Algebraic Statistics for Computational Biology

"If you can't stand algebra, keep out of evolutionary biology"

– John Maynard Smith
[Smith, 1998, page ix]

Algebraic Statistics for Computational Biology

Edited by

Lior Pachter and Bernd Sturmfels

University of California at Berkeley

CAMBRIDGE
UNIVERSITY PRESS

32 Avenue of the Americas, New York NY 10013-2473, USA

Cambridge University Press is part of the University of Cambridge.

It furthers the University's mission by disseminating knowledge in the pursuit of education, learning and research at the highest international levels of excellence.

www.cambridge.org
Information on this title: www.cambridge.org/9780521857000

First published 2005

A catalogue record for this publication is available from the British Library

Library of Congress Cataloguing in Publication data

Algebraic statistics for computational biology / edited by Lior Pachter, Bernd Sturmfels.
p.   cm.
Includes bibliographical references and index.
ISBN 0-521-85700-7
1. Biometry. 2. Algebra. I. Pachter, Lior, 1973– II. Sturmfels, Bernd, 1962–
QH323.5 .A43 2005
572.8/6 – 22   2005050070

ISBN 978-0-521-85700-0 Hardback

Contents

Preface

The title of this book reflects who we are: a computational biologist and an algebraist who share a common interest in statistics. Our collaboration sprang from the desire to find a mathematical language for discussing biological sequence analysis, with the initial impetus being provided by the introductory workshop on *Discrete and Computational Geometry* at the Mathematical Sciences Research Institute (MSRI) held at Berkeley in August 2003. At that workshop we began exploring the similarities between tropical matrix multiplication and the Viterbi algorithm for hidden Markov models. Our discussions ultimately led to two articles [Pachter and Sturmfels, 2004a,b] which are explained and further developed in various chapters of this book.

In the fall of 2003 we held a graduate seminar on *The Mathematics of Phylogenetic Trees*. About half of the authors of the second part of this book participated in that seminar. It was based on topics from the books [Felsenstein, 2003, Semple and Steel, 2003] but we also discussed other projects, such as Michael Joswig's polytope propagation on graphs (now Chapter 6). That seminar got us up to speed on research topics in phylogenetics, and led us to participate in the conference on *Phylogenetic Combinatorics* which was held in July 2004 in Uppsala, Sweden. In Uppsala we were introduced to David Bryant and his statistical models for split systems (now Chapter 17).

Another milestone was the workshop on *Computational Algebraic Statistics*, held at the American Institute for Mathematics (AIM) at Palo Alto in December 2003. That workshop was built on the algebraic statistics paradigm, which is that statistical models for discrete data can be regarded as solutions to systems of polynomial equations. Our current understanding of algebraic statistical models, maximum likelihood estimation and expectation maximization was shaped by the excellent discussions and lectures at AIM.

These developments led us to offer a mathematics graduate course titled *Algebraic Statistics for Computational Biology* in the fall of 2004. The course was attended mostly by mathematics students curious about computational biology, but also by computer scientists, statisticians, and bioengineering students interested in understanding the mathematical foundations of bioinformatics. Participants ranged from postdocs to first-year graduate students and even one undergraduate. The format consisted of lectures by us on basic principles of algebraic statistics and computational biology, as well as student participation in the form of group projects and presentations. The class was divided into four sections, reflecting the four themes of algebra, statistics, computation and biology. Each group was assigned a handful of projects to pursue, with the goal of completing a written report by the end of the semester. In some cases the groups worked on the problems we suggested, but, more often than not, original ideas by group members led to independent research plans.

Halfway through the semester, it became clear that the groups were making fantastic progress, and that their written reports would contain many novel ideas and results. At that point, we thought about preparing a book. The first half of the book would be based on our own lectures, and the second half would consist of chapters based on the final term papers. A tight schedule was seen as essential for the success of such an undertaking, given that many participants would be leaving Berkeley and the momentum would be lost. It was decided that the book should be written by March 2005, or not at all.

We were fortunate to find a partner in Cambridge University Press, which agreed to work with us on our concept. We are especially grateful to our editor, David Tranah, for his strong encouragement, and his trust that our half-baked ideas could actually turn into a readable book. After all, we were proposing to write to a book with twenty-nine authors during a period of three months.

The project did become reality and the result is in your hands. It offers an accurate snapshot of what happened during our seminars at UC Berkeley in 2003 and 2004. Nothing more and nothing less. The choice of topics is certainly biased, and the presentation is undoubtedly very far from perfect. But we hope that it may serve as an invitation to biology for mathematicians, and as an invitation to algebra for biologists, statisticians and computer scientists. Following this preface, we have included a guide to the chapters and suggested entry points for readers with different backgrounds and interests. Additional information and supplementary material may be found on the book website at http://bio.math.berkeley.edu/ascb/

Many friends and colleagues provided helpful comments and inspiration during the project. We especially thank Elizabeth Allman, Ruchira Datta, Manolis Dermitzakis, Serkan Hoşten, Ross Lippert, John Rhodes and Amelia Taylor. Serkan Hoşten was also instrumental in developing and guiding research which is described in Chapters 15 and 18.

Most of all, we are grateful to our wonderful students and postdocs from whom we learned so much. Their enthusiasm and hard work have been truly amazing. You will enjoy meeting them in Part II.

Lior Pachter and Bernd Sturmfels
Berkeley, California, May 2005

Guide to the chapters

The introductory Chapters 1–4 can be studied as a unit or read in parts with specific topics in mind. Although there are some dependencies and shared examples, the individual chapters are largely independent of each other. Suggested introductory sequences of study for specific topics are:

- Algebraic statistics: 1.1, 1.2, 1.4, 1.5.
- Maximum likelihood estimation: 1.1, 1.2, 1.3, 3.3.
- Tropical geometry: 2.1, 3.4, 3.5.
- Gröbner bases: 3.1, 3.2, 2.5.
- Comparative genomics: 4.1, 4.2, 4.3, 4.4, 4.5, 2.5.
- Sequence alignment: 1.1, 1.2, 1.4, 2.1, 2.2, 2.3.
- Phylogenetics: 1.1, 1.2, 1.4, 2.4, 3.4, 3.5, 4.5.

Dependencies of the Part II chapters on Part I are summarized in the table below. This should help readers interested in reading a specific chapter or section to find the location of background material. Pointers are also provided to related chapters that may be of interest.

Chapter	Prerequisites	Further reading
5	1.4, 2.2, 2.3	6, 7, 8, 9
6	1.1, 1.2, 1.4, 2.3	5, 7, 8, 9
7	2.2, 2.3	8
8	1.1, 1.2, 1.4, 2.1, 2.2, 2.3	5, 7, 9
9	1.5, 2.2, 2.3, 4.4	5, 8
10	1.1, 1.2, 1.4	9, 11
11	1.1, 1.2, 1.3, 3.1, 3.2	12
12	1.3, 1.4	4.4, 11
13	1.1, 1.2, 1.4, 1.5	22
14	1.1, 1.2, 1.4, 1.5, 3.1	11, 16
15	1.4, 3.1, 3.2, 3.3, 4.5	16, 17, 18, 19, 20
16	1.4, 3.1, 3.2, 4.5	15, 19
17	1.1, 1.2, 1.4, 1.5, 2.4 4.5	15, 18, 19
18	2.4, 4.5	20
19	2.4, 3.1, 4.5	15, 18
20	2.4, 4.5	17
21	1.4, 2.5, 4.5	17, 19
22	1.4, 4	7, 13, 21

Acknowledgment of support

We were fortunate to receive support from many agencies and institutions while working on the book. The following list is an acknowledgment of support for the many research activities that formed part of the *Algebraic Statistics for Computational Biology* book project.

Niko Beerenwinkel was funded by Deutsche Forschungsgemeinschaft (DFG) under Grant No. BE 3217/1-1. David Bryant was supported by NSERC grant number 238975-01 and FQRNT grant number 2003-NC-81840. Marta Casanellas was partially supported by RyC program of "Ministerio de Ciencia y Tecnologia", BFM2003-06001 and BIO2000-1352-C02-02 of "Plan Nacional I+D" of Spain. Anat Caspi was funded through the Genomics Training Grant at UC Berkeley: NIH 5-T32-HG00047. Mark Contois was partially supported by NSF grant DEB-0207090. Mathias Drton was support by NIH grant R01-HG02362-03. Dan Levy was supported by NIH grant GM 68423 and NSF grant DMS 9971169. Radu Mihaescu was supported by the Hertz foundation. Raaz Sainudiin was partly supported by a joint DMS/NIGMS grant 0201037. Sagi Snir was supported by NIH grant R01-HG02362-03. Kevin Woods was supported by NSF Grant DMS 0402148. Eric Kuo, Seth Sullivant and Josephine Yu were supported by NSF graduate research fellowships.

Lior Pachter was supported by NSF CAREER award CCF 03-47992, NIH grant R01-HG02362-03 and a Sloan Research Fellowship. He also acknowledges support from the Programs for Genomic Application (NHLBI). Bernd Sturmfels was supported by NSF grant DMS 0200729 and the Clay Mathematics Institute (July 2004). He was the Hewlett–Packard Research Fellow at the Mathematical Sciences Research Institute (MSRI) Berkeley during the year 2003–2004 which allowed him to study computational biology.

Finally, we thank staff at the University of California at Berkeley, Universitat de Barcelona (2001SGR-00071), the Massachusetts Institute of Technology and MSRI for extending hospitality to visitors at various times during which the book was being written.

Part I

Introduction to the four themes

Part I of this book is devoted to outlining the basic principles of algebraic statistics and their relationship to computational biology. Although some of the ideas are complex, and their relationships intricate, the underlying philosophy of our approach to biological sequence analysis is summarized in the cartoon on the cover of the book. The fictional character is DiaNA, who appears throughout the book, and is the statistical surrogate for our biological intuition. In the cartoon, DiaNA is walking randomly on a graph and is tossing tetrahedral dice that can land on one of the letters A,C,G or T. A key feature of the tosses is that the outcome depends on the direction of her route. We, the observers, record the letters that appear on the successive throws, but are unable to see the path that DiaNA takes on her graph. Our goal is to guess DiaNA's path from the die roll outcomes. That is, we wish to make an inference about missing data from certain observations.

In this book, the observed data are DNA sequences. A standard problem of computational biology is to infer an optimal alignment for two given DNA sequences. We shall see that this problem is precisely our example of guessing DiaNA's path. In Chapter 4 we give an introduction to the relevant biological concepts, and we argue that our example is not just a toy problem but is fundamental for designing efficient algorithms for analyzing real biological data.

The tetrahedral shape of DiaNA's dice hint at convex polytopes. We shall see in Chapter 2 that polytopes are geometric objects which play a key role in statistical inference. Underlying the whole story is computational algebra, featured in Chapter 3. Algebra is a universal language with which to describe the process at the heart of DiaNA's randomness.

Chapter 1 offers a fairly self-contained introduction to algebraic statistics. Many concepts of statistics have a natural analog in algebraic geometry, and there is an emerging dictionary which bridges the gap between these disciplines:

Statistics		Algebraic Geometry
independence	=	Segre variety
log-linear model	=	toric variety
curved exponential family	=	manifold
mixture model	=	join of varieties
MAP estimation	=	tropicalization
… …	=	… … …

Table 0.1. *A glimpse of the statistics – algebraic geometry dictionary.*

This dictionary is far from being complete, but it already suggests that algorithmic tools from algebraic geometry, most notably Gröbner bases, can be used for computations in statistics that are of interest for computational biology applications. While we are well aware of the limitations of algebraic algorithms, we nevertheless believe that computational biologists might benefit from adding the techniques described in Chapter 3 to their tool box. In addition, we have found the algebraic point of view to be useful in unifying and developing many computational biology algorithms. For example, the results on parametric sequence alignment in Chapter 7 do not require the language of algebra to be understood or utilized, but were motivated by concepts such as the Newton polytope of a polynomial. Chapter 2 discusses discrete algorithms which provide efficient solutions to various problems of statistical inference. Chapter 4 is an introduction to the biology, where we return to many of the examples in Chapter 1, illustrating how the statistical models we have discussed play a prominent role in computational biology.

We emphasize that Part I serves mainly as an introduction and reference for the chapters in Part II. We have therefore omitted many topics which are rightfully considered to be an integral part of computational biology. In particular, we have restricted ourselves to the topic of biological sequence analysis, and within that domain have focused on eukaryotic genome analysis. Readers may be interested in referring to [Durbin *et al.*, 1998] or [Ewens and Grant, 2005], our favorite introductions to the area of biological sequence analysis. Also useful may be a text on molecular biology with an emphasis on genomics, such as [Brown, 2002]. Our treatment of computational algebraic geometry in Chapter 3 is only a sliver taken from a mature and developed subject. The excellent book by [Cox *et al.*, 1997] fills in many of the details missing in our discussions.

Because Part I covers a wide range of topics, a comprehensive list of prerequisites would include a background in computer science, familiarity with molecular biology, and introductory courses in statistics and abstract algebra. Direct experience in computational biology would also be desirable. Of course, we recognize that this is asking too much. Real-life readers may be experts in one of these subjects but completely unfamiliar with others, and we have taken this into account when writing the book.

Various chapters provide natural points of entry for readers with different backgrounds. Those wishing to learn more about genomes can start with Chapter 4, biologists interested in software tools can start with Section 2.5, and statisticians who wish to brush up their algebra can start with Chapter 3.

In summary, the book is not meant to serve as the definitive text for algebraic statistics or computational biology, but rather as a first invitation to biology for mathematicians, and conversely as a mathematical primer for biologists. In other words, it is written in the spirit of interdisciplinary collaboration that is highlighted in the article *Mathematics is Biology's Next Microscope, Only Better; Biology is Mathematics' Next Physics, Only Better* [Cohen, 2004].

1
Statistics

Lior Pachter

Bernd Sturmfels

Statistics is the science of data analysis. The data to be encountered in this book are derived from *genomes*. Genomes consist of long chains of DNA which are represented by sequences in the letters A, C, G or T. These abbreviate the four nucleic acids Adenine, Cytosine, Guanine and Thymine, which serve as fundamental building blocks in molecular biology.

What do statisticians do with their data? They build models of the process that generated the data and, in what is known as *statistical inference*, draw conclusions about this process. Genome sequences are particularly interesting data to draw conclusions from: they are the blueprint for life, and yet their function, structure, and evolution are poorly understood. Statistical models are fundamental for genomics, a point of view that was emphasized in [Durbin *et al.*, 1998].

The inference tools we present in this chapter look different from those found in [Durbin *et al.*, 1998], or most other texts on computational biology or mathematical statistics: ours are written in the language of abstract algebra. The algebraic language for statistics clarifies many of the ideas central to the analysis of discrete data, and, within the context of biological sequence analysis, unifies the main ingredients of many widely used algorithms.

Algebraic Statistics is a new field, less than a decade old, whose precise scope is still emerging. The term itself was coined by Giovanni Pistone, Eva Riccomagno and Henry Wynn, with the title of their book [Pistone *et al.*, 2000]. That book explains how polynomial algebra arises in problems from experimental design and discrete probability, and it demonstrates how computational algebra techniques can be applied to statistics.

This chapter takes some additional steps along the algebraic statistics path. It offers a self-contained introduction to algebraic statistical models, with the aim of developing inference tools relevant for studying genomes. Special emphasis will be placed on (hidden) Markov models and graphical models.

1.1 Statistical models for discrete data

Imagine a fictional character named DiaNA who produces sequences of letters over the four-letter alphabet $\{A, C, G, T\}$. An example of such a sequence is

$$\text{CTCACGTGATGAGAGCATTCTCAGACCGTGACGCGTGTAGCAGCGGCTC.} \qquad (1.1)$$

The sequences produced by DiaNA are called *DNA sequences*. DiaNA generates her sequences by some random process. When modeling this random process we make assumptions about part of its structure. The resulting *statistical model* is a family of probability distributions, one of which governs the process by which DiaNA generates her sequences. In this book we consider parametric statistical models, which are families of probability distributions that can be parameterized by finitely many parameters. One important task is to estimate DiaNA's parameters from the sequences she generates. Estimation is also called *learning* in the computer science literature.

DiaNA uses *tetrahedral dice* to generate DNA sequences. Each die has the shape of a tetrahedron, and its four faces are labeled with the letters A, C, G and T. If DiaNA rolls a fair die then each of the four letters will appear with the same probability 1/4. If she uses a loaded tetrahedral die then the four probabilities can be any four non-negative numbers that sum to one.

Example 1.1 Suppose that DiaNA uses three tetrahedral dice. Two of her dice are loaded and one die is fair. The probabilities of rolling the four letters are known to us. They are the numbers in the rows of the following table:

	A	C	G	T
first die	0.15	0.33	0.36	0.16
second die	0.27	0.24	0.23	0.26
third die	0.25	0.25	0.25	0.25

$$(1.2)$$

DiaNA generates each letter in her DNA sequence independently using the following process. She first picks one of her three dice at random, where her first die is picked with probability θ_1, her second die is picked with probability θ_2, and her third die is picked with probability $1 - \theta_1 - \theta_2$. The probabilities θ_1 and θ_2 are unknown to us, but we do know that DiaNA makes one roll with the selected die, and then she records the resulting letter, A, C, G or T.

In the setting of biology, the first die corresponds to DNA that is G + C rich, the second die corresponds to DNA that is G + C poor, and the third is a fair die. We got the specific numbers in the first two rows of (1.2) by averaging the rows of the two tables in [Durbin *et al.*, 1998, page 50] (for more on this example and its connection to CpG island identification see Chapter 4).

Suppose we are given the DNA sequence of length $N = 49$ shown in (1.1). One question that may be asked is whether the sequence was generated by DiaNA using this process, and, if so, which parameters θ_1 and θ_2 did she use?

Let p_A, p_C, p_G and p_T denote the probabilities that DiaNA will generate any of her four letters. The statistical model we have discussed is written in algebraic notation as

$$
\begin{aligned}
p_A &= -0.10 \cdot \theta_1 + 0.02 \cdot \theta_2 + 0.25, \\
p_C &= 0.08 \cdot \theta_1 - 0.01 \cdot \theta_2 + 0.25, \\
p_G &= 0.11 \cdot \theta_1 - 0.02 \cdot \theta_2 + 0.25, \\
p_T &= -0.09 \cdot \theta_1 + 0.01 \cdot \theta_2 + 0.25.
\end{aligned}
$$

Note that $p_A + p_C + p_G + p_T = 1$, and we get the three distributions in the rows of (1.2) by specializing (θ_1, θ_2) to $(1, 0)$, $(0, 1)$ and $(0, 0)$ respectively.

To answer our questions, we consider the *likelihood* of observing the particular data (1.1). Since each of the 49 characters was generated independently, that likelihood is the product of the probabilities of the individual letters:

$$
L \quad = \quad p_C p_T p_C p_A p_C p_G \cdots p_A \quad = \quad p_A^{10} \cdot p_C^{14} \cdot p_G^{15} \cdot p_T^{10}.
$$

This expression is the *likelihood function* of DiaNA's model for the data (1.1). To stress the fact that the parameters θ_1 and θ_2 are unknowns we write

$$
L(\theta_1, \theta_2) \quad = \quad p_A(\theta_1, \theta_2)^{10} \cdot p_C(\theta_1, \theta_2)^{14} \cdot p_G(\theta_1, \theta_2)^{15} \cdot p_T(\theta_1, \theta_2)^{10}.
$$

This likelihood function is a real-valued function on the triangle

$$
\Theta \quad = \quad \big\{ (\theta_1, \theta_2) \in \mathbb{R}^2 \ : \ \theta_1 > 0 \text{ and } \theta_2 > 0 \text{ and } \theta_1 + \theta_2 < 1 \big\}.
$$

In the *maximum likelihood* framework we estimate the parameter values that DiaNA used by those values which make the likelihood of observing the data as large as possible. Thus our task is to maximize $L(\theta_1, \theta_2)$ over the triangle Θ. It is equivalent but more convenient to maximize the *log-likelihood function*

$$
\begin{aligned}
\ell(\theta_1, \theta_2) \quad = \quad & \log\big(L(\theta_1, \theta_2)\big) \\
= \quad & 10 \cdot \log(p_A(\theta_1, \theta_2)) + 14 \cdot \log(p_C(\theta_1, \theta_2)) \\
& + 15 \cdot \log(p_G(\theta_1, \theta_2)) + 10 \cdot \log(p_T(\theta_1, \theta_2)).
\end{aligned}
$$

The solution to this optimization problem can be computed in closed form, by equating the two partial derivatives of the log-likelihood function to zero:

$$
\begin{aligned}
\frac{\partial \ell}{\partial \theta_1} &= \frac{10}{p_A} \cdot \frac{\partial p_A}{\partial \theta_1} + \frac{14}{p_C} \cdot \frac{\partial p_C}{\partial \theta_1} + \frac{15}{p_G} \cdot \frac{\partial p_G}{\partial \theta_1} + \frac{10}{p_T} \cdot \frac{\partial p_T}{\partial \theta_1} = 0, \\
\frac{\partial \ell}{\partial \theta_2} &= \frac{10}{p_A} \cdot \frac{\partial p_A}{\partial \theta_2} + \frac{14}{p_C} \cdot \frac{\partial p_C}{\partial \theta_2} + \frac{15}{p_G} \cdot \frac{\partial p_G}{\partial \theta_2} + \frac{10}{p_T} \cdot \frac{\partial p_T}{\partial \theta_2} = 0.
\end{aligned}
$$

Each of the two expressions is a rational function in (θ_1, θ_2), i.e. it can be written as a quotient of two polynomials. By clearing denominators and by applying the algebraic technique of *Gröbner bases* (Section 3.1), we can transform the two equations above into the equivalent equations

$$
\begin{aligned}
13003050 \cdot \theta_1 + 2744 \cdot \theta_2^2 - 2116125 \cdot \theta_2 - 6290625 &= 0, \\
134456 \cdot \theta_2^3 - 10852275 \cdot \theta_2^2 - 4304728125 \cdot \theta_2 + 935718750 &= 0.
\end{aligned} \tag{1.3}
$$

The second equation has a unique solution $\hat{\theta}_2$ between 0 and 1. The corresponding value of $\hat{\theta}_1$ is obtained by solving the first equation. We find

$$
(\hat{\theta}_1, \hat{\theta}_2) \quad = \quad \big(0.5191263945, \ 0.2172513326 \big).
$$

The log-likelihood function attains its maximum value at this point:

$$\ell(\hat{\theta}_1, \hat{\theta}_2) \quad = \quad -67.08253037.$$

The corresponding probability distribution

$$(\hat{p}_A, \hat{p}_C, \hat{p}_G, \hat{p}_T) \quad = \quad \big(\, 0.202432, 0.289358, 0.302759, 0.205451 \,\big) \qquad (1.4)$$

is very close to the empirical distribution

$$\frac{1}{49}(10, 14, 15, 10) \quad = \quad \big(\, 0.204082, 0.285714, 0.306122, 0.204082 \,\big). \qquad (1.5)$$

We conclude that the proposed model is a good fit for the data (1.1). To make this conclusion precise we would need to employ a technique like the χ^2 test [Bickel and Doksum, 2000], but we keep our little example informal and simply assert that our calculation suggests that DiaNA used the probabilities $\hat{\theta}_1$ and $\hat{\theta}_2$ for choosing among her dice.   □

We now turn to our general discussion of statistical models for discrete data. A *statistical model* is a family of probability distributions on some state space. In this book we assume that the state space is finite, but possibly quite large. We often identify the state space with the set of the first m positive integers,

$$[m] \quad := \quad \{1, 2, \ldots, m\}. \qquad (1.6)$$

A probability distribution on the set $[m]$ is a point in the *probability simplex*

$$\Delta_{m-1} \quad := \quad \big\{\, (p_1, \ldots, p_m) \in \mathbb{R}^m : \sum_{i=1}^{m} p_i = 1 \text{ and } p_j \geq 0 \text{ for all } j \,\big\}. \qquad (1.7)$$

The index $m - 1$ indicates the dimension of the simplex Δ_{m-1}. We write Δ for the simplex Δ_{m-1} when the underlying state space $[m]$ is understood.

Example 1.2 The state space for DiaNA's dice is the set $\{A, C, G, T\}$ which we identify with the set $[4] = \{1, 2, 3, 4\}$. The simplex Δ is a tetrahedron. The probability distribution associated with a fair die is the point $(\frac{1}{4}, \frac{1}{4}, \frac{1}{4}, \frac{1}{4})$, which is the centroid of the tetrahedron Δ. Equivalently, we may think about our model via the concept of a *random variable*: that is, a function X taking values in the state space $\{A, C, G, T\}$. Then the point corresponding to a fair die gives the probability distribution of X as $\text{Prob}(X = A) = \frac{1}{4}$, $\text{Prob}(X = C) = \frac{1}{4}$, $\text{Prob}(X = G) = \frac{1}{4}$, $\text{Prob}(X = T) = \frac{1}{4}$. All other points in the tetrahedron Δ correspond to loaded dice.   □

A statistical model for discrete data is a family of probability distributions on $[m]$. Equivalently, a statistical model is simply a subset of the simplex Δ. The ith coordinate p_i represents the probability of observing the state i, and in that capacity p_i must be a non-negative real number. However, when discussing algebraic computations (as in Chapter 3), we sometimes relax this requirement and allow p_i to be negative or even a complex number.

An *algebraic statistical model* arises as the image of a polynomial map

$$\mathbf{f} \;:\; \mathbb{R}^d \to \mathbb{R}^m \;, \quad \theta = (\theta_1, \ldots, \theta_d) \;\mapsto\; \big(f_1(\theta), f_2(\theta), \ldots, f_m(\theta)\big). \qquad (1.8)$$

The unknowns $\theta_1, \ldots, \theta_d$ represent the model parameters. In most cases of interest, d is much smaller than m. Each coordinate function f_i is a polynomial in the d unknowns, which means it has the form

$$f_i(\theta) \;\;=\;\; \sum_{a \in \mathbb{N}^d} c_a \cdot \theta_1^{a_1} \theta_2^{a_2} \cdots \theta_d^{a_d}, \qquad (1.9)$$

where all but finitely many of the coefficients $c_a \in \mathbb{R}$ are zero. We use $\mathbb{N}$ to denote the non-negative integers: that is, $\mathbb{N} = \{0, 1, 2, 3, \ldots\}$.

The parameter vector $(\theta_1, \ldots, \theta_d)$ ranges over a suitable non-empty open subset Θ of $\mathbb{R}^d$ which is called the *parameter space* of the model $\mathbf{f}$. We assume that the parameter space Θ satisfies the condition

$$f_i(\theta) > 0 \qquad \text{for all } i \in [m] \text{ and } \theta \in \Theta. \qquad (1.10)$$

Under these hypotheses, the following two conditions are equivalent:

$$\mathbf{f}(\Theta) \subseteq \Delta \qquad \Longleftrightarrow \qquad f_1(\theta) + f_2(\theta) + \cdots + f_m(\theta) \;=\; 1. \qquad (1.11)$$

This is an identity of polynomial functions, which means that all non-constant terms of the polynomials f_i cancel, and the constant terms add up to 1. If (1.11) holds, then our model is simply the set $\mathbf{f}(\Theta)$.

Example 1.3 DiaNA's model in Example 1.1 is a *mixture model* which mixes three distributions on $\{\mathsf{A}, \mathsf{C}, \mathsf{G}, \mathsf{T}\}$. Geometrically, the image of DiaNA's map

$$\mathbf{f} : \mathbb{R}^2 \to \mathbb{R}^4, \quad (\theta_1, \theta_2) \mapsto (p_\mathsf{A}, p_\mathsf{C}, p_\mathsf{G}, p_\mathsf{T})$$

is the plane in $\mathbb{R}^4$ which is cut out by the two linear equations

$$p_\mathsf{A} + p_\mathsf{C} + p_\mathsf{G} + p_\mathsf{T} \;=\; 1 \qquad \text{and} \qquad 11\,p_\mathsf{A} + 15\,p_\mathsf{G} \;=\; 17\,p_\mathsf{C} + 9\,p_\mathsf{T}. \qquad (1.12)$$

These two linear equations are *algebraic invariants* of the model. The plane they define intersects the tetrahedron Δ in the quadrangle whose vertices are

$$\left(0, 0, \frac{3}{8}, \frac{5}{8}\right), \;\; \left(0, \frac{15}{32}, \frac{17}{32}, 0\right), \;\; \left(\frac{9}{20}, 0, 0, \frac{11}{20}\right) \text{ and } \left(\frac{17}{28}, \frac{11}{28}, 0, 0\right). \qquad (1.13)$$

Inside this quadrangle is the triangle $\mathbf{f}(\Theta)$ whose vertices are the three rows of the table in (1.2). The point (1.4) lies in that triangle and is near (1.5). $\qquad\square$

Some statistical models are given by a polynomial map $\mathbf{f}$ for which (1.11) does not hold. If this is the case then we scale each vector in $\mathbf{f}(\Theta)$ by the positive quantity $\sum_{i=1}^m f_i(\theta)$. Regardless of whether (1.11) holds or not, our model is the family of all probability distributions on $[m]$ of the form

$$\frac{1}{\sum_{i=1}^m f_i(\theta)} \cdot \big(f_1(\theta), f_2(\theta), \ldots, f_m(\theta)\big) \qquad \text{where} \quad \theta \in \Theta. \qquad (1.14)$$

We generally try to keep things simple and assume that (1.11) holds. However, there are some cases, such as the general toric model in the next section, when the formulation in (1.14) is more natural. It poses no great difficulty to extend our theorems and algorithms from polynomials to rational functions.

Our *data* are typically given in the form of a sequence of observations

$$i_1, i_2, i_3, i_4, \ldots, i_N. \tag{1.15}$$

Each data point i_j is an element from our state space $[m]$. The integer N, which is the length of the sequence, is called the *sample size*. Assuming that the observations (1.15) are independent and identically distributed samples, we can summarize the data (1.15) in the *data vector* $u = (u_1, u_2, \ldots, u_m)$ where u_k is the number of indices $j \in [N]$ such that $i_j = k$. Hence u is a vector in $\mathbb{N}^m$ with $u_1 + u_2 + \cdots + u_m = N$. The *empirical distribution* corresponding to the data (1.15) is the scaled vector $\frac{1}{N}u$ which is a point in the probability simplex Δ. The coordinates u_i/N of this vector are the *observed relative frequencies* of the various outcomes.

We consider the model $\mathbf{f}$ to be a "good fit" for the data u if there exists a parameter vector $\theta \in \Theta$ such that the probability distribution $\mathbf{f}(\theta)$ is very close, in a statistically meaningful sense [Bickel and Doksum, 2000], to the empirical distribution $\frac{1}{N}u$. Suppose we draw N times at random (independently and with replacement) from the set $[m]$ with respect to the probability distribution $\mathbf{f}(\theta)$. Then the probability of observing the sequence (1.15) equals

$$L(\theta) \quad = \quad f_{i_1}(\theta) f_{i_2}(\theta) \cdots f_{i_N}(\theta) \quad = \quad f_1(\theta)^{u_1} \cdots f_m(\theta)^{u_m}. \tag{1.16}$$

This expression depends on the parameter vector θ as well as the data vector u. However, we think of u as being fixed and then L is a function from Θ to the positive real numbers. It is called the *likelihood function* to emphasize that it is a function that depends on θ. Note that any reordering of the sequence (1.15) leads to the same data vector u. Hence the probability of observing the data vector u is equal to

$$\frac{(u_1 + u_2 + \cdots + u_m)!}{u_1! u_2! \cdots u_m!} \cdot L(\theta). \tag{1.17}$$

The vector u plays the role of a *sufficient statistic* for the model $\mathbf{f}$. This means that the likelihood function $L(\theta)$ depends on the data (1.15) only through u. In practice one often replaces the likelihood function by its logarithm

$$\ell(\theta) \quad = \quad \log L(\theta) \quad = \quad u_1 \cdot \log(f_1(\theta)) + u_2 \cdot \log(f_2(\theta)) + \cdots + u_m \cdot \log(f_m(\theta)). \tag{1.18}$$

This is the *log-likelihood function*. Note that $\ell(\theta)$ is a function from the parameter space $\Theta \subset \mathbb{R}^d$ to the negative real numbers $\mathbb{R}_{<0}$.

The problem of *maximum likelihood estimation* is to maximize the likelihood function $L(\theta)$ in (1.16), or, equivalently, the scaled likelihood function (1.17), or, equivalently, the log-likelihood function $\ell(\theta)$ in (1.18). Here θ ranges over the parameter space $\Theta \subset \mathbb{R}^d$. Formally, we consider the optimization problem:

$$\text{Maximize } \ell(\theta) \quad \text{subject to } \theta \in \Theta. \tag{1.19}$$

A solution to this optimization problem is denoted $\hat{\theta}$ and is called a *maximum likelihood estimate* of θ with respect to the model $\mathbf{f}$ and the data u.

Sometimes, if the model satisfies certain properties, it may be that there always exists a unique maximum likelihood estimate $\hat{\theta}$. This happens for *linear models* and *toric models*, due to the concavity of their log-likelihood function, as we shall see in Section 1.2. For most statistical models, however, the situation is not as simple. First, a maximum likelihood estimate need not exist (since we assume Θ to be open). Second, even if $\hat{\theta}$ exists, there can be more than one global maximum, in fact, there can be infinitely many of them. And, third, it may be very difficult to find any one of these global maxima. In that case, one may content oneself with a local maximum of the likelihood function. In Section 1.3 we shall discuss the *EM algorithm* which is a numerical method for finding solutions to the maximum likelihood estimation problem (1.19).

1.2 Linear models and toric models

In this section we introduce two classes of models which, under weak conditions on the data, have the property that the likelihood function has exactly one local maximum $\hat{\theta} \in \Theta$. Since the parameter spaces of the models are convex, the *maximum likelihood estimate* $\hat{\theta}$ can be computed using any of the hill-climbing methods of convex optimization, such as the gradient ascent algorithm.

1.2.1 Linear models

An algebraic statistical model $\mathbf{f} : \mathbb{R}^d \to \mathbb{R}^m$ is called a *linear model* if each of its coordinate polynomials $f_i(\theta)$ is a linear function. Being a linear function means that there exist real numbers $a_{i1}, \ldots, a_{1d}$ and b_i such that

$$f_i(\theta) = \sum_{j=1}^{d} a_{ij}\theta_j + b_i. \tag{1.20}$$

The m linear functions $f_1(\theta), \ldots, f_m(\theta)$ have the property that their sum is the constant function 1. DiaNA's model studied in Example 1.1 is a linear model. For the data discussed in that example, the log-likelihood function $\ell(\theta)$ had a unique local maximum on the parameter triangle Θ. The following proposition states that this desirable property holds for every linear model.

Proposition 1.4 *For any linear model $\mathbf{f}$ and data $u \in \mathbb{N}^m$, the log-likelihood function $\ell(\theta) = \sum_{i=1}^{m} u_i \log(f_i(\theta))$ is concave. If the linear map $\mathbf{f}$ is one-to-one and all u_i are positive then the log-likelihood function is strictly concave.*

Proof Our assertion that the log-likelihood function $\ell(\theta)$ is concave states that the *Hessian matrix* $\left(\frac{\partial^2 \ell}{\partial \theta_j \, \partial \theta_k}\right)$ is negative semi-definite for every $\theta \in \Theta$. In other words, we need to show that every eigenvalue of this symmetric matrix is non-positive. The partial derivative of the linear function $f_i(\theta)$ in (1.20) with respect to the unknown θ_j is the constant a_{ij}. Hence the partial derivative of the log-likelihood function $\ell(\theta)$ equals

$$\frac{\partial \ell}{\partial \theta_j} = \sum_{i=1}^{m} \frac{u_i a_{ij}}{f_i(\theta)}. \tag{1.21}$$

Taking the derivative again, we get the following formula for the Hessian matrix

$$\left(\frac{\partial^2 \ell}{\partial \theta_j \, \partial \theta_k} \right) = -A^{\mathrm{T}} \cdot \mathrm{diag}\left(\frac{u_1}{f_1(\theta)^2}, \frac{u_2}{f_2(\theta)^2}, \ldots, \frac{u_m}{f_m(\theta)^2} \right) \cdot A. \tag{1.22}$$

Here A is the $m \times d$ matrix whose entry in row i and column j equals a_{ij}. This shows that the Hessian (1.22) is a symmetric $d \times d$ matrix each of whose eigenvalues is non-positive.

The argument above shows that $\ell(\theta)$ is a concave function. Moreover, if the linear map $\mathbf{f}$ is one-to-one then the matrix A has rank d. In that case, provided all u_i are strictly positive, all eigenvalues of the Hessian are strictly negative, and we conclude that $\ell(\theta)$ is strictly concave for all $\theta \in \Theta$. $\qquad \square$

The critical points of the likelihood function $\ell(\theta)$ of the linear model $\mathbf{f}$ are the solutions to the system of d equations in d unknowns which are obtained by equating (1.21) to zero. What we get are the *likelihood equations*

$$\sum_{i=1}^{m} \frac{u_i a_{i1}}{f_i(\theta)} = \sum_{i=1}^{m} \frac{u_i a_{i2}}{f_i(\theta)} = \cdots = \sum_{i=1}^{m} \frac{u_i a_{id}}{f_i(\theta)} = 0. \tag{1.23}$$

The study of these equations involves the combinatorial theory of hyperplane arrangements. Indeed, consider the m hyperplanes in d-space $\mathbb{R}^d$ which are defined by the equations $f_i(\theta) = 0$ for $i = 1, 2, \ldots, m$. The complement of this arrangement of hyperplanes in $\mathbb{R}^d$ is the set of parameter values

$$\mathcal{C} = \{ \theta \in \mathbb{R}^d : f_1(\theta) f_2(\theta) f_3(\theta) \cdots f_m(\theta) \neq 0 \}.$$

This set is the disjoint union of finitely many open convex polyhedra defined by inequalities $f_i(\theta) > 0$ or $f_i(\theta) < 0$. These polyhedra are called the *regions* of the arrangement. Some of these regions are bounded, and others are unbounded. The natural parameter space of the linear model coincides with exactly one bounded region. The other bounded regions would give rise to negative probabilities. However, they are relevant for the algebraic complexity of our problem. Let μ denote the number of bounded regions of the arrangement.

Theorem 1.5 (Varchenko's Formula) *If the u_i are positive, then the likelihood equations (1.23) of the linear model $\mathbf{f}$ have precisely μ distinct real solutions, one in each bounded region of the hyperplane arrangement $\{f_i = 0\}_{i \in [m]}$. All solutions have multiplicity one and there are no other complex solutions.*

This result first appeared in [Varchenko, 1995]. The connection to maximum likelihood estimation was explored in [Catanese *et al.*, 2005].

We already saw one instance of Varchenko's Formula in Example 1.1. The four lines defined by the vanishing of DiaNA's probabilities p_{A}, p_{C}, p_{G} or p_{T}

partition the (θ_1, θ_2)-plane into eleven regions. Three of these eleven regions are bounded: one is the quadrangle (1.13) in Δ and two are triangles outside Δ. Thus DiaNA's linear model has $\mu = 3$ bounded regions. Each region contains one of the three solutions of the transformed likelihood equations (1.3). Only one of these three regions is of statistical interest.

Example 1.6 Consider a one-dimensional $(d = 1)$ linear model $\mathbf{f} : \mathbb{R}^1 \to \mathbb{R}^m$. Here θ is a scalar parameter and each $f_i = a_i \theta + b_i$ $(a_i \neq 0)$ is a linear function in one unknown θ. We have $a_1 + a_2 + \cdots + a_m = 0$ and $b_1 + b_2 + \cdots + b_m = 1$. Assuming the m quantities $-b_i/a_i$ are all distinct, they divide the real line into $m - 1$ bounded segments and two unbounded half-rays. One of the bounded segments is $\Theta = \mathbf{f}^{-1}(\Delta)$. The derivative of the log-likelihood function equals

$$\frac{d\ell}{d\theta} = \sum_{i=1}^{m} \frac{u_i a_i}{a_i \theta + b_i}.$$

For positive u_i, this rational function has precisely $m - 1$ zeros, one in each of the bounded segments. The maximum likelihood estimate $\widehat{\theta}$ is the unique zero of $d\ell/d\theta$ in the statistically meaningful segment $\Theta = \mathbf{f}^{-1}(\Delta)$. □

Example 1.7 Many statistical models used in biology have the property that the polynomials $f_i(\theta)$ are multilinear. The concavity result of Proposition 1.4 is a useful tool for varying the parameters one at a time. Here is such a model with $d = 3$ and $m = 5$. Consider the trilinear map $\mathbf{f} : \mathbb{R}^3 \to \mathbb{R}^5$ given by

$$
\begin{aligned}
f_1(\theta) &= -24\theta_1\theta_2\theta_3 + 9\theta_1\theta_2 + 9\theta_1\theta_3 + 9\theta_2\theta_3 - 3\theta_1 - 3\theta_2 - 3\theta_3 + 1, \\
f_2(\theta) &= -48\theta_1\theta_2\theta_3 + 6\theta_1\theta_2 + 6\theta_1\theta_3 + 6\theta_2\theta_3, \\
f_3(\theta) &= 24\theta_1\theta_2\theta_3 + 3\theta_1\theta_2 - 9\theta_1\theta_3 - 9\theta_2\theta_3 + 3\theta_3, \\
f_4(\theta) &= 24\theta_1\theta_2\theta_3 - 9\theta_1\theta_2 + 3\theta_1\theta_3 - 9\theta_2\theta_3 + 3\theta_2, \\
f_5(\theta) &= 24\theta_1\theta_2\theta_3 - 9\theta_1\theta_2 - 9\theta_1\theta_3 + 3\theta_2\theta_3 + 3\theta_1.
\end{aligned}
$$

This is a small instance of the *Jukes–Cantor model* of phylogenetics. Its derivation and its relevance for computational biology will be discussed in detail in Section 4.5. Let us fix two of the parameters, say θ_1 and θ_2, and vary only the third parameter θ_3. The result is a linear model as in Example 1.6, with $\theta = \theta_3$. We compute the maximum likelihood estimate $\widehat{\theta}_3$ for this linear model, and then we replace θ_3 by $\widehat{\theta}_3$. Next fix the two parameters θ_2 and θ_3, and vary the third parameter θ_1. Thereafter, fix (θ_3, θ_1) and vary θ_2, etc. Iterating this procedure, we may compute a local maximum of the likelihood function. □

1.2.2 Toric models

Our second class of models with well-behaved likelihood functions are the *toric models*. These are also known as *log-linear models*, and they form an important class of *exponential families*. Let $A = (a_{ij})$ be a non-negative integer $d \times m$

matrix with the property that all column sums are equal:

$$\sum_{i=1}^{d} a_{i1} \;=\; \sum_{i=1}^{d} a_{i2} \;=\; \cdots \;=\; \sum_{i=1}^{d} a_{im}. \qquad (1.24)$$

The jth column vector a_j of the matrix A represents the monomial

$$\theta^{a_j} \;=\; \prod_{i=1}^{d} \theta_i^{a_{ij}} \qquad \text{for } j = 1, 2, \ldots, m.$$

Our assumption (1.24) says that these monomials all have the same degree. The toric model of A is the image of the orthant $\Theta = \mathbb{R}_{>0}^d$ under the map

$$\mathbf{f} : \mathbb{R}^d \to \mathbb{R}^m, \quad \theta \mapsto \frac{1}{\sum_{j=1}^{m} \theta^{a_j}} \cdot \left(\theta^{a_1}, \theta^{a_2}, \ldots, \theta^{a_m} \right). \qquad (1.25)$$

Note that we can scale the parameter vector without changing the image: $\mathbf{f}(\theta) = \mathbf{f}(\lambda \cdot \theta)$. Hence the dimension of the toric model $\mathbf{f}(\mathbb{R}_{>0}^d)$ is at most $d - 1$. In fact, the dimension of $\mathbf{f}(\mathbb{R}_{>0}^d)$ is one less than the rank of A. The denominator polynomial $\sum_{j=1}^{m} \theta^{a_j}$ is known as the *partition function*.

Sometimes we are also given positive constants $c_1, \ldots, c_m > 0$ and the map $\mathbf{f}$ is modified as follows:

$$\mathbf{f} : \mathbb{R}^d \to \mathbb{R}^m, \quad \theta \mapsto \frac{1}{\sum_{j=1}^{m} c_j \theta^{a_j}} \cdot \left(c_1 \theta^{a_1}, c_2 \theta^{a_2}, \ldots, c_m \theta^{a_m} \right). \qquad (1.26)$$

In a toric model, the logarithms of the probabilities are linear functions in the logarithms of the parameters θ_i. It is for this reason that statisticians refer to toric models as log-linear models. For simplicity we stick with the formulation (1.25) but the discussion would be analogous for (1.26). Throughout this book, we identify each toric model $\mathbf{f}$ with the corresponding integer matrix A.

Maximum likelihood estimation for the toric model (1.25) means solving the following optimization problem

$$\text{Maximize } p_1^{u_1} \cdots p_m^{u_m} \text{ subject to } (p_1, \ldots, p_m) \in \mathbf{f}(\mathbb{R}_{>0}^d). \qquad (1.27)$$

This problem is equivalent to

$$\text{Maximize } \theta^{Au} \text{ subject to } \theta \in \mathbb{R}_{>0}^d \text{ and } \sum_{j=1}^{m} \theta^{a_j} \;=\; 1. \qquad (1.28)$$

Here we are using multi-index notation for monomials in $\theta = (\theta_1, \ldots, \theta_d)$:

$$\theta^{Au} \;=\; \prod_{i=1}^{d} \prod_{j=1}^{m} \theta_i^{a_{ij} u_j} \;=\; \prod_{i=1}^{d} \theta_i^{a_{i1} u_1 + a_{i2} u_2 + \cdots + a_{im} u_m} \qquad \text{and} \qquad \theta^{a_j} \;=\; \prod_{i=1}^{d} \theta_i^{a_{ij}}.$$

Writing $b = Au$ for the sufficient statistic, our optimization problem (1.28) is

$$\text{Maximize } \theta^b \text{ subject to } \theta \in \mathbb{R}_{>0}^d \text{ and } \sum_{j=1}^{m} \theta^{a_j} \;=\; 1. \qquad (1.29)$$

Example 1.8 Let $d = 2$, $m = 3$, $A = \begin{pmatrix} 2 & 1 & 0 \\ 0 & 1 & 2 \end{pmatrix}$ and $u = (11, 17, 23)$. The sample size is $N = 51$. Our problem is to maximize the likelihood function $\theta_1^{39} \theta_2^{63}$ over all positive real vectors (θ_1, θ_2) that satisfy $\theta_1^2 + \theta_1 \theta_2 + \theta_2^2 = 1$. The unique solution $(\hat\theta_1, \hat\theta_2)$ to this problem has coordinates

$$\hat\theta_1 = \frac{1}{51} \sqrt{1428 - 51\sqrt{277}} = 0.4718898805 \quad \text{and}$$

$$\hat\theta_2 = \frac{1}{51} \sqrt{2040 - 51\sqrt{277}} = 0.6767378938.$$

The probability distribution corresponding to these parameter values is

$$\hat p = (\hat p_1, \hat p_2, \hat p_3) = (\hat\theta_1^2, \hat\theta_1 \hat\theta_2, \hat\theta_2^2) = (0.2227, 0.3193, 0.4580). \qquad \square$$

Proposition 1.9 *Fix a toric model A and data $u \in \mathbb{N}^m$ with sample size $N = u_1 + \cdots + u_m$ and sufficient statistic $b = Au$. Let $\hat p = \mathbf{f}(\hat\theta)$ be any local maximum for the equivalent optimization problems* (1.27),(1.28),(1.29). *Then*

$$A \cdot \hat p = \frac{1}{N} \cdot b. \tag{1.30}$$

Writing $\hat p$ as a column vector, we check that (1.30) holds in Example 1.8:

$$A \cdot \hat p = \begin{pmatrix} 2\hat\theta_1^2 + \hat\theta_1\hat\theta_2 \\ \hat\theta_1\hat\theta_2 + 2\hat\theta_2^2 \end{pmatrix} = \frac{1}{51} \cdot \begin{pmatrix} 39 \\ 63 \end{pmatrix} = \frac{1}{N} \cdot Au.$$

Proof We introduce a *Lagrange multiplier* λ. Every local optimum of (1.29) is a critical point of the following function in $d + 1$ unknowns $\theta_1, \ldots, \theta_d, \lambda$:

$$\theta^b + \lambda \cdot \left(1 - \sum_{j=1}^m \theta^{a_j} \right).$$

We apply the scaled gradient operator

$$\theta \cdot \nabla_\theta = \left(\theta_1 \frac{\partial}{\partial \theta_1}, \theta_2 \frac{\partial}{\partial \theta_2}, \ldots, \theta_d \frac{\partial}{\partial \theta_d} \right)$$

to the function above. The resulting critical equations for $\hat\theta$ and $\hat p$ state that

$$(\hat\theta)^b \cdot b = \lambda \cdot \sum_{j=1}^m (\hat\theta)^{a_j} \cdot a_j = \lambda \cdot A \cdot \hat p.$$

This says that the vector $A \cdot \hat p$ is a scalar multiple of the vector $b = Au$. Since the matrix A has the vector $(1, 1, \ldots, 1)$ in its row space, and since $\sum_{j=1}^m \hat p_j = 1$, it follows that the scalar factor which relates the sufficient statistic $b = A \cdot u$ to $A \cdot \hat p$ must be the sample size $\sum_{j=1}^m u_j = N$. $\qquad \square$

Given the matrix $A \in \mathbb{N}^{d \times m}$ and any vector $b \in \mathbb{R}^d$, we consider the set

$$P_A(b) = \left\{ p \in \mathbb{R}^m : A \cdot p = \frac{1}{N} \cdot b \text{ and } p_j > 0 \text{ for all } j \right\}.$$

This is a relatively open *polytope*. (See Section 2.3 for an introduction to polytopes). We shall prove that $P_A(b)$ is either empty or meets the toric model in precisely one point. This result was discovered and re-discovered many times by different people from various communities. In toric geometry, it goes under the keyword "moment map". In the statistical setting of exponential families, it appears in the work of Birch in the 1960s; see [Agresti, 1990, page 168].

Theorem 1.10 (Birch's Theorem) *Fix a toric model A and let $u \in \mathbb{N}_{>0}^m$ be a strictly positive data vector with sufficient statistic $b = Au$. The intersection of the polytope $P_A(b)$ with the toric model $\mathbf{f}(\mathbb{R}_{>0}^d)$ consists of precisely one point. That point is the maximum likelihood estimate $\widehat{p}$ for the data u.*

Proof Consider the *entropy function*

$$H \; : \; \mathbb{R}_{\geq 0}^m \; \rightarrow \; \mathbb{R}_{\geq 0} \; , \quad (p_1, \ldots, p_m) \; \mapsto \; -\sum_{i=1}^{m} p_i \cdot \log(p_i).$$

This function is well-defined for non-negative vectors because $p_i \cdot \log(p_i)$ is 0 for $p_i = 0$. The entropy function H is strictly concave in $\mathbb{R}_{>0}^m$, i.e.,

$$H\big(\lambda \cdot p + (1 - \lambda) \cdot q\big) \; > \; \lambda \cdot H(p) + (1 - \lambda) \cdot H(q) \quad \text{for } p \neq q \text{ and } 0 < \lambda < 1,$$

because the Hessian matrix $\big(\partial^2 H / \partial p_i \partial p_j\big)$ is a diagonal matrix with entries $-1/p_1, -1/p_2, \ldots, -1/p_m$. The restriction of the entropy function H to the relatively open polytope $P_A(b)$ is strictly concave as well, so it attains its maximum at a unique point $p^* = p^*(b)$ in the polytope $P_A(b)$.

For any vector $u \in \mathbb{R}^m$ which lies in the kernel of A, the directional derivative of the entropy function H vanishes at the point $p^* = (p_1^*, \ldots, p_m^*)$:

$$u_1 \cdot \frac{\partial H}{\partial p_1}(p^*) + u_2 \cdot \frac{\partial H}{\partial p_2}(p^*) + \cdots + u_m \cdot \frac{\partial H}{\partial p_m}(p^*) \;\; = \;\; 0. \tag{1.31}$$

Since the derivative of $x \cdot \log(x)$ is $\log(x) + 1$, and since $(1, 1, \ldots, 1)$ is in the row span of the matrix A, the equation (1.31) implies

$$0 \; = \; \sum_{j=1}^{m} u_j \cdot \log(p_j^*) + \sum_{j=1}^{m} u_j \; = \; \sum_{j=1}^{m} u_j \cdot \log(p_j^*) \quad \text{for all } u \in \text{kernel}(A). \tag{1.32}$$

This implies that $\big(\log(p_1^*), \log(p_2^*), \ldots, \log(p_m^*)\big)$ lies in the row span of A. Pick a vector $\eta^* = (\eta_1^*, \ldots, \eta_d^*)$ such that $\sum_{i=1}^{d} \eta_i^* a_{ij} = \log(p_j^*)$ for all j. If we set $\theta_i^* = \exp(\eta_i^*)$ for $i = 1, \ldots, d$ then

$$p_j^* \; = \; \prod_{i=1}^{d} \exp(\eta_i^* a_{ij}) \; = \; \prod_{i=1}^{d} (\theta_i^*)^{a_{ij}} \; = \; (\theta^*)^{a_j} \quad \text{for } j = 1, 2, \ldots, m.$$

This shows that $p^* = \mathbf{f}(\theta^*)$ for some $\theta^* \in \mathbb{R}_{>0}^d$, so p^* lies in the toric model. Moreover, if A has rank d then θ^* is uniquely determined (up to scaling) by $p^* = \mathbf{f}(\theta)$. We have shown that p^* is a point in the intersection $P_A(b) \cap \mathbf{f}(\mathbb{R}_{>0}^d)$.

It remains to be seen that there is no other point. Suppose that q lies in

$P_A(b) \cap \mathbf{f}(\mathbb{R}^d_{>0})$. Then (1.32) holds, so that q is a critical point of the entropy function H. Since the Hessian matrix is negative definite at q, this point is a maximum of the strictly concave function H, and therefore $q = p^*$.

Let $\widehat{\theta}$ be a maximum likelihood estimate for the data u and let $\widehat{p} = \mathbf{f}(\widehat{\theta})$ be the corresponding probability distribution. Proposition 1.9 tells us that $\widehat{p}$ lies in $P_A(b)$. The uniqueness property in the previous paragraph implies $\widehat{p} = p^*$ and, assuming A has rank d, we can further conclude $\widehat{\theta} = \theta^*$. $\qquad\square$

Example 1.11 (Example 1.8 continued) Let $d = 2$, $m = 3$ and $A = \begin{pmatrix} 2 & 1 & 0 \\ 0 & 1 & 2 \end{pmatrix}$. If b_1 and b_2 are positive reals then the polytope $P_A(b_1, b_2)$ is a line segment. The maximum likelihood point $\widehat{p}$ is characterized by the equations

$$2\widehat{p}_1 + \widehat{p}_2 = \frac{1}{N} \cdot b_1 \quad \text{and} \quad \widehat{p}_2 + 2\widehat{p}_3 = \frac{1}{N} \cdot b_2 \quad \text{and} \quad \widehat{p}_1 \cdot \widehat{p}_3 = \widehat{p}_2 \cdot \widehat{p}_2.$$

The unique positive solution to these equations equals

$$
\begin{aligned}
\widehat{p}_1 &= \tfrac{1}{N} \cdot \left(\tfrac{7}{12} b_1 + \tfrac{1}{12} b_2 - \tfrac{1}{12} \sqrt{b_1{}^2 + 14\, b_1\, b_2 + b_2{}^2} \right), \\
\widehat{p}_2 &= \tfrac{1}{N} \cdot \left(-\tfrac{1}{6} b_1 - \tfrac{1}{6} b_2 + \tfrac{1}{6} \sqrt{b_1{}^2 + 14\, b_1\, b_2 + b_2{}^2} \right), \\
\widehat{p}_3 &= \tfrac{1}{N} \cdot \left(\tfrac{1}{12} b_1 + \tfrac{7}{12} b_2 - \tfrac{1}{12} \sqrt{b_1{}^2 + 14\, b_1\, b_2 + b_2{}^2} \right).
\end{aligned}
$$
$\qquad\square$

The most classical example of a toric model in statistics is the *independence model* for two random variables. Let X_1 be a random variable on $[m_1]$ and X_2 a random variable on $[m_2]$. The two random variables are *independent* if

$$\text{Prob}(X_1 = i, X_2 = j) = \text{Prob}(X_1 = i) \cdot \text{Prob}(X_2 = j).$$

Using the abbreviation $p_{ij} = \text{Prob}(X_1 = i, X_2 = j)$, we rewrite this as

$$p_{ij} = \left(\sum_{\nu=1}^{m_2} p_{i\nu} \right) \cdot \left(\sum_{\mu=1}^{m_1} p_{\mu j} \right) \qquad \text{for all } i \in [m_1], j \in [m_2].$$

The independence model is a toric model with $m = m_1 \cdot m_2$ and $d = m_1 + m_2$. Let Δ be the $(m-1)$-dimensional simplex (with coordinates p_{ij}) consisting of all joint probability distributions. A point $p \in \Delta$ lies in the image of the map

$$\mathbf{f} : \mathbb{R}^d \to \mathbb{R}^m, \quad (\theta_1, \ldots, \theta_d) \mapsto \frac{1}{\sum_{ij} \theta_i \theta_{j+m_1}} \cdot \left(\theta_i \theta_{j+m_1} \right)_{i \in [m_1], j \in [m_2]}$$

if and only if X_1 and X_2 are independent, i.e., if and only if the $m_1 \times m_2$ matrix (p_{ij}) has rank one. The map $\mathbf{f}$ can be represented by a $d \times m$ matrix A whose entries are in $\{0, 1\}$, with precisely two 1s per column. Here is an example.

Example 1.12 As an illustration consider the independence model for a binary random variable and a ternary random variable ($m_1 = 2, m_2 = 3$). Here

$$A = \begin{array}{c} \\ \theta_1 \\ \theta_2 \\ \theta_3 \\ \theta_4 \\ \theta_5 \end{array} \begin{array}{c} p_{11} \ \ p_{12} \ \ p_{13} \ \ p_{21} \ \ p_{22} \ \ p_{23} \\ \left(\begin{array}{cccccc} 1 & 1 & 1 & 0 & 0 & 0 \\ 0 & 0 & 0 & 1 & 1 & 1 \\ 1 & 0 & 0 & 1 & 0 & 0 \\ 0 & 1 & 0 & 0 & 1 & 0 \\ 0 & 0 & 1 & 0 & 0 & 1 \end{array} \right) \end{array}$$

This matrix A encodes the rational map $\mathbf{f} : \mathbb{R}^5 \to \mathbb{R}^{2 \times 3}$ given by

$$(\theta_1, \theta_2; \theta_3, \theta_4, \theta_5) \ \mapsto \ \frac{1}{(\theta_1 + \theta_2)(\theta_3 + \theta_4 + \theta_5)} \cdot \begin{pmatrix} \theta_1\theta_3 & \theta_1\theta_4 & \theta_1\theta_5 \\ \theta_2\theta_3 & \theta_2\theta_4 & \theta_2\theta_5 \end{pmatrix}.$$

Note that $\mathbf{f}(\mathbb{R}^5_{>0})$ consists of all positive 2×3 matrices of rank 1 whose entries sum to 1. The effective dimension of this model is three, which is one less than the rank of A. We can represent this model with only three parameters $(\theta_1, \theta_3, \theta_4) \in (0, 1)^3$ by setting $\theta_2 = 1 - \theta_1$ and $\theta_5 = 1 - \theta_3 - \theta_4$. $\qquad \square$

Maximum likelihood estimation for the independence model is easy: the optimal parameters are the normalized row and column sums of the data matrix.

Proposition 1.13 *Let $u = (u_{ij})$ be an $m_1 \times m_2$ matrix of positive integers. Then the maximum likelihood parameters $\widehat{\theta}$ for these data in the independence model are given by the normalized row and column sums of the matrix:*

$$\widehat{\theta}_\mu \ = \ \frac{1}{N} \cdot \sum_{\nu \in [m_2]} u_{\mu\nu} \quad and \quad \widehat{\theta}_{\nu+m_1} \ = \ \frac{1}{N} \cdot \sum_{\mu \in [m_1]} u_{\mu\nu} \qquad for \ \mu \in [m_1], \ \nu \in [m_2].$$

Proof We present the proof for the case $m_1 = 2, m_2 = 3$ in Example 1.12. The general case is completely analogous. Consider the reduced parameterization

$$\mathbf{f}(\theta) \ = \ \begin{pmatrix} \theta_1\theta_3 & \theta_1\theta_4 & \theta_1(1 - \theta_3 - \theta_4) \\ (1 - \theta_1)\theta_3 & (1 - \theta_1)\theta_4 & (1 - \theta_1)(1 - \theta_3 - \theta_4) \end{pmatrix}.$$

The log-likelihood function equals

$$\ell(\theta) \ = \ (u_{11} + u_{12} + u_{13}) \cdot \log(\theta_1) \ + \ (u_{21} + u_{22} + u_{23}) \cdot \log(1 - \theta_1)$$
$$+ (u_{11} + u_{21}) \cdot \log(\theta_3) + (u_{12} + u_{22}) \cdot \log(\theta_4) + (u_{13} + u_{23}) \cdot \log(1 - \theta_3 - \theta_4).$$

Taking the derivative of $\ell(\theta)$ with respect to θ_1 gives

$$\frac{\partial \ell}{\partial \theta_1} \ = \ \frac{u_{11} + u_{12} + u_{13}}{\theta_1} - \frac{u_{21} + u_{22} + u_{23}}{1 - \theta_1}.$$

Setting this expression to zero, we find that

$$\widehat{\theta}_1 \ = \ \frac{u_{11} + u_{12} + u_{13}}{u_{11} + u_{12} + u_{13} + u_{21} + u_{22} + u_{23}} \ = \ \frac{1}{N} \cdot (u_{11} + u_{12} + u_{13}).$$

Similarly, by setting $\partial\ell/\partial\theta_3$ and $\partial\ell/\partial\theta_4$ to zero, we get

$$\widehat{\theta}_3 \ = \ \frac{1}{N} \cdot (u_{11} + u_{21}) \quad and \quad \widehat{\theta}_4 \ = \ \frac{1}{N} \cdot (u_{12} + u_{22}). \qquad \square$$

1.3 Expectation Maximization

In the last section we saw that linear models and toric models enjoy the property that the likelihood function has at most one local maximum. Unfortunately, this property fails for most other algebraic statistical models, including the ones that are actually used in computational biology. A simple example of a model whose likelihood function has multiple local maxima will be featured in this section. For many models that are neither linear nor toric, statisticians use a numerical optimization technique called *Expectation Maximization* (or *EM* for short) for maximizing the likelihood function. This technique is known to perform well on many problems of practical interest. However, it must be emphasized that EM is not guaranteed to reach a global maximum. Under some conditions, it will converge to a local maximum of the likelihood function, but sometimes even this fails, as we shall see in our little example.

We introduce Expectation Maximization for the following class of algebraic statistical models. Let $F = (f_{ij}(\theta))$ be an $m \times n$ matrix of polynomials (or rational functions, as in the toric case) in the unknown parameters $\theta = (\theta_1, \ldots, \theta_d)$. We assume that the sum of all the $f_{ij}(\theta)$ equals the constant 1, and there exists an open subset $\Theta \subset \mathbb{R}^d$ of admissible parameters such that $f_{ij}(\theta) > 0$ for all $\theta \in \Theta$. We identify the matrix F with the polynomial map $F : \mathbb{R}^d \to \mathbb{R}^{m \times n}$ whose coordinates are the $f_{ij}(\theta)$. Here $\mathbb{R}^{m \times n}$ denotes the mn-dimensional real vector space consisting of all $m \times n$ matrices. We shall refer to F as the *hidden model* or the *complete data model*.

The key assumption we make about the hidden model F is that it has an easy and reliable algorithm for solving the maximum likelihood problem (1.19). For instance, F could be a linear model or a toric model, so that the likelihood function has at most one local maximum in Θ, and this global maximum can be found efficiently and reliably using the techniques of convex optimization. For special toric models, such as the independence model and certain Markov models, there are simple explicit formulas for the maximum likelihood estimates. See Propositions 1.13, 1.17 and 1.18 for such formulas.

Consider the linear map which takes an $m \times n$ matrix to its vector of row sums

$$\rho : \mathbb{R}^{m \times n} \to \mathbb{R}^m, \quad G = (g_{ij}) \mapsto \left(\sum_{j=1}^n g_{1j}, \sum_{j=1}^n g_{2j}, \ldots, \sum_{j=1}^n g_{mj} \right).$$

The *observed model* is the composition $\mathbf{f} = \rho \circ F$ of the hidden model F and the *marginalization map* ρ. The observed model is the one we really care about:

$$\mathbf{f} : \mathbb{R}^d \to \mathbb{R}^m, \quad \theta \mapsto \left(\sum_{j=1}^n f_{1j}(\theta), \sum_{j=1}^n f_{2j}(\theta), \ldots, \sum_{j=1}^n f_{mj}(\theta) \right). \quad (1.33)$$

Hence $f_i(\theta) = \sum_{j=1}^m f_{ij}(\theta)$. The model $\mathbf{f}$ is also known as the *partial data model*. Suppose we are given a vector $u = (u_1, u_2, \ldots, u_m) \in \mathbb{N}^m$ of data for the observed model $\mathbf{f}$. Our problem is to maximize the likelihood function for these data with respect to the observed model:

$$\text{maximize} \quad L_{\text{obs}}(\theta) \;=\; f_1(\theta)^{u_1} f_2(\theta)^{u_2} \cdots f_m(\theta)^{u_m} \quad \text{subject to } \theta \in \Theta. \quad (1.34)$$

This is a hard problem, for instance, because of multiple local solutions. Suppose we have no idea how to solve (1.34). It would be much easier to solve the corresponding problem for the hidden model F instead:

$$\text{maximize} \quad L_{\text{hid}}(\theta) \;=\; f_{11}(\theta)^{u_{11}} \cdots f_{mn}(\theta)^{u_{mn}} \quad \text{subject to} \quad \theta \in \Theta. \quad (1.35)$$

The trouble is, however, that we do not know the *hidden data*, that is, we do not know the matrix $U = (u_{ij}) \in \mathbb{N}^{m \times n}$. All we know about the matrix U is that its row sums are equal to the data we do know, in symbols, $\rho(U) = u$.

The idea of the EM algorithm is as follows. We start with some initial guess of what the parameter vector θ might be. Then we make an estimate, given θ, of what we expect the hidden data U might be. This latter step is called the *expectation step* (or *E-step* for short). Note that the expected values for the hidden data vector do not have to be integers. Next we solve the problem (1.35) to optimality, using the easy and reliable subroutine which we assumed is available for the hidden model F. This step is called the *maximization step* (or *M-step* for short). Let θ^* be the optimal solution found in the M-step. We then replace the old parameter guess θ by the new and improved parameter guess θ^*, and we iterate the process $E \to M \to E \to M \to E \to M \to \cdots$ until we are satisfied. Of course, what needs to be shown is that the likelihood function increases during this process and that the sequence of parameter guesses θ converges to a local maximum of $L_{\text{obs}}(\theta)$. We state the EM procedure formally in Algorithm 1.14. As before, it is more convenient to work with log-likelihood functions than with likelihood functions, and we abbreviate

$$\ell_{\text{obs}}(\theta) \;:=\; \log\big(L_{\text{obs}}(\theta)\big) \quad \text{and} \quad \ell_{\text{hid}}(\theta) \;:=\; \log\big(L_{\text{hid}}(\theta)\big).$$

Algorithm 1.14 (EM Algorithm)

Input: An $m \times n$ matrix of polynomials $f_{ij}(\theta)$ representing the hidden model F and observed data $u \in \mathbb{N}^m$.
Output: A proposed maximum $\widehat{\theta} \in \Theta \subset \mathbb{R}^d$ of the log-likelihood function $\ell_{\text{obs}}(\theta)$ for the observed model **f**.

Step 0: Select a threshold $\epsilon > 0$ and select starting parameters $\theta \in \Theta$ satisfying $f_{ij}(\theta) > 0$ for all i, j.
E-Step: Define the *expected hidden data matrix* $U = (u_{ij}) \in \mathbb{R}^{m \times n}$ by

$$u_{ij} \quad := \quad u_i \cdot \frac{f_{ij}(\theta)}{\sum_{j=1}^{m} f_{ij}(\theta)} \quad = \quad \frac{u_i}{f_i(\theta)} \cdot f_{ij}(\theta).$$

M-Step: Compute the solution $\theta^* \in \Theta$ to the maximum likelihood problem (1.35) for the hidden model $F = (f_{ij})$.
Step 3: If $\ell_{\text{obs}}(\theta^*) - \ell_{\text{obs}}(\theta) > \epsilon$ then set $\theta := \theta^*$ and go back to the E-Step.
Step 4: Output the parameter vector $\widehat{\theta} := \theta^*$ and the corresponding probability distribution $\widehat{p} = f(\widehat{\theta})$ on the set $[m]$.

The justification for this algorithm is given by the following theorem.

Theorem 1.15 *The value of the likelihood function weakly increases during each iteration of the EM algorithm, in other words, if θ is chosen in the open set Θ prior to the E-step and θ^* is computed by one E-step and one M-step then $\ell_{\text{obs}}(\theta) \leq \ell_{\text{obs}}(\theta^*)$. If $\ell_{\text{obs}}(\theta) = \ell_{\text{obs}}(\theta^*)$ then θ^* is a critical point of the likelihood function ℓ_{obs}.*

Proof We use the following fact about the logarithm of a positive number x:

$$\log(x) \leq x - 1 \quad \text{with equality if and only if} \quad x = 1. \tag{1.36}$$

Let $u \in \mathbb{N}^n$ and $\theta \in \Theta$ be given prior to the E-step, let $U = (u_{ij})$ be the matrix computed in the E-step, and let $\theta^* \in \Theta$ be the vector computed in the subsequent M-step. We consider the difference between the values at θ^* and θ of the log-likelihood function of the observed model:

$$
\ell_{\text{obs}}(\theta^*) - \ell_{\text{obs}}(\theta) \;=\; \sum_{i=1}^{m} u_i \cdot \left[\log(f_i(\theta^*)) - \log(f_i(\theta)) \right]
$$

$$
= \sum_{i=1}^{m} \sum_{j=1}^{n} u_{ij} \cdot \left[\log(f_{ij}(\theta^*)) - \log(f_{ij}(\theta)) \right] \tag{1.37}
$$

$$
+ \sum_{i=1}^{m} u_i \cdot \left(\log\!\left(\frac{f_i(\theta^*)}{f_i(\theta)} \right) - \sum_{j=1}^{n} \frac{u_{ij}}{u_i} \cdot \log\!\left(\frac{f_{ij}(\theta^*)}{f_{ij}(\theta)} \right) \right).
$$

The double-sum in the middle equals $\ell_{\text{hid}}(\theta^*) - \ell_{\text{hid}}(\theta)$. This difference is non-negative because the parameter vector θ^* was chosen so as to maximize the log-likelihood function for the hidden model with data (u_{ij}). We next show that the last sum is non-negative as well. The parenthesized expression equals

$$
\log\!\left(\frac{f_i(\theta^*)}{f_i(\theta)} \right) - \sum_{j=1}^{n} \frac{u_{ij}}{u_i} \log\!\left(\frac{f_{ij}(\theta^*)}{f_{ij}(\theta)} \right) \;=\; \log\!\left(\frac{f_i(\theta^*)}{f_i(\theta)} \right) + \sum_{j=1}^{n} \frac{f_{ij}(\theta)}{f_i(\theta)} \log\!\left(\frac{f_{ij}(\theta)}{f_{ij}(\theta^*)} \right).
$$

We rewrite this expression as follows

$$
\sum_{j=1}^{n} \frac{f_{ij}(\theta)}{f_i(\theta)} \cdot \log\!\left(\frac{f_i(\theta^*)}{f_i(\theta)} \right) + \sum_{j=1}^{n} \frac{f_{ij}(\theta)}{f_i(\theta)} \cdot \log\!\left(\frac{f_{ij}(\theta)}{f_{ij}(\theta^*)} \right)
$$

$$
= \sum_{j=1}^{n} \frac{f_{ij}(\theta)}{f_i(\theta)} \cdot \log\!\left(\frac{f_i(\theta^*)}{f_{ij}(\theta^*)} \cdot \frac{f_{ij}(\theta)}{f_i(\theta)} \right). \tag{1.38}
$$

This last expression is non-negative. This can be seen as follows. Consider the non-negative quantities

$$
\pi_j = \frac{f_{ij}(\theta)}{f_i(\theta)} \quad \text{and} \quad \sigma_j = \frac{f_{ij}(\theta^*)}{f_i(\theta^*)} \qquad \text{for } j = 1, 2, \ldots, n.
$$

We have $\pi_1 + \cdots + \pi_n = \sigma_1 + \cdots + \sigma_n = 1$, so the vectors π and σ can be regarded as probability distributions on the set $[n]$. The expression (1.38) equals the *Kullback–Leibler distance* between these two probability distributions:

$$H(\pi||\sigma) \;\; = \;\; \sum_{j=1}^{n}(-\pi_j)\cdot\log\left(\frac{\sigma_j}{\pi_j}\right) \;\; \geq \;\; \sum_{j=1}^{n}(-\pi_j)\cdot\left(1 - \frac{\sigma_j}{\pi_j}\right) \;\; = \;\; 0. \quad (1.39)$$

The inequality follows from (1.36).

If $\ell_{\mathrm{obs}}(\theta) = \ell_{\mathrm{obs}}(\theta^*)$, then the two terms in (1.37) are both zero. Since equality holds in (1.39) if and only if $\pi = \sigma$, it follows that

$$\frac{f_{ij}(\theta)}{f_i(\theta)} = \frac{f_{ij}(\theta^*)}{f_i(\theta^*)} \qquad \text{for } i = 1, 2, \ldots, m, \; j = 1, 2, \ldots, n. \qquad (1.40)$$

Therefore,

$$0 \;\; = \;\; \frac{\partial\ell_{\mathrm{hid}}(\theta^*)}{\partial\theta_k^*} \;\; = \;\; \sum_{i=1}^{m}\sum_{j=1}^{n}\frac{u_{ij}}{f_{ij}(\theta^*)}\cdot\frac{\partial f_{ij}}{\partial\theta_k}(\theta^*) \;\; \stackrel{(1.40)}{=} \;\; \sum_{i=1}^{m}\sum_{j=1}^{n}\frac{u_i}{f_i(\theta^*)}\cdot\frac{\partial f_{ij}}{\partial\theta_k}(\theta^*)$$

$$= \;\; \sum_{i=1}^{m}\frac{u_i}{f_i(\theta^*)}\cdot\left(\frac{\partial}{\partial\theta_k}\sum_{j=1}^{n}f_{ij}\right)(\theta^*) \;\; = \;\; \frac{\partial\ell_{\mathrm{obs}}(\theta^*)}{\partial\theta_k^*}.$$

This means that θ^* is a critical point of ℓ_{obs}.   □

The remainder of this section is devoted to a simple example which will illustrate the EM algorithm and the issue of multiple local maxima for $\ell(\theta)$.

Example 1.16 Our data are two DNA sequences of length 40:

$$\begin{array}{l} \texttt{ATCACCAAACATTGGGATGCCTGTGCATTTGCAAGCGGCT} \\ \texttt{ATGAGTCTTAAACGCTGGCCATGTGCCATCTTAGACAGCG} \end{array} \qquad (1.41)$$

We wish to test the hypothesis that these two sequences were generated by DiaNA using one biased coin and four tetrahedral dice, each with four faces labeled by the letters A, C, G and T. Two of her dice are in her left pocket, and the other two dice are in her right pocket. Our model states that DiaNA generated each column of this alignment independently by the following process. She first tosses her coin. If the coin comes up heads, she rolls the two dice in her left pocket, and if the coin comes up tails she rolls the two dice in her right pocket. In either case DiaNA reads off the column of the alignment from the two dice she rolled. All dice have a different color, so she knows which of the dice correspond to the first and second sequences.

To represent this model algebraically, we introduce the vector of parameters

$$\theta \;\; = \;\; \left(\pi, \lambda_{\mathsf{A}}^1, \lambda_{\mathsf{C}}^1, \lambda_{\mathsf{G}}^1, \lambda_{\mathsf{T}}^1, \lambda_{\mathsf{A}}^2, \lambda_{\mathsf{C}}^2, \lambda_{\mathsf{G}}^2, \lambda_{\mathsf{T}}^2, \rho_{\mathsf{A}}^1, \rho_{\mathsf{C}}^1, \rho_{\mathsf{G}}^1, \rho_{\mathsf{T}}^1, \rho_{\mathsf{A}}^2, \rho_{\mathsf{C}}^2, \rho_{\mathsf{G}}^2, \rho_{\mathsf{T}}^2\right).$$

The parameter π represents the probability that DiaNA's coin comes up heads. The parameter λ_j^i represents the probability that the ith dice in her left pocket comes up with nucleotide j. The parameter ρ_j^i represents the probability that the ith dice in her right pocket comes up with nucleotide j. In total there are $d = 13$ free parameters because

$$\lambda_{\mathsf{A}}^i + \lambda_{\mathsf{C}}^i + \lambda_{\mathsf{G}}^i + \lambda_{\mathsf{T}}^i \;\; = \;\; \rho_{\mathsf{A}}^i + \rho_{\mathsf{C}}^i + \rho_{\mathsf{G}}^i + \rho_{\mathsf{T}}^i \;\; = \;\; 1 \qquad \text{for } i = 1, 2.$$

More precisely, the parameter space in this example is a product of simplices

$$\Theta \;=\; \Delta_1 \times \Delta_3 \times \Delta_3 \times \Delta_3 \times \Delta_3.$$

The model is given by the polynomial map

$$\mathbf{f} \;:\; \mathbb{R}^{13} \to \mathbb{R}^{4\times 4}, \; \theta \mapsto (f_{ij}) \quad \text{where} \quad f_{ij} \;=\; \pi\cdot\lambda_i^1\cdot\lambda_j^2 + (1-\pi)\cdot\rho_i^1\cdot\rho_j^2. \tag{1.42}$$

The image of $\mathbf{f}$ is an 11-dimensional algebraic variety in the 15-dimensional probability simplex Δ, more precisely, $\mathbf{f}(\Theta)$ consists of all non-negative 4×4 matrices of rank at most two having coordinate sum 1. The difference in dimensions (11 versus 13) means that this model is *non-identifiable*: the preimage $\mathbf{f}^{-1}(v)$ of a rank-2 matrix $v \in \mathbf{f}(\Theta)$ is a surface in the parameters space Θ.

Now consider the given alignment (1.41). Each pair of distinct nucleotides occurs in precisely two columns. For instance, the pair $(\mathtt{C},\mathtt{G})$ occurs in the third and fifth columns of (1.41). Each of the four identical pairs of nucleotides (namely $\mathtt{AA}$, $\mathtt{CC}$, $\mathtt{GG}$ and $\mathtt{TT}$) occurs in precisely four columns of the alignment. We summarize our data in the following 4×4 matrix of counts:

$$u \;=\; \begin{array}{c@{\quad}c} & \begin{array}{cccc} \mathtt{A} & \mathtt{C} & \mathtt{G} & \mathtt{T} \end{array} \\ \begin{array}{c} \mathtt{A} \\ \mathtt{C} \\ \mathtt{G} \\ \mathtt{T} \end{array} & \left(\begin{array}{cccc} 4 & 2 & 2 & 2 \\ 2 & 4 & 2 & 2 \\ 2 & 2 & 4 & 2 \\ 2 & 2 & 2 & 4 \end{array}\right) \end{array}. \tag{1.43}$$

Our goal is to find parameters θ which maximize the log-likelihood function

$$\ell_{\mathrm{obs}}(\theta) \;=\; 4 \cdot \sum_i \log(f_{ii}(\theta)) + 2 \cdot \sum_{i \neq j} \log(f_{ij}(\theta)) :$$

here the summation indices i, j range over $\{\mathtt{A},\mathtt{C},\mathtt{G},\mathtt{T}\}$. Maximizing $\ell_{\mathrm{obs}}(\theta)$ means finding a 4×4 matrix $\mathbf{f}(\theta)$ of rank 2 that is close (in the statistical sense of maximum likelihood) to the empirical distribution $(1/40) \cdot u$.

We apply the EM algorithm to this problem. The hidden data is the decomposition of the given alignment into two subalignments according to the contributions made by dice from DiaNA's left and right pocket respectively:

$$u_{ij} \;=\; u_{ij}^l + u_{ij}^r \quad \text{for all} \quad i,j \in \{\mathtt{A},\mathtt{C},\mathtt{G},\mathtt{T}\}.$$

The hidden model equals

$$F \;:\; \mathbb{R}^{13} \to \mathbb{R}^{2\times 4\times 4}, \; \theta \mapsto (f_{ij}^l, f_{ij}^r)$$

$$\text{where} \quad f_{ij}^l \;=\; \pi \cdot \lambda_i^1 \cdot \lambda_j^2 \quad \text{and} \quad f_{ij}^r \;=\; (1-\pi) \cdot \rho_i^1 \cdot \rho_i^2.$$

The hidden model consists of two copies of the independence model for two random variables on $\{\mathtt{A},\mathtt{C},\mathtt{G},\mathtt{T}\}$, one copy for left and the other copy for right. In light of Proposition 1.13, it is easy to maximize the hidden likelihood function $L_{\mathrm{hid}}(\theta)$: we just need to divide the row and column sums of the hidden data matrices by the grand total. This is the M-step in our algorithm.

The EM algorithm starts in Step 0 by selecting a vector of initial parameters

$$\theta = \big(\pi, (\lambda_{\mathtt{A}}^1, \lambda_{\mathtt{C}}^1, \lambda_{\mathtt{G}}^1, \lambda_{\mathtt{T}}^1), (\lambda_{\mathtt{A}}^2, \lambda_{\mathtt{C}}^2, \lambda_{\mathtt{G}}^2, \lambda_{\mathtt{T}}^2), (\rho_{\mathtt{A}}^1, \rho_{\mathtt{C}}^1, \rho_{\mathtt{G}}^1, \rho_{\mathtt{T}}^1), (\rho_{\mathtt{A}}^2, \rho_{\mathtt{C}}^2, \rho_{\mathtt{G}}^2, \rho_{\mathtt{T}}^2)\big). \tag{1.44}$$

Then the current value of the log-likelihood function equals

$$\ell_{\text{obs}}(\theta) \;=\; \sum_{ij} u_{ij} \cdot \log\bigl(\pi \cdot \lambda_i^1 \cdot \lambda_j^2 + (1 - \pi) \cdot \rho_i^1 \cdot \rho_j^2\bigr). \qquad (1.45)$$

In the E-step we compute the expected hidden data by the following formulas:

$$u_{ij}^l \;:=\; u_{ij} \cdot \frac{\pi \cdot \lambda_i^1 \cdot \lambda_j^2}{\pi \cdot \lambda_i^1 \cdot \lambda_j^2 + (1 - \pi) \cdot \rho_i^1 \cdot \rho_j^2} \qquad \text{for } i,j \in \{\mathsf{A}, \mathsf{C}, \mathsf{G}, \mathsf{T}\},$$

$$u_{ij}^r \;:=\; u_{ij} \cdot \frac{(1 - \pi) \cdot \rho_i^1 \cdot \rho_j^2}{\pi \cdot \lambda_i^1 \cdot \lambda_j^2 + (1 - \pi) \cdot \rho_i^1 \cdot \rho_j^2} \qquad \text{for } i,j \in \{\mathsf{A}, \mathsf{C}, \mathsf{G}, \mathsf{T}\}.$$

In the subsequent M-step we now compute the maximum likelihood parameters $\theta^* = \bigl(\pi^*, \lambda_\mathsf{A}^{1*}, \lambda_\mathsf{C}^{1*}, \ldots, \rho_\mathsf{T}^{2*}\bigr)$ for the hidden model F. This is done by taking row sums and column sums of the matrix (u_{ij}^l) and the matrix (u_{ij}^r), and by defining the next parameters π to be the relative total counts of these two matrices. In symbols, in the M-step we perform the following computations:

$$\pi^* = \tfrac{1}{N} \cdot \sum_{ij} u_{ij}^l,$$

$$\lambda_i^{1*} = \tfrac{1}{N} \cdot \sum_j u_{ij}^l \quad \text{and} \quad \rho_i^{1*} = \tfrac{1}{N} \cdot \sum_j u_{ij}^r \qquad \text{for} \quad i \in \{\mathsf{A}, \mathsf{C}, \mathsf{G}, \mathsf{T}\},$$

$$\lambda_j^{2*} = \tfrac{1}{N} \cdot \sum_i u_{ij}^l \quad \text{and} \quad \rho_j^{2*} = \tfrac{1}{N} \cdot \sum_i u_{ij}^r \qquad \text{for} \quad j \in \{\mathsf{A}, \mathsf{C}, \mathsf{G}, \mathsf{T}\}.$$

Here $N = \sum_{ij} u_{ij} = \sum_{ij} u_{ij}^l + \sum_{ij} u_{ij}^r$ is the sample size of the data.

After the M-step, the new value $\ell_{\text{obs}}(\theta^*)$ of the likelihood function is computed, using the formula (1.45). If $\ell_{\text{obs}}(\theta^*) - \ell_{\text{obs}}(\theta)$ is small enough then we stop and output the vector $\widehat{\theta} = \theta^*$ and the corresponding 4×4 matrix $\mathbf{f}(\widehat{\theta})$. Otherwise we set $\theta = \theta^*$ and return to the E-step.

Here are four numerical examples for the data (1.43) with sample size $N = 40$. In each of our experiments, the starting vector θ is indexed as in (1.44). Our choices of starting vectors are not meant to reflect the reality of computational statistics. In practice, one would chose the starting parameters much more generically, so as to avoid singularities and critical points of the likelihood function. Our only objective here is to give a first illustration of the algorithm.

Experiment 1: We pick uniform starting parameters

$$\theta \;=\; \bigl(0.5, \;\; (0.25, 0.25, 0.25, 0.25), \;\; (0.25, 0.25, 0.25, 0.25),$$
$$(0.25, 0.25, 0.25, 0.25), \;\; (0.25, 0.25, 0.25, 0.25)\bigr).$$

The parameter vector θ is a stationary point of the EM algorithm, so after one step we output $\widehat{\theta} = \theta$. The resulting estimated probability distribution on pairs of nucleotides is the uniform distribution

$$\mathbf{f}(\widehat{\theta}) \;=\; \frac{1}{16}\begin{pmatrix} 1 & 1 & 1 & 1 \\ 1 & 1 & 1 & 1 \\ 1 & 1 & 1 & 1 \\ 1 & 1 & 1 & 1 \end{pmatrix}, \qquad \ell_{\text{obs}}(\widehat{\theta}) \;=\; -110.903548889592\ldots$$

Here $\widehat{\theta}$ is a critical point of the log-likelihood function $\ell_{\text{obs}}(\theta)$ but it is not a local maximum. The Hessian matrix of $\ell_{\mathrm{obs}}(\theta)$ evaluated at $\widehat{\theta}$ has both positive and negative eigenvalues. The characteristic polynomial of the Hessian equals

$$z(z-64)(z-16)^2(z+16)^2(z+64)(z+80)^4(z+320)^2.$$

Experiment 2: We decrease the starting parameter λ_{A}^1 and we increase λ_{C}^1:

$$\theta \;=\; \big(\,0.5, \quad (0.2,0.3,0.25,0.25),\ (0.25,0.25,0.25,0.25),$$
$$(0.25,0.25,0.25,0.25),\ (0.25,0.25,0.25,0.25)\,\big).$$

Now the EM algorithm converges to a distribution which is a local maximum:

$$\mathbf{f}(\widehat{\theta}) \;=\; \frac{1}{54}\cdot\begin{pmatrix} 6 & 3 & 3 & 3 \\ 3 & 4 & 4 & 4 \\ 3 & 4 & 4 & 4 \\ 3 & 4 & 4 & 4 \end{pmatrix}, \qquad \ell_{\mathrm{obs}}(\widehat{\theta}) \;=\; -110.152332481077\ldots$$

The Hessian of $\ell_{\mathrm{obs}}(\theta)$ at $\widehat{\theta}$ has rank 11, and all eleven non-zero eigenvalues are distinct and negative.

Experiment 3: We next increase the starting parameter ρ_{A}^1 and we decrease ρ_{C}^1:

$$\theta \;=\; \big(\,0.5, \quad (0.2,0.3,0.25,0.25),\ (0.25,0.25,0.25,0.25),$$
$$(0.3,0.2,0.25,0.25),\ (0.25,0.25,0.25,0.25)\,\big).$$

The EM algorithm converges to a distribution which is a saddle point of ℓ_{obs}:

$$\mathbf{f}(\widehat{\theta}) \;=\; \frac{1}{48}\cdot\begin{pmatrix} 4 & 2 & 3 & 3 \\ 2 & 4 & 3 & 3 \\ 3 & 3 & 3 & 3 \\ 3 & 3 & 3 & 3 \end{pmatrix}, \qquad \ell_{\mathrm{obs}}(\widehat{\theta}) \;=\; -110.223952742410\ldots$$

The Hessian of $\ell_{\mathrm{obs}}(\theta)$ at $\widehat{\theta}$ has rank 11, with nine eigenvalues negative.

Experiment 4: Let us now try the following starting parameters:

$$\theta \;=\; \big(\,0.5, \quad (0.2,0.3,0.25,0.25),\ (0.25,0.2,0.3,0.25),$$
$$(0.25,0.25,0.25,0.25),\ (0.25,0.25,0.25,0.25)\,\big).$$

The EM algorithm converges to a probability distribution which is a local maximum of the likelihood function. This maximum is greater than the local maximum found previously in Experiment 2. The new winner is

$$\mathbf{f}(\widehat{\theta}) \;=\; \frac{1}{40}\cdot\begin{pmatrix} 3 & 3 & 2 & 2 \\ 3 & 3 & 2 & 2 \\ 2 & 2 & 3 & 3 \\ 2 & 2 & 3 & 3 \end{pmatrix}, \qquad \ell_{\mathrm{obs}}(\widehat{\theta}) \;=\; -110.098128348563\ldots$$

All 11 nonzero eigenvalues of the Hessian of $\ell_{\mathrm{obs}}(\theta)$ are distinct and negative.

We repeated this experiment many times with random starting values, and we never found a parameter vector that was better than the one found in Experiment 4. Based on these findings, we would like to conclude that the maximum value of the observed likelihood function is attained by our best solution:

$$\max \{ L_{\mathrm{obs}}(\theta) \; : \; \theta \in \Theta \} \;\; = \;\; \frac{2^{16} \cdot 3^{24}}{40^{40}} \;\; = \;\; e^{-110.0981283}. \tag{1.46}$$

Assuming that this conclusion is correct, let us discuss the set of all optimal solutions. Since the data matrix u is invariant under the action of the symmetric group on $\{\mathtt{A}, \mathtt{C}, \mathtt{G}, \mathtt{T}\}$, that group also acts on the set of optimal solutions. There are three matrices like the one found in Experiment 4:

$$\frac{1}{40} \cdot \begin{pmatrix} 3 & 3 & 2 & 2 \\ 3 & 3 & 2 & 2 \\ 2 & 2 & 3 & 3 \\ 2 & 2 & 3 & 3 \end{pmatrix}, \quad \frac{1}{40} \cdot \begin{pmatrix} 3 & 2 & 3 & 2 \\ 2 & 3 & 2 & 3 \\ 3 & 2 & 3 & 2 \\ 2 & 3 & 2 & 3 \end{pmatrix} \quad \text{and} \quad \frac{1}{40} \cdot \begin{pmatrix} 3 & 2 & 2 & 3 \\ 2 & 3 & 3 & 2 \\ 2 & 3 & 3 & 2 \\ 3 & 2 & 2 & 3 \end{pmatrix}. \tag{1.47}$$

The preimage of each of these matrices under the polynomial map $\mathbf{f}$ is a surface in the space of parameters θ, namely, it consists of all representations of a rank 2 matrix as a convex combination of two rank 1 matrices. The topology of such "spaces of explanations" were studied in [Mond *et al.*, 2003]. The result (1.46) indicates that the set of optimal solutions to the maximum likelihood problem is the disjoint union of three "surfaces of explanations".

But how do we know that (1.46) is actually true? Does running the EM algorithm $100,000$ times without converging to a parameter vector whose likelihood is larger constitute a mathematical proof? Can it be turned into a mathematical proof? Algebraic techniques for addressing such questions will be introduced in Section 3.3. For a numerical approach see Chapter 20. $\qquad \square$

1.4 Markov models

We now introduce Markov chains, hidden Markov models and Markov models on trees, using the algebraic notation of the previous sections. While our presentation is self-contained, readers may find it useful to compare with the (more standard) description of these models in [Durbin *et al.*, 1998] or other text books. A natural point of departure is the following toric model.

1.4.1 Toric Markov chains

We fix an alphabet Σ with l letters, and we fix a positive integer n. We shall define a toric model whose state space is the set Σ^n of all words of length n. The model is parameterized by the set Θ of positive $l \times l$ matrices. Thus the number of parameters is $d = l^2$ and the number of states is $m = l^n$.

Every toric model with d parameters and m states is represented by a $d \times m$ matrix A with integer entries as in Section 1.2. The $d \times m$ matrix which represents the toric Markov model will be denoted by $A_{l,n}$. Its rows are indexed by Σ^2 and its columns indexed by Σ^n. The entry of the matrix $A_{l,n}$ in the row indexed by the pair $\sigma_1\sigma_2 \in \Sigma^2$ and the column indexed by the word $\pi_1\pi_2 \cdots \pi_n \in \Sigma^n$ is the number of occurrences of the pair inside the word, i.e., the number of indices $i \in \{1, \ldots, n-1\}$ such that $\sigma_1\sigma_2 = \pi_i\pi_{i+1}$. We define the *toric Markov chain model* to be the toric model specified by the matrix $A_{l,n}$.

For a concrete example let us consider words of length $n = 4$ over the binary alphabet $\Sigma = \{0, 1\}$, so that $l = 2$, $d = 4$ and $m = 16$. The matrix $A_{2,4}$ is the following 4×16 matrix:

	0000	0001	0010	0011	0100	0101	0110	0111	1000	1001	1010	1011	1100	1101	1110	1111
00	3	2	1	1	1	0	0	0	2	1	0	0	1	0	0	0
01	0	1	1	1	1	2	1	1	0	1	1	1	0	1	0	0
10	0	0	1	0	1	1	1	0	1	1	2	1	1	1	1	0
11	0	0	0	1	0	0	1	2	0	0	0	1	1	1	2	3

We write $\mathbb{R}^{2 \times 2}$ for the space of 2×2 matrices

$$\theta = \begin{pmatrix} \theta_{00} & \theta_{01} \\ \theta_{10} & \theta_{11} \end{pmatrix}.$$

The parameter space $\Theta \subset \mathbb{R}^{2 \times 2}$ consists of all matrices θ whose four entries θ_{ij} are positive. The toric Markov chain model of length $n = 4$ for the binary alphabet ($l = 2$) is the image of $\Theta = \mathbb{R}^{2 \times 2}_{>0}$ under the monomial map

$$\mathbf{f}_{2,4} : \mathbb{R}^{2 \times 2} \rightarrow \mathbb{R}^{16}, \quad \theta \mapsto \frac{1}{\sum_{ijkl} p_{ijkl}} \cdot (p_{0000}, p_{0001}, \ldots, p_{1111}),$$

where $\quad p_{i_1 i_2 i_3 i_4} = \theta_{i_1 i_2} \cdot \theta_{i_2 i_3} \cdot \theta_{i_3 i_4} \quad$ for all $i_1 i_2 i_3 i_4 \in \{0, 1\}^4$.

The map $\mathbf{f}_{l,n} : \mathbb{R}^d \rightarrow \mathbb{R}^m$ is defined analogously for larger alphabets and longer sequences.

The toric Markov chain model $\mathbf{f}_{2,4}(\Theta)$ is a three-dimensional object inside the 15-dimensional simplex Δ which consists of all probability distributions on the state space $\{0, 1\}^4$. Algebraically, the simplex is specified by the equation

$$p_{0000} + p_{0001} + p_{0010} + p_{0011} + \cdots + p_{1110} + p_{1111} = 1, \qquad (1.48)$$

where the p_{ijkl} are unknowns which represent the probabilities of the 16 states. To understand the geometry of the toric Markov chain model, we examine the matrix $A_{2,4}$. The 16 columns of $A_{2,4}$ represent twelve distinct points in

$$\{ (u_{00}, u_{01}, u_{10}, u_{11}) \in \mathbb{R}^{2 \times 2} : u_{00} + u_{01} + u_{10} + u_{11} = 3 \} \simeq \mathbb{R}^3.$$

The convex hull of these twelve points is the three-dimensional polytope depicted in Figure 1.1. We refer to Section 2.3 for a general introduction to polytopes. Only eight of the twelve points are vertices of the polytope.

Polytopes like the one in Figure 1.1 are important for *maximum a posteriori inference*, which is discussed in more detail in Sections 1.5, 2.2 and 4.4. Polytopes of Markov chains are discussed in detail in Chapter 10.

The adjective "toric" is used for the toric Markov chain model $\mathbf{f}_{2,4}(\Theta)$ because $\mathbf{f}_{2,4}$ is a monomial map, and so its image is a *toric variety*. (An introduction to varieties is given in Section 3.1). Every variety is characterized by a finite list of polynomials that vanish on that variety. In the context of

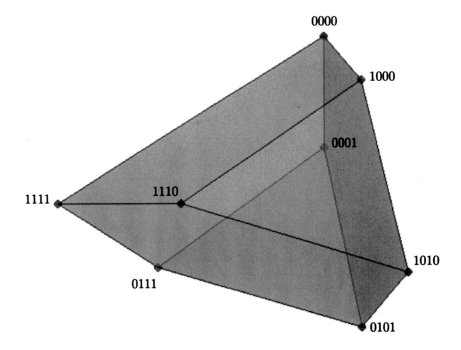

Fig. 1.1. The polytope of the toric Markov chain model $\mathbf{f}_{2,4}(\Theta)$.

statistics, these polynomials are called *model invariants*. A model invariant is an algebraic relation that holds for all probability distributions in the model. For a toric model these invariants can be derived from the geometry of its polytope. We explain this derivation for the toric Markov chain model $\mathbf{f}_{2,4}(\Theta)$.

The simplest model invariant is the equation (1.48). The other linear invariants come from the fact that the matrix $A_{2,4}$ has some repeated columns:

$$p_{0110} = p_{1011} = p_{1101} \quad \text{and} \quad p_{0010} = p_{0100} = p_{1001}. \qquad (1.49)$$

These relations state that $A_{2,4}$ is a configuration of only 12 distinct points. Next there are four relations which specify the location of the four non-vertices. Each of them is the midpoint on the segment between two of the eight vertices:

$$\begin{aligned} p_{0011}^2 &= p_{0001}p_{0111} & p_{1001}^2 &= p_{0001}p_{1010}, \\ p_{1100}^2 &= p_{1000}p_{1110} & p_{1101}^2 &= p_{0101}p_{1110}. \end{aligned} \qquad (1.50)$$

For instance, the first equation $p_{0011}^2 = p_{0001}p_{0111}$ corresponds to the following additive relation among the fourth, second and eighth column of $A_{2,4}$:

$$2 \cdot (1,1,0,1) \quad = \quad (2,1,0,0) + (0,1,0,2). \qquad (1.51)$$

The remaining eight columns of $A_{2,4}$ are vertices of the polytope depicted above. The corresponding probabilities satisfy the following relations

$$\begin{aligned} p_{0111}p_{1010} &= p_{0101}p_{1110}, & p_{0111}p_{1000} &= p_{0001}p_{1110}, & p_{0101}p_{1000} &= p_{0001}p_{1010}, \\ p_{0111}p_{1110}^2 &= p_{1010}p_{1111}^2, & p_{0111}^2p_{1110} &= p_{0101}p_{1111}^2, & p_{0001}p_{1000}^2 &= p_{0000}^2p_{1010}, \\ p_{0000}^2p_{0101} &= p_{0001}^2p_{1000}, & p_{0000}^2p_{1110}^3 &= p_{1000}^3p_{1111}^2, & p_{0000}^2p_{0111}^3 &= p_{0001}^3p_{1111}^2. \end{aligned}$$

These nine equations together with (1.48), (1.49) and (1.50) characterize the set of distributions $p \in \Delta$ that lie in the toric Markov chain model $\mathbf{f}_{2,4}(\Theta)$. Tools for computing such lists of model invariants will be presented in Chapter 3. Note that, just like in (1.51), each of the nine equations corresponds to the unique affine dependency among the vertices of a planar quadrangle formed by four of the eight vertices of the polytope in Figure 1.1.

1.4.2 Markov chains

The Markov chain model is a submodel of the toric Markov chain model. Let Θ_1 denote the subset of all matrices $\theta \in \mathbb{R}_{>0}^{l \times l}$ whose rows sum to one. The *Markov chain model* is the image of Θ_1 under the map $\mathbf{f}_{l,n}$. By a *Markov chain* we mean any point p in the model $\mathbf{f}_{l,n}(\Theta_1)$. This definition agrees with the familiar description of Markov chains in [Durbin *et al.*, 1998, Chapter 3], except that here we require the initial distribution at the first state to be uniform. This assumption is made to keep the exposition simple.

For instance, if $l = 2$ then the parameter space Θ_1 is a square. Namely, Θ_1 is the set of all pairs $(\theta_0, \theta_1) \in \mathbb{R}^2$ such that the following matrix is positive:

$$\theta = \begin{pmatrix} \theta_0 & 1 - \theta_0 \\ 1 - \theta_1 & \theta_1 \end{pmatrix}$$

The Markov chain model is the image of the square under the map $\mathbf{f}_{2,n}$. A Markov chain of length $n = 4$ is any probability distribution p of the form

$$p_{0000} = \frac{1}{2}\theta_0^3, \; p_{0001} = \frac{1}{2}\theta_0^2(1-\theta_0), \; p_{0010} = p_{1001} = p_{0100} = \frac{1}{2}\theta_0(1-\theta_0)(1-\theta_1),$$

$$p_{0011} = \frac{1}{2}\theta_0(1-\theta_0)\theta_1, \quad p_{0101} = \frac{1}{2}(1-\theta_0)^2(1-\theta_1), \quad p_{0111} = \frac{1}{2}(1-\theta_0)\theta_1^2,$$

$$p_{0110} = p_{1011} = p_{1101} = \frac{1}{2}(1-\theta_0)\theta_1(1-\theta_1), \quad p_{1010} = \frac{1}{2}(1-\theta_1)^2(1-\theta_0),$$

$$p_{1000} = \frac{1}{2}(1-\theta_1)\theta_0^2, \; p_{1100} = \frac{1}{2}\theta_1(1-\theta_1)\theta_0, \; p_{1110} = \frac{1}{2}\theta_1^2(1-\theta_1), \; p_{1111} = \frac{1}{2}\theta_1^3.$$

Thus the Markov chain model is the surface in the 15-dimensional simplex Δ given by this parameterization. It satisfies all the invariants of the toric Markov chain model plus some extra invariants due to the fact that probabilities must sum to 1, and the initial distribution is uniform. For example,

$$p_{0000} + p_{0001} + p_{0010} + p_{0011} + p_{0100} + p_{0101} + p_{0110} + p_{0111} = \frac{1}{2}.$$

We next discuss maximum likelihood estimation for Markov chains. Fix a data vector $u \in \mathbb{N}^{l^n}$ representing N observed sequences in Σ^n. The sufficient statistic $v = A_{l,n} \cdot u \in \mathbb{N}^{l^2}$ is regarded as an $l \times l$ matrix. The entry $v_{\sigma_1 i_2}$ in row σ_1 and column i_2 of the matrix v equals the number of occurrences of $\sigma_1 i_2 \in \Sigma^2$ as a consecutive pair in any of the N observed sequences.

Proposition 1.17 *The maximum likelihood estimate of the data $u \in \mathbb{N}^{l^n}$ in the Markov chain model is the $l \times l$ matrix $\widehat{\theta} = (\widehat{\theta}_{ij})$ in Θ_1 with coordinates*

$$\widehat{\theta}_{ij} \;=\; \frac{v_{ij}}{\sum_{s \in \Sigma} v_{is}} \qquad where \quad v = A_{l,n} \cdot u.$$

Proof The likelihood function for the toric Markov chain model equals

$$L(\theta) \;=\; \theta^{A_{l,n} \cdot u} \;=\; \theta^{v} \;=\; \prod_{ij \in \Sigma^2} \theta_{ij}^{v_{ij}}.$$

The log-likelihood function can be written as follows:

$$\ell(\theta) \;=\; \sum_{i \in \Sigma} \big(v_{i1} \cdot \log(\theta_{i1}) + v_{i2} \cdot \log(\theta_{i2}) + \cdots + v_{i,l-1} \cdot \log(\theta_{i,l-1}) + v_{il} \cdot \log(\theta_{il}) \big).$$

The log-likelihood function for the Markov chain model is obtained by restricting this function to the set Θ_1 of positive $l \times l$ matrices whose row sums are all equal to one. Therefore, $\ell(\theta)$ is the sum over all $i \in \Sigma$ of the expressions

$$v_{i1} \cdot \log(\theta_{i1}) + v_{i2} \cdot \log(\theta_{i2}) + \cdots + v_{i,l-1} \cdot \log(\theta_{i,l-1}) + v_{il} \cdot \log(1 - \sum_{s=1}^{l-1} \theta_{is}). \quad (1.52)$$

These expressions have disjoint sets of unknowns for different values of the index $i \in \Sigma$. To maximize $\ell(\theta)$ over Θ_1, it hence suffices to maximize the concave function (1.52) over the $(l-1)$-dimensional simplex consisting of all non-negative vectors $(\theta_{i1}, \theta_{i2}, \ldots, \theta_{i,l-1})$ of coordinate sum at most one. By equating the partial derivatives of (1.52) to zero, we see that the unique critical point has coordinates $\theta_{ij} = v_{ij}/(v_{i1} + v_{i2} + \cdots + v_{il})$ as desired. $\qquad \square$

We next introduce the *fully observed Markov model* that underlies the hidden Markov model considered in Subsection 1.4.3. We fix the sequence length n and we consider a first alphabet Σ with l letters and a second alphabet Σ' with l' letters. The observable states in this model are pairs $(\sigma, \tau) \in \Sigma^n \times (\Sigma')^n$ of words of length n. A sequence of N observations in this model is summarized in a matrix $u \in \mathbb{N}^{l^n \times (l')^n}$ where $u_{(\sigma,\tau)}$ is the number of times the pair (σ, τ) was observed. Hence, in this model, $m = (l \cdot l')^n$.

The fully observed Markov model is parameterized by a pair of matrices (θ, θ') where θ is an $l \times l$ matrix and θ' is an $l \times l'$ matrix. The matrix θ encodes a Markov chain as before: the entry θ_{ij} represents the probability of transitioning from state $i \in \Sigma$ to $j \in \Sigma$. The matrix θ' encodes the interplay between the two alphabets: the entry θ'_{ij} represents the probability of outputting symbol $j \in \Sigma'$ when the Markov chain is in state $i \in \Sigma$. As before in the Markov chain model, we restrict ourselves to positive matrices whose rows sum to one. To be precise, Θ_1 now denotes the set of pairs of matrices $(\theta, \theta') \in \mathbb{R}_{>0}^{l \times l} \times \mathbb{R}_{>0}^{l \times l'}$ whose row sums are equal to one. Hence $d = l(l+l'-2)$.

The fully observed Markov model is the restriction to Θ_1 of the toric model

$$F : \mathbb{R}^d \to \mathbb{R}^m, \quad (\theta, \theta') \mapsto p = (p_{\sigma, \tau})$$

where $\qquad p_{\sigma,\tau} \quad = \quad \frac{1}{l}\theta'_{\sigma_1\tau_1}\theta_{\sigma_1\sigma_2}\theta'_{\sigma_2\tau_2}\theta_{\sigma_2\sigma_3}\theta'_{\sigma_3\tau_3}\theta_{\sigma_3\sigma_4}\cdots\theta_{\sigma_{n-1}\sigma_n}\theta'_{\sigma_n\tau_n}.$ (1.53)

The computation of maximum likelihood estimates for this model is an extension of the method for Markov chains in Proposition 1.17. The role of the matrix $A_{l,n}$ for Markov chains is now played by the following linear map

$$ A \,:\, \mathbb{N}^{l^n \times (l')^n} \;\to\; \mathbb{N}^{l \times l} \oplus \mathbb{N}^{l \times l'}. $$

The image of the basis vector $e_{\sigma,\tau}$ corresponding to a single observation (σ,τ) under A is the pair of matrices (w, w'), where w_{rs} is the number of indices i such that $\sigma_i\sigma_{i+1} = rs$, and w'_{rt} is the number of indices i such that $\sigma_i\tau_i = rt$.

Let $u \in \mathbb{N}^{l^n \times (l')^n}$ be a matrix of data. The sufficient statistic is the pair of matrices $A \cdot u = (v, v')$. Here $v \in \mathbb{N}^{l \times l}$ and $v' \in \mathbb{N}^{l \times l'}$. The likelihood function $L_{\mathrm{hid}} : \Theta_1 \to \mathbb{R}$ of the fully observed Markov model is the monomial

$$ L_{\mathrm{hid}}(\theta) \quad = \quad \theta^v \cdot (\theta')^{v'}. $$

Proposition 1.18 *The maximum likelihood estimate for the data $u \in \mathbb{N}^{l^n \times (l')^n}$ in the fully observed Markov model is the matrix pair $(\widehat{\theta}, \widehat{\theta}') \in \Theta_1$ with*

$$ \widehat{\theta}_{ij} \quad = \quad \frac{v_{ij}}{\sum_{s\in\Sigma} v_{is}} \qquad and \qquad \widehat{\theta}'_{ij} \quad = \quad \frac{v'_{ij}}{\sum_{t\in\Sigma'} v'_{it}} \qquad (1.54) $$

Proof This is entirely analogous to the proof of Proposition 1.17, the point being that the log-likelihood function $\ell_{\mathrm{hid}}(\theta)$ decouples as a sum of expressions like (1.52), each of which is easy to maximize over the relevant simplex. $\qquad \square$

1.4.3 Hidden Markov models

The hidden Markov model $\mathbf{f}$ is derived from the fully observed Markov model F by summing out the first indices $\sigma \in \Sigma^n$. More precisely, consider the map

$$ \rho \,:\, \mathbb{R}^{l^n \times (l')^n} \;\longrightarrow\; \mathbb{R}^{(l')^n} $$

obtained by taking the column sums of a matrix with l^n rows and $(l')^n$ columns. The *hidden Markov model* is the algebraic statistical model defined by composing the fully observed Markov model F with the marginalization map ρ:

$$ \mathbf{f} \;=\; \rho \circ F \,:\, \Theta_1 \subset \mathbb{R}^d \;\longrightarrow\; \mathbb{R}^{(l')^n}. \qquad (1.55) $$

Here, $d = l(l + l' - 2)$ and it is natural to write $\mathbb{R}^d = \mathbb{R}^{l(l-1)} \times \mathbb{R}^{l(l'-1)}$ since the parameters are pairs of matrices (θ, θ'). The explicit formula for the coordinates of the map $\mathbf{f}$ in the case $n = 4$ is given in Example 1.36.

Remark 1.19 The hidden Markov model is a polynomial map $\mathbf{f}$ from the parameter space $\mathbb{R}^{l(l-1)} \times \mathbb{R}^{l(l'-1)}$ to the probability space $\mathbb{R}^{(l')^n}$. The degree of $\mathbf{f}$ in the entries of θ is $n - 1$, and the degree of $\mathbf{f}$ in the entries of θ' is n.

The notation in the definition in (1.55) is consistent with our discussion of the Expectation Maximization (EM) algorithm in Section 1.3. Thus we can find maximum likelihood estimates for the hidden Markov model by applying the EM algorithm to $\mathbf{f} = \rho \circ F$. See Chapter 12 for details.

Remark 1.20 The *Baum–Welch algorithm* is the special case of the EM algorithm obtained by applying EM to the hidden Markov model $\mathbf{f} = \rho \circ F$. A key ingredient in the Baum–Welch algorithm is the sum-product algorithm which allows for rapid evaluation of the coordinate polynomials f_i of the HMM.

The Baum–Welch algorithm in general, and Remark 1.20 in particular, are discussed in Section 11.6 of [Durbin *et al.*, 1998].

Example 1.21 Consider the *occasionally dishonest casino* which is featured as running example in [Durbin *et al.*, 1998]. In that casino they use a fair die most of the time, but occasionally they switch to a loaded die. Our two alphabets are $\Sigma = \{\text{fair}, \text{loaded}\}$ and $\Sigma' = \{1, 2, 3, 4, 5, 6\}$ for the six possible outcomes of rolling a die. Suppose a particular game involves rolling the dice $n = 4$ times. This hidden Markov model has $d = 12$ parameters, appearing in

$$
\theta \;=\; \begin{array}{c} \\ \text{fair} \\ \text{loaded} \end{array} \begin{array}{cc} \text{fair} & \text{loaded} \\ \left(\begin{array}{cc} x & 1-x \\ 1-y & y \end{array} \right) \end{array} \qquad \text{and}
$$

$$
\theta' \;=\; \begin{array}{c} \\ \text{fair} \\ \text{loaded} \end{array} \begin{array}{cccccc} 1 & 2 & 3 & 4 & 5 & 6 \\ \left(\begin{array}{cccccc} f_1 & f_2 & f_3 & f_4 & f_5 & 1-\sum_{i=1}^{5} f_i \\ l_1 & l_2 & l_3 & l_4 & l_5 & 1-\sum_{j=1}^{5} l_j \end{array} \right). \end{array}
$$

Presumably, the fair die is really fair, so that $f_1 = f_2 = f_3 = f_4 = f_5 = 1/6$, but, to be on the safe side, let us here keep the f_i as unknown parameters.

This hidden Markov model (HMM) has $m = 6^4 = 1,296$ possible outcomes, namely, all the words $\tau = \tau_1 \tau_2 \tau_3 \tau_4$ in $(\Sigma')^4$. The coordinates of the map $\mathbf{f} : \mathbb{R}^{12} \to \mathbb{R}^{1296}$ in (1.55) are polynomials of degree $7 = 3 + 4$:

$$
p_{\tau_1 \tau_2 \tau_3 \tau_4} \;=\; \frac{1}{2} \cdot \sum_{\sigma_1 \in \Sigma} \sum_{\sigma_2 \in \Sigma} \sum_{\sigma_3 \in \Sigma} \sum_{\sigma_4 \in \Sigma} \theta'_{\sigma_1 \tau_1} \theta_{\sigma_1 \sigma_2} \theta'_{\sigma_2 \tau_2} \theta_{\sigma_2 \sigma_3} \theta'_{\sigma_3 \tau_3} \theta_{\sigma_3 \sigma_4} \theta'_{\sigma_4 \tau_4}.
$$

Thus our HMM is specified by a list of $1,296$ polynomials p_τ in the twelve unknowns. The sum of all the polynomials is 1. Each one has degree three in the two unknowns x, y and degree four in the ten unknowns $f_1, f_2, \ldots, l_5$.

Suppose we observe the game N times. These observations are our data. The sufficient statistic is the vector $(u_\tau) \in \mathbb{N}^{1296}$, where $u_\tau = u_{\tau_1 \tau_2 \tau_3 \tau_4}$ counts the number of times the output sequence $\tau = \tau_1 \tau_2 \tau_3 \tau_4$ was observed. Hence $\sum_{\tau \in (\Sigma')^4} u_\tau = N$. The goal of EM is to maximize the log-likelihood function

$$
\ell(x, y, f_1, \ldots, f_5, l_1, \ldots, l_5) \;=\; \sum_{\tau \in \Sigma'^4} u_{\tau_1 \tau_2 \tau_3 \tau_4} \cdot \log(p_{\tau_1 \tau_2 \tau_3 \tau_4}),
$$

where (x, y) ranges over a square, $(f_1, \ldots, f_5)$ runs over a 5-simplex, and so does $(l_1, \ldots, l_5)$. Our parameter space $\Theta_1 \subset \mathbb{R}^{12}$ is the product of the square and the two 5-simplices. The Baum–Welch algorithm (i.e., the EM algorithm for the HMM) aims to maximize ℓ over the 12-dimensional polytope Θ_1. $\quad\square$

1.4.4 Tree models

Markov chains and hidden Markov models are special instances of *tree models*, a class of models which we discuss next. We begin by defining the *fully observed tree model*, from which we then derive the *hidden tree model*. These models relate to each other in the same way that the hidden Markov model is the composition of the fully observed Markov model with a marginalization map.

Let T be a rooted tree with n leaves. We write $N(T)$ for the set of all nodes of T. This set includes the root, which is denoted $\mathbf{r}$, and the leaves, which are indexed by $[n] = \{1, 2, \ldots, n\}$. The set $E(T)$ of edges of T is a subset of $N(T) \times N(T)$. Every edge is directed away from the root $\mathbf{r}$. We use the abbreviation kl for edges $(k, l) \in E(T)$. Every node $i \in N(T)$ represents a random variable which takes values in a finite alphabet Σ_i. Our tree models are parameterized by a collection of matrices θ^{kl}, one for each edge $kl \in E(T)$. The rows of the matrix θ^{kl} are indexed by Σ_k, and the columns are indexed by Σ_l. As before, we restrict ourselves to positive matrices whose rows sum to one. Let Θ_1 denote the collection of tuples $\left(\theta^{kl}\right)_{kl \in E(T)}$ of such matrices. The dimension of the parameter space Θ_1 is therefore $d = \sum_{kl \in E(T)} |\Sigma_k|(|\Sigma_l| - 1)$.

The *fully observed tree model* is the restriction to Θ_1 of the monomial map

$$F_T : \mathbb{R}^d \to \mathbb{R}^m, \quad \theta = \left(\theta^{kl}\right)_{kl \in E(T)} \mapsto p = (p_\sigma)$$

$$p_\sigma = \frac{1}{|\Sigma_{\mathbf{r}}|} \cdot \prod_{kl \in E(T)} \theta^{kl}_{\sigma_k \sigma_l}. \tag{1.56}$$

Here $m = \prod_{i \in N(T)} |\Sigma_i|$. The state space of this model is the Cartesian product of the sets Σ_i. A state is a vector $\sigma = (\sigma_i)_{i \in N(T)}$ where $\sigma_i \in \Sigma_i$. The factor $1/|\Sigma_{\mathbf{r}}|$ means that we are assuming the root distribution to be uniform. This assumption is made only to keep the exposition simple at this stage in the book. In the later sections and chapters on evolutionary models, the root distribution will be allowed to vary arbitrarily or be partially fixed in other ways.

The fully observed tree model F_T is (the restriction to Θ_1 of) a toric model. There is an easy formula for computing maximum likelihood parameters in this model. The formula and its derivation is similar to that in Proposition 1.18.

The *hidden tree model* $\mathbf{f}_T$ is obtained from the fully observed tree model F_T by summing over the internal nodes of the tree. Hidden tree models are therefore defined on a restricted state space corresponding only to leaves of the tree. The state space of the hidden tree model is $\Sigma_1 \times \Sigma_2 \times \cdots \times \Sigma_n$, the product of the alphabets associated with the leaves of T. The cardinality of the state space is $m' = |\Sigma_1| \cdot |\Sigma_2| \cdots |\Sigma_n|$. There is a natural linear *marginalization map* $\rho_T : \mathbb{R}^m \rightarrow \mathbb{R}^{m'}$ which takes real-valued functions on $\prod_{i \in N(T)} \Sigma_i$ to real-valued functions on $\prod_{i=1}^n \Sigma_i$. We now have $\mathbf{f}_T = \rho_T \circ F_T$.

Proposition 1.22 *The hidden tree model* $\mathbf{f}_T : \mathbb{R}^d \rightarrow \mathbb{R}^{m'}$ *is a multilinear polynomial map. Each of its coordinates has total degree* $|E(T)|$, *but is linear when regarded as a function of the entries of each matrix* θ^{kl} *separately.*

The model $\mathbf{f}_T$ described here is also known as the *general Markov model* on the tree T, relative to the given alphabets Σ_i. The adjective "general" refers to the fact that the matrices θ^{kl} are distinct and their entries obey no constraints beyond positivity and rows summing to one. In most applications of tree models, the parameters $(\theta^{kl})_{kl \in E(T)}$ are specialized in some manner, either by requiring that some matrices are identical or by specializing each individual matrix θ^{kl} to have fewer than $|\Sigma_k| \cdot (|\Sigma_l| - 1)$ free parameters.

Example 1.23 The hidden Markov model is a (specialization of the) hidden tree model, where the tree T is the *caterpillar tree* depicted in Figure 1.2.

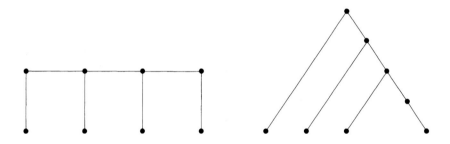

Fig. 1.2. Two views of the caterpillar tree.

In the HMM there are only two distinct alphabets: $\Sigma_i = \Sigma$ for $i \in N(T) \backslash [n]$ and $\Sigma_i = \Sigma'$ for $i \in [n]$. The matrices θ^{kl} are all square and identical along the non-terminal edges of the tree. A second matrix is used for all terminal edges. $\square$

Maximum likelihood estimation for the hidden tree model can be done with the EM algorithm, as described in Section 1.3. Indeed, the hidden tree model is the composition $\mathbf{f}_T = \rho_T \circ F_T$ of an easy toric model F_T and the marginalization map ρ_T, so Algorithm 1.14 is directly applicable to this situation.

Tree models used in phylogenetics have the same alphabet Σ on each edge, but the transition matrices remain distinct and independent. The two alphabets most commonly used are $\Sigma = \{0, 1\}$ and $\Sigma = \{A, C, G, T\}$. We present one example for each alphabet. In both cases, the tree T is the *claw tree*, which has no internal nodes other than the root: $N(T) = \{1, 2, \ldots, n, \mathbf{r}\}$.

Example 1.24 Let $\Sigma = \{0, 1\}$ and T the claw tree with $n = 6$ leaves. The hidden tree model $\mathbf{f}_T$ has $d = 12$ parameters. It has $m = 64$states which are indexed by binary strings $i_1 i_2 i_3 i_4 i_5 i_6 \in \Sigma^6$. The model $\mathbf{f}_T(\Theta_1)$ is the 12-dimensional variety in the 63-simplex given by the parameterization

$$p_{i_1 i_2 i_3 i_4 i_5 i_6} = \frac{1}{2}\theta_{0i_1}^{r1}\theta_{0i_2}^{r2}\theta_{0i_3}^{r3}\theta_{0i_4}^{r4}\theta_{0i_5}^{r5}\theta_{0i_6}^{r6} + \frac{1}{2}\theta_{1i_1}^{r1}\theta_{1i_2}^{r2}\theta_{1i_3}^{r3}\theta_{1i_4}^{r4}\theta_{1i_5}^{r5}\theta_{1i_6}^{r6}.$$

If the root distribution is unspecified then $d = 13$ and the parameterization is

$$p_{i_1 i_2 i_3 i_4 i_5 i_6} = \lambda\theta_{0i_1}^{r1}\theta_{0i_2}^{r2}\theta_{0i_3}^{r3}\theta_{0i_4}^{r4}\theta_{0i_5}^{r5}\theta_{0i_6}^{r6} + (1-\lambda)\theta_{1i_1}^{r1}\theta_{1i_2}^{r2}\theta_{1i_3}^{r3}\theta_{1i_4}^{r4}\theta_{1i_5}^{r5}\theta_{1i_6}^{r6}. \quad (1.57)$$

$\square$

Example 1.25 Let $\Sigma = \{A, C, G, T\}$ and let T be the claw tree with $n = 3$ leaves. The hidden tree model $\mathbf{f}_T$ has $m = 64$ states which are the triples $ijk \in \Sigma^3$. Writing $\lambda = (\lambda_A, \lambda_C, \lambda_G, \lambda_T)$ for the root distribution, we have

$$p_{ijk} = \lambda_A\theta_{Ai}^{r1}\theta_{Aj}^{r2}\theta_{Ak}^{r3} + \lambda_C\theta_{Ci}^{r1}\theta_{Cj}^{r2}\theta_{Ck}^{r3} + \lambda_G\theta_{Gi}^{r1}\theta_{Gj}^{r2}\theta_{Gk}^{r3} + \lambda_T\theta_{Ti}^{r1}\theta_{Tj}^{r2}\theta_{Tk}^{r3}. \quad (1.58)$$

If λ is unspecified then this model has $d = 12 + 12 + 12 + 3 = 39$ parameters. If the root distribution is uniform, i.e. $\lambda = (\frac{1}{4}, \frac{1}{4}, \frac{1}{4}, \frac{1}{4})$, then $d = 36 = 12 + 12 + 12$. We note that the small Jukes–Cantor model in Example 1.7 is the three-dimensional submodel of this 36-dimensional model obtained by setting

$$\theta^{r\nu} = \begin{array}{c} \\ A \\ C \\ G \\ T \end{array} \begin{array}{cccc} A & C & G & T \\ \begin{pmatrix} 1-3\theta_\nu & \theta_\nu & \theta_\nu & \theta_\nu \\ \theta_\nu & 1-3\theta_\nu & \theta_\nu & \theta_\nu \\ \theta_\nu & \theta_\nu & 1-3\theta_\nu & \theta_\nu \\ \theta_\nu & \theta_\nu & \theta_\nu & 1-3\theta_\nu \end{pmatrix} \end{array} \quad \text{for } \nu \in \{1, 2, 3\},$$

and $\lambda_A = \lambda_C = \lambda_G = \lambda_T = \frac{1}{4}$. The number of states drops from $m = 64$ in Example 1.25 to $m = 5$ in Example 1.7 since many of the probabilities p_{ijk} become equal under this specialization. The algebraic geometry of Examples 1.24 and 1.25 is discussed in Section 3.2. $\square$

A key statistical problem associated with hidden tree models is *model selection*. The general model selection problem is as follows: suppose we have a data vector $u = (u_1, \ldots, u_m)$, a collection of models $\mathbf{f}^1, \ldots, \mathbf{f}^k$ where $\mathbf{f}^i : \mathbb{R}^{d_i} \to \mathbb{R}^m$, and we would like to select a "good" model for the data. In the case where $d_1 = \cdots = d_m$, we may select the model $\mathbf{f}^i$ whose likelihood function attains the largest value of all. This problem arises for hidden tree models where there the leaf set $[n]$ and data are fixed, but we would like to select from among all phylogenetic trees on $[n]$ that tree which maximizes the likelihood of the data. Since the number of trees grows exponentially when n increases, this approach leads to combinatorial explosion. In applications to biology, this explosion is commonly dealt with by using the distance-based techniques in Section 2.4. Hidden tree models are studied in detail in Chapters 15 through 20.

1.5 Graphical models

Almost all the statistical models we have discussed in the previous four sections are instances of *graphical models*. Discrete graphical models are certain algebraic statistical models for joint probability distributions of n random variables $X_1, X_2, \ldots, X_n$ which can be specified in two possible ways:

- by a parameterization $\mathbf{f} : \mathbb{R}^d \to \mathbb{R}^m$ (with polynomial coordinates as before),
- by a collection of conditional independence statements.

Our focus in this section is on the latter representation, and its connection to the former via a result of statistics known as the *Hammersley–Clifford Theorem*, which concerns conditional independence statements derived from graphs. The graphs that underlie graphical models are key to developing efficient inference algorithms, an important notion which is the final topic of this section and is the basis for applications of graphical models to problems in biology.

We assume that each random variable X_i takes its values in a finite alphabet Σ_i. The common state space of all models to be discussed in this section is therefore the Cartesian product of the alphabets:

$$\prod_{i=1}^{n} \Sigma_i \quad = \quad \Sigma_1 \times \Sigma_2 \times \cdots \times \Sigma_n \qquad (1.59)$$

and the number of states is $m = \prod_{i=1}^{n} |\Sigma_i|$. This number is fixed throughout this section. A probability distribution on the state space (1.59) corresponds to an n-dimensional table $(p_{i_1 i_2 \cdots i_n})$. We think of $p_{i_1 i_2 \cdots i_n}$ as an unknown which represents the probability of the event $X_1 = i_1, X_2 = i_2, \ldots, X_n = i_n$.

A *conditional independence statement* about $X_1, X_2, \ldots, X_n$ has the form

$$A \text{ is independent of } B \text{ given } C \qquad (\text{in symbols: } A \perp\!\!\!\perp B \,|\, C), \qquad (1.60)$$

where A, B, C are pairwise disjoint subsets of $\{X_1, X_2, \ldots, X_n\}$. If C is the empty set then (1.60) reads "A is independent of B" and is denoted by $A \perp\!\!\!\perp B$.

Remark 1.26 The independence statement (1.60) translates into a set of quadratic equations in the unknowns $p_{i_1 \cdots i_n}$. The equations are indexed by

$$\binom{\prod_{X_i \in A} \Sigma_i}{2} \times \binom{\prod_{X_j \in B} \Sigma_j}{2} \times \prod_{X_k \in C} \Sigma_k. \qquad (1.61)$$

An element of the set (1.61) is a triple consisting of two distinct elements a and a' in $\prod_{X_i \in A} \Sigma_i$, two distinct elements b and b' in $\prod_{X_j \in B} \Sigma_j$, and an element c in $\prod_{X_k \in C} \Sigma_k$. The independence condition $A \perp\!\!\!\perp B \,|\, C$ is equivalent to the statement that, for all triples $\{a, a'\}, \{b, b'\}$ and $\{c\}$,

$$\text{Prob}(A = a, B = b, C = c) \cdot \text{Prob}(A = a', B = b', C = c)$$
$$- \text{Prob}(A = a', B = b, C = c) \cdot \text{Prob}(A = a, B = b', C = c) \quad = \quad 0.$$

To get our quadrics indexed by (1.61), we translate each of the probabilities above into a linear form in the unknowns $p_{i_1 i_2 \cdots i_n}$. Namely, $\text{Prob}(A = a, B = b, C = c)$ is replaced by a marginalization which is the sum of all $p_{i_1 i_2 \cdots i_n}$ that satisfy

- for all $X_\alpha \in A$, the X_α-coordinate of a equals i_α,

- for all $X_\beta \in B$, the X_β-coordinate of b equals i_β, and
- for all $X_\gamma \in C$, the X_γ-coordinate of c equals i_γ.

We define $\mathcal{Q}_{A \perp\!\!\!\perp B \,|\, C}$ to be the set of quadratic forms in the unknowns $p_{i_1 i_2 \cdots i_n}$ which result from this substitution. Thus $\mathcal{Q}_{A \perp\!\!\!\perp B \,|\, C}$ is indexed by (1.61).

We illustrate the definition of the set of quadrics $\mathcal{Q}_{A \perp\!\!\!\perp B \,|\, C}$ with an example:

Example 1.27 Let $n = 3$ and $i_1 = i_2 = i_3 = \{0, 1\}$, so that $(p_{i_1 i_2 i_3})$ is a $2 \times 2 \times 2$ table whose eight entries are unknowns. The independence statement $\{X_2\}$ *is independent of* $\{X_3\}$ *given* $\{X_1\}$ describes the pair of quadrics

$$\mathcal{Q}_{X_2 \perp\!\!\!\perp X_3 \,|\, X_1} \;=\; \{\, p_{000} p_{011} - p_{001} p_{010},\; p_{100} p_{111} - p_{101} p_{110} \,\}. \qquad (1.62)$$

The statement $\{X_2\}$ *is independent of* $\{X_3\}$ corresponds to a single quadric

$$\mathcal{Q}_{X_2 \perp\!\!\!\perp X_3} \;=\; \{\, (p_{000} + p_{100})(p_{011} + p_{111}) - (p_{001} + p_{101})(p_{010} + p_{110}) \,\}. \qquad (1.63)$$

The set $\mathcal{Q}_{X_1 \perp\!\!\!\perp \{X_2, X_3\}}$ representing the statement $\{X_1\}$ *is independent of* $\{X_2, X_3\}$ consists of the six 2×2 subdeterminants of the 2×4 matrix

$$\begin{pmatrix} p_{000} & p_{001} & p_{010} & p_{011} \\ p_{100} & p_{101} & p_{110} & p_{111} \end{pmatrix}. \qquad (1.64)$$

Each of these three statements specifies a model, which is a subset of the 7-simplex Δ with coordinates $p_{i_1 i_2 i_3}$. The model (1.62) has dimension five, the model (1.63) has dimension six, and the model (1.64) has dimension four. $\square$

In general, we write $V_\Delta(A \perp\!\!\!\perp B \,|\, C)$ for the family of all joint probability distributions that satisfy the quadratic equations in $\mathcal{Q}_{A \perp\!\!\!\perp B \,|\, C}$. The model $V_\Delta(A \perp\!\!\!\perp B \,|\, C)$ is a subset of the $(m-1)$-dimensional probability simplex Δ.

Consider any finite collection of conditional independence statements (1.60):

$$\mathcal{M} \;=\; \{\, A^{(1)} \perp\!\!\!\perp B^{(1)} \,|\, C^{(1)},\; A^{(2)} \perp\!\!\!\perp B^{(2)} \,|\, C^{(2)},\; \ldots,\; A^{(m)} \perp\!\!\!\perp B^{(m)} \,|\, C^{(m)} \,\}.$$

We write $\mathcal{Q}_\mathcal{M}$ for the set of quadratic forms representing these statements:

$$\mathcal{Q}_\mathcal{M} \;=\; \mathcal{Q}_{A^{(1)} \perp\!\!\!\perp B^{(1)} \,|\, C^{(1)}} \cup \mathcal{Q}_{A^{(2)} \perp\!\!\!\perp B^{(2)} \,|\, C^{(2)}} \cup \cdots \cup \mathcal{Q}_{A^{(m)} \perp\!\!\!\perp B^{(m)} \,|\, C^{(m)}}.$$

The common zero set of these quadratic forms in the simplex Δ equals

$$V_\Delta(\mathcal{M}) \;=\; V_\Delta(A^{(1)} \perp\!\!\!\perp B^{(1)} \,|\, C^{(1)}) \cap \cdots \cap V_\Delta(A^{(m)} \perp\!\!\!\perp B^{(m)} \,|\, C^{(m)}).$$

We call $V_\Delta(\mathcal{M})$ the *conditional independence model* of $\mathcal{M}$. This model is the family of joint probability distributions which satisfy all the statements in $\mathcal{M}$.

Example 1.28 Let $n = 3$ and $i_1 = i_2 = i_3 = \{0, 1\}$. Consider the model

$$\mathcal{M} \;=\; \{\, X_1 \perp\!\!\!\perp X_2 \,|\, X_3,\; X_1 \perp\!\!\!\perp X_3 \,|\, X_2 \,\}.$$

These two independence statements translate into four quadratic forms:

$$\mathcal{Q}_\mathcal{M} \;=\; \{\, p_{000} p_{110} - p_{010} p_{100},\; p_{001} p_{111} - p_{011} p_{101},$$
$$p_{000} p_{101} - p_{001} p_{100},\; p_{010} p_{111} - p_{011} p_{110} \,\}.$$

The model $V_\Delta(\mathcal{M})$ consists of three components. Two of them are tetrahedra which are faces of the 7-dimensional simplex Δ. These two tetrahedra are

$$X_2 = X_3: \quad \{\, p \in \Delta \;:\; p_{001} = p_{010} = p_{101} = p_{110} = 0 \,\},$$
$$X_2 \neq X_3: \quad \{\, p \in \Delta \;:\; p_{000} = p_{011} = p_{100} = p_{111} = 0 \,\}.$$

Only the third component meets the interior of the simplex. That component is the four-dimensional model $V_\Delta(X_1 \perp\!\!\!\perp \{X_2, X_3\})$ which consists of all distributions $p \in \Delta$ for which the 2×4 matrix in (1.64) has rank one. This analysis shows that for strictly positive probability distributions we have

$$X_1 \perp\!\!\!\perp X_2 \,|\, X_3 \quad \text{and} \quad X_1 \perp\!\!\!\perp X_3 \,|\, X_2 \quad \text{implies} \quad X_1 \perp\!\!\!\perp \{X_2, X_3\}. \qquad (1.65)$$

$\square$

We are now prepared to define graphical models, starting with the undirected case. Let G be an undirected graph with vertices $X_1, X_2, \ldots, X_n$. Let $\mathcal{M}_G$ denote the set of all conditional independence statements

$$X_i \perp\!\!\!\perp X_j \,|\, \{X_1, \ldots, X_n\} \backslash \{X_i, X_j\}, \qquad (1.66)$$

where (X_i, X_j) runs over all pairs of nodes that are not connected by an edge in G. In what follows we let Δ^0 denote the open probability simplex of dimension $m - 1$. The *Markov random field* (or *undirected graphical model* or *Markov network*) defined by the graph G is the model $V_{\Delta^0}(\mathcal{M}_G)$. This is the set of all strictly positive distributions which satisfy the statements in $\mathcal{M}_G$.

In the literature on graphical models, the set $\mathcal{M}_G$ is known as the *pairwise Markov property* on the graph G. There are also two larger sets of conditional independence statements that can be derived from the graph, called the *local Markov property* and the *global Markov property* [Lauritzen, 1996], which specify the same model $V_{\Delta^0}(\mathcal{M}_G)$ in the open simplex Δ^0. For simplicity, we restrict our presentation to the pairwise Markov property (1.66).

Example 1.29 Let $n = 4$ and G the 4-chain graph (Figure 1.3). The graph G is drawn with the random variables labeling the nodes, and shaded nodes indicating that all random variables are observed.

Fig. 1.3. Graph of the 4-chain Markov random field.

There are 3 pairs of nodes not connected by an edge, so that

$$\mathcal{M}_G \;=\; \{X_1 \perp\!\!\!\perp X_3 \,|\, \{X_2, X_4\}, \; X_1 \perp\!\!\!\perp X_4 \,|\, \{X_2, X_3\}, \; X_2 \perp\!\!\!\perp X_4 \,|\, \{X_1, X_3\}\}.$$

For binary alphabets Σ_i the set $\mathcal{Q}_{\mathcal{M}_G}$ consists of the twelve quadratic forms

$$p_{0010}p_{1000} - p_{0000}p_{1010}, \quad p_{0001}p_{1000} - p_{0000}p_{1001}, \quad p_{0001}p_{0100} - p_{0000}p_{0101},$$

$$p_{0011}p_{1001} - p_{0001}p_{1011}, \quad p_{0011}p_{1010} - p_{0010}p_{1011}, \quad p_{0011}p_{0110} - p_{0010}p_{0111},$$

$$p_{0110}p_{1100} - p_{0100}p_{1110}, \quad p_{0101}p_{1100} - p_{0100}p_{1101}, \quad p_{1001}p_{1100} - p_{1000}p_{1101},$$

$$p_{0111}p_{1101} - p_{0101}p_{1111}, \quad p_{0111}p_{1110} - p_{0110}p_{1111}, \quad p_{1011}p_{1110} - p_{1010}p_{1111}. \quad \square$$

Every Markov random field $V_{\Delta^0}(\mathcal{M}_G)$ is, in fact, a toric model specified parametrically by a matrix A_G with entries in $\{0,1\}$. The columns of the matrix A_G are indexed by $\prod_{i=1}^{n} \Sigma_i$. The rows are indexed by all the possible assignments to the maximal cliques in G. A *clique* in G is a collection of nodes any two of which are connected by an edge. If the graph G contains no triangles (as in Example 1.29) then the maximal cliques are just the edges.

An entry in the matrix A_G is 1 if the states corresponding to the column agree with the assignments specified by the row and is 0 otherwise. Returning to Example 1.29, the matrix A_G has 16 columns, and 12 rows. The rows are indexed by tuples $(i, j, \sigma_i, \sigma_j)$ where $\{X_i, X_j\}$ is an edge of the graph G and $\sigma_i \in \Sigma_i$ and $\sigma_j \in \Sigma_j$. The nonzero entries of A_G are therefore given by rows $(i, j, \sigma_i, \sigma_j)$ and columns $\pi_1 \pi_2 \cdots \pi_n$ where $\sigma_i = \pi_i$ and $\sigma_j = \pi_j$:

	0000	0001	0010	0011	0100	0101	0110	0111	1000	1001	1010	1011	1100	1101	1110	1111
00··	1	1	1	1	0	0	0	0	0	0	0	0	0	0	0	0
01··	0	0	0	0	1	1	1	1	0	0	0	0	0	0	0	0
10··	0	0	0	0	0	0	0	0	1	1	1	1	0	0	0	0
11··	0	0	0	0	0	0	0	0	0	0	0	0	1	1	1	1
·00·	1	1	0	0	0	0	0	0	1	1	0	0	0	0	0	0
·01·	0	0	1	1	0	0	0	0	0	0	1	1	0	0	0	0
·10·	0	0	0	0	1	1	0	0	0	0	0	0	1	1	0	0
·11·	0	0	0	0	0	0	1	1	0	0	0	0	0	0	1	1
··00	1	0	0	0	1	0	0	0	1	0	0	0	1	0	0	0
··01	0	1	0	0	0	1	0	0	0	1	0	0	0	1	0	0
··10	0	0	1	0	0	0	1	0	0	0	1	0	0	0	1	0
··11	0	0	0	1	0	0	0	1	0	0	0	1	0	0	0	1

Each of the 12 rows corresponds to pairs in $i_1 \times i_2$ or $i_2 \times i_3$ or $i_3 \times i_4$. For instance, the label $\cdot 01 \cdot$ of the sixth row represents $(i, j, \sigma_i, \sigma_j) = (2, 3, 0, 1)$. We note that each of the twelve quadrics in Example 1.29 corresponds to a vector in the kernel of the matrix A_G. For instance, the quadric $p_{0010}p_{1000} - p_{0000}p_{1010}$ corresponds to the following vector in the kernel of A_G:

0000	0001	0010	0011	0100	0101	0110	0111	1000	1001	1010	1011	1100	1101	1110	1111

$$\begin{pmatrix} -1 & 0 & 1 & 0 & 0 & 0 & 0 & 0 & 1 & 0 & -1 & 0 & 0 & 0 & 0 & 0 \end{pmatrix}.$$

The relationship between $\mathcal{M}_G$ and the matrix A_G generalizes as follows:

Theorem 1.30 (Undirected Hammersley–Clifford) *The Markov random field $V_{\Delta^0}(\mathcal{M}_G)$ coincides with the toric model specified by the matrix A_G.*

Proof See [Lauritzen, 1996] and [Geiger *et al.*, 2005]. □

Markov random fields are toric because their defining conditional independence statements $A \perp\!\!\!\perp B \,|\, C$ have the property that

$$A \cup B \cup C = \{X_1, X_2, \ldots, X_n\}. \tag{1.67}$$

This property ensures that all the quadrics in $\mathcal{Q}_{A \perp\!\!\!\perp B \,|\, C}$ are differences of two monomials of the form $p_{\ldots}p_{\ldots} - p_{\ldots}p_{\ldots}$. If the property (1.67) does not hold, then the quadrics have more terms and the models are generally not toric.

Remark 1.31 It is important to note that the conditional independence statements for a Markov random field are based on pairs of random variables not joined by an edge in the graph. This should be contrasted with the parameters in the toric model, where there are sets of parameters for each maximal clique in the graph. These parameters do not, in general, have an interpretation as marginal or conditional probabilities. They are sometimes called *potentials*.

We now define directed graphical models which are generally not toric. We also return to the closed simplex Δ. Let D be an acyclic directed graph with nodes $X_1, X_2, \ldots, X_n$. For any node X_i, let $\mathrm{pa}(X_i)$ denote the set of *parents* of X_i in D and let $\mathrm{nd}(X_i)$ denote the set of *non-descendants* of X_i in D which are not parents of X_i. The *directed graphical model* of D is described by the following set of independence statements:

$$\mathcal{M}_D = \big\{ X_i \perp\!\!\!\perp \mathrm{nd}(X_i) \,|\, \mathrm{pa}(X_i) \;:\; i = 1, 2, \ldots, n \big\}.$$

The directed graphical model $V_\Delta(\mathcal{M}_D)$ admits a polynomial parameterization, which amounts to a directed version of the Hammersley–Clifford theorem. Before stating this parameterization in general, we first discuss a small example.

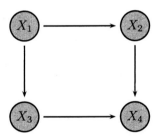

Fig. 1.4. The directed graphical model in Example 1.32.

Example 1.32 Let D be the directed graph with nodes $1, 2, 3, 4$ and four edges $(1, 2), (1, 3), (2, 4), (3, 4)$. Then $\mathcal{M}_D = \big\{ X_2 \perp\!\!\!\perp X_3 \,|\, X_1, \; X_4 \perp\!\!\!\perp X_1 \,|\, \{X_2, X_3\} \big\}$. The quadrics associated with this directed graphical model are

$$\mathcal{Q}_{\mathcal{M}_D} = \big\{ (p_{0000} + p_{0001})(p_{0110} + p_{0111}) - (p_{0010} + p_{0011})(p_{0100} + p_{0101}),$$
$$(p_{1000} + p_{1001})(p_{1110} + p_{1111}) - (p_{1010} + p_{1011})(p_{1100} + p_{1101}),$$
$$p_{0000}p_{1001} - p_{0001}p_{1000}, \; p_{0010}p_{1011} - p_{0011}p_{1010},$$
$$p_{0100}p_{1101} - p_{0101}p_{1100}, \; p_{0110}p_{1111} - p_{0111}p_{1110} \big\}.$$

The model $V_\Delta(\mathcal{M}_D)$ is nine-dimensional inside the 15-dimensional simplex Δ.

We present this model as the image of a polynomial map $F_D : \mathbb{R}^9 \to \mathbb{R}^{16}$. The vector of $9 = 2^0 + 2^1 + 2^1 + 2^2$ parameters for this model is written

$$\theta = (a, b_0, b_1, c_0, c_1, d_{00}, d_{00}, d_{10}, d_{11}).$$

The letters a, b, c, d correspond to the random variables X_1, X_2, X_3, X_4 in this order. The parameters represent the probabilities of each node given its parents. For instance, the parameter d_{10} is the probability of the event "$X_4 = 0$ given $X_2 = 1$ and $X_3 = 0$". The coordinates of the map $\mathbf{f} : \theta \mapsto p$ are

$$
\begin{aligned}
p_{0000} &= a \cdot b_0 \cdot c_0 \cdot d_{00} \\
p_{0001} &= a \cdot b_0 \cdot c_0 \cdot (1 - d_{00}) \\
p_{0010} &= a \cdot b_0 \cdot (1 - c_0) \cdot d_{01} \\
p_{0011} &= a \cdot b_0 \cdot (1 - c_0) \cdot (1 - d_{01}) \\
p_{0100} &= a \cdot (1 - b_0) \cdot c_0 \cdot d_{10} \\
p_{0101} &= a \cdot (1 - b_0) \cdot c_0 \cdot (1 - d_{10}) \\
p_{0110} &= a \cdot (1 - b_0) \cdot (1 - c_0) \cdot d_{11} \\
p_{0111} &= a \cdot (1 - b_0) \cdot (1 - c_0) \cdot (1 - d_{11}) \\
p_{1000} &= (1 - a) \cdot b_1 \cdot c_1 \cdot d_{00} \\
p_{1001} &= (1 - a) \cdot b_1 \cdot c_1 \cdot (1 - d_{00}) \\
p_{1010} &= (1 - a) \cdot b_1 \cdot (1 - c_1) \cdot d_{01} \\
p_{1011} &= (1 - a) \cdot b_1 \cdot (1 - c_1) \cdot (1 - d_{01}) \\
p_{1100} &= (1 - a) \cdot (1 - b_1) \cdot c_1 \cdot d_{10} \\
p_{1101} &= (1 - a) \cdot (1 - b_1) \cdot c_1 \cdot (1 - d_{10}) \\
p_{1110} &= (1 - a) \cdot (1 - b_1) \cdot (1 - c_1) \cdot d_{11} \\
p_{1111} &= (1 - a) \cdot (1 - b_1) \cdot (1 - c_1) \cdot (1 - d_{11}).
\end{aligned}
$$

Note that the six quadrics in $\mathcal{Q}_{\mathcal{M}_D}$ are zero for these expressions, and also

$$\sum_{i=0}^{1} \sum_{j=0}^{1} \sum_{k=0}^{1} \sum_{l=0}^{1} p_{ijkl} = 1.$$

$\square$

Let us return to our general discussion, where D is an acyclic directed graph on n nodes, each associated with a finite alphabet Σ_i. We introduce a parameter $\theta_{(\nu,\pi)}$ for each element $(\nu, \sigma) \in \Sigma_i \times \prod_{j \in \mathrm{pa}(i)} \Sigma_j$, where i ranges over all nodes. These parameters are supposed to satisfy the linear equations

$$\sum_{\nu \in \Sigma_i} \theta_{(\nu, \sigma)} = 1 \qquad \text{for all} \quad \sigma \in \prod_{j \in \mathrm{pa}(X_i)} \Sigma_j. \tag{1.68}$$

Thus the total number of free parameters is

$$d = \sum_{i=1}^{n} (|\Sigma_i| - 1) \cdot \prod_{j \in \mathrm{pa}(X_i)} |\Sigma_j|. \tag{1.69}$$

These constraints are imposed because the parameters represent conditional probabilities:

$$\theta_{(\nu,\sigma)} \quad = \quad \Pr\big(X_i = \nu \,|\, \mathrm{pa}(X_i) = \sigma\big).$$

With the directed acyclic graph D we associate the following monomial map:

$$F_D \;:\; \mathbb{R}^d \;\to\; \mathbb{R}^m \,,\; \theta \;\mapsto\; p,$$

$$\text{where} \quad p_\sigma \;=\; \textstyle\prod_{i=1}^{n} \theta_{(\sigma_i,\sigma|_{\mathrm{pa}(X_i)})} \quad \text{for all} \quad \sigma \in \textstyle\prod_{i=1}^{n} \Sigma_i.$$

Here $\sigma|_{\mathrm{pa}(X_i)}$ denotes the restriction of the vector σ to $\prod_{j \in \mathrm{pa}(X_i)} \Sigma_j$. Let Θ_1 be the set of positive parameter vectors $\theta \in \mathbb{R}^d$ which satisfy (1.68). The following theorem generalizes the result derived for the graph in Example 1.32.

Theorem 1.33 (Directed Hammersley–Clifford) *The directed graphical model $V_\Delta(\mathcal{M}_D)$ equals the image of the parameter space Θ_1 under the map F_D.*

Proof We refer the reader to Theorem 3.27 in [Lauritzen, 1996] and Theorem 3 in [Garcia *et al.*, 2004]. $\qquad\square$

Remark 1.34 Suppose that $D = T$ is a rooted tree with all edges directed away from the root $\mathbf{r}$. The directed graphical model $V_\Delta(\mathcal{M}_D)$ is precisely the fully observed tree model (with arbitrary root distributions), and the parameterization F_D specializes to the one given in (1.56). It is known that the model $V_\Delta(\mathcal{M}_T)$ does not depend on the location of the root $\mathbf{r}$, and, in fact, the model coincides with the Markov random field $V_\Delta(\mathcal{M}_G)$, where G denotes the undirected tree.

Statistical inference for graphical models leads to the problem of computing

$$\sum_{\sigma \in S} p_\sigma, \tag{1.70}$$

where S ranges over certain subsets of $\prod_{i=1}^{n} \Sigma_i$.

In the case where $S = \prod_{i=1}^{n} \Sigma_i$, the sum (1.70) is precisely the partition function of the graphical model. If S is not equal to the entire product of the alphabets, then S is typically a set of all strings σ where some coordinates σ_i have been fixed. Here the inference problem involves a marginalization, which we think of as evaluating one coordinate polynomial of a partial data model. The evaluation of the sum (1.70) may be performed using ordinary arithmetic, or with the *tropical semiring*, using *min* instead of $+$, and $+$ instead of $\times$, and replacing p_σ with the negative of its logarithm (Sections 2.1 and 2.2).

All of these variants of the inference problem are important statistically – and relevant for biological applications. For example, if some of the variables X_i of a Markov random field or directed graphical model D are hidden, then this gives rise to a marginalization map ρ_D and to a hidden model $\mathbf{f}_D = \rho_D \circ F_D$. Evaluating one coordinate of the polynomial map $\mathbf{f}_D$ tropically, also known as *maximum a posteriori (MAP) inference*, is therefore exactly the tropical evaluation of a subsum of the partition function. The case of trees is of particular interest in biological sequence analysis. More examples are discussed in Chapter 2. Connections to biology are developed in Chapter 4.

Remark 1.35 If inference with a graphical model involves computing the partition function tropically, then the model is referred to as *discriminative*. In the case where specific coordinates are selected before summing (1.70), then the model is *generative*. These terms are used in statistical learning theory [Jordan, 2005].

Inference can be computationally nontrivial for two reasons. In order to compute the partition function, the number of terms in the sum is equal to m which can be very large since many applications of graphical models require that the models have large numbers of random variables. One may easily encounter $n = 200$ binary random variables, in which case

$$m = 1606938044258990275541962092341162602522202993782792835301376.$$

The success of graphical models has been due to the possibility of efficient inference for many models of interest. The organizing principle is the generalized distributive law which gives a recursive decomposition of (1.70) according to the graph underlying the model.

Rather than explaining the details of the generalized distributive law in general, we illustrate its origins and application with the hidden Markov model:

Example 1.36 Recall that the hidden Markov model is a polynomial map $\mathbf{f}$ from the parameter space $\mathbb{R}^{l(l-1)} \times \mathbb{R}^{l(l'-1)}$ to the probability space $\mathbb{R}^{(l')^n}$. Consider the case $n = 4$. If we treat the hidden Markov model as a special case of the tree model (compare Figure 1.5 with Figure 1.2), allowing for different parameters on each edge, then a coordinate polynomial is

$$p_{j_1 j_2 j_3 j_4} = \frac{1}{|\Sigma|} \cdot \sum_{i_1 \in \Sigma} \sum_{i_2 \in \Sigma} \sum_{i_3 \in \Sigma} \sum_{i_4 \in \Sigma} \theta_{i_1 j_1}^{X_1 Y_1} \theta_{i_1 i_2}^{X_1 X_2} \theta_{i_2 j_2}^{X_2 Y_2} \theta_{i_2 i_3}^{X_2 X_3} \theta_{\sigma_3 j_3}^{X_3 Y_3} \theta_{\sigma_3 i_4}^{X_3 X_4} \theta_{i_4 j_4}^{X_4 Y_4}.$$

This sum $p_{j_1 j_2 j_3 j_4}$ can be rewritten as follows:

$$\frac{1}{|\Sigma|} \cdot \sum_{i_1 \in \Sigma} \theta_{i_1 j_1}^{X_1 Y_1} \left(\sum_{i_2 \in \Sigma} \theta_{i_1 i_2}^{X_1 X_2} \theta_{i_2 j_2}^{X_2 Y_2} \left(\sum_{i_3 \in \Sigma} \theta_{i_2 i_3}^{X_2 X_3} \theta_{i_3 j_3}^{X_3 Y_3} \left(\sum_{i_4 \in \Sigma} \theta_{i_3 i_4}^{X_3 X_4} \theta_{i_4 j_4}^{X_4 Y_4} \right) \right) \right).$$

The graph for the hidden Markov model is shown in Figure 1.5. Note that the unshaded nodes correspond to random variables which are summed in the marginalization map, thus resulting in one sum for each unshaded node.

This connection between graphs and recursive decompositions is made precise by the *junction tree algorithm* [Cowell *et al.*, 1999] (also known as *belief propagation* [Pearl, 1988], the *sum-product algorithm* [Kschischang *et al.*, 2001] or the *generalized distributive law* [Aji and McEliece, 2000]). Note that in terms of algorithmic complexity, the latter formulation, while equivalent to the first, requires only $O(n)$ additions and multiplications for an HMM of length n in order to compute $p_{j_1 j_2 \cdots j_n}$. The naive formulation requires $O(l^n)$ additions. $\qquad \square$

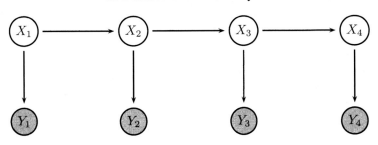

Fig. 1.5. Graph of the hidden Markov model.

The inference problem for graphical models can be formulated as an instance of a more general *marginalization of a product function* (MPF) problem. Formally, suppose that we have n indeterminates $x_1, \ldots, x_n$ taking on values in finite sets $A_1, \ldots, A_n$. Let R be a commutative semiring and $\alpha_i : A_1 \times A_2 \cdots \times A_n \to R$ $(i = 1, \ldots, m)$ be functions with values in R. The MPF problem is to evaluate, for a set $S = \{j_1, \ldots, j_r\} \subset [n]$,

$$\beta(S) \quad = \quad \bigoplus_{x_{j_1} \in A_{j_1}, \ldots, x_{j_r} \in A_{j_r}} \bigodot_{i=1}^{M} \alpha_i(x_1, \ldots, x_n)$$

where $\oplus$ is addition and $\odot$ is multiplication in the semi-ring R. Two important semirings which make their appearance in the next chapter are the *tropical semiring* (or *min-plus algebra*, in Section 2.1) and the *polytope algebra* (in Section 2.3).

2

Computation

Lior Pachter

Bernd Sturmfels

Many of the algorithms used for biological sequence analysis are *discrete algorithms*, i.e., the key feature of the problems being solved is that some optimization needs to be performed on a finite set. Discrete algorithms are complementary to *numerical algorithms*, such as Expectation Maximization, Singular Value Decomposition and Interval Arithmetic, which make their appearance in later chapters. They are also distinct from *algebraic algorithms*, such as the Buchberger Algorithm, which is discussed in Section 3.1. In what follows we introduce discrete algorithms and mathematical concepts which are relevant for biological sequence analysis. The final section of this chapter offers an annotated list of the computer programs which are used throughout the book. The list ranges over all three themes (discrete, algebraic, numerical) and includes software tools which are useful for research in computational biology.

Some discrete algorithms arise naturally from algebraic statistical models, which are characterized by finitely many polynomials, each with finitely many terms. Inference methods for drawing conclusions about missing or hidden data depend on the combinatorial structure of the polynomials in the algebraic representation of the models. In fact, many widely used dynamic programming methods, such as the *Needleman–Wunsch algorithm* for sequence alignment, can be interpreted as evaluating polynomials, albeit with *tropical arithmetic*.

The combinatorial structure of a polynomial, or polynomial map, is encoded in its *Newton polytope*. Thus every algebraic statistical model has a Newton polytope, and it is the structure of this polytope which governs dynamic programming related to that model. Computing the entire polytope is what we call *parametric inference*. This computation can be done efficiently in the *polytope algebra* which is a natural generalization of tropical arithmetic. In Section 2.4 we study the combinatorics of one of the central objects in genome analysis, viz., phylogenetic trees, with an emphasis on the *neighbor-joining algorithm*.

2.1 Tropical arithmetic and dynamic programming

Dynamic programming was introduced by Bellman in the 1950s to solve sequential decision problems with a compositional cost structure. Dynamic programming offers efficient methods for progressively building a set of scores or probabilities in order to solve a problem, and many discrete algorithms for biological sequence analysis are based on the principles of dynamic programming.

A convenient algebraic structure for stating various dynamic programming algorithms is the *tropical semiring* $(\mathbb{R} \cup \{\infty\}, \oplus, \odot)$. The tropical semiring consists of the real numbers $\mathbb{R}$, together with an extra element ∞, and with the arithmetic operations of addition and multiplication redefined as follows:

$$x \oplus y := \min(x, y) \quad \text{and} \quad x \odot y := x + y.$$

In other words, the *tropical sum* of two real numbers is their minimum, and the *tropical product* of two numbers is their sum. Here are some examples of how to do arithmetic in this strange number system. The tropical sum of 3 and 7 is 3. The tropical product of 3 and 7 equals 10. We write this as follows:

$$3 \oplus 7 = 3 \quad \text{and} \quad 3 \odot 7 = 10.$$

Many of the familiar axioms of arithmetic remain valid in the tropical semiring. For instance, both addition and multiplication are *commutative*:

$$x \oplus y = y \oplus x \quad \text{and} \quad x \odot y = y \odot x.$$

The *distributive law* holds for tropical addition and tropical multiplication:

$$x \odot (y \oplus z) = x \odot y \oplus x \odot z.$$

Both arithmetic operations have a *neutral element*. Infinity is the neutral element for addition and zero is the neutral element for multiplication:

$$x \oplus \infty = x \quad \text{and} \quad x \odot 0 = x.$$

The tropical *addition table* and the tropical *multiplication table* look like this:

$\oplus$	1	2	3	4	5	6	7
1	1	1	1	1	1	1	1
2	1	2	2	2	2	2	2
3	1	2	3	3	3	3	3
4	1	2	3	4	4	4	4
5	1	2	3	4	5	5	5
6	1	2	3	4	5	6	6
7	1	2	3	4	5	6	7

$\odot$	1	2	3	4	5	6	7
1	2	3	4	5	6	7	8
2	3	4	5	6	7	8	9
3	4	5	6	7	8	9	10
4	5	6	7	8	9	10	11
5	6	7	8	9	10	11	12
6	7	8	9	10	11	12	13
7	8	9	10	11	12	13	14

Although tropical addition and multiplication are straightforward, subtraction is tricky. There is no tropical "10 minus 3" because the equation $3 \oplus x = 10$ has no solution x. In this book we use addition $\oplus$ and multiplication $\odot$ only.

Example 2.1 It is important to keep in mind that 0 is the multiplicatively neutral element. For instance, the *tropical binomial coefficients* are all 0, as in

$$
\begin{aligned}
(x \oplus y)^3 &= (x \oplus y) \odot (x \oplus y) \odot (x \oplus y) \\
&= 0 \odot x^3 \oplus 0 \odot x^2 y \oplus 0 \odot xy^2 \oplus 0 \odot y^3.
\end{aligned}
$$

The zero coefficients can be dropped in this identity, and we conclude

$$
(x \oplus y)^3 = x^3 \oplus x^2 y \oplus xy^2 \oplus y^3 = x^3 \oplus y^3.
$$

This identity is verified by noting that

$$
3 \cdot \min\{x, y\} = \min\{3x, 2x + y, x + 2y, 3y\} = \min\{3x, 3y\}
$$

holds for all real numbers x and y. $\square$

The familiar linear algebra operations of adding and multiplying vectors and matrices make perfect sense over the tropical semiring. For instance, the tropical scalar product in $\mathbb{R}^3$ of a row vector with a column vector is the scalar

$$
\begin{aligned}
(u_1, u_2, u_3) \odot (v_1, v_2, v_3)^{\mathrm{T}} &= u_1 \odot v_1 \oplus u_2 \odot v_2 \oplus u_3 \odot v_3 \\
&= \min\{u_1 + v_1, \, u_2 + v_2, \, u_3 + v_3\}.
\end{aligned}
$$

Here is the product of a column vector and a row vector of length three:

$$
\begin{aligned}
&(u_1, u_2, u_3)^T \odot (v_1, v_2, v_3) \\
&= \begin{pmatrix} u_1 \odot v_1 & u_1 \odot v_2 & u_1 \odot v_3 \\ u_2 \odot v_1 & u_2 \odot v_2 & u_2 \odot v_3 \\ u_3 \odot v_1 & u_3 \odot v_2 & u_3 \odot v_3 \end{pmatrix} = \begin{pmatrix} u_1 + v_1 & u_1 + v_2 & u_1 + v_3 \\ u_2 + v_1 & u_2 + v_2 & u_2 + v_3 \\ u_3 + v_1 & u_3 + v_2 & u_3 + v_3 \end{pmatrix}.
\end{aligned}
$$

Any matrix which can be expressed as such a product has *tropical rank one*.

To see why tropical arithmetic is relevant for discrete algorithms we consider the problem of finding shortest paths in a weighted directed graph. This is a standard problem of dynamic programming. Let G be a directed graph with n nodes which are labeled by $1, 2, \ldots, n$. Every directed edge (i, j) in G has an associated length d_{ij} which is a non-negative real number. If (i, j) is not an edge of G then we set $d_{ij} = +\infty$. We represent the weighted directed graph G by its $n \times n$ adjacency matrix $D_G = (d_{ij})$ whose off-diagonal entries are the edge lengths d_{ij}. The diagonal entries of D_G are zero, i.e., $d_{ii} = 0$ for all i.

If G is an undirected graph with edge lengths, then we can represent G as a directed graph with two directed edges (i, j) and (j, i) for each undirected edge $\{i, j\}$. In that special case, D_G is a symmetric matrix, and we can think of $d_{ij} = d_{ji}$ as the distance between node i and node j. For a general directed graph G, the adjacency matrix D_G will not be symmetric. Consider the result of tropically multiplying the $n \times n$ matrix D_G with itself $n - 1$ times:

$$
D_G^{\odot n-1} = D_G \odot D_G \odot \cdots \odot D_G. \tag{2.1}
$$

This is an $n \times n$ matrix with entries in $\mathbb{R}_{\geq 0} \cup \{+\infty\}$.

Proposition 2.2 *Let G be a weighted directed graph on n nodes with $n \times n$ adjacency matrix D_G. Then the entry of the matrix $D_G^{\odot n-1}$ in row i and column j equals the length of a shortest path from node i to node j in G.*

Proof Let $d_{ij}^{(r)}$ denote the minimum length of any path from node i to node j which uses at most r edges in G. Thus $d_{ij}^{(1)} = d_{ij}$ for any two nodes i and j. Since the edge weights d_{ij} were assumed to be non-negative, a shortest path from node i to node j visits each node of G at most once. In particular, any such shortest path in the directed graph G uses at most $n - 1$ directed edges. Hence the length of a shortest path from i to j equals $d_{ij}^{(n-1)}$.

For $r \geq 2$ we have the following recursive formula for the lengths of these shortest paths:

$$d_{ij}^{(r)} \quad = \quad \min\{d_{ik}^{(r-1)} + d_{kj} : k = 1, 2, \ldots, n\}. \tag{2.2}$$

Using tropical arithmetic, this formula can be rewritten as follows:

$$\begin{aligned} d_{ij}^{(r)} \quad &= \quad d_{i1}^{(r-1)} \odot d_{1j} \;\oplus\; d_{i2}^{(r-1)} \odot d_{2j} \;\oplus\; \cdots \;\oplus\; d_{in}^{(r-1)} \odot d_{nj}. \\ &= \quad (d_{i1}^{(r-1)}, d_{i2}^{(r-1)}, \ldots, d_{in}^{(r-1)}) \odot (d_{1j}, d_{2j}, \ldots, d_{nj})^T. \end{aligned}$$

From this it follows, by induction on r, that $d_{ij}^{(r)}$ coincides with the entry in row i and column j of the $n \times n$ matrix $D_G^{\odot r}$. Indeed, the right hand side of the recursive formula is the tropical product of row i of $D_G^{\odot r-1}$ and column j of D_G, which is the (i, j) entry of $D_G^{\odot r}$. In particular, $d_{ij}^{(n-1)}$ coincides with the entry in row i and column j of $D_G^{\odot n-1}$. This proves the claim. □

The iterative evaluation of the formula (2.2) is known as the *Floyd–Warshall Algorithm* [Floyd, 1962, Warshall, 1962] for finding shortest paths in a weighted digraph. Floyd–Warshall simply means performing the matrix multiplication

$$D_G^{\odot r} \quad = \quad D_G^{\odot r-1} \odot D_G \qquad \text{for } r = 2, \ldots, n - 1.$$

Example 2.3 Let G be the complete bi-directed graph on $n = 4$ nodes with

$$D_G \quad = \quad \begin{pmatrix} 0 & 1 & 3 & 7 \\ 2 & 0 & 1 & 3 \\ 4 & 5 & 0 & 1 \\ 6 & 3 & 1 & 0 \end{pmatrix}.$$

The first and second tropical power of this matrix are found to be

$$D_G^{\odot 2} \quad = \quad \begin{pmatrix} 0 & 1 & 2 & 4 \\ 2 & 0 & 1 & 2 \\ 4 & 4 & 0 & 1 \\ 5 & 3 & 1 & 0 \end{pmatrix} \qquad \text{and} \qquad D_G^{\odot 3} \quad = \quad \begin{pmatrix} 0 & 1 & 2 & 3 \\ 2 & 0 & 1 & 2 \\ 4 & 4 & 0 & 1 \\ 5 & 3 & 1 & 0 \end{pmatrix}.$$

The entries in $D_G^{\odot 3}$ are the lengths of the shortest paths in the graph G.

The tropical computation above can be related to the following matrix computation in ordinary arithmetic. Let ϵ denote an indeterminate, and let $A_G(\epsilon)$ be the $n \times n$ matrix whose entries are the monomials $\epsilon^{d_{ij}}$. In our example,

$$A_G(\epsilon) \quad = \quad \begin{pmatrix} 1 & \epsilon^1 & \epsilon^3 & \epsilon^7 \\ \epsilon^2 & 1 & \epsilon^1 & \epsilon^3 \\ \epsilon^4 & \epsilon^5 & 1 & \epsilon^1 \\ \epsilon^6 & \epsilon^3 & \epsilon^1 & 1 \end{pmatrix}.$$

Now compute the third power of this matrix in ordinary arithmetic

$$A_G(\epsilon)^3 \quad = \quad \begin{pmatrix} 1 + 3\epsilon^3 + \cdots & 3\epsilon + \epsilon^4 + \cdots & 3\epsilon^2 + 3\epsilon^3 + \cdots & \epsilon^3 + 6\epsilon^4 + \cdots \\ 3\epsilon^2 + 4\epsilon^5 + \cdots & 1 + 3\epsilon^3 + \cdots & 3\epsilon + \epsilon^3 + \cdots & 3\epsilon^2 + 3\epsilon^3 + \cdots \\ 3\epsilon^4 + 2\epsilon^6 + \cdots & 3\epsilon^4 + 6\epsilon^5 + \cdots & 1 + 3\epsilon^2 + \cdots & 3\epsilon + \epsilon^3 + \cdots \\ 6\epsilon^5 + 3\epsilon^6 + \cdots & 3\epsilon^3 + \epsilon^5 + \cdots & 3\epsilon + \epsilon^3 + \cdots & 1 + 3\epsilon^2 + \cdots \end{pmatrix}.$$

The entry of $A_G(\epsilon)^3$ in row i and column j is a polynomial in ϵ which represents the lengths of all paths from node i to node j using at most three edges. The lowest exponent appearing in this polynomial is the (i, j)-entry in the matrix $D_G^{\odot 3}$. This is a general phenomenon, summarized informally as follows:

$$\text{tropical} \quad = \quad \lim_{\epsilon \to 0} \log\big(\text{classical}(\epsilon)\big) \qquad (2.3)$$

This process of passing from classical arithmetic to tropical arithmetic is referred to as *tropicalization*. In the later sections of Chapter 3, we discuss the tropicalization of algebraic-geometric objects such as curves and surfaces. $\square$

We shall give two more examples of how tropical arithmetic ties in naturally with familiar algorithms in discrete mathematics. The first example concerns the dynamic programming approach to *integer linear programming*. The general integer linear programming problem can be stated as follows. Let $A = (a_{ij})$ be a $d \times n$ matrix of non-negative integers, let $w = (w_1, \ldots, w_n)$ be a row vector with real entries, and let $b = (b_1, \ldots, b_d)^T$ be a column vector with non-negative integer entries. Our task is to find a non-negative integer column vector $u = (u_1, \ldots, u_n)$ which solves the following optimization problem:

$$\text{Minimize } w \cdot u \text{ subject to } u \in \mathbb{N}^n \text{ and } A \cdot u = b. \qquad (2.4)$$

Let us further assume that all columns of the matrix A sum to the same number α and that $b_1 + \cdots + b_d = m \cdot \alpha$. This assumption is convenient because it ensures that all feasible solutions $u \in \mathbb{N}^n$ of (2.4) satisfy $u_1 + \cdots + u_n = m$. We refer to Chapter 13 for a discussion of integer linear programming in the context of applying Markov random fields for inference in computational biology.

We can solve the integer programming problem (2.4) using tropical arithmetic as follows. Let $q_1, \ldots, q_d$ be indeterminates and consider the expression

$$w_1 \odot q_1^{a_{11}} \odot q_2^{a_{21}} \odot \cdots \odot q_d^{a_{d1}} \quad \oplus \quad \cdots \quad \oplus \quad w_n \odot q_1^{a_{1n}} \odot q_2^{a_{2n}} \odot \cdots \odot q_d^{a_{dn}}. \qquad (2.5)$$

Proposition 2.4 *The optimal value of* (2.4) *is the coefficient of the monomial* $q_1^{b_1} q_2^{b_2} \cdots q_d^{b_d}$ *in the mth power, evaluated tropically, of the expression* (2.5).

The proof of this proposition is not difficult and is similar to that of Proposition 2.2. The process of taking the mth power of the tropical polynomial (2.5) can be regarded as solving the shortest path problem in a certain graph. This is precisely the dynamic programming approach to integer linear programming, as described in [Schrijver, 1986]. Prior to the result by [Lenstra, 1983] that integer linear programming can be solved in polynomial time for fixed dimensions, the dynamic programming method provided a polynomial-time algorithm under the assumption that the integers in A are bounded.

Example 2.5 Let $d = 2$, $n = 5$ and consider the instance of (2.4) given by

$$A = \begin{pmatrix} 4 & 3 & 2 & 1 & 0 \\ 0 & 1 & 2 & 3 & 4 \end{pmatrix}, \quad b = \begin{pmatrix} 5 \\ 7 \end{pmatrix} \quad \text{and} \quad w = (2, 5, 11, 7, 3).$$

Here we have $\alpha = 4$ and $m = 3$. The matrix A and the cost vector w are encoded by a tropical polynomial as in (2.5):

$$f = 2q_1^4 \oplus 5q_1^3 q_2 \oplus 11q_1^2 q_2^2 \oplus 7q_1 q_2^3 \oplus 3q_2^4.$$

The third power of this polynomial, evaluated tropically, is equal to

$$\begin{aligned} f \odot f \odot f = \ & 6q_1^{12} \oplus 9q_1^{11}q_2 \oplus 12q_1^{10}q_2^2 \oplus 11q_1^9 q_2^3 \oplus 7q_1^8 q_2^4 \oplus 10q_1^7 q_2^5 \oplus 13q_1^6 q_2^6 \\ & \oplus 12q_1^5 q_2^7 \oplus 8q_1^4 q_2^8 \oplus 11q_1^3 q_2^9 \oplus 17q_1^2 q_2^{10} \oplus 13q_1 q_2^{11} \oplus 9q_2^{12}. \end{aligned}$$

The coefficient 12 of $q_1^5 q_2^7$ in this tropical polynomial is the optimal value. An optimal solution to this integer programming problem is $u = (1, 0, 0, 1, 1)^{\mathrm{T}}$. $\square$

Our final example concerns the notion of the determinant of an $n \times n$ matrix $Q = (q_{ij})$. Since there is no negation in tropical arithmetic, the *tropical determinant* is the same as the *tropical permanent*, namely, it is the sum over the diagonal products obtained by taking all $n!$ permutations π of $\{1, 2, \ldots, n\}$:

$$\text{tropdet}(Q) \quad := \quad \bigoplus_{\pi \in S_n} q_{1\pi(1)} \odot q_{2\pi(2)} \odot \cdots \odot q_{n\pi(n)}. \tag{2.6}$$

Here S_n denotes the *symmetric group* of permutations of $\{1, 2, \ldots, n\}$. The evaluation of the tropical determinant is the classical *assignment problem* of combinatorial optimization. Consider a company which has n jobs and n workers, and each job needs to be assigned to exactly one of the workers. Let q_{ij} be the cost of assigning job i to worker j. The company wishes to find the cheapest assignment $\pi \in S_n$. The optimal total cost is the following minimum:

$$\min\{q_{1\pi(1)} + q_{2\pi(2)} + \cdots + q_{n\pi(n)} \ : \ \pi \in S_n\}.$$

This number is precisely the tropical determinant of the matrix $Q = (q_{ij})$.

Remark 2.6 The tropical determinant solves the assignment problem.

In the assignment problem we need to find the minimum over $n!$ quantities, which appears to require exponentially many operations. However, there is a well-known polynomial-time algorithm for solving this problem. The method was introduced in [Kuhn, 1955] and is known as the *Hungarian Assignment Method*. It maintains a price for each job and an (incomplete) assignment of workers and jobs. At each iteration, the method chooses an unassigned worker and computes a shortest augmenting path from this person to the set of jobs. The total number of arithmetic operations is $O(n^3)$.

In classical arithmetic, the evaluation of determinants and the evaluation of permanents are in different complexity classes. The determinant of an $n \times n$ matrix can be computed in $O(n^3)$ steps, namely by *Gaussian elimination*, while computing the permanent of an $n \times n$ matrix is a fundamentally harder problem (it is #P-complete [Valiant, 1979]). It would be interesting to explore whether the Hungarian Method can be derived from some version of Gaussian Elimination by the principle of tropicalization (2.3).

To see what we mean, consider a 3×3 matrix $A(\epsilon)$ whose entries are polynomials in the indeterminate ϵ. For each entry we list the term of lowest order:

$$A(\epsilon) \quad = \quad \begin{pmatrix} a_{11}\epsilon^{q_{11}} + \cdots & a_{12}\epsilon^{q_{12}} + \cdots & a_{13}\epsilon^{q_{13}} + \cdots \\ a_{21}\epsilon^{q_{21}} + \cdots & a_{22}\epsilon^{q_{22}} + \cdots & a_{23}\epsilon^{q_{23}} + \cdots \\ a_{31}\epsilon^{q_{31}} + \cdots & a_{32}\epsilon^{q_{32}} + \cdots & a_{33}\epsilon^{q_{33}} + \cdots \end{pmatrix}.$$

Suppose that the a_{ij} are sufficiently general non-zero real numbers, so that no cancellation occurs in the lowest-order coefficient when we expand the determinant of $A(\epsilon)$. Writing Q for the 3×3 matrix with entries q_{ij}, we have

$$\det(A(\epsilon)) \quad = \quad \alpha \cdot \epsilon^{\text{tropdet}(Q)} + \cdots \qquad \text{for some } \alpha \in \mathbb{R} \backslash \{0\}.$$

Thus the tropical determinant of Q can be extracted from this expression by taking the logarithm and letting ϵ tend to zero, as suggested by (2.3).

The reader may have wondered where the adjective "tropical" comes from. The algebraic structure $(\mathbb{R} \cup \{\infty\}, \oplus, \odot)$, which is also known as the *min-plus algebra*, has been invented (or re-invented) many times by many people. One of its early developers, in the 1960s, was the Brazilian mathematician Imre Simon. Simon's work was followed up on by French scholars [Pin, 1998], who coined the term "tropical semiring" for the min-plus algebra, in honor of their Brazilian colleague. Hence "tropical" stands for the French view of Brazil. Currently, many mathematicians are working on tropical mathematics and they are exploring a wide range of applications [Litvinov, 2005].

2.2 Sequence alignment

A fundamental task in computational biology is the alignment of DNA or protein sequences. Since biological sequences arising in practice are usually fairly long, researchers have developed highly efficient algorithms for finding optimal alignments. Although in some cases heuristics are used to reduce the combinatorial complexity, most of the algorithms are based on, or incorporate the dynamic programming principle (see the books by [Waterman, 1995, Gusfield, 1997]). What we hope to accomplish in this section is to explain what algebraic statistics and tropical arithmetic have to do with discrete algorithms used for sequence alignment.

First, we give a self-contained explanation of the Needleman–Wunsch algorithm [Needleman and Wunsch, 1970] for aligning biological sequences. Then we explain an algebraic statistical model for pairs of sequences, namely the *pair hidden Markov model*, and we use Needleman–Wunsch to illustrate how dynamic programming algorithms arise naturally from the tropicalization of this model.

We begin by specifying the sequence alignment problem in precise terms. Fix a finite alphabet Σ with l letters, for instance, $\Sigma = \{0, 1, \ldots, l-1\}$. If $l = 4$ then the alphabet of choice is $\Sigma = \{A, C, G, T\}$. Suppose we are given two sequences $\sigma^1 = \sigma_1^1 \sigma_2^1 \cdots \sigma_n^1$ and $\sigma^2 = \sigma_1^2 \sigma_2^2 \cdots \sigma_m^2$ over the alphabet Σ. The sequence lengths n and m may be different. Our aim is to measure the complexity of transforming the sequence σ^1 into the sequence σ^2 by changes to individual characters, insertion of new characters, or deletion of existing characters. Such changes are called *edits*. The *sequence alignment problem* is to find the shortest sequence of edits that relates the two sequences σ^1 and σ^2.

Such sequences of edits are called *alignments*. The shortest sequence of edits relating σ^1 and σ^2 consists of at most $n + m$ edits, and therefore it is a finite problem to identify the best alignment: one can exhaustively enumerate all edit sequences and then pick the shortest one. However, exhaustive search can be improved on considerably. We shall present a dynamic programming algorithm for solving the alignment problem which requires only $O(nm)$ steps.

Each alignment of the pair (σ^1, σ^2) is represented by a string h over the *edit alphabet* $\{H, I, D\}$. These letters stand for **h**omology, **i**nsertion and **d**eletion; this terminology is explained in more detail in Chapter 4. We call the string h the *edit string* of the alignment. An I in the edit string represents an insertion in the first sequence σ^1, a D in the edit string is a deletion in the first sequence σ^1, and an H is a character change, including the "identity change". Writing $\#H, \#I$ and $\#D$ for the number of instances of H, I and D in an edit string for an alignment of the pair (σ^1, σ^2), we find that

$$\#H + \#D = n \quad \text{and} \quad \#H + \#I = m. \tag{2.7}$$

Example 2.7 Let $n = 7$ and $m = 9$ and consider the sequences $\sigma^1 = \text{ACGTAGC}$ and $\sigma^2 = \text{ACCGAGACC}$. Then the following table shows an alignment of σ^1 and σ^2 with $\#H = 6$, $\#I = 3$ and $\#D = 1$. The first row is the edit string:

$$
\begin{array}{cccccccccc}
H & H & I & H & I & H & H & I & D & H \\
A & C & - & G & - & T & A & - & G & C \\
A & C & C & G & A & G & A & C & - & C
\end{array}
\tag{2.8}
$$

Although the alignment has length ten, it represents the transformation of σ^1 into σ^2 by five edit steps which are performed from the left to the right. This transformation is uniquely encoded by the edit string $HHIHIHHIDH$. □

Proposition 2.8 *A string over the edit alphabet* $\{H, I, D\}$ *represents an align-*

ment of an n-letter sequence σ^1 and an m-letter sequence σ^2 if and only if the equations (2.7) hold.

Proof As we perform the edits from the left to the right, every letter in σ^1 either corresponds to a letter in σ^2, in which case we record an H in the edit string, or it gets deleted, in which case we record a D. This shows the first identity in (2.7). The second identity holds because every letter σ^2 either corresponds to a letter in σ^1, in which case there is an H in the edit string, or it has been inserted, in which case we record an I in the edit string. Any string over $\{H, I, D\}$ with (2.7), when read from left to right, produces a valid sequence of edits that transforms σ^1 into σ^2. □

We write $\mathcal{A}_{n,m}$ for the set of all strings over $\{H, I, D\}$ that satisfy (2.7). We call $\mathcal{A}_{n,m}$ the *set of all alignments* of the sequences σ^1 and σ^2, in spite of the fact that it only depends on n and m rather than the specific sequences σ^1 and σ^2. Each element h in $\mathcal{A}_{n,m}$ corresponds to a pair of sequences (μ^1, μ^2) over the alphabet $\Sigma \cup \{-\}$ such that μ^1 consists of a copy of σ^1 together with inserted "$-$" characters, and similarly μ^2 is a copy of σ^2 with inserted "$-$" characters. The cardinalities of the sets $\mathcal{A}_{n,m}$ are the *Delannoy numbers* [Stanley, 1999, §6.3]. They can be computed by a generating function.

Proposition 2.9 *The cardinality of the set $\mathcal{A}_{n,m}$ of all alignments can be computed as the coefficient of $x^m y^n$ in the generating function $1/(1-x-y-xy)$.*

Proof Consider the expansion of the given generating function

$$\frac{1}{1-x-y-xy} = \sum_{m=0}^{\infty} \sum_{n=0}^{\infty} a_{m,n} x^m y^n.$$

The coefficients are characterized by the linear recurrence

$$a_{m,n} = a_{m-1,n} + a_{m,n-1} + a_{m-1,n-1} \text{ with } a_{0,0} = 1,\ a_{m,-1} = a_{-1,n} = 0. \quad (2.9)$$

The same recurrence is valid for the cardinality of $\mathcal{A}_{n,m}$. Indeed, for $m+n \geq 1$, every string in $\mathcal{A}_{n,m}$ is either a string in $\mathcal{A}_{n-1,m-1}$ followed by an H, or a string in $\mathcal{A}_{n-1,m}$ followed by an I, or it is a string in $\mathcal{A}_{n,m-1}$ followed by a D. Also, $\mathcal{A}_{0,0}$ has only one element, namely the empty string, and $\mathcal{A}_{n,m}$ is the empty set if $m < 0$ or $n < 0$. Hence the numbers $a_{m,n}$ and $\#\mathcal{A}_{n,m}$ satisfy the same initial conditions and the same recurrence (2.9), so they must be equal. □

In light of the recurrence (2.9), it is natural to introduce the following graph.

Definition 2.10 The *alignment graph* $\mathcal{G}_{n,m}$ is the directed graph on the set of nodes $\{0, 1, \ldots, n\} \times \{0, 1, \ldots, m\}$ and three classes of directed edges as follows: there are edges labeled by I between pairs of nodes $(i, j) \to (i, j+1)$, there are edges labeled by D between pairs of nodes $(i, j) \to (i+1, j)$, and there are edges labeled by H between pairs of nodes $(i, j) \to (i+1, j+1)$.

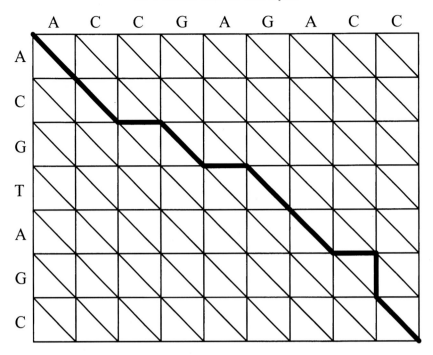

Fig. 2.1. The alignment (2.8) shown as a path in the alignment graph $\mathcal{G}_{7,9}$.

Remark 2.11 The set $\mathcal{A}_{n,m}$ of all alignments is in bijection with the set of paths from the node $(0,0)$ to the node (n,m) in the alignment graph $\mathcal{G}_{n,m}$.

We have introduced three equivalent combinatorial objects: strings over $\{H, I, D\}$ satisfying (2.7), sequence pairs (μ^1, μ^2) that are equivalent to σ^1, σ^2 with the possible insertion of "$-$" characters, and paths in the alignment graph $\mathcal{G}_{n,m}$. All three represent alignments, and they are useful in designing algorithms for finding good alignments. In order to formalize what "good" means, we need to give scores to alignments. A *scoring scheme* is a pair of maps

$$
\begin{aligned}
w &: \Sigma \cup \{-\} \times \Sigma \cup \{-\} \rightarrow \mathbb{R}, \\
w' &: \{H, I, D\} \times \{H, I, D\} \rightarrow \mathbb{R}.
\end{aligned}
$$

Scoring schemes induce weights on alignments of sequences as follows. Fix the two given sequences σ^1 and σ^2 over the alphabet $\Sigma = \{\mathtt{A}, \mathtt{C}, \mathtt{G}, \mathtt{T}\}$. Each alignment is given by an edit string h over $\{H, I, D\}$. We write $|h|$ for the length of h. The edit string h determines the two sequences μ^1 and μ^2 of length $|h|$ over $\Sigma \cup \{-\}$. The weight of the alignment h is defined to be

$$
W(h) \quad := \quad \sum_{i=1}^{|h|} w(\mu_i^1, \mu_i^2) + \sum_{i=2}^{|h|} w'(h_{i-1}, h_i). \tag{2.10}
$$

We represent a scoring scheme (w, w') by a pair of matrices. The first one is

$$
w \quad = \quad
\begin{pmatrix}
w_{\mathtt{A,A}} & w_{\mathtt{A,C}} & w_{\mathtt{A,G}} & w_{\mathtt{A,T}} & w_{\mathtt{A,-}} \\
w_{\mathtt{C,A}} & w_{\mathtt{C,C}} & w_{\mathtt{C,G}} & w_{\mathtt{C,T}} & w_{\mathtt{C,-}} \\
w_{\mathtt{G,A}} & w_{\mathtt{G,C}} & w_{\mathtt{G,G}} & w_{\mathtt{G,T}} & w_{\mathtt{G,-}} \\
w_{\mathtt{T,A}} & w_{\mathtt{T,C}} & w_{\mathtt{T,G}} & w_{\mathtt{T,T}} & w_{\mathtt{T,-}} \\
w_{\mathtt{-,A}} & w_{\mathtt{-,C}} & w_{\mathtt{-,G}} & w_{\mathtt{-,T}} &
\end{pmatrix}.
\tag{2.11}
$$

Here the lower right entry $w_{-,-}$ is left blank because it is never used in computing the weight of an alignment. The second matrix is a 3×3 matrix:

$$
w' \quad = \quad
\begin{pmatrix}
w'_{H,H} & w'_{H,I} & w'_{H,D} \\
w'_{I,H} & w'_{I,I} & w'_{I,D} \\
w'_{D,H} & w'_{D,I} & w'_{D,D}
\end{pmatrix}
\tag{2.12}
$$

Thus the total number of parameters in the alignment problem is $24 + 9 = 33$. We identify the space of parameters with $\mathbb{R}^{33}$. Each alignment $h \in \mathcal{A}_{n,m}$ of a pair of sequences (σ^1, σ^2) gives rise to a linear functional $W(h)$ on $\mathbb{R}^{33}$.

For instance, the weight of the alignment $h = HHIHIHHIDH$ of our sequences $\sigma^1 = \mathtt{ACGTAGC}$ and $\sigma^2 = \mathtt{ACCGAGACC}$ is the linear functional

$$
\begin{aligned}
W(h) \quad = \quad & 2 \cdot w_{\mathtt{A,A}} + 2 \cdot w_{\mathtt{C,C}} + w_{\mathtt{G,G}} + w_{\mathtt{T,G}} + 2 \cdot w_{\mathtt{-,C}} + w_{\mathtt{-,A}} + w_{\mathtt{G,-}} \\
& + 2 \cdot w'_{H,H} + 3 \cdot w'_{H,I} + 2 \cdot w'_{I,H} + w'_{I,D} + w'_{D,H}.
\end{aligned}
$$

Suppose we are given two input sequences σ^1 and σ^2 of lengths n and m over the alphabet Σ. Suppose further that we are given a fixed scoring scheme (w, w'). The *global alignment problem* is to compute an alignment $h \in \mathcal{A}_{n,m}$ whose weight $W(h)$ is minimal among all alignments in $\mathcal{A}_{n,m}$. In the computational biology literature, it is more common to use "maximal" instead of "minimal", but, of course, that is equivalent if we replace (w, w') by $(-w, -w')$.

In the following discussion let us simplify the problem and assume that $w' = 0$, so the weight of an alignment is the linear functional $W(h) = \sum_{i=1}^{|h|} w(\mu_i^1, \mu_i^2)$ on $\mathbb{R}^{24}$. The problem instance (σ^1, σ^2, w) induces weights on the edges of the alignment graph $\mathcal{G}_{n,m}$ as follows. The weight of the edge $(i, j) \to (i+1, j)$ is $w(\sigma_{i+1}^1, -)$, the weight of the edge $(i, j) \to (i, j+1)$ is $w(-, \sigma_{j+1}^2)$, and the weight of the edge $(i, j) \to (i+1, j+1)$ is $w(\sigma_{i+1}^1, \sigma_{j+1}^2)$. This gives a graph-theoretic reformulation of the global alignment problem.

Remark 2.12 The global alignment problem is equivalent to finding the minimum weight path from $(0, 0)$ to (n, m) in the alignment graph $\mathcal{G}_{n,m}$.

Thus the global alignment problem is equivalent to finding shortest paths in a weighted graph. Proposition 2.2 gave a general dynamic programming algorithm for the shortest path problem, the Floyd–Warshall algorithm, which amounts to multiplying matrices in tropical arithmetic. For the specific graph arising in the global alignment problem, this translates into an $O(nm)$ dynamic programming algorithm, called the *Needleman–Wunsch algorithm*.

Algorithm 2.13 (Needleman–Wunsch)

Input: Two sequences $\sigma^1 \in \Sigma^n$, $\sigma^2 \in \Sigma^m$ and a scoring scheme $w \in \mathbb{R}^{24}$.

Output: An alignment $h \in \mathcal{A}_{n.m}$ whose weight $W(h)$ is minimal.

Initialization: Create an $(n+1) \times (m+1)$ matrix M whose rows are indexed by $\{0, 1, \ldots, n\}$ and whose columns are indexed by $\{0, 1, \ldots, m\}$. Set $M[0,0] = 0$.

$$\text{Set} \quad M[i,0] := M[i-1,0] + w(\sigma^1_i, -) \quad \text{for} \quad i = 1, \ldots, n$$
$$\text{and} \quad M[0,j] := M[0,j-1] + w(-, \sigma^2_j) \quad \text{for} \quad j = 1, \ldots, m.$$

Loop: For $i = 1, \ldots, n$ and $j = 1, \ldots, m$ set

$$M[i,j] \quad := \quad \min \left\{ \begin{array}{l} M[i-1,j-1] + w(\sigma^1_i, \sigma^2_j) \\ M[i-1,j] + w(\sigma^1_i, -) \\ M[i,j-1] + w(-, \sigma^2_j) \end{array} \right.$$

Color one or more of the three edges which are adjacent to and directed towards (i,j), and which attain the minimum.

Backtrack: Trace an optimal path in the backward direction from (n,m) to $(0,0)$. This is done by following an arbitrary sequence of colored edges.

Output: Edge labels in $\{H, I, D\}$ of an optimal path in the forward direction.

The more general case when the 3×3 matrix w' is not zero can be modeled by replacing each interior node in $\mathcal{G}_{n,m}$ by a complete bipartite graph $K_{3,3}$ whose edge weights are $w'_{HH}, w'_{H,I}, \ldots, w'_{DD}$. These $9(m-1)(n-1)$ new edges represent transitions between the different states in $\{H, I, D\}$. The resulting graph is denoted $\mathcal{G}'_{n,m}$ and called the *extended alignment graph*. Figure 2.2 illustrates what happens to a node of $\mathcal{G}_{n,m}$ when passing to $\mathcal{G}'_{n,m}$.

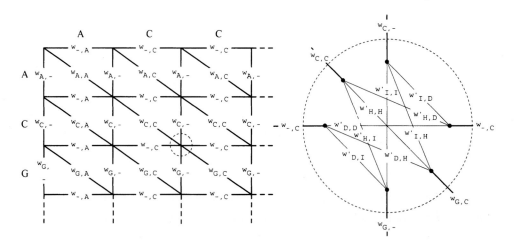

Fig. 2.2. Creating the extended alignment graph by replacing vertices with $K_{3,3}$.

The minimum weight path in $\mathcal{G}'_{n,m}$ is found by a variant of the Needleman–Wunsch algorithm. In the following example we stick to the case $w' = 0$.

Example 2.14 Consider the sequences $\sigma^1 = $ ACGTAGC and $\sigma^2 = $ ACCGAGACC from Example 2.7. According to Proposition 2.9, the number of alignments is

$$\#\mathcal{A}_{7,9} \quad = \quad 224,143.$$

We assume $w' = 0$. The alignment graph $\mathcal{G}_{7,9}$ is depicted in Figure 2.1.

For any particular choice of a scoring scheme $w \in \mathbb{R}^{24}$, the Needleman–Wunsch algorithm easily finds an optimal alignment. Consider the example

$$w \quad = \quad \begin{pmatrix} -91 & 114 & 31 & 123 & x \\ 114 & -100 & 125 & 31 & x \\ 31 & 125 & -100 & 114 & x \\ 123 & 31 & 114 & -91 & x \\ x & x & x & x \end{pmatrix},$$

where the *gap penalty* x is an unknown number between 150 and 200. The 16 specified parameter values in the matrix w are the ones used in the `blastz` alignment program scoring matrix [Schwartz *et al.*, 2003]. For $x \geq 169.5$ an optimal alignment is

$$\begin{pmatrix} h \\ \mu^1 \\ \mu^2 \end{pmatrix} = \begin{pmatrix} H & D & H & H & D & H & H & H & H \\ A & - & C & G & - & T & A & G & C \\ A & C & C & G & A & G & A & C & C \end{pmatrix} \text{ with } W(h) = 2x - 243.$$

If the gap penalty x is below 169.5 then an optimal alignment is

$$\begin{pmatrix} h \\ \mu^1 \\ \mu^2 \end{pmatrix} = \begin{pmatrix} H & D & H & H & I & H & H & D & D & H \\ A & - & C & G & T & A & G & - & - & C \\ A & C & C & G & - & A & G & A & C & C \end{pmatrix} \text{ with } W(h) = 4x - 582.$$

After verifying this computation, the reader may now wish to vary all 24 parameters in the matrix w and run the Needleman–Wunsch algorithm many times. How does the resulting optimal alignment change? How many of the $224,143$ alignments occur for some choice of scoring scheme w? Is there a scoring scheme $w \in \mathbb{R}^{24}$ which makes the alignment (2.8) optimal? Such questions form the subject of *parametric alignment* [Waterman *et al.*, 1992, Gusfield *et al.*, 1994, Gusfield, 1997] which is the topic of Chapter 7. □

We now shift gears and present the *pair hidden Markov model* for alignments. This is an algebraic statistical model which depends on two integers n and m:

$$\mathbf{f} \; : \; \mathbb{R}^{33} \; \rightarrow \; \mathbb{R}^{4^{n+m}}. \tag{2.13}$$

The 4^{n+m} states are the pairs (σ^1, σ^2) of sequences of length n and m. The $33 = 24 + 9$ parameters are written as a pair of matrices (θ, θ') where

$$\theta = \begin{pmatrix} \theta_{A,A} & \theta_{A,C} & \theta_{A,G} & \theta_{A,T} & \theta_{A,-} \\ \theta_{C,A} & \theta_{C,C} & \theta_{C,G} & \theta_{C,T} & \theta_{C,-} \\ \theta_{G,A} & \theta_{G,C} & \theta_{G,G} & \theta_{G,T} & \theta_{G,-} \\ \theta_{T,A} & \theta_{T,C} & \theta_{T,G} & \theta_{T,T} & \theta_{T,-} \\ \theta_{-,A} & \theta_{-,C} & \theta_{-,G} & \theta_{-,T} \end{pmatrix}, \quad \theta' = \begin{pmatrix} \theta'_{H,H} & \theta'_{H,I} & \theta'_{H,D} \\ \theta'_{I,H} & \theta'_{I,I} & \theta'_{I,D} \\ \theta'_{D,H} & \theta'_{D,I} & \theta'_{D,D} \end{pmatrix}. \tag{2.14}$$

In order to be statistically meaningful these parameters have to be non-negative and satisfy six independent linear equations. That is, they must lie in

$$\Theta \quad = \quad \Delta_{15} \times \Delta_3 \times \Delta_3 \times \Delta_2 \times \Delta_2 \times \Delta_2 \quad \subset \quad \mathbb{R}^{33}.$$

The parameter space Θ is the product of six simplices of dimensions $15, 3, 3, 2, 2$ and 2. The big simplex Δ_{15} consists of all non-negative 4×4 matrices $(\theta_{ij})_{i,j\in\Sigma}$ whose entries sum to 1. The two tetrahedra Δ_3 come from requiring that

$$\theta_{-,A} + \theta_{-,C} + \theta_{-,G} + \theta_{-,T} \quad = \quad \theta_{A,-} + \theta_{C,-} + \theta_{G,-} + \theta_{T,-} \quad = \quad 1.$$

The three triangles Δ_2 come from requiring that

$$\theta'_{H,H} + \theta'_{H,I} + \theta'_{H,D} \quad = \quad \theta'_{I,H} + \theta'_{I,I} + \theta'_{I,D} \quad = \quad \theta'_{D,H} + \theta'_{D,I} + \theta'_{D,D} \quad = \quad 1.$$

The coordinate f_{σ^1,σ^2} of the pair hidden Markov model $\mathbf{f}$ represents the probability of observing the pair of sequences (σ^1, σ^2). This is the polynomial

$$f_{\sigma^1,\sigma^2} \quad = \quad \sum_{h\in\mathcal{A}_{n,m}} \prod_{i=1}^{|h|} \theta_{\mu_i^1,\mu_i^2} \cdot \prod_{i=2}^{|h|} \theta'_{h_{i-1},h_i}. \tag{2.15}$$

Here (μ^1, μ^2) is the pair of sequences over $\Sigma \cup \{-\}$ which corresponds to h. The following observation is crucial for understanding parametric alignment.

Proposition 2.15 *The objective function of the sequence alignment problem is the tropicalization of a coordinate polynomial f_{σ^1,σ^2} of the pair HMM.*

Proof The tropicalization of the polynomial (2.15) is obtained by replacing the outer sum by a tropical sum $\oplus$ and the inner products by tropical products $\odot$. We replace each unknown $\theta_{\dots}$ by the corresponding unknown $w_{\dots}$, which we think of as the negated logarithm of $\theta_{\dots}$. The result is the tropical polynomial

$$\text{trop}(f_{\sigma^1,\sigma^2}) \quad = \quad \bigoplus_{h\in\mathcal{A}_{n,m}} \bigodot_{i=1}^{|h|} w_{\mu_i^1,\mu_i^2} \cdot \bigodot_{i=2}^{|h|} w'_{h_{i-1},h_i}. \tag{2.16}$$

The tropical product inside the tropical sum is precisely the weight $W(h)$ of the alignment h or (μ^1, μ^2) as defined in (2.10). Hence (2.16) is equivalent to

$$\text{trop}(f_{\sigma^1,\sigma^2}) \quad = \quad \min_{h\in\mathcal{A}_{n,m}} W(h).$$

Evaluating the tropical polynomial is therefore equivalent to finding an optimal alignment of the two sequences σ^1 and σ^2. Stated in the language of pair HMMs, evaluating (2.16) is equivalent to finding a Viterbi sequence. $\square$

Remark 2.16 Since the logarithm of a probability is always negative, the correspondence in Proposition 2.15 only accounts for scoring schemes in which the weights have the same sign. Scoring schemes in which the weights have mixed signs, as in Example 2.14, result from associating $w_{\dots}$ with the *log-odds ratio* $\log(\theta_{\dots}/\tilde{\theta}_{\dots})$ where the $\tilde{\theta}_{\dots}$ are additional new parameters.

Remark 2.17 The tropicalization of the hidden Markov model (Section 1.4.3) is given by polynomials analogous to those in (2.16):

$$\text{trop}(f_{\tau_1 \ldots \tau_n}) \quad = \quad \bigoplus_{\sigma_1, \ldots, \sigma_n} \bigodot_{i=1}^{n} w'_{\sigma_i \tau_i} \bigodot_{i=2}^{n} w_{\sigma_{i-1} \sigma_i} \qquad (2.17)$$

where $w_{\ldots}$ is the negated logarithm of the transition probability $\theta_{\ldots}$ and $w'_{\ldots}$ is the negated logarithm of the output probability $\theta'_{\ldots}$. Evaluating the tropical polynomial is known as the *Viterbi algorithm* [Viterbi, 1967] and a minimizing sequence $\hat{\sigma}_1 \cdots \hat{\sigma}_n$ is known as a *Viterbi sequence* for the HMM. The extension to arbitrary hidden tree models is straightforward; in that case the tropical evaluation of a coordinate of the model is known as *Felsenstein's algorithm* [Felsenstein, 1981].

It is an instructive exercise to show that the sum of the polynomials f_{σ^1, σ^2} over all 4^{n+m} pairs of sequences (σ^1, σ^2) simplifies to 1 when (θ, θ') lies in Θ. The key idea is to derive a recursive decomposition of the polynomial f_{σ^1, σ^2} by grouping together all summands with fixed last factor pair $\theta'_{h_{|h|-1}, h_{|h|}} \theta_{\mu^1_{|h|}, \mu^2_{|h|}}$. This recursive decomposition is equivalent to performing dynamic programming along the extended alignment graph $\mathcal{G}'_{n,m}$. The variant of the Needleman–Wunsch algorithm on the graph $\mathcal{G}'_{n,m}$ is precisely the efficient evaluation of the tropical polynomial $\text{trop}(f_{\sigma^1, \sigma^2})$ using the same recursive decomposition.

We explain this circle of ideas for the simpler case of Algorithm 2.13 where

$$w' \quad = \quad \begin{pmatrix} 0 & 0 & 0 \\ 0 & 0 & 0 \\ 0 & 0 & 0 \end{pmatrix}$$

To be precise, we shall implement dynamic programming on the alignment graph $\mathcal{G}_{n,m}$ as the efficient computation of a (tropical) polynomial. In terms of the pair HMM, this means that we are fixing all entries of the 3×3 matrix θ' to be identical. Let us consider the following two possible specializations:

$$\theta' = \begin{pmatrix} 1/3 & 1/3 & 1/3 \\ 1/3 & 1/3 & 1/3 \\ 1/3 & 1/3 & 1/3 \end{pmatrix} \quad \text{and} \quad \theta' = \begin{pmatrix} 1 & 1 & 1 \\ 1 & 1 & 1 \\ 1 & 1 & 1 \end{pmatrix}$$

The first specialization is the statistically meaningful one, but it leads to more complicated formulas in the coefficients. For that reason we use the second specialization in our implementation. We write g_{σ^1, σ^2} for the polynomial in the 24 unknowns $\theta_{\ldots}$ obtained from f_{σ^1, σ^2} by setting each of the 9 unknowns $\theta_{\ldots}$ to 1. The following short `Maple` code computes the polynomial $g_{\text{s1,s2}}$ for

```
s1 := [A,C,G]: s2 := [A,C,C]:

T := array([ [ tAA, tAC, tAG, tAT, t_A ],
             [ tCA, tCC, tCG, tCT, t_C ],
             [ tGA, tGC, tGG, tGT, t_G ],
             [ tTA, tTC, tTG, tTT, t_T ],
```

[tA_, tC_, tG_, tT_, 0]]):

This represents the matrix θ with tAA $= \theta_{AA}$, tAC $= \theta_{AC}, \ldots$ etc. We initialize

```
n := nops(s1): m := nops(s2):
u1 := subs({A=1,C=2,G=3,T=4},s1):
u2 := subs({A=1,C=2,G=3,T=4},s2):
M := array([],0..n,0..m): M[0,0] := 1:
for i from 1 to n do
M[i,0] := M[i-1,0] * T[u1[i],5]:
od:
for j from 1 to m do
M[0,j] := M[0,j-1] * T[5,u2[j]]:
od:
```

We then perform a loop precisely as in Algorithm 2.13, with tropical arithmetic on real numbers replaced by ordinary arithmetic on polynomials.

```
for i from 1 to n do
 for j from 1 to m do
    M[i,j] := M[i-1,j-1] * T[u1[i],u2[j]] + M[i-1, j ] *
              T[u1[i], 5 ] + M[ i ,j-1] * T[ 5 ,u2[j]]:
 od:
od:
lprint(M[n,m]);
```

Our Maple code produces a recursive decomposition of the polynomial $g_{\text{ACG,ACC}}$:

```
((tAA+2*tA_*t_A)*tCC+(tA_*tAC+tA_*tC_*t_A+(tAA+2*tA_*t_A)*tC_)
*t_C+(t_A*tCA+(tAA+2*tA_*t_A)*t_C+t_A*t_C*tA_)*tC_)*tGC+((tA_*
tAC+tA_*tC_*t_A+(tAA+2*tA_*t_A)*tC_)*tCC+(tA_*tC_*tAC+tA_*tC_^2
*t_A+(tA_*tAC+tA_*tC_*t_A+(tAA+2*tA_*t_A)*tC_)*tC_)*t_C+((tAA+
2*tA_*t_A)*tCC+(tA_*tAC+tA_*tC_*t_A+(tAA+2*tA_*t_A)*tC_)*t_C+
(t_A*tCA+(tAA+2*tA_*t_A)*t_C+t_A*t_C*tA_)*tC_)*tC_)*t_G+((t_A*
tCA+(tAA+2*tA_*t_A)*t_C+t_A*t_C*tA_)*tGC+((tAA+2*tA_*t_A)*tCC+
(tA_*tAC+tA_*tC_*t_A+(tAA+2*tA_*t_A)*tC_)*t_C+(t_A*tCA+(tAA+2*
tA_*t_A)*t_C+t_A*t_C*tA_)*tC_)*t_G+(t_A*t_C*tGA+(t_A*tCA+(tAA+
2*tA_*t_A)*t_C+t_A*t_C*tA_)*t_G+t_A*t_C*t_G*tA_)*tC_)*tC_
```

The expansion of this polynomial has 14 monomials. The sum of its coefficients is $\#\mathcal{A}_{3,3} = 63$. Next we run same code for the sequences of Example 2.7:

s1 := [A,C,G,T,A,G,C]: s2 := [A,C,C,G,A,G,A,C,C]:

The expansion of the resulting polynomial $g_{s1,s2}$ has $1,615$ monomials, and the sum of its coefficients equals $\#\mathcal{A}_{7,9} = 224,143$. Each monomial in $g_{s1,s2}$ represents a family of alignments h all of which have the same $W(h)$. We have chosen a simple example to illustrate the main points, but the method shown can be used for computing the polynomials associated to much longer sequence pairs. We summarize our discussion of sequence alignment as follows:

Remark 2.18 The Needleman–Wunsch algorithm is the tropicalization of the pair hidden Markov model for sequence alignment. Similarly, the Viterbi algorithm is the tropicalization of the hidden Markov model, and the Felsenstein algorithm is the tropicalization of the hidden tree model.

In order to answer parametric questions, such as the ones raised at the end of Example 2.14, we need to better understand the combinatorial structure encoded in the polynomials f_{σ^1,σ^2} and g_{σ^1,σ^2}. The key to unraveling this combinatorial structure lies in the study of polytopes, which is our next topic.

2.3 Polytopes

In this section we review basic facts about convex polytopes and algorithms for computing them, and we explain how they relate to algebraic statistical models. Every polynomial and every polynomial map has an associated polytope, called its *Newton polytope*. This allows us to replace tropical arithmetic by the *polytope algebra*, which is useful for solving parametric inference problems.

As a motivation for the mathematics in this section, let us give a sneak preview of Newton polytopes arising from the pair HMM for sequence alignment.

Example 2.19 Consider the following 14 points v_i in 11-dimensional space:

$$
\begin{aligned}
v_1 &= (0,0,1,0,0,2,0,0,1,1,1) & 20\,\theta_{\text{A}-}\,\theta_{-\text{A}}\,\theta^2_{\text{C}-}\,\theta_{-\text{C}}\,\theta_{-\text{G}} \\
v_2 &= (1,0,0,0,0,2,0,0,0,1,1) & 6\,\theta_{\text{AA}}\,\theta^2_{\text{C}-}\,\theta_{-\text{C}}\,\theta_{-\text{G}} \\
v_3 &= (0,0,1,0,1,1,0,0,1,0,1) & 7\,\theta_{\text{A}-}\,\theta_{-\text{A}}\,\theta_{\text{CC}}\,\theta_{\text{C}-}\,\theta_{-\text{G}} \\
v_4 &= (0,0,1,0,0,1,0,1,1,1,0) & 9\,\theta_{\text{A}-}\,\theta_{-\text{A}}\,\theta_{\text{C}-}\,\theta_{-\text{C}}\,\theta_{\text{GC}} \\
v_5 &= (0,1,1,0,0,1,0,0,0,1,1) & 4\,\theta_{\text{A}-}\,\theta_{\text{AC}}\,\theta_{\text{C}-}\,\theta_{-\text{C}}\,\theta_{-\text{G}} \\
v_6 &= (0,0,0,0,0,2,1,0,1,1,0) & \theta_{-\text{A}}\,\theta^2_{\text{C}-}\,\theta_{-\text{C}}\,\theta_{\text{GA}} \\
v_7 &= (0,0,0,1,0,2,0,0,1,0,1) & 3\,\theta_{-\text{A}}\,\theta^2_{\text{C}-}\,\theta_{\text{CA}}\,\theta_{-\text{G}} \\
v_8 &= (1,0,0,0,1,1,0,0,0,0,1) & 3\,\theta_{\text{AA}}\,\theta_{\text{CC}}\,\theta_{\text{C}-}\,\theta_{-\text{G}} \\
v_9 &= (1,0,0,0,0,1,0,1,0,1,0) & 3\,\theta_{\text{AA}}\,\theta_{\text{C}-}\,\theta_{-\text{C}}\,\theta_{\text{GC}} \\
v_{10} &= (0,0,1,0,1,0,0,1,1,0,0) & 2\,\theta_{\text{A}-}\,\theta_{-\text{A}}\,\theta_{\text{CC}}\,\theta_{\text{GC}} \\
v_{11} &= (0,1,1,0,1,0,0,0,0,0,1) & \theta_{\text{A}-}\,\theta_{\text{CC}}\,\theta_{\text{AC}}\,\theta_{-\text{G}} \\
v_{12} &= (0,1,1,0,0,0,0,1,0,1,0) & \theta_{\text{A}-}\,\theta_{\text{AC}}\,\theta_{-\text{C}}\,\theta_{\text{GC}} \\
v_{13} &= (0,0,0,1,0,1,0,1,1,0,0) & 2\,\theta_{-\text{A}}\,\theta_{\text{C}-}\,\theta_{\text{CA}}\,\theta_{\text{GC}} \\
v_{14} &= (1,0,0,0,1,0,0,1,0,0,0) & \theta_{\text{AA}}\,\theta_{\text{CC}}\,\theta_{\text{GC}}
\end{aligned}
$$

To the right of each point v_i is the corresponding monomial in the unknowns $\theta_{\text{AA}}, \theta_{\text{AC}}, \theta_{\text{A}-}, \theta_{\text{CA}}, \theta_{\text{CC}}, \theta_{\text{C}-}, \theta_{\text{GA}}, \theta_{\text{GC}}, \theta_{-\text{A}}, \theta_{-\text{C}}, \theta_{-\text{G}}$. The jth coordinate in v_i equals the exponent of the jth unknown. The sum of these 14 monomials is the polynomial $g_{\text{ACG,ACC}}$ computed by the `Maple` code at the end of Section 2.2.

The 14 points v_i span a six-dimensional linear space in $\mathbb{R}^{11}$, and it is their location inside that space which determines which alignment is optimal. For instance, the gapless alignment (H, H, H) which corresponds to the last monomial $\theta_{AA}\,\theta_{CC}\,\theta_{GC}$ is optimal if and only if the scoring scheme w satisfies

$$w_{C-} + w_{-G} \geq w_{GC}, \qquad w_{A-} + w_{AC} + w_{-G} \geq w_{AA} + w_{GC},$$

$$w_{C-} + w_{-C} \geq w_{CC}, \qquad w_{A-} + w_{AC} + w_{-C} \geq w_{AA} + w_{CC},$$

$$w_{A-} + w_{-A} \geq w_{AA}, \qquad w_{-A} + w_{C-} + w_{CA} \geq w_{AA} + w_{CC},$$

$$\text{and} \quad w_{-A} + 2w_{C-} + w_{-C} + w_{GA} \geq w_{AA} + w_{CC} + w_{GC}.$$

In this section we introduce the geometry behind such derivations. $\qquad\square$

Given any points $v_1, \ldots, v_n$ in $\mathbb{R}^d$, their *convex hull* is the set

$$P \quad = \quad \left\{ \sum_{i=1}^{n} \lambda_i v_i \in \mathbb{R}^d \ : \ \lambda_1, \ldots, \lambda_n \geq 0 \ \text{and} \ \sum_{i=1}^{n} \lambda_i = 1 \right\}. \qquad (2.18)$$

Any subset of $\mathbb{R}^d$ of this form is called a *convex polytope* or just a *polytope*, for short. The *dimension* of the polytope P is the dimension of its affine span $\{ \sum_{i=1}^{n} \lambda_i v_i \in \mathbb{R}^d \ : \ \sum_{i=1}^{n} \lambda_i = 1 \}$. We can also represent a polytope as a finite intersection of closed half-spaces. Let A be a real $m \times d$ matrix and let $b \in \mathbb{R}^m$. Each row of A and corresponding entry of b defines a half-space in $\mathbb{R}^d$. Their intersection is the following set which may be bounded or unbounded:

$$P \quad = \quad \{ x \in \mathbb{R}^d \ : \ A \cdot x \geq b \}. \qquad (2.19)$$

Any subset of $\mathbb{R}^d$ of this form is called a *convex polyhedron*.

Theorem 2.20 *Convex polytopes are precisely the bounded convex polyhedra.*

Proof A proof (and lots of information on polytopes) can be found in the books [Grünbaum, 2003] and [Ziegler, 1995]. This theorem is known as the *Weyl–Minkowski Theorem*. $\qquad\square$

Thus every polytope can be represented either in the form (2.18) or in the form (2.19). These representations are known as *V-polytopes* and *H-polytopes*. Transforming one into the other is a fundamental algorithmic task in geometry.

Example 2.21 Let P be the standard cube of dimension $d = 3$. As an H-polytope the cube is the solution to $m = 6$ linear inequalities

$$P \quad = \quad \{ (x, y, z) \in \mathbb{R}^3 \ : \ 0 \leq x \leq 1, \ 0 \leq y \leq 1, \ 0 \leq z \leq 1 \},$$

and as a V-polytope the cube is the convex hull of $n = 8$ points

$$P = \mathrm{conv}\{(0,0,0), (0,0,1), (0,1,0), (0,1,1), (1,0,0), (1,0,1), (1,1,0), (1,1,1)\}.$$

$\qquad\square$

Closely related computational tasks are to make the V-representation (2.18) irredundant by removing points v_i, and to make the H-representation (2.19) irredundant by removing halfspaces, each while leaving the set P unchanged. To understand the underlying geometry, we need to define faces of polytopes.

Given a polytope $P \subset \mathbb{R}^d$ and a vector $w \in \mathbb{R}^d$, consider the set of all points in P at which the linear functional $x \mapsto x \cdot w$ attains its minimum. It is denoted

$$\text{face}_w(P) \quad = \quad \{\, x \in P \;:\; x \cdot w \leq y \cdot w \text{ for all } y \in P \,\}. \qquad (2.20)$$

Let $w^* = \min\{x \cdot w : x \in P\}$. Then we can write (2.20) equivalently as

$$\text{face}_w(P) \quad = \quad \{\, x \in P \;:\; x \cdot w \leq w^* \,\}.$$

This shows that $\text{face}_w(P)$ is a bounded polyhedron, and hence it is a polytope by Theorem 2.20. Every polytope of this form is called a *face* of P. In particular P is a face of itself, obtained by taking $w = 0$. A face of dimension zero consists of a single point and is called a *vertex* of P. A face of dimension one is called an *edge*, a face of dimension $\dim(P) - 1$ is called a *facet*, and a face of dimension $\dim(P) - 2$ is called a *ridge*. The cube in Example 2.21 has 27 faces. Of these, there are 8 vertices, 12 edges ($=$ ridges), 6 facets, and the cube itself.

We write $f_i(P)$ for the number of i-dimensional faces of a polytope P. The vector $f(P) = \big(f_0(P), f_1(P), f_2(P), \ldots, f_{d-1}(P)\big)$ is called the *f-vector* of P. So, the three-dimensional cube P has the f-vector $f(P) = (8, 12, 6)$. Its dual polytope P^*, which is the octahedron, has the f-vector $f(P^*) = (6, 12, 8)$.

Let P be a polytope and F a face of P. The *normal cone* of P at F is

$$N_P(F) \quad = \quad \{\, w \in \mathbb{R}^d \;:\; \text{face}_w(P) = F \,\}.$$

This is a relatively open convex polyhedral cone in $\mathbb{R}^d$. Its dimension satisfies

$$\dim N_P(F) \quad = \quad d - \dim(F).$$

In particular, if $F = \{v\}$ is a vertex of P then its normal cone $N_P(v)$ is d-dimensional and consists of all linear functionals w that are minimized at v.

Example 2.22 Let P be the convex hull of the points $v_1, \ldots, v_{14}$ in Example 2.19. The normal cone $N_P(v_{14})$ consists of all weights for which the gapless alignment (H, H, H) is optimal. It is characterized by the seven inequalities on the previous page. $\qquad\qquad\square$

The collection of all cones $N_P(F)$ as F runs over all faces of P is denoted $\mathcal{N}(P)$ and is called the *normal fan* of P. Thus the normal fan $\mathcal{N}(P)$ is a partition of $\mathbb{R}^d$ into cones. The cones in $\mathcal{N}(P)$ are in bijection with the faces of P. For instance, if P is the 3-cube then $\mathcal{N}(P)$ is the partition of $\mathbb{R}^3$ into cones with constant sign vectors. Hence $\mathcal{N}(P)$ is combinatorially isomorphic to the octahedron P^*. Figure 2.3 shows a two-dimensional example.

Our next result ties in the faces of a polytope P with its irredundant representations. Let a_i be one of the row vectors of the matrix A in (2.19) and let b_i be the corresponding entry in the vector b. This defines the face

$$\text{face}_{a_i}(P) \quad = \quad \{\, x \in P \;:\; a_i \cdot x = b_i \,\}.$$

Proposition 2.23 *The V-representation (2.18) of the polytope P is irredundant if and only if v_i is a vertex of P for $i = 1, \ldots, n$. The H-representation (2.19) is irredundant if and only if $\text{face}_{a_i}(P)$ is a facet of P for $i = 1, \ldots, m$.*

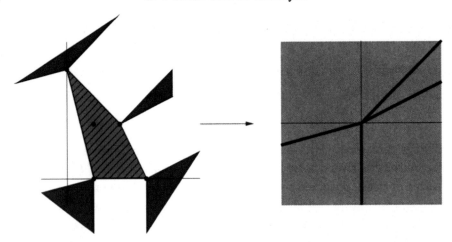

Fig. 2.3. The (negated) normal fan of a quadrangle in the plane.

A comprehensive software system for computing with polytopes is the program **polymake**. We show the use of **polymake** by computing the polytope of the toric Markov chain model $\mathbf{f}_{2,4}(\Theta)$. This model has $m = 16$ states and $d = 4$ parameters. We create an input file named **foo** which looks like this:

```
POINTS
1 3 0 0 0
1 2 1 0 0
1 1 1 1 0
1 1 1 0 1
1 1 1 1 0
1 0 2 1 0
1 0 1 1 1
1 0 1 0 2
1 2 0 1 0
1 1 1 1 0
1 0 1 2 0
1 0 1 1 1
1 1 0 1 1
1 0 1 1 1
1 0 0 1 2
1 0 0 0 3
```

These 16 points are the columns of the 4×16-matrix in Subsection 1.4.1. The extra character 1 is prepended for technical reasons. We run the command

```
polymake foo VERTICES
```

Then the system responds by listing the eight vertices of this polytope

```
VERTICES
1 3 0 0 0
```

```
1 2 1 0 0
1 0 2 1 0
1 0 1 0 2
1 2 0 1 0
1 0 1 2 0
1 0 0 1 2
1 0 0 0 3
```

Furthermore, on the file `foo` itself we find the irredundant H-representation

```
FACETS
1 0 -1 1 0
0 1 0 0 0
1 0 1 -1 0
0 0 1 0 0
0 0 0 1 0
3 -1 -1 -1 0
```

```
AFFINE_HULL
-3 1 1 1 1
```

This output tells us that our polytope is defined by the one linear equation $x_1 + x_2 + x_3 + x_4 = 3$ and the six linear inequalities

$$x_2 - x_3 \leq 1, \; x_1 \geq 0, \; x_3 - x_2 \leq 1, \; x_2 \geq 0, \; x_3 \geq 0, \; x_1 + x_2 + x_3 \leq 3.$$

Indeed, the command `DIM` confirms that the polytope is three-dimensional:

```
polymake foo DIM
DIM
3
```

The f-vector of our polytope coincides with that of the three-dimensional cube

```
polymake foo F_VECTOR
F_VECTOR
8 12 6
```

But our polytope is not a cube at all. Inspecting the updated file `foo` reveals that its facets are two triangles, two quadrangles and two pentagons:

```
VERTICES_IN_FACETS
{1 2 3}
{2 3 5 6 7}
{4 5 6}
{0 4 6 7}
{0 1 3 7}
{0 1 2 4 5}
```

This is the polytope depicted in Figure 1.1. We return to our general discussion.

Let $\mathcal{P}_d$ denote the set of all polytopes in $\mathbb{R}^d$. There are two natural operations, namely, addition $\oplus$ and multiplication $\odot$, defined on the set $\mathcal{P}_d$. The resulting structure is the *polytope algebra* $(\mathcal{P}_d, \oplus, \odot)$. So, if $P, Q \in \mathcal{P}_d$ are polytopes then their sum $P \oplus Q$ is the convex hull of the union of P and Q:

$$
\begin{aligned}
P \oplus Q \quad &:= \quad \mathrm{conv}(P \cup Q) \\
&= \quad \{\, \lambda p + (1 - \lambda)q \in \mathbb{R}^d \; : \; p \in P, \, q \in Q, \, 0 \le \lambda \le 1 \,\}.
\end{aligned}
$$

The product in the polytope algebra is defined to be the *Minkowski sum*:

$$
\begin{aligned}
P \odot Q \quad &:= \quad P + Q \\
&= \quad \{\, p + q \in \mathbb{R}^d \; : \; p \in P, \, q \in Q \,\}.
\end{aligned}
$$

It follows from the Weyl–Minkowski Theorem that both $P \oplus Q$ and $P \odot Q$ are polytopes in $\mathbb{R}^d$. The polytope algebra $(\mathcal{P}_d, \oplus, \odot)$ satisfies many of the familiar axioms of arithmetic. Clearly, addition and multiplication are commutative. But it is also the case that the distributive law holds for polytopes:

Proposition 2.24 *If P, Q, R are polytopes in $\mathbb{R}^d$ then*

$$
(P \oplus Q) \odot R \quad = \quad (P \odot R) \oplus (Q \odot R). \tag{2.21}
$$

Proof Consider points $p \in P$, $q \in Q$ and $r \in R$. For $0 \le \lambda \le 1$ note that

$$
(\lambda p + (1 - \lambda)q) + r \quad = \quad \lambda(p + r) + (1 - \lambda)(q + r).
$$

The left hand side represents an arbitrary point in the left hand side of (2.21), and similarly for the right hand side. □

Example 2.25 *(The tropical semiring revisited)* Let us consider the algebra $(\mathcal{P}_1, \oplus, \odot)$ of all polytopes on the real line $(d = 1)$. Each element of $\mathcal{P}_1$ is a segment $[a, b]$ where $a < b$ are real numbers. The arithmetic operations are

$$
\begin{aligned}
[a, b] \oplus [c, d] \quad &= \quad [\,\min(a, c), \max(b, d)\,], \\
[a, b] \odot [c, d] \quad &= \quad [\, a + c, \, b + d \,].
\end{aligned}
$$

Thus the one-dimensional polytope algebra is essentially the same as the tropical semiring $(\mathbb{R}, \oplus, \odot)$. Or, stated differently, the polytope algebra $(\mathcal{P}_d, \oplus, \odot)$ is a natural higher-dimensional generalization of the tropical semiring. □

One of the main connections between polytopes and algebraic statistics is via the Newton polytopes of the polynomials which parameterize a model. Consider the polynomial

$$
f \quad = \quad \sum_{i=1}^{n} c_i \cdot \theta_1^{v_{i1}} \theta_2^{v_{i2}} \cdots \theta_d^{v_{id}}, \tag{2.22}
$$

where c_i is a non-zero real number and $v_i = (v_{i1}, v_{i2}, \ldots, v_{id}) \in \mathbb{N}^d$ for $i =$

$1, 2, \ldots, n$. We define the *Newton polytope* of the polynomial f as the convex hull of all exponent vectors that appear in the expansion (2.22) of f:

$$\mathrm{NP}(f) \quad := \quad \mathrm{conv}\{v_1, v_2, \ldots, v_n\} \quad \subset \quad \mathbb{R}^d. \tag{2.23}$$

Hence the Newton polytope $\mathrm{NP}(f)$ is precisely the V-polytope in (2.18). The operation of taking Newton polytopes respects the arithmetic operations:

Theorem 2.26 *Let f and g be polynomials in $\mathbb{R}[\theta_1, \ldots, \theta_d]$. Then*

$$\mathrm{NP}(f \cdot g) = \mathrm{NP}(f) \odot \mathrm{NP}(g) \quad and \quad \mathrm{NP}(f + g) \subseteq \mathrm{NP}(f) \oplus \mathrm{NP}(g).$$

If all coefficients of f and g are positive then $\mathrm{NP}(f + g) = \mathrm{NP}(f) \oplus \mathrm{NP}(g)$.

Proof Let $f = \sum_{i=1}^{n} c_i \cdot \theta^{v_i}$ be as in (2.22) and let $g = \sum_{j=1}^{n'} c'_j \cdot \theta^{v'_j}$. For any $w \in \mathbb{R}^d$ let $\mathrm{in}_w(f)$ denote the *initial form* of f. This is the subsum of all terms $c_i \theta^{v_i}$ such that $v_i \cdot w$ is minimal. Then the following identity holds:

$$\mathrm{NP}\big(\mathrm{in}_w(f)\big) \quad = \quad \mathrm{face}_w\big(\mathrm{NP}(f)\big). \tag{2.24}$$

The initial form of a product is the product of the initial forms:

$$\mathrm{in}_w(f \cdot g) \quad = \quad \mathrm{in}_w(f) \cdot \mathrm{in}_w(g). \tag{2.25}$$

For generic $w \in \mathbb{R}^d$, the initial form (2.25) is a monomial $\theta^{v_i + v'_j}$, and its coefficient in $f \cdot g$ is the product of the corresponding coefficients in f and g. Finally, the face operator $\mathrm{face}_w(\cdot)$ is a linear map on the polytope algebra:

$$\mathrm{face}_w\big(\mathrm{NP}(f) \odot \mathrm{NP}(g)\big) \quad = \quad \mathrm{face}_w\big(\mathrm{NP}(f)\big) \odot \mathrm{face}_w\big(\mathrm{NP}(g)\big). \tag{2.26}$$

Combining the three identities (2.24), (2.25) and (2.26), for w generic, shows that the polytopes $\mathrm{NP}(f \cdot g)$ and $\mathrm{NP}(f) \odot \mathrm{NP}(g)$ have the same set of vertices.

For the second identity, note that $\mathrm{NP}(f) \oplus \mathrm{NP}(g)$ is the convex hull of $\{v_1, \ldots, v_n, v'_1, \ldots, v'_{n'}\}$. Every term of $f + g$ has its exponent in this set, so this convex hull contains $\mathrm{NP}(f+g)$. If all coefficients are positive then equality holds because there is no cancellation when forming the sum $f + g$. □

Example 2.27 Consider the polynomials $f = (x+1)(y+1)(z+1)$ and $g = (x + y + z)^2$. Then $\mathrm{NP}(f)$ is a cube and $\mathrm{NP}(g)$ is a triangle. The Newton polytope $\mathrm{NP}(f+g)$ of their sum is the *bipyramid* with vertices $(0,0,0)$, $(2,0,0)$, $(0,2,0)$, $(0,0,2)$, $(1,1,1)$. The Newton polytope $\mathrm{NP}(f \cdot g)$ of their product is the Minkowski sum of the cube with the triangle. It has 15 vertices. □

Newton polytopes allow us to transfer constructions from the algebraic setting of polynomials to the geometric setting of polytopes. The following example illustrates this. Suppose we are given a 4×4 matrix of polynomials,

$$A(x,y,z) \quad = \quad \begin{pmatrix} a_{11}(x,y,z) & a_{12}(x,y,z) & a_{13}(x,y,z) & a_{14}(x,y,z) \\ a_{21}(x,y,z) & a_{22}(x,y,z) & a_{23}(x,y,z) & a_{24}(x,y,z) \\ a_{31}(x,y,z) & a_{32}(x,y,z) & a_{33}(x,y,z) & a_{34}(x,y,z) \\ a_{41}(x,y,z) & a_{42}(x,y,z) & a_{43}(x,y,z) & a_{44}(x,y,z) \end{pmatrix},$$

and suppose we are interested in the Newton polytope of its determinant $\det(A(x,y,z))$. One possible way to compute this Newton polytope is to evaluate the determinant, list all terms that occur in that polynomial, and then compute the convex hull. However, assuming that the coefficients of the $a_{ij}(x,y,z)$ are such that no cancellations occur, it is more efficient to do the arithmetic directly at the level of Newton polytopes. Namely, we replace each matrix entry by its Newton polytope $P_{ij} = \text{NP}(a_{ij})$, consider the 4×4 matrix of polytopes (P_{ij}), and compute its determinant in the polytope algebra. Just like in the tropical semiring (2.6), here the determinant equals the permanent:

$$\det \begin{pmatrix} P_{11} & P_{12} & P_{13} & P_{14} \\ P_{21} & P_{22} & P_{23} & P_{24} \\ P_{31} & P_{32} & P_{33} & P_{34} \\ P_{41} & P_{42} & P_{43} & P_{44} \end{pmatrix} = \bigoplus_{\sigma \in S_4} P_{1\sigma(1)} \odot P_{2\sigma(2)} \odot P_{3\sigma(3)} \odot P_{4\sigma(4)}.$$

This determinant of polytopes represents a parameterized family of assignment problems. Indeed, suppose the cost q_{ij} of assigning job i to worker j depends piecewise-linearly on a vector of three parameters $w = (w_x, w_y, w_z)$, namely

$$q_{ij} = \min\{w \cdot p : p \in P_{ij}\}.$$

Thus the cost q_{ij} is determined by solving the linear programming problem with polytope P_{ij}. The *parametric assignment problem* would be to solve the assignment problem simultaneously for all vectors $w \in \mathbb{R}^3$. In other words, we wish to preprocess the problem specification so that the cost of an optimal assignment can be computed rapidly. This preprocessing amounts to computing the irredundant V-representation of the polytope obtained from the determinant. Then the cost of an optimal assignment can be computed as follows:

$$\min\{w \cdot p : p \in \det((P_{ij})_{1 \leq i,j \leq 4})\}.$$

Our discussion furnishes a higher-dimensional generalization of Remark 2.6:

Remark 2.28 The parametric assignment problem is solved by computing the determinant of the matrix of polytopes (P_{ij}) in the polytope algebra.

We can similarly define the *parametric shortest path problem* on a directed graph. The weight of each edge is now a polytope P_{ij} in $\mathbb{R}^d$, and for a specific parameter vector $w \in \mathbb{R}^d$ we recover the scalar edge weights by linear programming on that polytope: $d_{ij} = \min\{w \cdot p : p \in P_{ij}\}$. Then the shortest path from i to j is given by $d_{ij}^{(n-1)} = \min\{w \cdot p : p \in P_{ij}^{(n-1)}\}$, where $P_{ij}^{(n-1)}$ is the (i,j)-entry in the $(n-1)$-th power of the matrix (P_{ij}). Here matrix multiplication is carried out in the polytope algebra $(\mathcal{P}_d, \oplus, \odot)$.

The Hungarian algorithm for assignments and the Floyd–Warshall algorithm for shortest paths can be extended to the parametric setting. Provided the number d of parameters is fixed, these algorithms still run in polynomial time. The efficient computation of such polytopes by dynamical programming using polytope algebra arithmetic along a graph is referred to as *polytope propagation* (see Chapters 5–8). We close this section by revisiting the case of alignments.

Remark 2.29 The problem of parametric alignment of two DNA sequences σ^1 and σ^2 is to compute the Newton polytopes $\mathrm{NP}(f_{\sigma^1,\sigma^2})$ of the corresponding coordinate polynomial f_{σ^1,σ^2} of the pair hidden Markov model (2.13).

If some of the scores have been specialized then we compute Newton polytopes of polynomials in fewer unknowns. For instance, if $w' = 0$ then our task is to compute the Newton polytope $\mathrm{NP}(g_{\sigma^1,\sigma^2})$ of the specialized polynomial g_{σ^1,σ^2}. This can be done efficiently by running the Needleman–Wunsch Algorithm 2.13 in the polytope algebra and is the topic of Chapters 5–7.

Example 2.30 Returning to Example 2.19, we observe that the 14 points $v_1, \ldots, v_{14}$ are the vertices of the Newton polytope $P = \mathrm{NP}(g_{\mathsf{ACG},\mathsf{ACC}})$. It is important to note that all of the 14 points corresponding to monomials in $g_{\mathsf{ACG},\mathsf{ACC}}$ are in fact vertices of P, which means that every possible alignment of ACG and ACC is an optimal alignment for some choice of parameters.

The polytope P is easily computed in `polymake`, which confirms that the polytope is six-dimensional. The f-vector is $f(P) = (14, 51, 86, 78, 39, 10)$. These numbers have an interpretation in terms of alignments. For example, there is an edge between two vertices in the polytope if for two different optimal alignments (containing different numbers of matches, mismatches, and gaps) the parameter regions which yield the optimal alignments share a boundary. In other words, the fact that the polytope has 51 edges tells us that there are precisely 51 "parameter boundaries", where an infinitesimal change in parameters can result in a different optimal alignment. The normal cones and their defining inequalities (like the seven in Example 2.19) characterize these boundaries, thus offering a solution to the parametric alignment problem. □

2.4 Trees and metrics

One of the important mathematical structures that arises in biology is the phylogenetic tree [Darwin, 1859, Felsenstein, 2003, Semple and Steel, 2003]. A *phylogenetic tree* is a tree T together with a labeling of its leaves. The number of combinatorial types of phylogenetic trees with the same leaves grows exponentially (Lemma 2.33). In phylogenetics a typical problem is to select a tree, based on data, from the large number of possible choices.

This section introduces some basic concepts regarding the combinatorics of trees that are important for phylogeny. The notion of *tree space* is related to the tropicalization principle introduced in Section 2.1 and will be revisited in Section 3.5. A widely used algorithm in phylogenetics, the neighbor-joining algorithm, is a method for projecting a metric onto tree space. This algorithm draws on a number of ideas in phylogenetics and serves as the focus of our presentation in this section. We begin by discussing a number of different, yet combinatorially equivalent, characterizations of trees.

A *dissimilarity map* on $[n] = \{1, 2, \ldots, n\}$ is a function $d : [n] \times [n] \to \mathbb{R}$ such that $d(i, i) = 0$ and $d(i, j) = d(j, i) \geq 0$. The set of all dissimilarity maps on $[n]$ is a real vector space of dimension $\binom{n}{2}$, which we identify with $\mathbb{R}^{\binom{n}{2}}$. A dissimilarity map d is called a *metric on* $[n]$ if the triangle inequality holds:

$$d(i,j) \ \leq \ d(i,k) + d(k,j) \qquad \text{for } i,j,k \in [n]. \tag{2.27}$$

A dissimilarity map d can be written as a non-negative symmetric $n \times n$ matrix $D = (d_{ij})$ where $d_{ij} = d(i,j)$ and $d_{ii} = 0$. The triangle inequality (2.27) can be expressed by matrix multiplication where the arithmetic is tropical.

Remark 2.31 The matrix D represents a metric if and only if $D \odot D = D$.

Proof The entry of the matrix $D \odot D$ in row i and column j equals

$$d_{i1} \odot d_{1j} \ \oplus \ \cdots \ \oplus \ d_{in} \odot d_{nj} \quad = \quad \min\{\, d_{ik} + d_{kj} : 1 \leq k \leq n \,\}. \tag{2.28}$$

This quantity is less than or equal to $d_{ij} = d_{ii} \odot d_{ij} = d_{ij} \odot d_{jj}$, and it equals d_{ij} if and only if the triangle inequality $d_{ij} \leq d_{ik} + d_{kj}$ holds for all k. $\square$

Note that, for weighted graph G on n nodes, the matrix $D_G^{\odot n-1}$ represents the corresponding graph metric. This is the content of Proposition 2.2.

The set of all metrics on $[n]$ is a full-dimensional convex polyhedral cone in $\mathbb{R}^{\binom{n}{2}}$, called the *metric cone*. The metric cone has a distinguished subcone, known as the *cut cone*, which is the $\mathbb{R}_{\geq 0}$-linear span of all metrics arising from all *splits* $\{A,B\}$ of $[n]$ into two disjoint non-empty subsets A and B. The *split metric* $d_{\{A,B\}}$ is defined by

$$
\begin{aligned}
d_{\{A,B\}}(i,j) \ &= \ 0 \ \text{ if } \ i,j \in A \ \text{ or } \ i,j \in B, \\
d_{\{A,B\}}(i,j) \ &= \ 1 \ \text{ if } \ i \in A, j \in B \ \text{ or } \ i \in B, j \in A. \tag{2.29}
\end{aligned}
$$

The cut cone is strictly contained in the metric cone if $n \geq 6$. This and many other results on metrics can be found in [Deza and Laurent, 1997].

A metric d is a *tree metric* if there exists a tree T with n leaves, labeled by $[n] = \{1,2,\ldots,n\}$, and a non-negative length for each edge of T, such that the length of the unique path from leaf x to leaf y equals $d(x,y)$ for all $x,y \in [n]$. We sometimes write d_T for the tree metric d which is derived from the tree T.

Example 2.32 Let $n = 4$ and consider the metric d given by the matrix

$$D \ = \ \begin{pmatrix} 0 & 1.1 & 1.0 & 1.4 \\ 1.1 & 0 & 0.3 & 1.3 \\ 1.0 & 0.3 & 0 & 1.2 \\ 1.4 & 1.3 & 1.2 & 0 \end{pmatrix}.$$

The metric d is a tree metric, as can be verified by examining Figure 2.4. $\square$

The *space of trees* [Billera *et al.*, 2001] is the following subset of the metric cone:

$$\mathcal{T}_n \ = \ \{\, d_T \ : \ d_T \text{ is a tree metric} \,\} \quad \subset \quad \mathbb{R}^{\binom{n}{2}}. \tag{2.30}$$

The structure of $\mathcal{T}_n$ is best understood by separating the combinatorial types

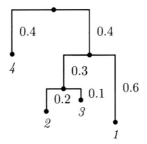

Fig. 2.4. The metric in Example 2.32 is a tree metric.

of trees from the lengths of the edges. A tree T is *trivalent* if every interior node is adjacent to three edges. A trivalent tree T has $n - 2$ interior nodes and $2n - 3$ edges. We can create any tree on $n + 1$ leaves by attaching the new leaf to any of the $2n - 3$ edges of T. By induction on n, we derive:

Lemma 2.33 *The number of combinatorial types of unrooted trivalent trees on a fixed set of n leaves is the* Schröder *number*

$$(2n - 5)!! \quad = \quad 1 \cdot 3 \cdot 5 \cdot \cdots \cdot (2n - 7) \cdot (2n - 5). \quad (2.31)$$

Each edge of a tree T corresponds to a split $\{A, B\}$ of the leaf set $[n]$ into two disjoint subsets A and B. Two splits $\{A_1, B_1\}$ and $\{A_2, B_2\}$ are *compatible* if at least one of the four intersections $A_1 \cap A_2$, $A_1 \cap B_2$, $B_1 \cap A_2$, and $B_1 \cap B_2$ is empty. We have the following easy combinatorial lemma:

Lemma 2.34 *If $\{A_1, B_1\}$ and $\{A_2, B_2\}$ are splits corresponding to two edges on a tree T with leaf set $[n]$ then $\{A_1, B_1\}$ and $\{A_2, B_2\}$ are compatible.*

Let $\{A_1, B_1\}$ and $\{A_2, B_2\}$ be two distinct compatible splits. We say that A_1 is *mixed* with respect to $\{A_2, B_2\}$ if $A_1 \cap A_2$ and $A_1 \cap B_2$ are both nonempty. Otherwise A_1 is *pure* with respect to $\{A_2, B_2\}$. Of the two components A_1 and B_1 exactly one is pure and the other is mixed with respect to the other split $\{A_2, B_2\}$. Let Splits(T) denote the collection of all $2n - 3$ splits (A, B) arising from T. For instance, if $n = 4$ and T is the tree in Figure 2.4 then

$$\text{Splits}(T) \quad = \quad \{\{1, 234\}, \{14, 23\}, \{123, 4\}, \{134, 2\}, \{124, 3\}\}. \quad (2.32)$$

Theorem 2.35 (Splits Equivalence Theorem) *A collection S of splits is pairwise compatible if and only if there exists a tree T such that $S = \text{Splits}(T)$. Moreover, if such a tree T exists then it is unique.*

Proof If there are no splits then the tree is a single node. Otherwise, we proceed by induction. Consider the set of splits $S' = S \backslash \{\{A, B\}\}$ where $\{A, B\}$ is a split in S. There is a unique tree T' corresponding to the set of splits S'. Any split in S' has one pure and one mixed component with respect to $\{A, B\}$. We orient the corresponding edge e of T' so that it is directed from the pure component to the mixed component. We claim that no node in T' can have out-degree ≥ 2. If this was the case there would be a split with a component that is both pure and mixed with respect to (A, B). Thus every node of T' has out-degree either 0 or 1. Since the number of nodes is one more than the number of edges, we conclude that the directed tree T' has a unique sink v'. Replace v' with two new nodes v_A and v_B and add a new edge between them as indicated in Figure 2.5. The result is the unique tree T with $S = \mathrm{Splits}(T)$.

$\square$

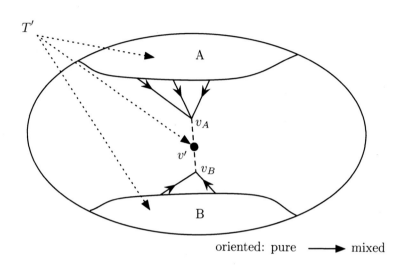

Fig. 2.5. Proof of the Splits Equivalence Theorem.

We next establish the classical *four-point condition* which characterizes membership in tree space $\mathcal{T}_n$. The proof is based on the notion of a *quartet*, which for any phylogenetic tree T is a subtree spanned by four leaves i, j, k, l. If $\{\{i, j\}, \{k, l\}\}$ is a split of that subtree then we denote the quartet by $(ij; kl)$.

Theorem 2.36 (The four-point condition) *A metric d is a tree metric if and only if, for any four leaves u, v, x, y, the maximum of the three numbers $d(u, v) + d(x, y)$, $d(u, x) + d(v, y)$ and $d(u, y) + d(v, x)$ is attained at least twice.*

Proof If $d = d_T$ for some tree then for any quartet $(uv; xy)$ of T it is clear that

$$d(u, v) + d(x, y) \ \leq \ d(u, x) + d(v, y) \ = \ d(u, y) + d(v, x).$$

Hence the "only if" direction holds.

We prove the converse using induction. The result holds trivially for trees with three leaves. Suppose that the number of leaves $n > 3$ and that the result holds for all metrics with fewer than n leaves. Let d be a metric on $[n] = \{1, 2, \ldots, n\}$ which satisfies the four-point condition.

Choose a triplet i, j, k that maximizes $d(i, k) + d(j, k) - d(i, j)$. By the induction hypothesis there is a tree T' on $[n] \backslash i$ that realizes d restricted to $[n] \backslash i$. Let λ be the length of the edge e of T' adjacent to j. We subdivide e

by attaching the leaf i next to the leaf j. The edge adjacent to i is assigned length $\lambda_i = (d(i,j) + d(i,k) - d(j,k))/2$, the edge adjacent to j is assigned length $\lambda_j = (d(i,j) + d(j,k) - d(i,k))/2$ and remaining part of e is assigned length $\lambda - \lambda_j$. We claim that the resulting tree T has non-negative edge weights and satisfies $d = d_T$. Clearly $d(x,y) = d_T(x,y)$ for all $x,y \in [n] \setminus i$.

Let l be any other leaf of T'. By the choice of i,j,k we have $d(i,k) + d(j,k) - d(i,j) \geq d(k,l) + d(i,k) - d(i,l)$ and $d(i,k) + d(j,k) - d(i,j) \geq d(k,l) + d(j,k) - d(j,l)$. The four-point condition then gives

$$d(i,j) + d(k,l) \leq d(i,k) + d(j,l) = d(i,l) + d(j,k).$$

The four-point condition also implies that $\lambda_i \geq 0$ and $\lambda_j \geq 0$. It remains to show that $\lambda - \lambda_j > 0$. This can be established by choosing a leaf l of T' such that $\lambda = (d(j,k) + d(j,l) - d(k,l))/2$, in which case

$$\lambda - \lambda_j = (d(i,k) + d(j,l) - d(i,j) - d(k,l))/2 \geq 0.$$

Thus T has non-negative edge weights. We have $d_T(i,j) = \lambda_i + \lambda_j = d(i,j)$ and $d_T(i,l) = d_T(j,l) - \lambda_j + \lambda_i = d(j,l) + d(i,k) - d(j,k) = d(i,l)$ for $l \neq i$. $\qquad \square$

It can be shown that the set of split metrics

$$\left\{ d_{\{A,B\}} : (A,B) \in \mathrm{Splits}(T) \right\} \tag{2.33}$$

is linearly independent in $\mathbb{R}^{\binom{n}{2}}$. We wrote the tree metric d_T uniquely as a linear combination of this set of split metrics. Let $\mathcal{C}_T$ denote the non-negative span of the set (2.33). The cone $\mathcal{C}_T$ is isomorphic to the orthant $\mathbb{R}_{\geq 0}^{2n-3}$.

Proposition 2.37 *The space of trees $\mathcal{T}_n$ is the union of the $(2n-5)!!$ orthants $\mathcal{C}_T$. More precisely, $\mathcal{T}_n$ is a simplicial fan of pure dimension $2n-3$ in $\mathbb{R}^{\binom{n}{2}}$.*

We return to tree space (and its relatives) in Section 3.5, where we show that $\mathcal{T}_n$ can be interpreted as a Grassmannian in tropical algebraic geometry.

The relevance of tree space to efficient statistical computation is this: suppose that our data consists of measurements of the frequency of occurrence of the different words in $\{A, C, G, C\}^n$ as columns of an alignment on n DNA sequences. As discussed in Section 1.4, we would like to select a tree model. In principle, we could compute the MLE for each of the $(2n-5)!!$ trees; however, this approach has a number of difficulties. First, even for a single tree the MLE computation is very difficult, even if we are satisfied with a reasonable local maximum of the likelihood function. Even if the MLE computation were feasible, a naive approach to model selection requires examining all exponentially many (in n) trees. One popular way to avoid these problems is the "distance based approach" which is to collapse the data to a dissimilarity map and then to obtain a tree via a projection onto tree space (see 4.26). The projection of choice for most biologists is the *neighbor-joining algorithm* which provides an easy-to-compute map from the metric cone onto

$\mathcal{T}_n$. The algorithm is based on Theorem 2.36 and the Cherry-Picking Theorem [Saitou and Nei, 1987, Studier and Keppler, 1988].

Fix a dissimilarity map d on the set $[n]$. For any $a_1, a_2, b_1, b_2 \in [n]$ we set

$$w(a_1 a_2; b_1 b_2) :=$$
$$\tfrac{1}{4}[d(a_1, b_1) + d(a_1, b_2) + d(a_2, b_1) + d(a_2, b_2) - 2[d(a_1, a_2) + d(b_1, b_2)]].$$

The function w provides a natural "weight" for quartets when d is a tree metric. The following result on quartets is proved by inspecting a tree with four leaves.

Lemma 2.38 *If d is a tree metric with $(a_1 a_2; b_1 b_2)$ a quartet in the tree, then $w(a_1 a_2; b_1 b_2) = -2 \cdot w(a_1 b_1; a_2 b_2)$, and this number is the length of the path which connects the path between a_1 and a_2 with the path between b_1 and b_2.*

A *cherry* of a tree is a pair of leaves which are both adjacent to the same node. The following theorem gives a criterion for identifying cherries.

Theorem 2.39 (Cherry-Picking Theorem) *If d is a tree metric on $[n]$ and*

$$Z_d(i,j) = \sum_{k,l \in [n] \setminus \{i,j\}} w(ij; kl) \tag{2.34}$$

then any pair of leaves that maximizes $Z_d(i,j)$ is a cherry in the tree.

Proof Suppose that i, j is not a cherry in the tree. Without loss of generality, we may assume that either there is a leaf k forming a cherry with i, or neither i nor j form a cherry with any leaf. In the first case, observe that

$$Z_d(i,k) - Z_d(i,j) = \sum_{x,y \neq i,j,k} (w(ik; xy) - w(ij; xy))$$
$$+ \sum_{x \neq i,j,k} (w(ik; xj) - w(ij; xk)) > 0.$$

Here we are using Lemma 2.38. In the latter case, there must be cherries (k, l) and (p, q) arranged as in Figure 2.6. Without loss of generality, we assume that the cherry (k, l) has the property that the number of leaves in $T \setminus e$ in the same component as k is less than or equal to the number of leaves in $T \setminus e'$ in the same component as p. We now compare $Z_d(k, l)$ with $Z_d(i, j)$:

$$Z_d(k,l) - Z_d(i,j) = \sum_{x,y \neq i,j,k,l} (w(kl; xy) - w(ij; xy))$$
$$+ \sum_{x \neq i,j,k,l} (w(kl; xj) + w(kl; ix) - w(ij; xl) - w(ij; kx)).$$

The two sums are each greater than 0. In the first sum, we need to evaluate all possible positions for x and y within the tree. If, for example, x and y lie in the component of $T \setminus \{x, y\}$ that contains i and j then it is clear that $w(kl; xy) - w(ij; xy) > 0$. If x and y lie in the same component of $T \setminus e$ as leaf k, then it may be that $w(kl; xy) - w(ij; xy) < 0$; however for each such pair

x, y there will be another pair that lies in the same component of $T \setminus e'$ as leaf p. The deficit for the former pair will be less than the surplus provided by the second. The remaining cases follow directly from Lemma 2.38. □

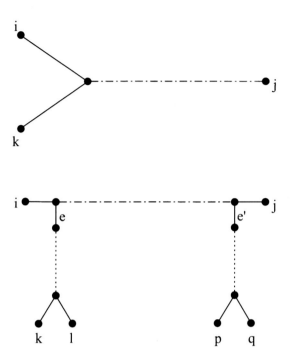

Fig. 2.6. Cases in the proof of the Cherry-Picking Theorem.

Theorem 2.39 is conceptually simple and useful, and we will see that it is useful for understanding the neighbor-joining algorithm. It is however not computationally efficient because $O(n^2)$ additions are necessary just to find one cherry. An equivalent, but computationally superior, formulation is:

Corollary 2.40 *Let d be a tree metric on* $[n]$. *For every pair* $i, j \in [n]$ *set*

$$Q_d(i,j) \quad = \quad (n-2) \cdot d(i,j) - \sum_{k \neq i} d(i,k) - \sum_{k \neq j} d(j,k). \qquad (2.35)$$

Then the pair $x, y \in [n]$ *that minimizes* $Q_d(x,y)$ *is a cherry in the tree.*

Proof Let $\tau = \sum_{x,y \in [n]} d(x,y)$. A direct calculation reveals the identity

$$Z_d(i,j) \quad = \quad -\frac{1}{2} \cdot \tau - \frac{n-2}{2} \cdot Q_d(i,j).$$

Thus maximizing $Z_d(x,y)$ is equivalent to minimizing $Q_d(x,y)$. □

The neighbor-joining algorithm makes use of the Cherry-Picking Theorem by peeling off cherries to recursively build a tree:

Algorithm 2.41 (Neighbor-joining algorithm)
Input: A dissimilarity map d on the set $[n]$.

Output: A phylogenetic tree T whose tree metric d_T is "close" to d.

Step 1:   Construct the $n \times n$ matrix Q_d whose (i,j)-entry is given by the formula (2.35), and identify the minimum off-diagonal entry $Q_d(x, y)$.

Step 2: Remove x, y from the tree, thereby creating a new leaf z. For each leaf k among the remaining $n - 2$ leaves, set

$$d(z, k) \;=\; \frac{1}{2}\big(d(x, k) + d(y, k) - d(x, y)\big). \qquad (2.36)$$

This replaces the $n \times n$ matrix Q_d by an $(n - 1) \times (n - 1)$ matrix. Return to Step 1 until there are no more leaves to collapse.

Step 3: Output the tree T. The edge lengths of T are determined recursively: If (x, y) is a cherry connected to node z as in Step 2, then the edge from x to z has length $d(x, k) - d_T(z, k)$ and the edge from y to z has length $d(y, k) - d_T(z, k)$.

This neighbor-joining algorithm recursively constructs a tree T whose metric d_T is, we hope, close to the given metric d. If d is a tree metric to begin with then the method is guaranteed to reconstruct the correct tree. More generally, instead of estimating pairwise distances, one can attempt to (more accurately) estimate the sum of the branch lengths of subtrees of size $m \geq 3$.

For any positive integer $d \geq 2$, we define a *d-dissimilarity map* on $[n]$ to be a function $D : [n]^d \to \mathbb{R}$ such that $D(i_1, i_2, \ldots, i_d) = D(i_{\pi(1)}, i_{\pi(2)}, \ldots, i_{\pi(d)})$ for all permutations π on $\{1, \ldots, d\}$ and $D(i_1, i_2, \ldots, i_d) = 0$ if the taxa $i_1, i_2, \ldots, i_d$ are not distinct. The set of all d-dissimilarity maps on $[n]$ is a real vector space of dimension $\binom{n}{d}$ which we identify with $\mathbb{R}^{\binom{n}{d}}$. Every tree T gives rise to an d-dissimilarity map D_T as follows. We define $D_T(i_1, \ldots, i_d)$ to be the sum of all branch lengths in the subtree of T spanned by $i_1, \ldots, i_d \in [n]$.

The following theorem is a generalization of Corollary 2.40. It leads to a generalized neighbor-joining algorithm which provides a better approximation of the maximum likelihood tree and parameters. A proof is given in Chapter 18 together with an explanation of the relevance of algebraic techniques for maximum likelihood estimation.

Theorem 2.42 *Let T be a tree on $[n]$ and $d < n$. For any $i, j \in [n]$ set*

$$Q_T(i, j) \;=\; \left(\frac{n-2}{d-1}\right) \sum_{Y \in \binom{[n] \setminus \{i,j\}}{d-2}} D_T(i, j, Y) \;-\; \sum_{Y \in \binom{[n] \setminus \{i\}}{d-1}} D_T(i, Y) \;-\; \sum_{Y \in \binom{[n] \setminus \{j\}}{d-1}} D_T(j, Y).$$

Then the pair $x, y \in [n]$ that minimizes $Q_T(x, y)$ is a cherry in the tree T.

The subset of $\mathbb{R}^{\binom{n}{d}}$ consisting of all d-dissimilarity maps D_T arising from trees T is a polyhedral space which is the image of the tree space $\mathcal{T}_n$ under a linear map $\mathbb{R}^{\binom{n}{2}} \to \mathbb{R}^{\binom{n}{d}}$. This polyhedral space is related to the tropicalization of the Grassmannian $G_{d,n}$, which is discussed in Section 3.5, but the details of this relationship are still not fully understood and deserve further study.

2.5 Software

In this section we introduce the software packages which were used by the authors of this book. These programs were discussed in our seminar in the Fall of 2004 and they played a key role for the studies which are presented in Part II. The subsection on *Mathematical Software* describes packages traditionally used by mathematicians but which may actually be very useful for statisticians and biologists. The subsection on *Computational Biology Software* summarizes programs more traditionally used for biological sequence analysis. In each subsection the software packages are listed in alphabetic order by name. Short examples or pointers to such examples are included for each package. These illustrate how the software was used in our computations.

2.5.1 Mathematical software

We describe ten packages for mathematical calculations relevant for this book.

4TI2

Summary: A package for linear algebra over the non-negative integers (e.g. integer programming). Also very useful for studying toric models (Section 2.2) and for computing phylogenetic invariants (Chapter 15).

Example: To solve the linear equation $2x + 5y = 3u + 4v$ for non-negative integers x, y, u, v we create a file named `foo` with the following two lines

```
1 4
2 5 -3 -4
```

Running the command `hilbert foo` creates a 10×4 matrix on a file `foo.hil`:

```
10 4
2 0 0 1
3 0 2 0
1 1 1 1
2 1 3 0
1 2 0 3
0 2 2 1
1 2 4 0
0 3 1 3
0 3 5 0
0 4 0 5
```

Every solution to the equation is an $\mathbb{N}$-linear combination of these ten rows.
Availability: Freely available executable only.
Website: `http://www.4ti2.de/`

LAPACK/SVDPACK

Summary: LAPACK is a Fortran library for linear algebra, widely used for scientific computation. It provides routines for solving systems of simultaneous linear equations, least-squares solutions of linear systems of equations, eigenvalue problems, and singular value problems.

SVDPACK is a specialized package which implements four numerical (iterative) methods for computing the singular value decomposition (SVD) of large sparse matrices using double precision Fortran. It makes use of basic linear algebra subroutines in LAPACK. SVDLIBC is a C library based on SVDPACK.

Example: These libraries are used in Chapter 19 to construct phylogenetic trees from alignments of DNA sequences.

Website: `http://www.netlib.org/lapack/` and `http://www.netlib.org/svdpack/` (or `http://tedlab.mit.edu/~dr/SVDLIBC/` for SVDLIBC).

MACAULAY2

Summary: A software system supporting research in algebraic geometry and commutative algebra [Grayson and Stillman, 2002].

Example: We illustrate the computation of toric models in MACAULAY2. Consider the toric Markov chain of length $n = 4$ with alphabet $\Sigma = \{0, 1\}$ that appears in Subsection 1.4.2. The model is specified with the commands:

```
i1 : R = QQ[p0000,p0001,p0010,p0011, p0100,p0101,p0110,p0111,
              p1000,p1001,p1010,p1011, p1100,p1101,p1110,p1111];
```

```
i2 : S = QQ[a00,a01,a10,a11];
```

```
i3 : f = map(S,R,{ a00*a00*a00, a00*a00*a01, a00*a01*a10,
     a00*a01*a11, a01*a10*a00, a01*a10*a01, a01*a11*a10,
     a01*a11*a11, a10*a00*a00, a10*a00*a01, a10*a01*a10,
     a10*a01*a11, a11*a10*a00, a11*a10*a01, a11*a11*a10,
     a11*a11*a11});
```

```
o3 : RingMap S <--- R
```

We have used the indeterminates a00,a01,a10,a11 for the parameters

$$\theta = \begin{pmatrix} \theta_{00} & \theta_{01} \\ \theta_{10} & \theta_{11} \end{pmatrix}.$$

The labels i1,i2,i3 indicate input to the program, o3 is output generated by MACAULAY2. We compute a Gröbner basis for the ideal $I_{\mathbf{f}}$ (see Section 3.2):

```
i4 : time If = kernel(f); -- Used 0.88 seconds
```

```
o4 : Ideal of R
```

```
i5 : gb If
```

```
o5 = | p1011-p1101 p0110-p1101 p0100-p1001 p0010-p1001
       p1101p1110-p1010p1111 p0111p1110-p1101p1111
       p0011p1110-p1001p1111 p1101^2-p0101p1110
```

```
p1100p1101-p1001p1110 p0111p1101-p0101p1111
p1100^2-p1000p1110 p1001p1100-p1000p1101
p0111p1100-p1001p1111 p0101p1100-p1001p1101
p0011p1100-p0001p1110 p0001p1100-p0000p1101
p0111p1010-p0101p1110 p0011p1010-p1001p1101
p1001^2-p0001p1010 p1000p1001-p0000p1010
p0111p1001-p0011p1101 p0011p1001-p0001p1101
p0111p1000-p0001p1110 p0101p1000-p0001p1010
p0011p1000-p0000p1101 p0001p1000-p0000p1001
p0000p0101-p0001p1001 p0011^2-p0001p0111
p0101p1110^2-p1010p1101p1111 p1001p1110^2-p1010p1100p1111
p0001p1110^2-p1000p1101p1111 p0000p1010p1100-p1000^2p1101
p0000p0011p1101-p0001^2p1110 p0000p1110^2-p1000p1100p1111
p0000p1100p1110-p1000^2p1111 p0000p0111^2-p0001p0011p1111
p0000p0011p0111-p0001^2p1111 |
```

These are the constraints on probabilities listed at the end of Subsection 1.4.1.
Availability: Open source.
Website: http://www.math.uiuc.edu/Macaulay2/

MAGMA

Summary: A software package for computation with algebraic, geometric and combinatorial structures such as graphs, groups, rings and fields. It includes a new fast implementation of the Faugère F4 algorithm for computing Gröbner bases [Bosma *et al.*, 1997].
Example: We compute a Gröbner basis for the example in Section 3.1.

```
Q := RationalField();
P<p1,p2,p3> := PolynomialRing(Q, 3);
I := ideal<P | p1^4+p2^4-p3^4,p1^4+p2^4+p3^4-2*p1*p2*p3,
          p1+p2+p3-1>;
G := GroebnerBasis(I);
G;
[ p1 + p2 + p3 - 1,
  p2^4 - 2*p2^3 + 3*p2^2 - 2*p2*p3^4 + p2*p3^3 - p2*p3^2
  + 2*p2*p3 - 2*p2 - p3^5 + 4*p3^4 - 2*p3^3 + 3*p3^2
  - 2*p3 + 1/2, p2^2*p3 + p2*p3^2 - p2*p3 + p3^4,
  p3^7 - 2*p3^6 + 4*p3^5 - 4*p3^4 + 3*p3^3 - 2*p3^2 + 1/2*p3 ]
```

Availability: Commercial software.
Website: http://magma.maths.usyd.edu.au/magma/

MAPLE

Summary: General purpose platform for mathematical computations. MAPLE is a versatile and powerful system, which includes many toolboxes and routines for standard symbolic and numerical computations. It is also an intuitive high-level interpreted language, which is convenient for quick computations.

Example: In Section 2.2, a specific example is provided showing how to compute sequence alignment polynomials using `MAPLE`.
Availability: Commercial software.
Website: `http://www.maplesoft.com/`

MATHEMATICA

Summary: General purpose platform for mathematical computations.
Example: In Chapter 12 `MATHEMATICA` is used to plot the likelihood surface for various hidden Markov models.
Availability: Commercial software.
Website: `http://www.wolfram.com/`

MATLAB

Summary: A general purpose high level mathematics package, particularly suited towards numerical linear algebra computations. MATLAB is supported by numerous specialized toolboxes: the statistics toolbox and bioinformatics toolbox are useful for computational biology.
Example: The following example illustrates the use of the statistics toolbox for experimenting with hidden Markov models. The example shows how to set up a simple model with $l = 2$ and $l' = 4$, generate data from the model, and how to run basic inference routines.

```
S=[0.8 0.2; 0.1 0.9]

S = 0.8 0.2
    0.1 0.9

T=[0.25 0.25 0.25 0.25; 0.125 0.375 0.375 0.125]

T =  0.250 0.250 0.250 0.250
     0.125 0.375 0.375 0.125
```

These commands set up the matrices θ and θ'. In other words, we have fixed a point on the model. The command **hmmgenerate** generates data from the model, and also specifies the alphabets Σ and Σ' to be used:

```
DNAseq=hmmgenerate(100,S,T,'Statenames',{'exon','intron'},
       'Symbols',{'A','C','G','T'})

DNAseq =  Columns 1 through 14

  'G' 'C' 'C' 'C' 'G' 'A' 'C' 'G' 'T' 'C' 'T' 'A' 'C' 'C'
  ...
```

The probability of `DNAseq` given the model, i.e., the evaluation of the `DNAseq` coordinate polynomial, is done with

```
[PSTATES,logpseq]=hmmdecode[DNAseq,S,T,'Symbols',
                    {'A','C','G','T'}]
```

The matrix `PSTATES` returns the forward variables (see Chapter 12). The logarithm of the probability of the sequence is also returned:

```
logpseq =    -1.341061974974420e+02
```

The tropicalization of the coordinate polynomial is evaluated as follows:

```
STATES=hmmviterbi(DNAseq,S,T,'Statenames',{'exon','intron'},
              'Symbols',{'A','C','G','T'}}
```

```
STATES =    Columns 1 through 6

    'intron' 'intron' 'intron' 'intron' 'intron' 'intron'


    ...


    Columns 99 through 100

    'exon' 'exon'
```

The `MATLAB` statistics toolbox also has an implementation of the EM algorithm for hidden Markov models, using the command `hmmtrain`.
Availability: Commercial software.
Website: http://www.mathworks.com/

POLYMAKE

Summary: A collection of programs for building, manipulating, analyzing and otherwise computing with polytopes and related polyhedral objects [Gawrilow and Joswig, 2000, Gawrilow and Joswig, 2001]. `polymake` uses the software `Javaview` for displaying pictures of polytopes, such as Figure 1.1.
Example: Several computations with polytopes are shown in Section 2.3.
Availability: Open source.
Website: http://www.math.tu-berlin.de/polymake/

SINGULAR

Summary: A system for polynomial computations, commutative algebra, and computational algebraic geometry. Very useful for algebraic statistics.
Example: See Sections 2.1, 2.2 and 2.3 for various examples. For a reference on `SINGULAR` with many worked out examples see [Greuel and Pfister, 2002].
Availability: Free under the GNU (**GNU**'s **N**ot **U**nix) Public License.
Website: http://www.singular.uni-kl.de/

R

Summary: A statistical computing language and environment, similar in syntax and focus to the S language [Ross and Gentleman, 1996]. Mathematicians find R comparable to MATLAB. The BIOCONDUCTOR package for R provides support for bioinformatics [Gentleman *et al.*, 2004].

Example: The following R code was used to produce Figure 3.1:

```
# Hardy--Weinberg curve
p <- c(seq(0, 1, 0.001), seq(1, 100, 0.01))
z0 <- p^2/(1+p)^2
z1 <- 2*p/(1+p)^2
z2 <- 1/(1+p)^2
x.rec <- cbind((2*z0+z1)/sqrt(3), z1)


## plot the Hardy--Weinberg curve
plot(x.rec[,1], x.rec[,2], type='l', xlim=c(0, 2/sqrt(3)),
    ylim=c(0, 1), xlab='', ylab='', yaxt='n', xaxt='n')


# plot simplex
lines(x=c(0, 2/sqrt(3)), y=c(0, 0))
lines(x=c(0, 1/sqrt(3)), y=c(0, 1))
lines(x=c(1/sqrt(3), 2/sqrt(3)), y=c(1, 0))
```

Availability: Open source.

Website: http://www.r-project.org/

2.5.2 Computational biology software

The five software programs highlighted here were all used during the preparation of the book, and are mostly accessible through web servers.

BLAST

Summary: A tool for searching through large biological sequence databases for matches to a query [Altschul *et al.*, 1990].

Example: There are many different "flavors" of BLAST, which allow for querying databases of DNA or protein, automatic translation of the input sequence, and other similar modifications. In what follows we illustrate the use of the BLASTN tool. We begin by submitting the sequence

ATGGCGGAGTCTGTGGAGCGCCTGCAGCAGCGGGTCCAGGAGCTGGAGCGGGAACTT

taken from an example in Section 7.4, to the BLASTN website. There are a number of important variables that can be set for the search, for example: the low complexity filter removes repeated subsequences, such as TTTT...TTT from the search. The word size is the minimum size of an exact match necessary for BLAST to return a "hit". The Expect parameter sets the threshold at which to report "significant" hits. It is based on the Karlin–Altschul model used to calculate statistical significance [Karlin and Altschul, 1990]. The remaining choices during submission are which database to search against (the default is nr which consists of all non-redundant nucleotide sequences in GENBANK), and various options for formatting the output. We selected the default for all settings, with the exception of **Alignments** which was set to 100, i.e., we opted to receive up to 100 reported alignments rather than the default 50.

Upon submitting the query, **BLAST** takes a few seconds (or minutes), and returns a page with a graphic showing which parts of the submitted sequence matched sequences in the database, and a text part containing links to the database hits, as well as the alignments. In our example, the text output is

```
                                          Score   E
Sequences producing significant alignments: (bits) Value

Homo sapiens ubiquitin-activat...          113    3e-23
Homo sapiens ubiquitin-activat...          113    3e-23
Homo sapiens ThiFP1 mRNA,comple..          113    3e-23
Homo sapiens cDNA FLJ31676 fis,..          113    3e-23
Homo sapiens cDNA: FLJ23251 fis..          113    3e-23
Homo sapiens ubiquitin-activatin.          113    3e-23
Homo sapiens 3 BAC RP11-333H9(...          113    3e-23
full-length cDNA clone CSODIO66...         113    3e-23
Homosapiens Uba5 mRNA for Ubiq...          113    3e-23
Homo sapiens mRNA;cDN...                   113    3e-23
PREDICTED: Pan troglodytes sim...          105    8e-21
Pongo pygmaeus mRNA; cDNA DKFZp...          98    2e-18
Sus scrofa cloneClu_21888.scr.m...          72    1e-10
...  ...
```

The entries are preceded with a **GENBANK** identifier (and a link to the original sequence in the database). Below this are the actual alignments, for example:

```
gi|33942036|emb|AL928824.13|   Zebrafish DNA sequence from clone
                               CH211-105D18 in linkage group 6,
                               complete sequence
          Length = 189742

 Score = 38.2 bits (19), Expect = 1.6
 Identities = 19/19 (100%)
 Strand = Plus / Minus

Query: 37      caggagctggagcgggaac 55
               |||||||||||||||||||
Sbjct: 155640 caggagctggagcgggaac 155622
```

A handy reference on how to use **BLAST** is [Korf *et al.*, 2003]. There are many variants of **BLAST** that have been designed for specialized tasks, including **BLASTZ** [Schwartz *et al.*, 2003] for rapid local alignment or large genomic regions and **BLAT** [Kent, 2002] for fast mRNA/DNA alignments.

Availability: Open source.

Website: http://www.ncbi.nlm.nih.gov/blast/

MAVID

Summary: A multiple alignment program designed for large genomic sequences [Bray and Pachter, 2004].

Example: Sequences can be submitted in multi-FASTA format through the website or the program can be downloaded for standalone use. A sequence in FASTA format begins with a single-line description, followed by lines of sequence data. The description line is distinguished from the sequence data by a greater-than (">") symbol in the first column. Sequences are expected to be represented in the standard IUB/IUPAC nucleic acid code, with these exceptions: lower-case letters are accepted and are mapped into upper-case; any characters other than A, C, G or T are converted into N which represents "unknown". Multi-FASTA format consists of alternating description lines followed by sequence data. It is important that each ">" symbol appear on a new line. For example:

```
>human
AGTGAGACACGACGAGCCTACTATCAGGACGAGAGCAGGAGAGTGATGATGAGTAGCG
CACAGCGACGATCATCACGAGAGAGTAAGAAGCAGTGATGATGTAGAGCGACGAGAGC
ACAGCGGCGACTACTACTAGG
>mouse
AGTGTGTCTCGTCGTGCCTACTTTCAGGACGAGAGCAGGTGAGTGTTGATGAGTTGCG
CTCTGCGACGTTCATCTCGAGTGAGTTAGAAAGTGAAGGTATAACACAAGGTGTGAAG
GCAGTGATGATGTAGAGCGACGAGAGCACAGCGGCGGGATGATATATCTAGGAGGATG
CCCAATTTTTTTTT
>platypus
CTCTGCGGCGTTCGTCTCGGGTGGGTTGGGGGGTGGGGGTGTGGCGCAAGGTGTGAAG
CACGACGACGATCTACGACGAGCGAGTGATGAGAGTGATGAGCGACGACGAGCACTAG
AAGCGACGACTACTATCGACGAGCAGCCGAGATGATGATGAAAGAGAGAGA
```

The MAVID program can align sequences that are much longer than the ones above (including alignments of sequences up to megabases long). Once the multi-FASTA file has been prepared, it is uploaded to the website. Consider for example, 13 sequences from the Cystic Fibrosis gene region (CFTR): human chimp, baboon, cow, pig, cat, dog, mouse, rat, chicken, zebra fish, fugu fish and tetraodon fish. This region is one of the ENCODE regions (see Chapter 21). The result of the MAVID run, including the original sequences, is too large to include here, but is stored on a website, in this case:

http://baboon.math.berkeley.edu/mavid/examples/zoo.target1/.

The website contains the alignment in multi-FASTA format (MAVID.mfa), as well as in PHYLIP format (MAVID.phy). A phylogenetic tree inferred from the alignment using neighbor-joining is also included:

The tree agrees well with the known phylogeny of the species, with the exception of the rodent placement. This issue is discussed in Section 21.4.

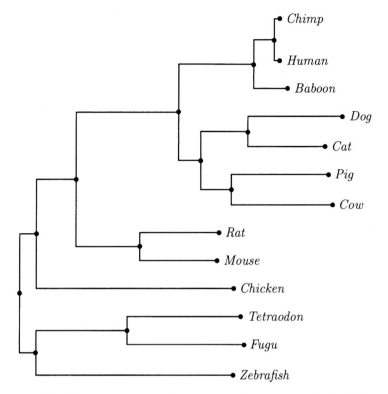

Fig. 2.7. Neighbor-joining tree from MAVID alignment of the CFTR region.

Availability: Open source.
Website: http://baboon.math.berkeley.edu/mavid/

PAML

Summary: Software for **P**hylogenetic **A**nalysis by **M**aximum **L**ikelihood. Consists of a collection of programs for estimating rate matrices and branch lengths for different tree models.
Example: Chapter 21 contains examples showing how to use PAML with different model assumptions (e.g. Jukes–Cantor, HKY85).
Availability: Open source.
Website: http://abacus.gene.ucl.ac.uk/software/paml.html

PHYLIP

Summary: A collection of programs for inferring phylogenies. This software has been continuously developed since 1981, and includes many routines utilities for manipulating and working with trees [Felsenstein, 2004].
Availability: Open source.
Example: PHYLIP reads alignments in a format which looks like this:

```
   5 10
human    AAGTGA
mouse    CAA--A
```

```
rat     AGCA-G
dog     G-AGCT
chicken T-ACCA
```

The first number in the first row is the number of sequences, and the second number if the number of columns in the alignment. Any of a number of routines can then be called, for example dnaml which constructs a tree.

Website: http://evolution.genetics.washington.edu/phylip.html

SPLITSTREE

Summary: Implementation of the neighbor-net algorithm, as well as split decomposition, neighbor-joining and other related methods. Includes a versatile visualization tool for splits graphs.

Availability: Open source.

Example: The "rodent problem" (Chapter 21) is a question about the placement of the rodents on the vertebrate tree. Figure 2.8 shows how SPLITSTREE can suggest different splits for a dataset. The rodents can be grouped with the chicken, or with the primates. The data used was obtained as explained in Section 21.3, and the model used was the HKY85 model (see Section 4.5). For more on splits networks see Chapter 17.

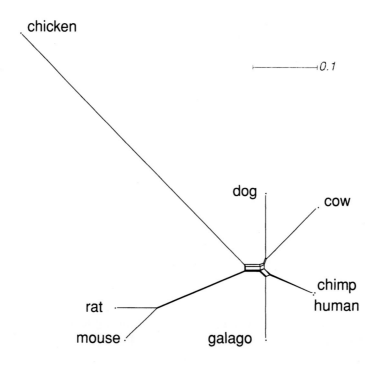

Fig. 2.8. Neighbor-net splits graph for the vertebrates.

Website:
http://www-ab.informatik.uni-tuebingen.de/software/splits/

3

Algebra

Lior Pachter

Bernd Sturmfels

The philosophy of algebraic statistics is that *statistical models are algebraic varieties*. We encountered many such models in Chapter 1. The purpose of this chapter is to give an elementary introduction to the relevant algebraic concepts, with examples drawn from statistics and computational biology.

Algebraic varieties are zero sets of systems of polynomial equations in several unknowns. These geometric objects appear in many contexts. For example, in genetics, the familiar *Hardy–Weinberg curve* is an algebraic variety (see Figure 3.1). In statistics, the distributions corresponding to independent random variables form algebraic varieties, called *Segre varieties*, that are well known to mathematicians. There are many questions one can ask about a system of polynomial equations; for example whether the solution set is empty, nonempty but finite, or infinite. Gröbner bases can be used to answer these questions.

Algebraic varieties can be described in two different ways, either by equations or parametrically. Each of these representations is useful. We encountered this duality in the *Hammersley–Clifford Theorem* which says that a graphical model can be described by conditional independence statements or by a polynomial parameterization. Clearly, methods for switching between these two representations are desirable. We discuss such methods in Section 3.2.

The study of systems of polynomial equations is the main focus of a central area in mathematics called *algebraic geometry*. This is a rich, beautiful, and well-developed subject, at whose heart lies a deep connection between algebra and geometry. In algebraic geometry, it is customary to study varieties over the field $\mathbb{C}$ of complex numbers even if the given polynomials have their coefficients in a subfield of $\mathbb{C}$ such as the real numbers $\mathbb{R}$ or the rational numbers $\mathbb{Q}$. This perspective leads to an algebraic approach to maximum likelihood estimation which may be unfamiliar to statisticians and is explained in Section 3.3.

Algebraic geometry makes sense also over the tropical semiring $(\mathbb{R}, \oplus, \odot)$. In that setting, algebraic varieties are piecewise-linear spaces. An important example for biology is the *space of trees* which will be discussed in Section 3.5.

3.1 Varieties and Gröbner bases

We write $\mathbb{Q}[p] = \mathbb{Q}[p_1, p_2, \ldots, p_m]$ for the set of all polynomials in m unknowns $p_1, p_2, \ldots, p_m$ with coefficients in the field $\mathbb{Q}$ of rational numbers. The set $\mathbb{Q}[p]$ has the structure of a $\mathbb{Q}$-vector space and also that of a ring. We call $\mathbb{Q}[p]$ the *polynomial ring*. A distinguished $\mathbb{Q}$-linear basis of $\mathbb{Q}[p]$ is the set of *monomials*

$$\{ p_1^{i_1} p_2^{i_2} \cdots p_m^{i_m} \quad : \quad i_1, i_2, \ldots, i_m \in \mathbb{N} \}. \tag{3.1}$$

To write down polynomials in a systematic way, we need to order the monomials. A *monomial order* is a total order $\prec$ on the set (3.1) which satisfies:

(i) the monomial $1 = p_1^0 p_2^0 \cdots p_m^0$ is smaller than all other monomials, and

(ii) if $p_1^{i_1} \cdots p_m^{i_m} \prec p_1^{j_1} \cdots p_m^{j_m}$ then $p_1^{i_1 + k_1} \cdots p_m^{i_m + k_m} \prec p_1^{j_1 + k_1} \cdots p_m^{j_m + k_m}$.

For polynomials in one unknown ($m = 1$) there is only one monomial order,

$$1 \prec p_1 \prec p_1^2 \prec p_1^3 \prec p_1^4 \prec \cdots,$$

but in several unknowns ($m \geq 2$) there are infinitely many monomial orders. One example is the *lexicographic monomial order* $\prec_{\text{lex}}$ which is defined as follows: $p_1^{i_1} \cdots p_m^{i_m} \prec_{\text{lex}} p_1^{j_1} \cdots p_m^{j_m}$ if the leftmost non-zero entry in the vector $(j_1 - i_1, j_2 - i_2, \ldots, j_m - i_m)$ is positive. In this section, all polynomials are written with their monomials in decreasing $\prec_{\text{lex}}$ order. The first monomial, or *initial monomial*, is often underlined: it is the $\prec_{\text{lex}}$ largest monomial appearing with non-zero coefficients in that polynomial. Here are three examples of polynomials in $\mathbb{Q}[p_1, p_2, p_3]$, each with its terms sorted in lexicographic order:

$$
\begin{aligned}
f_1 &= 4\underline{p_1 p_3} - p_2^2, \\
f_2 &= \underline{p_1^2} - 2p_1 + p_2^2 - 4p_2 - p_3^2 + 6p_3 - 2, \\
f_3 &= \underline{p_1 p_2^2} + p_1 p_3^3 + p_1 + p_2^3 + p_2^2 + p_2 p_3^3 + p_2 + p_3^3 + 1.
\end{aligned}
$$

What we are interested in is the geometry of these polynomials. The zero set of each of them is a surface in three-dimensional space $\mathbb{R}^3$. For instance, $\{f_2 = 0\}$ is the sphere of radius 4 around the point with coordinates $(1, 2, 3)$, and $\{f_3 = 0\}$ is the union of a plane with a parabolic surface. The surface $\{f_1 = 0\}$ is a quadratic cone: its intersection with the probability triangle is known as the *Hardy–Weinberg curve* in statistical genetics (Figure 3.1).

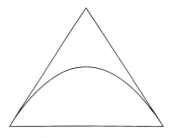

Fig. 3.1. The Hardy–Weinberg curve.

In our applications, the unknown p_i represents the probability of the ith event among m possible ones. But for now think of p_i just as a formal symbol.

Every polynomial $f \in \mathbb{Q}[p_1, \ldots, p_m]$ in m unknowns defines a *hypersurface*

$$V(f) \quad = \quad \{ (z_1, \ldots, z_m) \in \mathbb{C}^m : f(z_1, \ldots, z_m) = 0 \}.$$

Note that $V(f)$ is defined over the complex numbers $\mathbb{C}$. If S is any subset of $\mathbb{C}^m$ then we write $V_S(f) := V(f) \cap S$ for the part of the hypersurface that lies in S. For instance, $V_{\mathbb{R}^m}(f)$ is the set of solutions to $f = 0$ over the real numbers, and $V_\Delta(f)$ is the set of solutions to $f = 0$ in the probability simplex

$$\Delta \quad = \quad \{ (z_1, \ldots, z_m) \in \mathbb{R}^m : \sum_{i=1}^m z_i = 1 \text{ and } z_1, z_2, \ldots, z_m \geq 0 \}.$$

A polynomial is *homogeneous* if all of its monomials $p_1^{i_1} p_2^{i_2} \cdots p_m^{i_m}$ have the same *total degree* $i_1 + i_2 + \cdots + i_m$. The following three polynomials in $\mathbb{Q}[p_1, p_2, p_3]$ have total degree four. The first two are homogeneous but the third is not:

$$\begin{aligned} g_1 &= p_1^4 + p_2^4 - p_3^4, \\ g_2 &= p_1^4 + p_2^4 + p_3^4, \\ g_3 &= p_1^4 + p_2^4 + p_3^4 - 2p_1 p_2 p_3. \end{aligned}$$

All three of $V(g_1), V(g_2)$ and $V(g_3)$ are complex surfaces in $\mathbb{C}^3$, and $V_{\mathbb{R}^3}(g_1)$ and $V_{\mathbb{R}^3}(g_3)$ are real surfaces in $\mathbb{R}^3$, but $V_{\mathbb{R}^3}(g_2)$ is just the point $(0,0,0)$. Restricting to the probability triangle Δ, we see that $V_\Delta(g_2) = \emptyset$, while $V_\Delta(g_1)$ and $V_\Delta(g_3)$ are curves in the triangle Δ.

To understand why algebraic geometers prefer to work over the complex numbers $\mathbb{C}$ rather than over the real numbers $\mathbb{R}$, let us consider polynomials in one unknown p. For $a_0, a_1, \ldots, a_s \in \mathbb{Q}$ with $a_s \neq 0$ consider

$$f(p) \quad = \quad a_s \cdot p^s + a_{s-1} \cdot p^{s-1} + \cdots + a_2 \cdot p^2 + a_1 \cdot p + a_0.$$

Recall that the following basic result holds over the complex numbers:

Theorem 3.1 (Fundamental Theorem of Algebra) *If f is a polynomial of degree s then $V(f)$ consists of s complex numbers, counting multiplicities.*

By contrast, the number of real roots of $f(p)$, i.e. the cardinality of $V_\mathbb{R}(f)$, does depend on the particular coefficients $a_0, a_1, \ldots, a_s$. It can range anywhere between 0 and s, and the dependence is very complicated. So, the reason we use $\mathbb{C}$ is quite simple: *It is easier to do algebraic geometry over the complex numbers $\mathbb{C}$ than over the real numbers $\mathbb{R}$.* In algebraic statistics, we postpone issues of real numbers and inequalities as long as we can get away with it. But of course, at the end of the day, we are dealing with parameters and probabilities, and those are real numbers which are constrained by inequalities.

Let $\mathcal{F}$ be an arbitrary subset of the polynomial ring $\mathbb{Q}[p_1, \ldots, p_m]$. We define its *variety* $V(\mathcal{F})$ as the intersection of the hypersurfaces $V(f)$ where f ranges over $\mathcal{F}$. Similarly, $V_S(\mathcal{F}) = \bigcap_{f \in \mathcal{F}} V_S(f)$ for any subset $S \subset \mathbb{C}^m$. Using the example above, the variety $V(\{g_1, g_3\})$ is a curve in three-dimensional space $\mathbb{C}^3$. That curve meets the probability triangle Δ in precisely two points:

$$V_\Delta(\{g_1, g_3\}) = \{(0.4117, 0.1735, 0.4149), (0.1735, 0.4117, 0.4149)\}. \quad (3.2)$$

These two points are found by first computing the variety $V(\{ g_1, g_3, p_1 + p_2 + p_3 - 1 \})$. We did this by running the following sequence of six commands in the computer algebra package `Singular`. See Section 2.5 for software references.

```
ring R = 0, (p1,p2,p3), lp;
```

Here `lp` specifies the lexicographic monomial order.

```
ideal I = (p1^4+p2^4-p3^4,p1^4+p2^4+p3^4-2*p1*p2*p3,p1+p2+p3-1);
ideal G = groebner(I); G; LIB "solve.lib"; solve(G,10);
```

For an explanation of these commands, and a discussion of how to solve polynomial systems in general, see Section 2.5 of [Sturmfels, 2002]. Running this `Singular` code shows that $V(\{ g_1, g_3, p_1 + p_2 + p_3 - 1 \})$ consists of 16 distinct points (which is consistent with *Bézout's Theorem* [Cox et al., 1997]). Only two of the 16 points have all their coordinates real. They lie in the triangle Δ.

Algebraists feel uneasy about floating point numbers. For a specific numerical example consider the common third coordinate of the two points in $V_\Delta(\{g_1, g_3\})$. When an algebraist sees the floating point number

$$\widehat{p}_3 = 0.4148730882\ldots, \quad (3.3)$$

(s)he will want to know whether $\widehat{p}_3$ can be expressed in terms of radicals.

Indeed, the floating point coordinates produced by the algorithms in this book are usually *algebraic numbers*. An algebraic number has a *degree* which is the degree of its minimal polynomial over $\mathbb{Q}$. For instance, our floating point number $\widehat{p}_3$ is an algebraic number of degree six. Its minimal polynomial equals

$$f(p_3) = 2 \cdot p_3^6 - 4 \cdot p_3^5 + 8 \cdot p_3^4 - 8 \cdot p_3^3 + 6 \cdot p_3^2 - 4 \cdot p_3 + 1.$$

This polynomial appears in the output of the command `G;` in our `Singular` program. Most algebraists would probably prefer the following description (3.4) of our number over the description given earlier in (3.3):

$$\widehat{p}_3 = \text{the smaller of the two real roots of the polynomial } f(p_3). \quad (3.4)$$

The other real root is 0.7845389895 but this does not appear in $V_\Delta(\{g_1, g_3\})$. Our number $\widehat{p}_3$ cannot be written in terms of radicals over $\mathbb{Q}$. This is because the *Galois group* of the polynomial $f(p_3)$ is the symmetric group on six letters, which is not a solvable group. To see this, run the following in `Maple`:

```
galois( 2*p3^6-4*p3^5+8*p3^4-8*p3^3+6*p3^2-4*p3+1, p3);
```

In summary, algorithms used in algebraic statistics produce floating numbers, and these numbers are often algebraic numbers, which means they have a well-defined *algebraic degree over* $\mathbb{Q}$. In algebraic statistics, we are sensitive to this intrinsic measure of complexity of the real numbers we are dealing with.

The command `ideal G = groebner(I);` in our `Singular` code computes the lexicographic *Gröbner basis* for the *ideal* generated by the three given polynomials. In what follows, we give a very brief introduction to these notions. For further details, the reader is referred to any of the numerous textbooks on computational algebraic geometry, such as [Cox *et al.*, 1997, Schenck, 2003].

Let $\mathcal{F} \subset \mathbb{Q}[p] = \mathbb{Q}[p_1, \ldots, p_m]$. The *ideal generated* by $\mathcal{F}$ is the set $\langle \mathcal{F} \rangle$ consisting of all polynomial linear combinations of the elements in $\mathcal{F}$. In symbols,

$$\langle \mathcal{F} \rangle = \{\, h_1 f_1 + \cdots + h_r f_r \ : \ f_1, \ldots, f_r \in \mathcal{F} \text{ and } h_1, \ldots, h_r \in \mathbb{Q}[p] \,\}.$$

An ideal I in $\mathbb{Q}[p]$ is any set of the form $I = \langle \mathcal{F} \rangle$. It is quite possible for two different subsets $\mathcal{F}$ and $\mathcal{F}'$ of $\mathbb{Q}[p]$ to generate the same ideal I, i.e.,

$$\langle \mathcal{F} \rangle = \langle \mathcal{F}' \rangle.$$

This equation means that every polynomial in $\mathcal{F}$ is a $\mathbb{Q}[p]$-linear combination of elements in $\mathcal{F}'$, and vice versa. If this holds then the two varieties coincide:

$$V(\mathcal{F}) = V(\mathcal{F}').$$

Hilbert's basis theorem implies that every variety is the intersection of finitely many hypersurfaces:

Theorem 3.2 (Hilbert's basis theorem) *Every infinite set $\mathcal{F}$ of polynomials in $\mathbb{Q}[p]$ has a finite subset $\mathcal{F}' \subset \mathcal{F}$ such that $\langle \mathcal{F} \rangle = \langle \mathcal{F}' \rangle$.*

This theorem is often stated in the following form:

Every ideal in a polynomial ring is finitely generated.

Let us now fix a term order $\prec$. Every polynomial $f \in \mathbb{Q}[p]$ has a unique *initial monomial* denoted $\mathrm{in}_\prec(f)$. The initial monomial of f is the $\prec$-largest monomial $p^a = p_1^{a_1} p_2^{a_2} \cdots p_m^{a_m}$ which appears with non-zero coefficient in the expansion of f. Let I be an ideal in $\mathbb{Q}[p]$. Then its *initial ideal* $\mathrm{in}_\prec(I)$ is the ideal generated by the initial monomials of all the polynomials in I,

$$\mathrm{in}_\prec(I) = \langle \mathrm{in}_\prec(f) \ : \ f \in I \rangle.$$

A finite subset $\mathcal{G}$ of an ideal I is a *Gröbner basis* with respect to the monomial order $\prec$ if the initial monomials of elements in $\mathcal{G}$ generate the initial ideal:

$$\mathrm{in}_\prec(I) = \langle \mathrm{in}_\prec(g) \ : \ g \in \mathcal{G} \rangle. \tag{3.5}$$

As we have defined it in (3.5), there is no minimality requirement for being a Gröbner basis. If $\mathcal{G}$ is a Gröbner basis for I then we can augment $\mathcal{G}$ by any additional elements from I and the resulting set is still a Gröbner basis. To remedy this non-minimality, we make one more definition. We say that $\mathcal{G}$ is a *reduced Gröbner basis* if the following three additional conditions hold:

(i) For each $g \in \mathcal{G}$, the coefficient of $\mathrm{in}_\prec(g)$ in g is 1.
(ii) The set $\{\, \mathrm{in}_\prec(g) \ : \ g \in \mathcal{G} \,\}$ minimally generates $\mathrm{in}_\prec(I)$.
(iii) No trailing term of any $g \in \mathcal{G}$ lies in $\mathrm{in}_\prec(I)$.

For a fixed term order $\prec$, every ideal I in $\mathbb{Q}[p_1, \ldots, p_m]$ has a unique reduced Gröbner basis $\mathcal{G}$. This reduced Gröbner basis is finite, and it can be computed from an arbitrary generating set $\mathcal{F}$ of I by the *Buchberger algorithm*. Any Gröbner basis generates the ideal for which it is a Gröbner basis, so in particular, the reduced Gröbner basis satisfies $\langle \mathcal{G} \rangle = \langle \mathcal{F} \rangle = I$.

We will present the Buchberger algorithm towards the end of this section. First, we discuss some applications to the study of algebraic varieties. Recall that varieties are the solution sets of polynomial equations in several unknowns. Here we take polynomials with rational coefficients, and we consider a finite set of them $\mathcal{F} \subset \mathbb{Q}[p_1, \ldots, p_m]$. Recall that the variety of $\mathcal{F}$ is the set of all common zeros of $\mathcal{F}$ over the field of complex numbers. As above it is denoted

$$V(\mathcal{F}) \quad = \quad \big\{ (z_1, \ldots, z_m) \in \mathbb{C}^m : f(z_1, \ldots, z_m) = 0 \text{ for all } f \in \mathcal{F} \big\}.$$

The variety does not change if we replace $\mathcal{F}$ by another set of polynomials that generates the same ideal in $\mathbb{Q}[p_1, \ldots, p_m]$. In particular, the reduced Gröbner basis $\mathcal{G}$ for the ideal $\langle \mathcal{F} \rangle$ specifies the same variety:

$$V(\mathcal{F}) \quad = \quad V(\langle \mathcal{F} \rangle) \quad = \quad V(\langle \mathcal{G} \rangle) \quad = \quad V(\mathcal{G}).$$

The advantage of $\mathcal{G}$ is that it reveals geometric properties of the variety which are not visible from the given polynomials $\mathcal{F}$. A most basic question which one might ask about the variety $V(\mathcal{F})$ is whether it is non-empty: does the given system of equations $\mathcal{F}$ have any solution over the complex numbers?

Theorem 3.3 (Hilbert's Nullstellensatz) *The variety $V(\mathcal{F})$ is empty if and only if the reduced Gröbner basis $\mathcal{G}$ of the ideal $\langle \mathcal{F} \rangle$ equals $\{1\}$.*

Example 3.4 Consider a set of three polynomials in two unknowns:

$$\mathcal{F} \quad = \quad \{ \theta_1^2 + \theta_1\theta_2 - 10, \ \theta_1^3 + \theta_1\theta_2^2 - 25, \ \theta_1^4 + \theta_1\theta_2^3 - 70 \}.$$

Running the Buchberger algorithm on the input $\mathcal{F}$, we find that $\mathcal{G} = \{1\}$, so the three given polynomials have no common zero (θ_1, θ_2) in $\mathbb{C}^2$. We now change the constant term of the middle polynomial as follows:

$$\mathcal{F} \quad = \quad \{ \theta_1^2 + \theta_1\theta_2 - 10, \ \theta_1^3 + \theta_1\theta_2^2 - \underline{26}, \ \theta_1^4 + \theta_1\theta_2^3 - 70 \}.$$

The reduced Gröbner basis of $\langle \mathcal{F} \rangle$ is $\mathcal{G} = \{ \theta_1 - 2, \ \theta_2 - 3 \}$. Thus, the variety of $\mathcal{F}$ consists of a single point in $\mathbb{C}^2$, namely, $V(\mathcal{F}) = V(\mathcal{G}) = \{ (2,3) \}$. $\qquad \square$

Our next question is *how many zeros* does a given system of equations have? To answer this we need one more definition. Given a fixed ideal I in $\mathbb{Q}[p_1, \ldots, p_m]$ and a fixed term order $\prec$, a monomial $p^a = p_1^{a_1} \cdots p_m^{a_m}$ is called *standard* if it is not in the initial ideal $\text{in}_\prec(I)$. The number of standard monomials is finite if and only if every unknown p_i appears to some power among the generators of the initial ideal. For example, if $\text{in}_\prec(I) = \langle p_1^3, p_2^4, p_3^5 \rangle$ then there are 60 standard monomials, but if $\text{in}_\prec(I) = \langle p_1^3, p_2^4, p_1 p_3^4 \rangle$ then the set of standard monomials is infinite (because every power of p_3 is standard).

Theorem 3.5 *The variety $\mathcal{V}(I)$ is finite if and only if the set of standard monomials is finite. In this case, the number of standard monomials equals the cardinality of $\mathcal{V}(I)$, when zeros are counted with multiplicity.*

In the case of one unknown p, this result is the Fundamental Theorem of Algebra (Theorem 3.1), which states that the variety $\mathcal{V}(f)$ of a polynomial $f \in \mathbb{Q}[p]$ of degree s consists of s complex numbers. Indeed, in this case $\{f\}$ is a Gröbner basis for its ideal $I = \langle f \rangle$, we have $\mathrm{in}_{\prec}(I) = \langle p^s \rangle$, and there are precisely s standard monomials: $1, p, p^2, \ldots, p^{s-1}$. Thus we can regard Theorem 3.5 as the *Multidimensional Fundamental Theorem of Algebra*.

Example 3.6 Consider the system of three polynomials in three unknowns

$$\mathcal{F} = \left\{ p_1^4 + p_2^4 - p_3^4, \ p_1^4 + p_2^4 + p_3^4 - 2p_1p_2p_3, \ p_1 + p_2 + p_3 - 1 \right\}.$$

A Gröbner basis for the purely lexicographic order $p_1 > p_2 > p_3$ is

$$\mathcal{G} = \big\{ \underline{p_1} + p_2 + p_3 - 1, \ \underline{p_2^2 p_3} + p_2 p_3^2 - p_2 p_3 + p_3^4, \ \underline{2p_3^7} - 4p_3^6 + 8p_3^5 + \cdots,$$
$$\underline{2p_2^4} + 4p_2^3 p_3 - 4p_2^3 + 6p_2^2 p_3^2 - 10p_2^2 p_3 + 6p_2^2 + 4p_2 p_3^3 - 10p_2 p_3^2 + \cdots \big\}.$$

The underlined initial monomials show that there are 16 standard monomials:

$$1, \ p_2, \ p_2^2, \ p_2^3, \ p_3, \ p_3^2, \ p_3^3, \ p_3^4, \ p_3^5, \ p_3^6, \ p_2 p_3, \ p_2 p_3^2, \ p_2 p_3^3, \ p_2 p_3^4, \ p_2 p_3^5, \ p_2 p_3^6.$$

Theorem 3.5 says $V(\mathcal{F})$ consists of 16 points. Two of them appear in (3.2). $\square$

Our criterion in Theorem 3.5 for deciding whether a variety is finite generalizes to the following formula for the dimension of a variety. A subset S of the set of unknowns $\{p_1, p_2, \ldots, p_m\}$ is a *standard set* if every monomial $\prod_{p_j \in S} p_j^{a_j}$ in those unknowns is standard. Equivalently, $\mathrm{in}_{\prec}(I) \cap \mathbb{C}[p_j : j \in S] = \{0\}$.

Theorem 3.7 (Dimension Formula) *The dimension of an algebraic variety $\mathcal{V}(I) \subset \mathbb{C}^m$ is the maximal cardinality of any standard set for the ideal I.*

For a proof of this combinatorial dimension formula, and many other basic results on Gröbner basis, we refer to [Cox *et al.*, 1997].

Example 3.8 Let $I \subset \mathbb{Q}[p_1, p_2, p_3]$ be the ideal generated by the Hardy–Weinberg polynomial $f_1 = 4\underline{p_1 p_3} - p_2^2$. The maximal standard sets for I in the lexicographic monomial order are $\{p_1, p_2\}$ and $\{p_2, p_3\}$. Both have cardinality two. Hence the variety $V(f_1)$ has dimension two: it is a surface in $\mathbb{C}^3$. $\square$

Another basic result states that the set of standard monomials is a $\mathbb{Q}$-vector space basis for the *residue ring* $\mathbb{Q}[p_1, \ldots, p_m]/I$. The image of any polynomial h in this residue ring can be expressed uniquely as a $\mathbb{Q}$-linear combination of standard monomials. This expression is the *normal form* of h modulo the Gröbner basis $\mathcal{G}$. The process of computing the normal form is the *division algorithm*. In the case of one unknown p, where $I = \langle f \rangle$ and f has degree s, the division algorithm writes any polynomial $h \in \mathbb{Q}[p]$ as a unique $\mathbb{Q}$-linear combination of the standard monomials $1, p, p^2, \ldots, p^{s-1}$. The division algorithm works relative to a Gröbner basis in any number m of unknowns.

Example 3.9 Let $\mathbb{Q}[p]$ be the polynomial ring in 16 unknowns, denoted

$$p \;=\; \begin{pmatrix} p_{\mathrm{AA}} & p_{\mathrm{AC}} & p_{\mathrm{AG}} & p_{\mathrm{AT}} \\ p_{\mathrm{CA}} & p_{\mathrm{CC}} & p_{\mathrm{CG}} & p_{\mathrm{CT}} \\ p_{\mathrm{GA}} & p_{\mathrm{GC}} & p_{\mathrm{GG}} & p_{\mathrm{GT}} \\ p_{\mathrm{TA}} & p_{\mathrm{TC}} & p_{\mathrm{TG}} & p_{\mathrm{TT}} \end{pmatrix}.$$

Recall that DiaNA's model in Example 1.16 for generating two independent DNA sequences has the parameterization

$$p_{ij} \;=\; \pi \cdot \lambda_i^1 \cdot \lambda_j^2 \;+\; (1-\pi) \cdot \rho_i^1 \cdot \rho_j^2 \qquad \text{where } i,j \in \{\mathrm{A},\mathrm{C},\mathrm{G},\mathrm{T}\}. \quad (3.6)$$

Since *"statistical models are algebraic varieties"*, this model can be represented as a variety $V(I)$ in $\mathbb{C}^{4\times4}$. The homogeneous ideal $I \subset \mathbb{Q}[p]$ corresponding to the model (3.6) is generated by the sixteen 3×3-minors of the 4×4-matrix p. These sixteen determinants form a reduced Gröbner basis for I:

$$\mathcal{G} = \big\{\, \underline{p_{\mathrm{AA}}p_{\mathrm{CC}}p_{\mathrm{GG}}} - p_{\mathrm{AA}}p_{\mathrm{CG}}p_{\mathrm{GC}} - p_{\mathrm{AC}}p_{\mathrm{CA}}p_{\mathrm{GG}} + p_{\mathrm{AC}}p_{\mathrm{CG}}p_{\mathrm{GA}} + p_{\mathrm{AG}}p_{\mathrm{CA}}p_{\mathrm{GC}} - p_{\mathrm{AG}}p_{\mathrm{CC}}p_{\mathrm{GA}},$$

$$\underline{p_{\mathrm{AA}}p_{\mathrm{CC}}p_{\mathrm{GT}}} - p_{\mathrm{AA}}p_{\mathrm{CT}}p_{\mathrm{GC}} - p_{\mathrm{AC}}p_{\mathrm{CA}}p_{\mathrm{GT}} + p_{\mathrm{AC}}p_{\mathrm{CT}}p_{\mathrm{GA}} + p_{\mathrm{AT}}p_{\mathrm{CA}}p_{\mathrm{GC}} - p_{\mathrm{AT}}p_{\mathrm{CC}}p_{\mathrm{GA}},$$

$$\cdots\cdots\cdots \qquad \cdots\cdots\cdots \qquad \cdots\cdots\cdots$$

$$\underline{p_{\mathrm{CC}}p_{\mathrm{GG}}p_{\mathrm{TT}}} - p_{\mathrm{CC}}p_{\mathrm{GT}}p_{\mathrm{TG}} - p_{\mathrm{CG}}p_{\mathrm{GC}}p_{\mathrm{TT}} + p_{\mathrm{CG}}p_{\mathrm{GT}}p_{\mathrm{TC}} + p_{\mathrm{CT}}p_{\mathrm{GC}}p_{\mathrm{TG}} - p_{\mathrm{CT}}p_{\mathrm{GG}}p_{\mathrm{TC}} \big\}.$$

Indeed, it is known [Sturmfels, 1990] that, for any $a,b,c \in \mathbb{N}$, the $a \times a$-minors of a $b \times c$-matrix of unknowns form a reduced Gröbner basis with respect to a term order that makes the main diagonal terms of any determinant highest. We are looking at the case $a = 3, b = c = 4$. The variety $V(\mathcal{G}) = V(I)$ consists of all complex 4×4-matrices of rank ≤ 2. We can compute the dimension of this variety using Theorem 3.7. There are twenty maximal standard sets for I. They all have cardinality 12. One such standard set is

$$S \;=\; \big\{\, p_{\mathrm{AA}},\, p_{\mathrm{AC}},\, p_{\mathrm{AG}},\, p_{\mathrm{AT}},\; p_{\mathrm{CA}},\, p_{\mathrm{CC}},\, p_{\mathrm{CG}},\, p_{\mathrm{CT}},\; p_{\mathrm{GA}},\, p_{\mathrm{GC}},\; p_{\mathrm{TA}},\, p_{\mathrm{TC}} \,\big\}.$$

Indeed, none of the monomials in these twelve unknowns lies in the initial ideal

$$\mathrm{in}_{\prec}(I) \;=\; \big\langle\, p_{\mathrm{AA}}p_{\mathrm{CC}}p_{\mathrm{GG}},\, p_{\mathrm{AA}}p_{\mathrm{CC}}p_{\mathrm{GT}},\, p_{\mathrm{AA}}p_{\mathrm{CG}}p_{\mathrm{GT}},\, \ldots,\, p_{\mathrm{CC}}p_{\mathrm{GG}}p_{\mathrm{TT}} \,\big\rangle.$$

Theorem 3.7 implies that the variety $V(I)$ has dimension $|S| = 12$, and its intersection with the probability simplex, $V_{\Delta}(I)$, has dimension 11.

To illustrate the division algorithm, we consider the non-standard monomial

$$h \;=\; p_{\mathrm{AA}} \cdot p_{\mathrm{CC}} \cdot p_{\mathrm{GG}} \cdot p_{\mathrm{TT}}$$

The normal form of h modulo $\mathcal{G}$ is the following sum of 12 standard monomials:

$$\mathrm{nf}_{\mathcal{G}}(h) \;=\; p_{\mathrm{AA}}p_{\mathrm{CT}}p_{\mathrm{GG}}p_{\mathrm{TC}} + p_{\mathrm{AC}}p_{\mathrm{CA}}p_{\mathrm{GT}}p_{\mathrm{TG}} - p_{\mathrm{AC}}p_{\mathrm{CT}}p_{\mathrm{GG}}p_{\mathrm{TA}} + p_{\mathrm{AG}}p_{\mathrm{CC}}p_{\mathrm{GA}}p_{\mathrm{TT}}$$

$$- p_{\mathrm{AG}}p_{\mathrm{CC}}p_{\mathrm{GT}}p_{\mathrm{TA}} - p_{\mathrm{AG}}p_{\mathrm{CT}}p_{\mathrm{GA}}p_{\mathrm{TC}} + p_{\mathrm{AG}}p_{\mathrm{CT}}p_{\mathrm{GC}}p_{\mathrm{TA}} - p_{\mathrm{AT}}p_{\mathrm{CA}}p_{\mathrm{GG}}p_{\mathrm{TC}}$$

$$- p_{\mathrm{AT}}p_{\mathrm{CC}}p_{\mathrm{GA}}p_{\mathrm{TG}} + p_{\mathrm{AT}}p_{\mathrm{CG}}p_{\mathrm{GA}}p_{\mathrm{TC}} - p_{\mathrm{AT}}p_{\mathrm{CG}}p_{\mathrm{GC}}p_{\mathrm{TA}} + 2 \cdot p_{\mathrm{AT}}p_{\mathrm{CC}}p_{\mathrm{GG}}p_{\mathrm{TA}}$$

As functions restricted to the set of probability distributions of the form (3.6), the monomial h and the polynomial $\mathrm{nf}_{\mathcal{G}}(h)$ are equal. $\qquad\square$

Our assertion in Example 3.9 that the 3×3-minors form a Gröbner basis raises the following question. Given a fixed term order $\prec$, how can one test whether a given set of polynomials $\mathcal{G}$ is a Gröbner basis or not?

The answer is given by the following criterion [Buchberger, 1965]. Consider any two polynomials g and g' in $\mathcal{G}$ and form their *S-polynomial* $m'g - mg'$. Here m and m' are monomials of smallest possible degree such that $m' \cdot \text{in}_\prec(g) = m \cdot \text{in}_\prec(g')$. The S-polynomial $m'g - mg'$ lies in the ideal $\langle \mathcal{G} \rangle$. We apply the division algorithm modulo the tentative Gröbner basis $\mathcal{G}$ to the input $m'g - mg'$. The resulting normal form $\text{nf}_{\mathcal{G}}(m'g - mg')$ is a $\mathbb{Q}$-linear combination of monomials none of which is divisible by an initial monomial from $\mathcal{G}$ (note that it may depend on the order of the tentative Gröbner basis). A necessary condition for $\mathcal{G}$ to be a Gröbner basis is that this result be zero:

$$\text{nf}_{\mathcal{G}}(m'g - mg') \;=\; 0 \qquad \text{for all } g, g' \in \mathcal{G}. \tag{3.7}$$

Theorem 3.10 (Buchberger's Criterion) *A finite set of polynomials $\mathcal{G} \subset \mathbb{Q}[p_1, \ldots, p_m]$ is a Gröbner basis for its ideal $\langle \mathcal{G} \rangle$ if and only if (3.7) holds, that is, if and only if all S-polynomials have normal form zero.*

So, to check that the set $\mathcal{G}$ of the sixteen 3×3-determinants in Example 3.9 is indeed a Gröbner basis, it suffices to compute the normal forms of all $\binom{16}{2}$ pairwise S-polynomials, such as

$$p_{TG} \cdot (p_{AA}p_{CC}p_{GG} - p_{AA}p_{CG}p_{GC} - \cdots) - p_{GG} \cdot (p_{AA}p_{CC}p_{TG} - p_{AA}p_{CG}p_{TC} - \cdots)$$

$$= \qquad -p_{AA}p_{CG}p_{GC}p_{TG} + p_{AA}p_{CG}p_{GG}p_{TC} + p_{AC}p_{CG}p_{GA}p_{TG} - p_{AC}p_{CG}p_{GG}p_{TA}$$

$$+ p_{AG}p_{CA}p_{GC}p_{TG} - p_{AG}p_{CA}p_{GG}p_{TC} - p_{AG}p_{CC}p_{GA}p_{TG} + p_{AG}p_{CC}p_{GG}p_{TA}$$

The normal form of this expression modulo $\mathcal{G}$ is zero, as promised.

We are now prepared to state the algorithm for computing Gröbner bases.

Algorithm 3.11 (Buchberger's Algorithm)
Input: A finite set $\mathcal{F}$ of polynomials in $\mathbb{Q}[p_1, p_2, \ldots, p_m]$ and a term order $\prec$.
Output: The reduced Gröbner basis $\mathcal{G}$ of the ideal $\langle \mathcal{F} \rangle$ with respect to $\prec$.
Step 1: Apply Buchberger's Criterion to see whether $\mathcal{F}$ is already a Gröbner basis. If yes go to Step 3.
Step 2: If no, we found a non-zero polynomial $\text{nf}_{\mathcal{G}}(m'g - mg')$. Enlarge the set $\mathcal{F}$ by adding this non-zero polynomial and go back to Step 1.
Step 3: Transform the Gröbner basis $\mathcal{F}$ to a reduced Gröbner basis $\mathcal{G}$.

This loop between Steps 1 and 2 will terminate after finitely many iterations because at each stage the ideal generated by the current initial monomials get strictly bigger. By Hilbert's Basis Theorem, every strictly increasing sequence of ideals $\mathbb{Q}[p_1, \ldots, p_m]$ must stabilize eventually.

The Gröbner basis $\mathcal{F}$ produced in Steps 1 and 2 is usually not reduced, so in Step 3 we perform *auto-reduction* to make $\mathcal{F}$ reduced. To achieve the three conditions in the definition of reduced Gröbner basis, here is what Step 3 does. First, each polynomial in $\mathcal{F}$ is divided by its leading coefficient to achieve condition 1. Next, one removes redundant polynomials to achieve condition 2. Finally, each polynomial is replaced by its normal form with respect to $\mathcal{F}$ to achieve condition 3. The resulting set $\mathcal{G}$ satisfies all three conditions.

We illustrate Buchberger's algorithm for a very simple example with $m = 1$:

$$\mathcal{F} = \left\{ \underline{p^2} + 3p - 4, \ \underline{p^3} - 5p + 4 \right\}.$$

This set is not a Gröbner basis because the S-polynomial

$$p \cdot (p^2 + 3p - 4) - 1 \cdot (p^3 - 5p + 4) \quad = \quad 3p^2 + p - 4$$

has the non-zero normal form

$$3p^2 + p - 4 - 3 \cdot (p^2 + 3p - 4) \quad = \quad -8p + 8.$$

The new set $\mathcal{F} \cup \{-8p + 8\}$ now passes the test imposed by Buchberger's Criterion: it is a Gröbner basis. The resulting reduced Gröbner basis equals $\mathcal{G} = \{ p - 1 \}$. In particular, we conclude $V(\mathcal{F}) = \{1\} \subset \mathbb{C}$.

Remark 3.12 If $\mathcal{F} \subset \mathbb{Q}[p]$ is a set of polynomials in one unknown p then the reduced Gröbner basis $\mathcal{G}$ of the ideal $\langle \mathcal{F} \rangle$ consists of only one polynomial g. The polynomial g is the greatest common divisor of $\mathcal{F}$.

Buchberger's algorithm is therefore a generalization of the Euclidean algorithm for polynomials in one unknown. Likewise, the Buchberger Algorithm simulates Gaussian elimination if we apply it to a set $\mathcal{F}$ of linear polynomials. We can thus think of Gröbner bases as a *Euclidean algorithm for multivariate polynomials* or a *Gaussian elimination for non-linear equations*.

In summary, Gröbner bases and the Buchberger Algorithm for finding them are fundamental notions in computational algebraic geometry. They also furnish the engine for more advanced algorithms for algebraic varieties. Polynomial models are ubiquitous across the life sciences, including settings quite different from those in this book. For example, they are used in systems biology [Laubenbacher and Stigler, 2004] and for finding equilibria in chemical reaction networks [Craciun and Feinberg, 2005, Gatermann and Wolfrum, 2005]. Computer programs for algebraic geometry include CoCoA, Macaulay2 and Singular (see Section 2.5). All three are free and easy to use.

3.2 Implicitization

Consider the polynomial map which represents an algebraic statistical model:

$$\mathbf{f} \ : \ \mathbb{C}^d \ \rightarrow \ \mathbb{C}^m. \tag{3.8}$$

Here the ambient spaces are taken over the complex numbers, but the coordinates $f_1, \ldots, f_m$ of the map $\mathbf{f}$ are polynomials with rational coefficients, i.e., $f_1, \ldots, f_m \in \mathbb{Q}[\theta_1, \ldots, \theta_d]$. These assumptions are consistent with our discussion in the previous section. We start out by investigating the following basic question: is the image of a polynomial map $\mathbf{f}$ really an algebraic variety?

Example 3.13 Consider the following map from the plane into three-space:

$$\mathbf{f} : \mathbb{C}^2 \to \mathbb{C}^3, \ (\theta_1, \theta_2) \mapsto \left(\theta_1^2, \theta_1 \cdot \theta_2, \theta_1 \cdot \theta_2 \right)$$

The image of $\mathbf{f}$ is a dense subset of a plane in three-space, namely, it is

$$
\begin{aligned}
\mathbf{f}(\mathbb{C}^2) &= \left\{ (p_1, p_2, p_3) \in \mathbb{C}^3 \ : \ p_2 = p_3 \text{ and } (p_1 = 0 \text{ implies } p_2 = 0) \right\} \\
&= \left(V(p_2 - p_3) \setminus V(p_1, p_2 - p_3) \right) \ \cup \ V(p_1, p_2, p_3).
\end{aligned}
$$

Thus the image of $\mathbf{f}$ is not an algebraic variety, but its closure is: $\overline{\mathbf{f}(\mathbb{C}^2)} = V(p_2 - p_3)$. The set $\mathbf{f}(\mathbb{C}^2)$ is a Boolean combination of algebraic varieties. □

The following general theorem holds in algebraic geometry. It can be derived from the *Closure Theorem* in Section 3.2 of [Cox *et al.*, 1997].

Theorem 3.14 *The image of a polynomial map* $\mathbf{f} : \mathbb{C}^d \to \mathbb{C}^m$ *is a Boolean combination of algebraic varieties in* $\mathbb{C}^m$. *The topological closure* $\overline{\mathbf{f}(\mathbb{C}^d)}$ *of the image* $\mathbf{f}(\mathbb{C}^d)$ *in* $\mathbb{C}^m$ *is an algebraic variety.*

The statements in this theorem are not true if we replace the complex numbers $\mathbb{C}$ by the real numbers $\mathbb{R}$. This can already be seen for the map $\mathbf{f}$ in Example 3.13. The image of this map over the reals equals

$$\mathbf{f}(\mathbb{R}^2) = \left\{ (p_1, p_2, p_3) \in \mathbb{R}^3 \ : \ p_2 = p_3 \text{ and } (p_1 > 0 \text{ or } p_1 = p_2 = p_3 = 0) \right\}.$$

The closure of the image is a half-plane in $\mathbb{R}^3$, which is not an algebraic variety.

$$\overline{\mathbf{f}(\mathbb{R}^2)} = \left\{ (p_1, p_2, p_3) \in \mathbb{R}^3 \ : \ p_2 = p_3 \text{ and } p_1 \geq 0 \right\}.$$

It is instructive to carry this example a little further and compute the images of various subsets Θ of $\mathbb{R}^2$. For instance, what is the image $\mathbf{f}(\Theta)$ of the square $\Theta = \{0 \leq \theta_1, \theta_2 \leq 1\}$? For answering such questions in general, we need algorithms for solving *polynomial inequalities over the real numbers*. Such algorithms exist in *real algebraic geometry*, which is an active area of research. However, real algebraic geometry lies beyond what we hope to explain in this book. In this chapter, we restrict ourselves to the much simpler setting of *polynomial equations over the complex numbers*. For an introduction to algorithms in real algebraic geometry see [Basu *et al.*, 2003].

We shall adopt the following convention: By *the image of the polynomial map* $\mathbf{f}$ in (3.8) we shall mean the algebraic variety $\overline{\mathbf{f}(\mathbb{C}^d)}$ in $\mathbb{C}^m$. Thus we disregard potential points p in $\overline{\mathbf{f}(\mathbb{C}^d)} \setminus \mathbf{f}(\mathbb{C}^d)$. This is not to say they are not important. In fact, in a statistical model for a biological problem, such boundary points p might represent probability distributions we really care about. If so, we need to refine our techniques. For the discussion in this chapter, however, we keep the algebra as simple as possible and refer to $\overline{\mathbf{f}(\mathbb{C}^d)}$ as the image of $\mathbf{f}$.

Let $I_{\mathbf{f}}$ denote the set of all polynomials in $\mathbb{Q}[p_1, \ldots, p_m]$ that vanish on the set $\mathbf{f}(\mathbb{C}^d)$. Thus $I_{\mathbf{f}}$ is the ideal which represents the variety $\overline{\mathbf{f}(\mathbb{C}^d)}$. A polynomial $h \in \mathbb{Q}[p_1, \ldots, p_m]$ lies in the ideal $I_{\mathbf{f}}$ if and only if

$$h\big(f_1(t), f_2(t), \ldots, f_m(t)\big) = 0 \qquad \text{for all } t = (t_1, t_2, \ldots, t_d) \in \mathbb{R}^d. \tag{3.9}$$

The ideal $I_{\mathbf{f}}$ is a *prime ideal*. This means that if a factorizable polynomial $h = h' \cdot h''$ satisfies (3.9) then one of its factors h' or h'' will also satisfy (3.9). In the condition (3.9) we can replace $\mathbb{R}^d$ by any open subset $\Theta \subset \mathbb{R}^d$ and we get an equivalent condition. Thus $I_{\mathbf{f}}$ equals the set of all polynomials that vanish on the points $\mathbf{f}(t)$ where t runs over the parameter space Θ. The polynomials in the prime ideal $I_{\mathbf{f}}$ are known as *model invariants* in algebraic statistics. For instance, for DiaNA's model in Example 3.9, the model invariants include the 3×3-minors of the 4×4-matrix of probabilities.

The computational task resulting from our discussion is called *implicitization*. Given m polynomials $f_1, \ldots, f_m$ in $\mathbb{Q}[\theta_1, \ldots, \theta_d]$ that represent a polynomial map $\mathbf{f} : \mathbb{C}^d \to \mathbb{C}^m$, implicitization seeks to compute a finite set $\mathcal{F}$ of polynomials in $\mathbb{Q}[p_1, p_2, \ldots, p_m]$ such that $\langle \mathcal{F} \rangle = I_{\mathbf{f}}$. Actually, it would be preferable to have a Gröbner basis $\mathcal{G}$ of the ideal $I_{\mathbf{f}}$. Our point of view is this:

"compute the image of a polynomial map $\mathbf{f}$ "

means   "compute generators of the prime ideal $I_{\mathbf{f}}$ "

Example 3.15 We compute the images of five different maps $\mathbf{f} : \mathbb{C}^2 \to \mathbb{C}^3$:

(a) If $\mathbf{f} = (\theta_1^2, \theta_1\theta_2, \theta_1\theta_2)$ then $I_{\mathbf{f}} = \langle p_2 - p_3 \rangle$. This is Example 3.13.

(b) If $\mathbf{f} = (\theta_1^2, 2\theta_1\theta_2, \theta_2^2)$ then $I_{\mathbf{f}} = \langle p_2^2 - 4p_1p_3 \rangle$ = Hardy–Weinberg.

(c) If $\mathbf{f} = (\theta_1^5, \theta_1\theta_2, \theta_2^4)$ then $I_{\mathbf{f}} = \langle p_1^4 p_3^5 - p_2^{20} \rangle$.

(d) If $\mathbf{f} = (\theta_1^5 + \theta_1\theta_2, \theta_1^5 + \theta_2^4, \theta_1\theta_2 + \theta_2^4)$ then we get the same ideal in new coordinates: $I_{\mathbf{f}} = \langle 2^{11}(p_1 + p_2 - p_3)^4(p_2 + p_3 - p_1)^5 - (p_1 + p_3 - p_2)^{20} \rangle$.

(e) If $\mathbf{f} = (\theta_1^2 + \theta_2^2, \theta_1^3 + \theta_2^3, \theta_1^4 + \theta_2^4)$ then we actually have to do a computation to find $I_{\mathbf{f}} = \langle p_1^6 - 4p_1^3p_2^2 - 4p_2^4 + 12p_1p_2^2p_3 - 3p_1^2p_3^2 - 2p_3^3 \rangle$.

□

The last ideal $I_{\mathbf{f}}$ was computed in **Singular** using the following six commands:

```
ring s=0, (p1,p2,p3),lp;
ring r=0, (t1,t2), lp;
map f = s, t1^2+t2^2, t1^3+t2^3, t1^4+t2^4;
ideal i0 = 0;
setring s;
preimage(r,f,i0);
```

It should be tried and then redone with the third line replaced as follows:

```
map f = s, t1^5+t1*t2, t1^5+t2^4, t1*t2+t2^4;
```

This produces the surface of degree 20 in Example 3.15 (d). The output is very large, and underlines the importance of identifying a coordinate change that will simplify a computation. This will be crucial for the applications to phylogenetics discussed in Chapter 15 and 16.

In order to understand the way Gröbner basis software (such as **Singular**) computes images of polynomial maps, we need to think about the ideal $I_{\mathbf{f}}$ in the following algebraic manner. Our polynomial map $\mathbf{f} : \mathbb{C}^d \to \mathbb{C}^m$ induces the map between polynomial rings in the opposite direction:

$$\mathbf{f}^* : \mathbb{Q}[p_1, p_2, \ldots, p_m] \to \mathbb{Q}[\theta_1, \ldots, \theta_d],$$
$$h(p_1, p_2, \ldots, p_m) \mapsto h(f_1(\theta), f_2(\theta), \ldots, f_m(\theta)).$$

The map $\mathbf{f}^*$ is a *ring homomorphism*, which means that $\mathbf{f}^*(h' + h'') = \mathbf{f}^*(h') + \mathbf{f}^*(h'')$ and $\mathbf{f}^*(h' \cdot h'') = \mathbf{f}^*(h') \cdot \mathbf{f}^*(h'')$. Thus the ring homomorphism is uniquely specified by saying that $\mathbf{f}^*(p_i) = f_i(\theta)$ for all i. The *kernel* of $\mathbf{f}^*$ is the set $(\mathbf{f}^*)^{-1}(0)$ of all polynomials $h \in \mathbb{Q}[p_1, \ldots, p_m]$ that get mapped to zero by $\mathbf{f}^*$.

Proposition 3.16 *The kernel of* $\mathbf{f}^*$ *equals the prime ideal* $I_{\mathbf{f}} \subset \mathbb{Q}[p_1, \ldots, p_m]$.

Proof A polynomial h satisfies $\mathbf{f}^*(h) = 0$ if and only if the condition (3.9) holds. Thus h lies in kernel$(\mathbf{f}^*) = (\mathbf{f}^*)^{-1}(0)$ if and only if h lies in $I_{\mathbf{f}}$. □

If I is any ideal in $\mathbb{Q}[\theta_1, \ldots, \theta_d]$ then its pre-image $(\mathbf{f}^*)^{-1}(I)$ is an ideal in $\mathbb{Q}[p_1, \ldots, p_m]$. The next theorem characterizes the variety in $\mathbb{C}^m$ of this ideal.

Theorem 3.17 *The variety of* $(\mathbf{f}^*)^{-1}(I)$ *is the closure in* $\mathbb{C}^m$ *of the image of the variety* $V(I) \subset \mathbb{C}^d$ *under the map* $\mathbf{f}$; *in symbols,*

$$V\big((\mathbf{f}^*)^{-1}(I)\big) = \overline{\mathbf{f}\big(V(I)\big)} \subset \mathbb{C}^m. \tag{3.10}$$

Proof We can view I as an ideal in the enlarged polynomial ring

$$\mathbb{Q}[\theta_1, \ldots \theta_d, p_1, p_2, \ldots, p_m].$$

Inside this big polynomial ring we consider the ideal

$$J = I + \langle p_1 - f_1(\theta), p_2 - f_2(\theta), \ldots, p_m - f_m(\theta) \rangle. \tag{3.11}$$

The ideal J represents the graph of the restricted map $\mathbf{f} : V(I) \to \mathbb{C}^m$. Indeed, that graph is precisely the variety $V(J) \subset \mathbb{C}^{d+m}$. The desired image $\mathbf{f}(V(I))$ is obtained by projecting the graph $V(J)$ onto the space $\mathbb{C}^m$ with coordinates $p_1, \ldots, p_m$. Algebraically, this corresponds to computing the *elimination ideal*

$$(\mathbf{f}^*)^{-1}(I) = J \cap \mathbb{Q}[p_1, \ldots, p_m]. \tag{3.12}$$

Now use the Closure Theorem in [Cox *et al.*, 1997, Sect. 3.2]. □

The `Singular` code displayed earlier is designed for the setup in (3.10). The `map` command specifies a homomorphism `f` from the second polynomial ring `s` to the first polynomial ring `r`, and the `preimage` command computes the pre-image of an ideal `i0` in `r` under the homomorphism `f`. The computation inside `Singular` is done by cleverly executing the two steps (3.11) and (3.12).

Example 3.18 We compute the image of the hyperbola $V(\theta_1\theta_2 - 1)$ under the map $\mathbf{f}$ in Example 3.15 (e) by replacing the line `ideal i0 = 0 ;` with the new line `ideal i0 = t1*t2-1 ;` in our code. The image of the hyperbola in three-space is a curve that is the intersection of two quadratic surfaces:

$$(\mathbf{f}^*)^{-1}(\langle\theta_1\theta_2 - 1\rangle) \;\; = \;\; \langle p_1p_3 - p_1 - p_2^2 + 2,\; p_1^2 - p_3 - 2\rangle \;\; \subset \;\; \mathbb{Q}[p_1, p_2, p_3].$$

$\square$

Example 3.19 Consider the *hidden Markov model* of Subsection 1.4.3 where $n = 3$ and both the hidden and observed states are binary $(l = l' = 2)$. The parameterization (1.55) is a map $\mathbf{f} : \mathbb{R}^4 \to \mathbb{R}^8$ which we enter into `Singular`:

```
ring s = 0, (p000, p001, p010, p011, p100, p101, p110, p111),lp;
ring r = 0, ( x,y, u,v ), lp;
map f = s, x^2*u^3 + x*(1-x)*u^2*(1-v) +
  (1-x)*(1-y)*u^2*(1-v) + (1-x)*y*u*(1-v)^2 + (1-y)*x*(1-v)*u^2 +
  (1-y)*(1-x)*(1-v)^2*u + y*(1-y)*(1-v)^2*u + y^2*(1-v)^3,
x^2*u^2*(1-u) + x*(1-x)*u^2*v + (1-x)*(1-y)*u*(1-v)*(1-u) +
  (1-x)*y*u*(1-v)*v + (1-y)*x*(1-v)*u*(1-u) +
  (1-y)*(1-x)*(1-v)*u*v + y*(1-y)*(1-v)^2*(1-u) + y^2*(1-v)^2*v,
x^2*u^2*(1-u) + x*(1-x)*u*(1-u)*(1-v) + (1-x)*(1-y)*u^2*v +
  (1-x)*y*u*(1-v)*v + (1-y)*x*(1-v)*u*(1-u) +
  (1-y)*(1-x)*(1-v)^2*(1-u) + y*(1-y)*(1-v)*v*u +y^2*(1-v)^2*v, ·
x^2*u*(1-u)^2 + x*(1-x)*u*(1-u)*v + (1-x)*(1-y)*u*v*(1-u) +
  (1-x)*y*u*v^2 + (1-y)*x*(1-v)*(1-u)^2 + y^2*(1-v)*v^2 +
  (1-y)*(1-x)*(1-v)*(1-u)*v + y*(1-y)*(1-v)*v*(1-u),
x^2*u^2*(1-u) + x*(1-x)*u*(1-u)*(1-v) +
  (1-x)*(1-y)*u*(1-v)*(1-u) + (1-x)*y*(1-u)*(1-v)^2 +
  (1-y)*x*v*u^2 + (1-y)*(1-x)*(1-v)*u*v + y*(1-y)*(1-v)*v*u +
y^2*(1-v)^2*v, x^2*u*(1-u)^2 + x*(1-x)*u*(1-u)*v +
  (1-x)*(1-y)*(1-u)^2*(1-v) + (1-x)*y*(1-u)*(1-v)*v +
  (1-y)*x*v*u*(1-u) + (1-y)*(1-x)*v^2*u + y*(1-y)*(1-v)*v*(1-u) +
y^2*(1-v)*v^2, x^2*u*(1-u)^2 + x*(1-x)*(1-u)^2*(1-v) +
  (1-x)*(1-y)*u*v*(1-u) + (1-x)*y*(1-u)*(1-v)*v +
  (1-y)*x*v*u*(1-u) + (1-y)*(1-x)*(1-v)*(1-u)*v + y*(1-y)*v^2*u +
y^2*(1-v)*v^2, x^2*(1-u)^3 + x*(1-x)*(1-u)^2*v +
  (1-x)*(1-y)*(1-u)^2*v + (1-x)*y*(1-u)*v^2 + (1-y)*x*v*(1-u)^2 +
  (1-y)*(1-x)*v^2*(1-u) + y*(1-y)*v^2*(1-u) + y^2*v^3;
```

Here the eight probabilities have been scaled by a factor of two (the initial distribution is uniform), and the model parameters are abbreviated

$$\theta_{00} = x, \quad \theta_{01} = 1 - x, \quad \theta_{10} = 1 - y, \quad \theta_{11} = y,$$
$$\theta'_{00} = u, \quad \theta'_{01} = 1 - u, \quad \theta'_{10} = 1 - v, \quad \theta'_{11} = v.$$

The model invariants of the hidden Markov model can now be computed using

```
ideal i0 = 0; setring s; preimage(r,f,i0);
```

This computation will be discussed in Chapter 11. Suppose we are interested (for some strange reason) in the submodel obtained by equating the transition matrix θ with the inverse of the output matrix θ'. The invariants of this two-dimensional submodel are found by the method of Theorem 3.17, using

```
ideal i = x*u + x*v - x - v, y*u + y*v - y - u ; setring s;
preimage(r,f,i);
```

The extension of these computations to longer chains ($n \geq 4$) becomes prohibitive. Off-the-shelf implementations in any Gröbner basis package will always run out of steam quickly when the instances get bigger. More specialized linear algebra techniques need to be employed in order to compute invariants of larger statistical model and Chapter 11 is devoted to these issues. $\quad\square$

We next discuss an implicitization problem which concerns an algebraic variety known as the *Grassmannian*. In our discussion of the *space of phylogenetic trees* in Section 3.5, we shall argue that the Grassmannian is a valuable geometric tool for understanding and designing algorithms for phylogenetic trees. Let $\mathbb{Q}[\theta]$ be the polynomial ring in the unknown entries of the $2 \times n$-matrix

$$\theta = \begin{pmatrix} \theta_{11} & \theta_{12} & \theta_{13} & \cdots & \theta_{1n} \\ \theta_{21} & \theta_{22} & \theta_{23} & \cdots & \theta_{2n} \end{pmatrix}.$$

Let $\mathbb{Q}[p] = \mathbb{Q}[p_{ij} : 1 \leq i < j \leq n]$ be the polynomial ring in the unknowns

$$\{p_{12}, p_{13}, p_{23}, p_{14}, p_{24}, p_{34}, p_{15}, \ldots, p_{n-1,n}\}. \tag{3.13}$$

Consider the ring homomorphism $\mathbf{f}^* : \mathbb{Q}[p] \to \mathbb{Q}[\theta]$, $p_{ij} \mapsto \theta_{1i}\theta_{2j} - \theta_{1j}\theta_{2i}$. The corresponding polynomial map $\mathbf{f} : \mathbb{C}^{2 \times n} \mapsto \mathbb{C}^{\binom{n}{2}}$ takes a $2 \times n$-matrix to the vector of 2×2-subdeterminants of θ. The image of this map is the *Grassmannian*, denoted $G_{2,n} = \mathbf{f}(\mathbb{C}^{2n})$. The Grassmannian is an algebraic variety, i.e., it is closed: $\mathbf{f}(\mathbb{C}^{2n}) = \overline{\mathbf{f}(\mathbb{C}^{2n})}$. The prime ideal of the Grassmannian is denoted $I_{2,n} = \text{kernel}(\mathbf{f}^*)$. This ideal has a nice Gröbner basis:

Theorem 3.20 *The ideal* $I_{2,n}$ *is generated by the quadratic polynomials*

$$\underline{p_{ik}p_{jl}} - p_{ij}p_{kl} - p_{il}p_{jk} \qquad (1 \leq i < j < k < l \leq n). \tag{3.14}$$

These form the reduced Gröbner basis when the underlined terms are leading.

Proof See Theorem 3.1.7 and Proposition 3.7.4 in [Sturmfels, 1993]. $\quad\square$

The dimension of the Grassmannian $G_{2,n}$ is computed using Theorem 3.7. The initial ideal $\text{in}_\prec(I_{2,n}) = \langle p_{ik} \cdot p_{jl} : 1 \leq i < j < k < l \leq n \rangle$ can be visualized as follows. Draw a convex n-gon with vertices labeled $1, 2, 3, \ldots, n$. We identify the unknown p_{ij} with the line segment connecting the vertex i and the vertex j. The generators of $\text{in}_\prec(I_{2,n})$ are the pairs of line segments that cross each other. Consider an arbitrary monomial in $\mathbb{Q}[p]$:

$$m = \prod_{(i,j) \in S} p_{ij}^{a_{ij}} \qquad \text{(where } a_{ij} > 0 \text{ for all } (i,j) \in S).$$

This monomial m is standard if and only if m does not lie in $\mathrm{in}_{\prec}(I_{2,n})$ if and only if the set S contains no crossing diagonals if and only if the line segments in S form a subdivision of the n-gon. Hence a subset S of (3.13) is a maximal standard set if and only if the edges in S form a triangulation of the n-gon.

The number of triangulations S of the n-gon is the *Catalan number* $\frac{1}{n+1}\binom{2n}{n}$ [Stanley, 1999]. Each triangulation S has precisely $|S| = 2n - 3$ edges.

Corollary 3.21 *The Grassmannian $G_{2,n} = V(I_{2,n})$ has dimension $2n - 3$.*

This derivation illustrates the Dimension Formula (Theorem 3.7). The fact that $G_{2,n}$ has dimension $2n - 3$ can also be seen directly, without reference to any Gröbner basis.

The ideal $I_{2,n}$ is known as the *Plücker ideal*, and the quadratic polynomials in (3.14) are known as the *Plücker relations*. The Plücker ideal $I_{2,n}$ has two natural generalizations. First, we can replace θ by a $d \times n$-matrix of unknowns (for any $d < n$) and we can define the Plücker ideal $I_{d,n}$ by taking the algebraic relations among the $d \times d$-minors of θ. Thus $I_{d,n}$ is a prime ideal in the polynomial ring in $\binom{n}{d}$ unknowns $\mathbb{Q}[p] = \mathbb{Q}\left[p_{i_1 \cdots i_d} : 1 \leq i_1 < \cdots < i_d \leq n \right]$. The corresponding variety in $\mathbb{C}^{\binom{n}{d}}$ is the Grassmannian $G_{d,n} = V(I_{d,n})$.

In algebraic geometry, it is natural to regard $G_{d,n}$ as a subvariety of the projective space of dimension $\binom{n}{d} - 1$. The points of the projective variety $G_{d,n}$ are in bijection with the d-dimensional linear subspaces of the n-dimensional vector space $\mathbb{C}^n$. Here $p = \mathbf{f}(\theta) \in \mathbb{C}^{\binom{n}{d}}$ corresponds to the row space of the matrix θ. Theorem 3.20 generalizes to this situation: the ideal $I_{d,n}$ is generated by quadratic polynomials known as the Plücker relations. Among these are the *three-term Plücker relations* which are derived from (3.14):

$$p_{\nu_1 \cdots \nu_{d-2} ik} \cdot p_{\nu_1 \cdots \nu_{d-2} jl} - p_{\nu_1 \cdots \nu_{d-2} ij} \cdot p_{\nu_1 \cdots \nu_{d-2} kl} - p_{\nu_1 \cdots \nu_{d-2} il} \cdot p_{\nu_1 \cdots \nu_{d-2} jk}. \quad (3.15)$$

The three-term Plücker relations are not quite enough to generate $I_{d,n}$.

The second natural generalization of the ideal $I_{2,n}$ is based on the identity

$$\left(p_{ik} p_{jl} - p_{ij} p_{kl} - p_{il} p_{jk} \right)^2 \;=\; \det \begin{pmatrix} 0 & p_{ij} & p_{ik} & p_{il} \\ -p_{ij} & 0 & p_{jk} & p_{jl} \\ -p_{ik} & -p_{jk} & 0 & p_{kl} \\ -p_{il} & -p_{jl} & -p_{kl} & 0 \end{pmatrix}. \quad (3.16)$$

This is a skew-symmetric 4×4-matrix with unknown entries p_{ij}. The square root of the determinant of a skew-symmetric $2k \times 2k$-matrix is a polynomial of degree k known as its *Pfaffian*. Hence the Plücker relation (3.14) is the Pfaffian of the matrix in (3.16). Skew-symmetric matrices of odd size are singular, so the determinant of a skew-symmetric $(2k + 1) \times (2k + 1)$-matrix is zero.

Remark 3.22 By Theorem 3.20 and equation (3.16), the Plücker ideal $I_{2,n}$ is generated by the 4×4-subPfaffians of an indeterminate skew-symmetric $n \times n$-matrix (p_{ij}).

Let $I_{2,n,k}$ be the ideal generated by the $2k \times 2k$-subPfaffians of a skew-symmetric $n \times n$-matrix p_{ij}. Thus $I_{2,n,2} = I_{2,n}$, and $I_{2,6,3}$ is generated by

$$\underline{p_{14}p_{25}p_{36}} - p_{13}p_{24}p_{56} + p_{14}p_{23}p_{56} - p_{15}p_{26}p_{34} + p_{15}p_{24}p_{36}$$
$$-p_{15}p_{23}p_{46} + p_{16}p_{23}p_{45} - p_{16}p_{24}p_{35} + p_{16}p_{25}p_{34} - p_{12}p_{34}p_{56}$$
$$+p_{12}p_{36}p_{45} - p_{12}p_{35}p_{46} + p_{13}p_{46}p_{25} - p_{13}p_{45}p_{26} + p_{14}p_{26}p_{35}$$

$$= \quad \det{}^{1/2} \begin{pmatrix} 0 & p_{12} & p_{13} & p_{14} & p_{15} & p_{16} \\ -p_{12} & 0 & p_{23} & p_{24} & p_{25} & p_{26} \\ -p_{13} & -p_{23} & 0 & p_{34} & p_{35} & p_{36} \\ -p_{14} & -p_{24} & -p_{34} & 0 & p_{45} & p_{46} \\ -p_{15} & -p_{25} & -p_{35} & -p_{45} & 0 & p_{56} \\ -p_{16} & -p_{26} & -p_{36} & -p_{46} & -p_{56} & 0 \end{pmatrix}. \qquad (3.17)$$

We introduce $k - 1$ matrices

$$\theta^{(s)} \quad = \quad \begin{pmatrix} \theta_{11}^{(s)} & \theta_{12}^{(s)} & \theta_{13}^{(s)} & \cdots & \theta_{1n}^{(s)} \\ \theta_{21}^{(s)} & \theta_{22}^{(s)} & \theta_{23}^{(s)} & \cdots & \theta_{2n}^{(s)} \end{pmatrix} \qquad (s = 1, 2, \ldots, k - 1).$$

The $2(k - 1)n$ entries of these $k - 1$ matrices are the parameters for the map

$$\mathbf{g} : (\mathbb{C}^{2 \times n})^{k-1} \to \mathbb{C}^{\binom{n}{2}}, \ (\theta^{(1)}, \ldots, \theta^{(k-1)}) \mapsto \mathbf{f}(\theta^{(1)}) + \cdots + \mathbf{f}(\theta^{(k-1)}). \ (3.18)$$

Theorem 3.23 *The image of the map* $\mathbf{g}$ *is the variety defined by the $2k \times 2k$-subPfaffians. We have $I_{\mathbf{g}} = I_{2,n,k}$ and image$(\mathbf{g}) = V(I_{2,n,k})$. In particular, the Pfaffian ideal $I_{2,n,k}$ is prime.*

The variety $V(I_{2,n,k})$ consists of all skew-symmetric $n \times n$ matrices of rank less than $2k$. Geometrically, $V(I_{2,n,k})$ is the $(k - 1)$st *secant variety of the Grassmannian*. Indeed, the passage from the polynomial map $\mathbf{f}$ to the polynomial map $\mathbf{g}$ in (3.18) corresponds to the geometric construction of passing from a projective variety to its $(k-1)$th *secant variety*. For a proof of Theorem 3.23 we refer to [DeConcini *et al.*, 1982].

The Gröbner basis in Theorem 3.20 generalizes from $I_{2,n}$ to $I_{2,n,k}$, and so does its convex n-gon interpretation. The initial ideal $\mathrm{in}_{\prec}(I_{2,n,k})$ for a suitable term order $\prec$ is generated by the k-element sets of pairwise crossing diagonals (see [Dress *et al.*, 2002]). As an example consider the 15 monomials in the cubic Pfaffian given above (this is the case $k = 3, n = 6$). The underlined initial monomial is the only one that represents three pairwise crossing diagonals.

There are many biologically important models for which a complete description of the prime ideal of model invariants has not yet been established. Consider the two hidden tree models in Examples 1.24 and 1.25. When taken with unspecified root distributions, these models are given by polynomial maps

$$\mathbf{f} : \mathbb{C}^{13} \to \mathbb{C}^{64} \qquad \text{and} \qquad \mathbf{f}' : \mathbb{C}^{39} \to \mathbb{C}^{64}.$$

The corresponding prime ideals $I_{\mathbf{f}}$ and $I_{\mathbf{f}}$ have the following conjectural description (where we disregard the linear form $\sum p_{..} - 1$ and consider the ideals of all homogeneous polynomials vanishing on the models):

Conjecture 3.24 *The ideal I_f is generated by homogeneous polynomials of degree 3, and $I_{f'}$ is generated by homogeneous polynomials of degree 5 and 9.*

This conjecture represents the limit of our knowledge on what is called the *naive Bayes model* in statistics and *secant varieties of Segre varieties* in geometry. For known results and further background see Chapters 15 and 16, [Allman and Rhodes, 2004b, Section 6], and [Garcia *et al.*, 2004, Section 7].

Example 3.25 Here we make the first part of Conjecture 3.24 precise by describing an explicit set of cubics which are believed to generate the kernel of

$$\mathbf{f}^* : \mathbb{Q}[p] \rightarrow \mathbb{Q}[\theta, \lambda]$$
$$p_{i_1 i_2 i_3 i_4 i_5 i_6} \mapsto \lambda \theta_{0i_1}^{r1} \theta_{0i_2}^{r2} \theta_{0i_3}^{r3} \theta_{0i_4}^{r4} \theta_{0i_5}^{r5} \theta_{0i_6}^{r6} + (1-\lambda)\theta_{1i_1}^{r1} \theta_{1i_2}^{r2} \theta_{1i_3}^{r3} \theta_{1i_4}^{r4} \theta_{1i_5}^{r5} \theta_{1i_6}^{r6}.$$

Consider any split of $\{1, 2, 3, 4, 5, 6\}$ into two subsets A and B of size at least two. We can write the $2 \times 2 \times 2 \times 2 \times 2 \times 2$ table $(p_{i_1 i_2 i_3 i_4 i_5 i_6})$ as an ordinary two-dimensional matrix where the rows are indexed by functions $A \rightarrow \{0, 1\}$ and the columns are indexed by functions $B \rightarrow \{0, 1\}$. These matrices have rank at most two for all distributions in the model, hence their 3×3-subdeterminants lie in the ideal I_f. It is conjectured that I_f is generated by these 3×3 determinants. For example, the 8×8-matrix for $A = \{1, 2, 3\}$ and $B = \{4, 5, 6\}$ equals

$$\begin{pmatrix}
p_{000000} & p_{000001} & p_{000010} & p_{000011} & p_{000100} & p_{000101} & p_{000110} & p_{000111} \\
p_{001000} & p_{001001} & p_{001010} & p_{001011} & p_{001100} & p_{001101} & p_{001110} & p_{001111} \\
p_{010000} & p_{010001} & p_{010010} & p_{010011} & p_{010100} & p_{010101} & p_{010110} & p_{010111} \\
p_{011000} & p_{011001} & p_{011010} & p_{011011} & p_{011100} & p_{011101} & p_{011110} & p_{011111} \\
p_{100000} & p_{100001} & p_{100010} & p_{100011} & p_{100100} & p_{100101} & p_{100110} & p_{100111} \\
p_{101000} & p_{101001} & p_{101010} & p_{101011} & p_{101100} & p_{101101} & p_{101110} & p_{101111} \\
p_{110000} & p_{110001} & p_{110010} & p_{110011} & p_{110100} & p_{110101} & p_{110110} & p_{110111} \\
p_{111000} & p_{111001} & p_{111010} & p_{111011} & p_{111100} & p_{111101} & p_{111110} & p_{111111}
\end{pmatrix}.$$

Chapter 19 presents a new algorithm for phylogenetic reconstruction based on the fact that such matrices have low rank for all splits (A, B) in the tree. It is based on an algebraic variant of neighbor-joining (Section 2.4), where cherries are picked based on singular value decompositions. $\square$

3.3 Maximum likelihood estimation

An algebraic statistical model is a map $\mathbf{f} : \mathbb{C}^d \rightarrow \mathbb{C}^m$ whose coordinate functions $f_1, \ldots, f_m$ are polynomials with rational coefficients in the parameters $\theta = (\theta_1, \ldots, \theta_d)$. The parameter space Θ is an open subset of $\mathbb{R}^d$ such that $\mathbf{f}(\Theta) \subseteq \mathbb{R}_{\geq 0}^m$. If we make the extra assumption that $f_1 + \cdots + f_m - 1$ is the zero polynomial then $\mathbf{f}(\Theta)$ is a family of probability distributions on the state space $[m] = \{1, \ldots, m\}$. A given data set is summarized in a vector $u = (u_1, \ldots, u_m)$ of positive integers. The problem of *maximum likelihood estimation* is to find a parameter vector $\hat{\theta}$ in Θ which best explains the data u. This leads to the

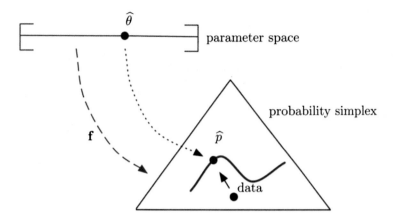

Fig. 3.2. The geometry of maximum likelihood estimation.

problem of maximizing the log-likelihood function

$$\ell_u(\theta) \;=\; \sum_{i=1}^{m} u_i \cdot \log(f_i(\theta)). \tag{3.19}$$

The geometry of maximum likelihood estimation is illustrated in Figure 3.2. The polynomial map **f** maps the low-dimensional parameter space into a very high-dimensional probability simplex. The image is the statistical model. The empirical distribution derived from the data vector u is a point in the probability simplex, and its maximum likelihood estimate $\widehat{p}$ is a point in the model. If these two points are close to each other then the model is a good fit for the data. Assuming that the model is identifiable (i.e., the map **f** is locally one-to-one), we can compute the unique parameter vector $\widehat{\theta}$ which maps to $\widehat{p}$.

Every local and global maximum $\widehat{\theta}$ in Θ of the log-likelihood function (3.19) is a solution of the *critical equations*

$$\frac{\partial \ell_u}{\partial \theta_1} \;=\; \frac{\partial \ell_u}{\partial \theta_2} \;=\; \cdots \;=\; \frac{\partial \ell_u}{\partial \theta_d} \;=\; 0. \tag{3.20}$$

The derivative of $\ell_u(\theta)$ with respect to the unknown θ_i is the rational function

$$\frac{\partial \ell_u}{\partial \theta_i} \;=\; \frac{u_1}{f_1(\theta)} \frac{\partial f_1}{\partial \theta_i} + \frac{u_2}{f_2(\theta)} \frac{\partial f_2}{\partial \theta_i} + \cdots + \frac{u_m}{f_m(\theta)} \frac{\partial f_m}{\partial \theta_i}. \tag{3.21}$$

The problem to be studied in this section is computing all solutions $\theta \in \mathbb{C}^d$ of the critical equations (3.20). Since (3.21) is a rational function, this set of critical points is an algebraic variety outside the locus where the denominators of these rational functions are zero. Hence the closure of the set of critical points of ℓ_u is an algebraic variety in $\mathbb{C}^d$, called the *likelihood variety* of the model **f** with respect to the data u.

In order to compute the likelihood variety we proceed as follows. We introduce m new unknowns $z_1, \ldots, z_m$ where z_i represents the inverse of $f_i(\theta)$. The polynomial ring $\mathbb{Q}[\theta, z] = \mathbb{Q}[\theta_1, \ldots, \theta_d, z_1, \ldots, z_m]$ is our "big ring", as opposed to the "small ring" $\mathbb{Q}[\theta] = \mathbb{Q}[\theta_1, \ldots, \theta_d]$ which is a subring of $\mathbb{Q}[\theta, z]$. We introduce an ideal generated by $m + d$ polynomials in the big ring $\mathbb{Q}[\theta, z]$:

$$J_u := \left\langle z_1 f_1(\theta) - 1, \ldots, z_m f_m(\theta) - 1, \sum_{j=1}^m u_j z_j \frac{\partial f_j}{\partial \theta_1}, \ldots, \sum_{j=1}^m u_j z_j \frac{\partial f_j}{\partial \theta_d} \right\rangle.$$

A point $(\theta, z) \in \mathbb{C}^{d+m}$ lies in the variety $V(J_u)$ of this ideal if and only if θ is a critical point of the log-likelihood function with $f_j(\theta) \neq 0$ and $z_j = 1/f_j(\theta)$ for all j. We next compute the elimination ideal in the small ring:

$$I_u = J_u \cap \mathbb{Q}[\theta_1, \ldots, \theta_d]. \tag{3.22}$$

We call I_u the *likelihood ideal* of the model $\mathbf{f}$ with respect to the data u. A point $\theta \in \mathbb{C}^d$ with all $f_j(\theta)$ non-zero lies in $V(I_u)$ if and only if θ is a critical point of the log-likelihood function $\ell_u(\theta)$. Thus $V(I_u)$ is the likelihood variety.

The algebraic approach to solving our optimization problem is this: Compute the variety $V(I_u) \subset \mathbb{C}^d$, intersect it with the pre-image $\mathbf{f}^{-1}(\Delta)$ of the $(m-1)$-dimensional probability simplex Δ, and identify all local maxima among the points in $V(I_u) \cap \mathbf{f}^{-1}(\Delta)$. We demonstrate this for an example.

Example 3.26 Let $d = 2$ and $m = 5$ and consider the following model:

```
ring bigring = 0, (t1,t2,z1,z2,z3,z4,z5), dp;   number t3 = 1/10;
poly f1 = -24*t1*t2*t3+9*t1*t2+9*t1*t3+9*t2*t3-3*t1-3*t2-3*t3+1;
poly f2 = -48*t1*t2*t3 + 6*t1*t2 + 6*t1*t3 + 6*t2*t3 ;
poly f3 =  24*t1*t2*t3 + 3*t1*t2 - 9*t1*t3 - 9*t2*t3 + 3*t3 ;
poly f4 =  24*t1*t2*t3 - 9*t1*t2 + 3*t1*t3 - 9*t2*t3 + 3*t2 ;
poly f5 =  24*t1*t2*t3 - 9*t1*t2 - 9*t1*t3 + 3*t2*t3 + 3*t1 ;
```

We use `Singular` notation with $\theta_1 = $ `t1` and $\theta_2 = $ `t2`. This map $\mathbf{f} : \mathbb{C}^2 \to \mathbb{C}^5$ is the submodel of the Jukes–Cantor model in Examples 1.7 and 4.26 obtained by fixing the third parameter θ_3 to be $1/10$. Suppose the given data are

```
int u1 = 31; int u2 = 5; int u3 = 7; int u4 = 11; int u5 = 13;
```

We specify the ideal J_u in the big ring $\mathbb{Q}[\theta_1, \theta_2, z_1, z_2, z_3, z_4, z_5]$:

```
ideal Ju = z1*f1-1, z2*f2-1, z3*f3-1, z4*f4-1, z5*f5-1,
u1*z1*diff(f1,t1)+u2*z2*diff(f2,t1)+u3*z3*diff(f3,t1)
   +u4*z4*diff(f4,t1)+u5*z5*diff(f5,t1),
u1*z1*diff(f1,t2)+u2*z2*diff(f2,t2)+u3*z3*diff(f3,t2)
   +u4*z4*diff(f4,t2)+u5*z5*diff(f5,t2);
```

Next we carry out the elimination step in (3.22) to get the likelihood ideal I_u:

```
ideal Iu = eliminate( Ju, z1*z2*z3*z4*z5 );
ring smallring = 0, (t1,t2), dp;
ideal Iu = fetch(bigring,Iu); Iu;
```

The likelihood ideal I_u is generated by six polynomials in (θ_1, θ_2) with large integer coefficients. Its variety $V(I_u) \subset \mathbb{C}^2$ is a finite set (it has dimension zero) consisting of 16 points. This is seen with the commands

```
ideal G = groebner(Iu); dim(G); vdim(G);
```

The numerical solver in `Singular` computes the 16 points in $V(I_u)$:

```
ideal G = groebner(Iu); LIB "solve.lib"; solve(G,20);
```

Eight of the 16 points in $V(I_u)$ have real coordinates, but only one of these corresponds to a probability distribution in our model. That solution is

$$\hat{\theta}_1 = 0.089827, \quad \hat{\theta}_2 = 0.073004, \qquad (\hat{\theta}_3 = 0.1 \text{ had been fixed}) \qquad (3.23)$$

The corresponding probability distribution in $\Delta \subset \mathbb{R}^5$ equals

$$\mathbf{f}(\hat{\theta}) \quad = \quad (\, 0.401336, 0.105568, 0.188864, 0.136975, 0.167257 \,),$$

and the values of the log-likelihood function at this distribution is

$$\ell_u(\hat{\theta}) \quad = \quad -96.32516427.$$

To determine the nature of this critical point we examine the Hessian matrix

$$\begin{pmatrix} \partial^2 \ell_u / \partial \theta_1^2 & \partial^2 \ell_u / \partial \theta_1 \theta_2 \\ \partial^2 \ell_u / \partial \theta_1 \theta_2 & \partial^2 \ell_u / \partial \theta_2^2 \end{pmatrix}.$$

At the point $\theta = \hat{\theta}$ with coordinates (3.23) the Hessian matrix equals

$$\begin{pmatrix} -2002.481296 & -624.2571252 \\ -624.2571252 & -2067.301931 \end{pmatrix}.$$

Both eigenvalues of this symmetric matrix are negative, and hence $\hat{\theta}$ is a local maximum of the likelihood function. We note that the maximum likelihood parameters $\hat{\theta}$ would be biologically meaningful in the context of Section 4.5. $\square$

An important question for computational statistics is this: What happens to the maximum likelihood estimate $\hat{\theta}$ when the model $\mathbf{f}$ is fixed but the data u vary? This variation is continuous in u because the log-likelihood function $\ell_u(\theta)$ is well-defined for all real vectors $u \in \mathbb{R}^m$. While the u_i are positive integers when dealing with data, there is really no mathematical reason for assuming that the u_i are integers. The problem (3.19) and the algebraic approach explained in Example 3.26 make sense for any real vector $u \in \mathbb{R}^m$.

If the model is algebraic (i.e. the f_i are polynomials or rational functions) then the maximum likelihood estimate $\hat{\theta}$ is an algebraic function of the data u. Being an *algebraic function* means that each coordinate $\hat{\theta}_i$ of the vector $\hat{\theta}$ is one of the zeroes of a polynomial of the following form in one unknown θ_i:

$$a_r(u) \cdot \theta_i^r + a_{r-1}(u) \cdot \theta_i^{r-1} + \cdots + a_2(u) \cdot \theta_i^2 + a_1(u) \cdot \theta_i + a_0(u). \quad (3.24)$$

Here each coefficient $a_i(u)$ is a polynomial in $\mathbb{Q}[u_1, \ldots, u_m]$, and the leading coefficient $a_r(u)$ is non-zero. We can further assume that the polynomial (3.24) is irreducible as an element of $\mathbb{Q}[u_1, \ldots, u_m, \theta_i]$. This means that the *discriminant* of (3.24) with respect to θ_i is a non-zero polynomial in $\mathbb{Q}[u_1, \ldots, u_m]$. We say that a vector $u \in \mathbb{R}^m$ is *generic* if u is not a zero of that discriminant polynomial for all $i \in \{1, \ldots, m\}$. The generic vectors u are dense in $\mathbb{R}^m$.

Definition 3.27 The *maximum likelihood degree* (or *ML degree*) of an algebraic statistical model is the number of complex critical points of the log-likelihood function $\ell_u(\theta)$ for a generic vector $u \in \mathbb{R}^m$.

For nice models $\mathbf{f}$ and most data u, the following three things will happen:

(i) The variety $V(I_u)$ is finite, so the ML degree is a positive integer.
(ii) The ideal $\mathbb{Q}[\theta_i] \cap I_u$ is prime and is generated by the polynomial (3.24).
(iii) The ML degree equals the degree r of the polynomial (3.24).

The ML degree measures the algebraic complexity of the process of maximum likelihood estimation for an algebraic statistical model $\mathbf{f}$. In particular, the ML degree of $\mathbf{f}$ is an upper bound for the number of critical points in Θ of the likelihood function, and hence an upper bound for the number of local maxima. For instance, for the specialized Jukes–Cantor model in Example 3.26, the ML degree is 16, and the maximum likelihood estimate $\widehat{\theta} = (\widehat{\theta}_1, \widehat{\theta}_2)$ is an algebraic function of degree 16 of the data $(u_1, \ldots, u_5)$. Hence, for any specific $u \in \mathbb{N}^5$, the number of local maxima is bounded above by 16. In general, the following upper bound for the ML degree is available.

Theorem 3.28 *Let* $\mathbf{f} : \mathbb{C}^d \to \mathbb{C}^m$ *be an algebraic statistical model whose coordinates are polynomials* $f_1, \ldots, f_m$ *of degrees* $b_1, \ldots, b_m$ *in the d unknowns* $\theta_1, \ldots, \theta_d$. *If the maximum likelihood degree of the model $\mathbf{f}$ is finite then it is less than or equal to the coefficient of* z^d *in the rational generating function*

$$\frac{(1-z)^d}{(1-zb_1)(1-zb_2)\cdots(1-zb_m)}. \tag{3.25}$$

Equality holds if the coefficients of the polynomials $f_1, f_2, \ldots, f_m$ *are generic.*

Proof See [Catanese *et al.*, 2005]. □

The ML degree is infinite when $V(I_u)$ is infinite. This happens when the model $\mathbf{f}$ is not identifiable; for instance in DiaNA's Example 1.16.

Example 3.29 Consider any statistical model which is parameterized by $m = 5$ quadratic polynomials f_i in $d = 2$ parameters θ_1 and θ_2. The formula in Theorem 3.28 says that the ML degree of the model $\mathbf{f} = (f_1, f_2, f_3, f_4, f_5)$ is at most the coefficient of z^2 in the generating function

$$\frac{(1-z)^2}{(1-2z)^5} = 1 + 8z + \underline{41}z^2 + 170z^3 + 620z^4 + \cdots.$$

Hence the ML degree is 41 if the f_i are generic and for special quadrics f_i it is ≤ 41. An instance is Example 3.26, where the model was given by five special quadrics in two unknowns, and the ML degree was $16 < 41$.

We can derive the number 41 directly from the critical equations

$$\frac{u_1}{f_1}\frac{\partial f_1}{\partial \theta_1} + \frac{u_2}{f_2}\frac{\partial f_2}{\partial \theta_1} + \cdots + \frac{u_5}{f_5}\frac{\partial f_5}{\partial \theta_1} = \frac{u_1}{f_1}\frac{\partial f_1}{\partial \theta_2} + \frac{u_2}{f_2}\frac{\partial f_2}{\partial \theta_2} + \cdots + \frac{u_5}{f_5}\frac{\partial f_5}{\partial \theta_2} = 0.$$

We claim that these equations have 41 solutions. Clearing denominators gives

$$u_1 \frac{\partial f_1}{\partial \theta_i} f_2 f_3 f_4 f_5 + u_2 \frac{\partial f_2}{\partial \theta_i} f_1 f_3 f_4 f_5 + \cdots + u_5 \frac{\partial f_5}{\partial \theta_i} f_1 f_2 f_3 f_4 = 0. \qquad (3.26)$$

for $i = 1, 2$. Each of these two equations specifies a curve of degree 9 in the plane $\mathbb{C}^2$. By Bézout's Theorem, these two curves intersect in $81 = 9 \cdot 9$ points. However, of these 81 points, precisely $40 = \binom{5}{2} \cdot 2 \cdot 2$ are accounted for by pairwise intersecting the given quadratic curves:

$$f_i(\theta_1, \theta_2) \quad = \quad f_j(\theta_1, \theta_2) \quad = \quad 0 \qquad (1 \leq i < j \leq 5).$$

Each of the four solutions to this system will also solve (3.26). After removing those four, we are left with $41 = 81 - 40$ solutions to the two critical equations. Theorem 3.28 says that a similar argument works not just for plane curves but for algebraic varieties in any dimension. $\qquad \square$

For some applications it is advantageous to replace the unconstrained optimization problem (3.19) by the *constrained optimization problem*:

$$\text{Maximize} \quad u_1 \cdot \log(p_1) + \cdots + u_m \cdot \log(p_m) \quad \text{subject to} \quad p \in \mathbf{f}(\Theta). \quad (3.27)$$

The image of $\mathbf{f}$ can be computed, in the sense discussed in the previous section, using the algebraic techniques of implicitization. Let $I_{\mathbf{f}} \subset \mathbb{Q}[p_1, p_2, \ldots, p_m]$ denote the prime ideal consisting of all polynomials that vanish on the image of the map $\mathbf{f} : \mathbb{C}^d \to \mathbb{C}^m$. Then we can replace $\mathbf{f}(\Theta)$ by $V_\Delta(I_{\mathbf{f}})$ in (3.27).

Algebraic geometers prefer to work with homogeneous polynomials and projective spaces rather than non-homogeneous polynomials and affine spaces. For that reason we prefer the ideal $P_{\mathbf{f}}$ generated by all *homogeneous* polynomials in $I_{\mathbf{f}}$. The homogeneous ideal $P_{\mathbf{f}}$ represents the model $\mathbf{f}$ just as well because

$$I_{\mathbf{f}} \quad = \quad P_{\mathbf{f}} + \langle p_1 + p_2 + \cdots + p_m - 1 \rangle \qquad \text{and} \qquad V_\Delta(I_{\mathbf{f}}) = V_\Delta(P_{\mathbf{f}}).$$

For more details see Chapter 14. Note that Conjecture 3.24 is all about the homogeneous ideals $P_{\mathbf{f}}$ and $P_{\mathbf{f}'}$.

Example 3.30 The homogeneous ideal for the model in Example 3.26 equals

$$P_{\mathbf{f}} \quad = \quad \langle\, 4p_2^2 - 3p_2p_3 - 6p_2p_4 - 6p_2p_5 + 2p_3p_4 + 2p_3p_5 + 10p_4p_5 \,,$$
$$6p_1 + 3p_2 - 4p_3 - 2p_4 - 2p_5 \,\rangle.$$

Thus $V_\Delta(P_{\mathbf{f}})$ is a quadratic surface which lies in a (four-dimensional) hyperplane. Our computation in Example 3.26 was aimed at finding the critical points of the function $p_1^{u_1} p_2^{u_2} p_3^{u_3} p_4^{u_4} p_5^{u_5}$ in the surface $V_\Delta(P_{\mathbf{f}})$. This constrained optimization problem has 16 complex critical points for generic $u_1, \ldots, u_5$, i.e., the maximum likelihood degree of this quadratic surface is equal to 16. $\qquad \square$

Once we are working with statistical models in their implicitized form, we can work without referring to the map $\mathbf{f}$. Since $V_\Delta(I_{\mathbf{f}}) = V_\Delta(P_{\mathbf{f}})$, we may in fact suppose that the model is given as an arbitrary homogeneous ideal $P \subset \mathbb{Q}[p_1, \ldots, p_m]$. The *MLE problem* for P is

$$\text{Maximize } u_1 \cdot \log(p_1) + \cdots + u_m \cdot \log(p_m) \quad \text{subject to } p \in V_\Delta(P). \quad (3.28)$$

Since P is homogeneous, we can regard $V(P)$ as a variety in complex projective $(m-1)$-space $\mathbb{P}^{m-1}$. Let $\text{Sing}(P)$ denote the singular locus of the projective variety $V(P)$, and let $V_{\mathbb{C}^*}(P)$ be the set of points in $V(P)$ all of whose coordinates are non-zero. We define the *likelihood locus* of P for the data u to be the set of all points in $V_{\mathbb{C}^*}(P) \backslash \text{Sing}(P)$ that are critical points of the function $\sum_{j=1}^m u_j \cdot \log(p_j)$. The *maximum likelihood degree* (or *ML degree*) of the ideal P is the cardinality of the likelihood locus when u is a generic vector in $\mathbb{R}^m$.

To characterize the likelihood locus algebraically, we use the technique of *Lagrange multipliers*. Suppose that $P = \langle g_1, g_2, \ldots, g_r \rangle$ where the g_i are homogeneous polynomials in $\mathbb{Q}[p_1, p_2, \ldots, p_m]$. We consider

$$u_1 \cdot \log(p_1) + \cdots + u_m \cdot \log(p_m) + \lambda_0 \left(1 - \sum_{i=1}^m p_i\right) + \lambda_1 g_1 + \cdots + \lambda_r g_r. \quad (3.29)$$

This is a function of the $m + r + 1$ unknowns $p_1, \ldots, p_m, \lambda_0, \lambda_1, \ldots, \lambda_r$. A point $p \in V_{\mathbb{C}^*}(P) \backslash \text{Sing}(P)$ lies in the likelihood locus if there exists $\lambda \in \mathbb{C}^{r+1}$ such that (p, λ) is a critical point of (3.29). Thus we can compute the likelihood locus (and hence the ML degree) from the ideal P using Gröbner basis elimination techniques. Details are described in [Hoşten *et al.*, 2005].

If the generators $g_1, \ldots, g_r$ of the homogeneous ideal P are chosen at random relative to their degrees $d_1, \ldots, d_r$ then we call $V(P)$ a *generic complete intersection*. The following formula for the ML degree is valid in this case.

Theorem 3.31 *Let $P = \langle g_1, \ldots, g_r \rangle$ be an ideal in $\mathbb{Q}[p_1, \ldots, p_m]$ where g_i is a homogeneous polynomial of degree d_i for $i = 1, \ldots, r$. Then the maximum likelihood degree of P is finite and is bounded above by*

$$\sum_{\substack{i_1 + i_2 + \cdots + i_r \leq m-1 \\ i_1 > 0, \ldots, i_r > 0}} d_1^{i_1} d_2^{i_2} \cdots d_r^{i_r}.$$

Equality holds when $V(P)$ is a generic complete intersection, that is, when the coefficients of the defining polynomials $g_1, g_2, \ldots, g_r$ are chosen generically.

Proof See [Hoşten *et al.*, 2005].   □

Example 3.32 Let $m = 5$, $r = 2$ and $P = \langle g_1, g_2 \rangle$ where g_1 and g_2 are random homogeneous polynomials of degrees d_1 and d_2 in $\mathbb{Q}[p_1, p_2, p_3, p_4, p_5]$. Then $V(P)$ is a surface in $\mathbb{P}^4$, and $V_\Delta(P)$ is either empty or is a surface in the 4-simplex Δ. The maximum likelihood degree of such a random surface equals

$$d_1 d_2 + d_1^2 d_2 + d_1 d_2^2 + d_1^3 d_2 + d_1^2 d_2^2 + d_1 d_2^3.$$

In particular, if $d_1 = 1$ and $d_2 = 2$, so that $V(P)$ is a quadratic surface in a hyperplane, then the ML degree is $2 + 2 + 4 + 2 + 4 + 8 = 22$. Compare this with the ML degree 16 of the surface in Example 3.30. Indeed, $P_{\mathbf{f}}$ has codimension 2 and is generated by two polynomials with $d_1 = 1$ and $d_2 = 2$. But the coefficients of the two generators of $P_{\mathbf{f}}$ are not generic enough, so the ML degree drops from 22 to 16 for the specific surface $V(P_{\mathbf{f}})$. □

3.4 Tropical geometry

In the first three sections of this chapter we introduced algebraic varieties and we showed how computations in algebraic geometry can be useful for statistical analysis. In this section we give an introduction to algebraic geometry in the piecewise-linear setting of the *tropical semiring* $(\mathbb{R} \cup \{\infty\}, \oplus, \odot)$.

We had our first encounter with the tropical universe in Chapter 2. While the emphasis there was on tropical *arithmetic* and its computational significance, here we aim to develop the elements of tropical *algebraic geometry*. We shall see that every algebraic variety can be tropicalized, and, since statistical models are algebraic varieties, statistical models can be tropicalized.

Let $q_1, \ldots, q_m$ be unknowns which represent elements in the tropical semiring $(\mathbb{R} \cup \{\infty\}, \oplus, \odot)$. A *monomial* is a product of these unknowns, where repetition is allowed. By commutativity, we can sort the product and write monomials in the usual notation, with the unknowns raised to an exponent, e.g.,

$$q_2 \odot q_1 \odot q_3 \odot q_1 \odot q_4 \odot q_2 \odot q_3 \odot q_2 \quad = \quad q_1^2 q_2^3 q_3^2 q_4. \tag{3.30}$$

When evaluating a tropical monomial in classical arithmetic we get a linear function in the unknowns. For instance, the monomial in (3.30) represents

$$q_2 + q_1 + q_3 + q_1 + q_4 + q_2 + q_3 + q_2 \quad = \quad 2q_1 + 3q_2 + 2q_3 + q_4.$$

Every linear function with integer coefficients arises in this manner. A *tropical polynomial* is a finite tropical linear combination of tropical monomials:

$$p(q_1, \ldots, q_m) \quad = \quad a \odot q_1^{i_1} q_2^{i_2} \cdots q_m^{i_m} \oplus b \odot q_1^{j_1} q_2^{j_2} \cdots q_m^{j_m} \oplus \cdots$$

Here the coefficients $a, b, \ldots$ are real numbers and the exponents $i_1, j_1, \ldots$ are non-negative integers. Every tropical polynomial represents a function $g : \mathbb{R}^m \to \mathbb{R}$. When evaluating this function in classical arithmetic, what we get is the minimum of a finite collection of linear functions, namely,

$$g(q_1, \ldots, q_m) \quad = \quad \min\big(a + i_1 q_1 + \cdots + i_m q_m, \, b + j_1 q_1 + \cdots + j_m q_m, \, \ldots\big).$$

This function $g : \mathbb{R}^m \to \mathbb{R}$ has the following three characteristic properties:

   (i) It is continuous;
  (ii) It is piecewise-linear, where the number of pieces is finite and each piece has rational slope;
 (iii) It is concave, i.e., $g\big((q + q')/2\big) \geq \frac{1}{2}(g(q) + g(q'))$ for all $q, q' \in \mathbb{R}^m$.

Every function $g : \mathbb{R}^m \to \mathbb{R}$ which satisfies these three properties can be represented as the minimum of a finite collection of linear functions. We conclude:

Proposition 3.33 *The tropical polynomials in m unknowns $q_1, \ldots, q_m$ are the piecewise-linear concave functions on $\mathbb{R}^m$ with non-negative integer coefficients.*

Example 3.34 Let $m = 1$, so we are considering tropical polynomials in one variable q. A general cubic polynomial has the form

$$g(q) \quad = \quad a \odot q^3 \,\oplus\, b \odot q^2 \,\oplus\, c \odot q \,\oplus\, d \qquad \text{where } a, b, c, d \in \mathbb{R}. \quad (3.31)$$

To graph this function we draw four lines in the (q, q') plane: $q' = 3q + a$, $q' = 2q + b$, $q' = q + c$ and the horizontal line $q' = d$. The value of $g(q)$ is the smallest q-value such that (q, q') is on one of these four lines, i.e., the graph of $g(q)$ is the lower envelope of the lines. All four lines actually contribute if

$$b - a \;\leq\; c - b \;\leq\; d - c. \quad (3.32)$$

These three values of q are the breakpoints where $g(q)$ fails to be linear, and the cubic has a corresponding factorization into three linear factors:

$$g(q) \quad = \quad a \odot (q \oplus (b - a)) \odot (q \oplus (c - b)) \odot (q \oplus (d - c)). \quad (3.33)$$

Generalizing this example, we can see that every tropical polynomial function in one unknown q can be written as a tropical product of tropical linear functions. This representation is essentially unique, i.e., the *Fundamental Theorem of Algebra* (Theorem 3.1) holds for tropical polynomials. $\qquad\square$

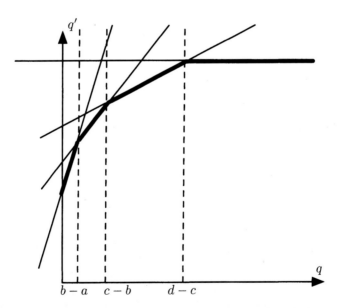

Fig. 3.3. The graph of a tropical cubic polynomial and its roots.

Example 3.35 The factorization of tropical polynomials in $m \geq 2$ unknowns into irreducible tropical polynomials is not unique. Here is a simple example

$$(0 \odot q_1 \oplus 0) \,\odot\, (0 \odot q_2 \oplus 0) \,\odot\, (0 \odot q_1 \odot q_2 \oplus 0)$$
$$= \quad (0 \odot q_1 \odot q_2 \oplus 0 \odot q_1 \oplus 0) \,\odot\, (0 \odot q_1 \odot q_2 \oplus 0 \odot q_2 \oplus 0),$$

keeping in mind that zero is the multiplicatively neutral element. This identity is equivalent to an identity in the *polytope algebra* (Section 2.3): a regular hexagon factors either into two triangles or into three line segments. □

A tropical polynomial function $g : \mathbb{R}^m \to \mathbb{R}$ is given as the minimum of a finite set of linear functions. We define the *tropical hypersurface* $\mathcal{T}(g)$ to be the set of all points $q \in \mathbb{R}^m$ at which this minimum is attained at least twice at q. Equivalently, a point $q \in \mathbb{R}^m$ lies in the hypersurface $\mathcal{T}(g)$ if and only if g is not linear at q. For example, if $m = 1$ and p is the cubic in (3.31) with the assumption (3.32), then $\mathcal{T}(g)$ is the set of breakpoints: $\{\, b - a, \, c - b, \, d - c \,\}$.

We next consider the case $m = 2$ of a tropical polynomial in two variables:

$$g(q_1, q_2) \quad = \quad \bigoplus_{(i,j)} c_{ij} \odot q_1^i \odot q_2^j.$$

Proposition 3.36 *The tropical curve $\mathcal{T}(g)$ is a finite graph which is embedded in the plane $\mathbb{R}^2$. It has both bounded and unbounded edges, all edge directions are rational, and $\mathcal{T}(g)$ satisfies the* zero tension condition.

The zero tension condition means the following. Consider any node p of the graph. Then the edges adjacent to p lie on lines with rational slopes. For each such line emanating from the origin consider the first non-zero lattice vector on that line. *Zero tension* at p means that the sum of these vectors is zero.

Here is a general method for drawing a tropical curve $\mathcal{T}(g)$ in the plane. Consider any term $\gamma \odot q_1^i \odot q_2^j$ appearing in the polynomial p. We represent this term by the point (γ, i, j) in $\mathbb{R}^3$, and we compute the convex hull of these points in $\mathbb{R}^3$. Now project the lower envelope of that convex hull into the plane under the map $\mathbb{R}^3 \to \mathbb{R}^2$, $(\gamma, i, j) \mapsto (i, j)$. The image is a planar convex polygon together with a distinguished subdivision Δ into smaller polygons. The tropical curve $\mathcal{T}(g)$ is a graph which is dual to this subdivision.

Example 3.37 Consider the general quadratic polynomial

$$g(q_1, q_2) \quad = \quad a \odot q_1^2 \oplus b \odot q_1 q_2 \oplus c \odot q_2^2 \oplus d \odot q_1 \oplus e \odot q_2 \oplus f.$$

Then Δ is a subdivision of the triangle with vertices $(0,0)$, $(0,2)$ and $(2,0)$. The lattice points $(0,1)$, $(1,0)$, $(1,1)$ are allowed to be used as vertices in these subdivisions. Assuming that $a, b, c, d, e, f \in \mathbb{R}$ satisfy

$$2b < a + c, \ 2d < a + f, \ 2e < c + f,$$

the subdivision Δ consists of four triangles, three interior edges and six boundary edges. The tropical quadratic curve $\mathcal{T}(g)$ has four vertices, three bounded edges and six half-rays (two northern, two eastern and two southwestern). In Figure 3.4, $\mathcal{T}(g)$ is shown in bold and the subdivision of Δ is in thin lines. □

It is known that tropical hypersurfaces $\mathcal{T}(g)$ intersect and interpolate like algebraic hypersurfaces do. For instance, two lines in the plane meet in one point, a line and a quadric meet in two points, two quadrics meet in four

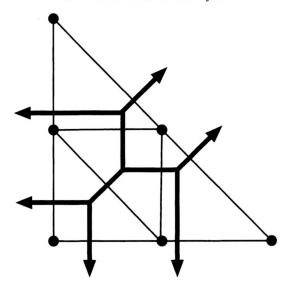

Fig. 3.4. The subdivision Δ and the tropical curve in Example 3.37.

points, etc ... Also, two general points lie on a unique line, five general points lie on a unique quadric, etc ... For a general discussion of Bézout's Theorem in tropical algebraic geometry, and for pictures illustrating these facts we refer to [Sturmfels, 2002, §9] and [Richter-Gebert *et al.*, 2003].

It is tempting to define tropical varieties as intersections of tropical hypersurfaces. But this is not quite the right definition to retain the desired properties from classical algebraic geometry. What we do instead is to utilize the field $\mathbb{Q}(\epsilon)$ of rational functions in one variable ϵ. We think of ϵ as a positive infinitesimal. Any non-zero rational function $c(\epsilon) \in \mathbb{Q}(\epsilon)$ has a series expansion

$$c(\epsilon) = \alpha_0 \epsilon^{i_0} + \alpha_1 \epsilon^{i_1} + \alpha_2 \epsilon^{i_2} + \cdots ,$$

where $i_0 < i_1 < \cdots \in \mathbb{Z}$, $\alpha_0, \alpha_1, \ldots \in \mathbb{Q}$, and $\alpha_0 \neq 0$. This series is unique. The *order* of the rational function $c(\epsilon)$ is the integer i_0.

Example 3.38 The following rational function in $\mathbb{Q}(\epsilon)$ has order$(c(\epsilon)) = -2$:

$$c(\epsilon) \quad = \quad \frac{3\epsilon^4 + 8\epsilon^5 + \epsilon^7 - \epsilon^{10}}{17\epsilon^6 - 11\epsilon^9 + 2\epsilon^{13}} \quad = \quad \frac{3}{17}\epsilon^{-2} + \frac{8}{17}\epsilon^{-1} + \frac{50}{289}\epsilon^1 + \frac{88}{289}\epsilon^2 + \cdots$$

$\square$

Let $\mathbb{Q}(\epsilon)[p_1, \ldots, p_m]$ be the ring of polynomials in m unknowns with coefficients in $\mathbb{Q}(\epsilon)$. For any (classical) polynomial

$$f \quad = \quad \sum_{i=1}^{s} c_i(\epsilon) \cdot p_1^{a_{1i}} p_2^{a_{2i}} \cdots p_m^{a_{mi}} \quad \in \mathbb{Q}(\epsilon)[p_1, \ldots, p_m], \qquad (3.34)$$

we define the *tropicalization* $g = \mathrm{trop}(f)$ to be the tropical polynomial

$$g \quad = \quad \bigoplus_{i=1}^{s} \mathrm{order}(c_i(\epsilon)) \odot q_1^{a_{1i}} \odot q_2^{a_{2i}} \odot \cdots \odot q_m^{a_{mi}}. \qquad (3.35)$$

In this manner, every polynomial f defines a tropical hypersurface $\mathcal{T}(g) = \mathcal{T}(\text{trop}(f))$ in $\mathbb{R}^m$. Recall that the function $g : \mathbb{R}^m \to \mathbb{R}$ is given as the minimum of s linear functions. The hypersurface $\mathcal{T}(g)$ consists of all points $q = (q_1, \ldots, q_m) \in \mathbb{R}^m$ where this minimum is attained at least twice, i.e.,

$$
\begin{aligned}
\text{order}(c_i(\epsilon)) + a_{1i}q_1 + \cdots + a_{mi}q_m &= \text{order}(c_j(\epsilon)) + a_{1j}q_1 + \cdots + a_{mj}q_m \\
\leq \quad \text{order}(c_k(\epsilon)) + a_{1k}q_1 + \cdots &+ a_{mk}q_m \quad \text{for } i \neq j \text{ and } k \in \{1, \ldots, s\} \backslash \{i, j\}.
\end{aligned}
$$

If, in addition to this condition, the leading coefficients α_0 of the series $c_i(\epsilon)$ and $c_j(\epsilon)$ are rational numbers of opposite sign then we say that q is a *positive point* of $\mathcal{T}(g)$. The subset of all positive points in $\mathcal{T}(g)$ is denoted $\mathcal{T}^+(g)$ and called the *positive tropical hypersurface* of the polynomial f.

If the polynomial f in (3.34) does not depend on ϵ at all, i.e., if $f \in \mathbb{Q}[p_1, \ldots, p_m]$, then $\text{order}(c_i(\epsilon)) = 0$ for all coefficients of its tropicalization g in (3.35), and the function $g : \mathbb{R}^m \to \mathbb{R}$ is given as the minimum of s linear functions with zero constant terms. Here is an example where this is the case.

Example 3.39 Let $m = 9$ and consider the determinant of a 3×3-matrix,

$$
f \quad = \quad p_{11}p_{22}p_{33} - p_{11}p_{23}p_{32} - p_{12}p_{21}p_{33} + p_{12}p_{23}p_{31} + p_{13}p_{21}p_{32} - p_{13}p_{22}p_{31}.
$$

Its tropicalization is the *tropical determinant*

$$
\begin{aligned}
g \quad = \quad & q_{11} \odot q_{22} \odot q_{33} \ \oplus \ q_{11} \odot q_{23} \odot q_{32} \ \oplus \ q_{12} \odot q_{21} \odot q_{33} \\
\oplus \ & q_{12} \odot q_{23} \odot q_{31} \ \oplus \ q_{13} \odot q_{21} \odot q_{32} \ \oplus \ q_{13} \odot q_{22} \odot q_{31}.
\end{aligned}
$$

Evaluating g at a 3×3-matrix (q_{ij}) means solving the *assignment problem* of finding a permutation σ of $\{1, 2, 3\}$ whose weight $q_{1\sigma_1} + q_{2\sigma_2} + q_{3\sigma_3}$ is minimal (Remark 2.6). The tropical hypersurface $\mathcal{T}(g)$ consists of all matrices $q \in \mathbb{R}^{3 \times 3}$ for which the minimum weight permutation is not unique. Working modulo the five-dimensional space of all 3×3-matrices (q_{ij}) with zero row sums and zero column sums, the tropical hypersurface $\mathcal{T}(g)$ is a three-dimensional polyhedral fan sitting in a four-dimensional space. If we intersect this fan with a 3-sphere around the origin, then we get a two-dimensional polyhedral complex consisting of six triangles and nine quadrangles. This complex consists of all 2-faces of the product of two triangles, labeled as in Figure 3.5. This complex is a bouquet of five 2-spheres. The positive tropical variety $\mathcal{T}^+(g)$ is the subcomplex consisting of the nine quadrangles shown in Figure 3.5. Note that $\mathcal{T}^+(g)$ is a torus. $\square$

Every *tropical algebraic variety* is derived from an ideal I in the polynomial ring $\mathbb{Q}(\epsilon)[p_1, \ldots, p_m]$. Namely, we define $\mathcal{T}(I)$ as the intersection of the tropical hypersurfaces $\mathcal{T}(\text{trop}(f))$ where f runs over the ideal I. Likewise, the *positive tropical variety* $\mathcal{T}^+(I)$ is the intersection of the positive tropical hypersurfaces $\mathcal{T}^+(\text{trop}(f))$ where f runs over I. In these definitions it suffices to let f run over a certain finite subset of the ideal I. Such a subset is called a *tropical basis* of I. From this finiteness property, it follows that $\mathcal{T}(I)$ and $\mathcal{T}^+(I)$ are *finite unions of convex polyhedra*. This means they are characterized by finite Boolean combinations of linear inequalities. Finding a tropical

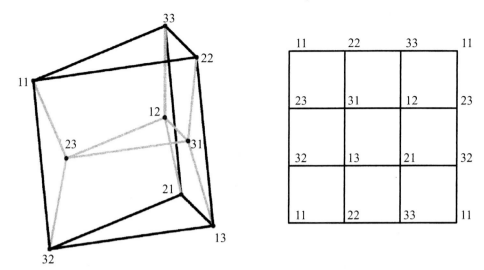

Fig. 3.5. The tropical 3×3 determinant.

basis from given generators of an ideal I and computing the polyhedra that make up its tropical variety is an active topic of research in tropical geometry.

Example 3.40 We consider the tropicalization of DiaNA's model in Example 1.16. The 3×3-minors of a 4×4-matrix of unknowns form a tropical basis for the ideal they generate. This follows from results in [Develin *et al.*, 2003]. The tropical variety $\mathcal{T}(I)$ consists of all 4×4-matrices of tropical rank at most two. The positive tropical variety $\mathcal{T}^+(I)$ is discussed in Example 3.43. ☐

Let $\mathbf{f} : \mathbb{C}^d \to \mathbb{C}^m$ be a polynomial map with coordinates $f_1, \ldots, f_m \in \mathbb{Q}[\theta_1, \ldots, \theta_d]$. We say that the map $\mathbf{f}$ is *positive* if each coefficient of each polynomial f_i is a positive real number. If this holds then $\mathbf{f}$ maps positive vectors in $\mathbb{R}^d$ to positive vectors in $\mathbb{R}^m$. We say that the map $\mathbf{f}$ is *surjectively positive* if $\mathbf{f}$ is positive and, in addition, $\mathbf{f}$ maps the positive orthant surjectively onto the positive points in the image; in symbols:

$$\mathbf{f}\big(\mathbb{R}^d_{>0}\big) \quad = \quad \text{image}(\mathbf{f}) \cap \mathbb{R}^m_{>0}. \tag{3.36}$$

Example 3.41 Let $d = 1, m = 2$ and $\mathbf{f} : \mathbb{R}^1 \mapsto \mathbb{R}^2, \theta \mapsto (\theta + 2, 2\theta + 1)$. The map $\mathbf{f}$ is positive. But $\mathbf{f}$ is not surjectively positive: for instance, the point $(7/4, 1/2)$ is in image$(\mathbf{f}) \cap \mathbb{R}^2_{>0}$ but not in $\mathbf{f}(\mathbb{R}^1_{>0})$.

On the other hand, if we take $\mathbf{f}' : \mathbb{R}^1 \mapsto \mathbb{R}^2, \theta \mapsto (\frac{1}{2}\theta + \frac{3}{2}, \theta)$ then $\mathbf{f}'$ is surjectively positive. Both maps have the same image, namely, image$(\mathbf{f}) =$ image$(\mathbf{f}')$ is the line $V(I_{\mathbf{f}}) \subset \mathbb{R}^2$ which is specified by the ideal

$$I_{\mathbf{f}} \ = \ I_{\mathbf{f}'} \ = \ \langle 2p_1 - p_2 - 3 \rangle.$$

The tropical variety $\mathcal{T}(I_{\mathbf{f}})$ is the curve defined by the tropical linear form

$$\text{trop}(2p_1 - p_2 - 3) \quad = \quad q_1 \oplus q_2 \oplus 0.$$

This tropical line is the union of three half-rays:

$$\mathcal{T}(I_{\mathbf{f}}) = \{(\varphi,0) : \varphi \in \mathbb{R}_{\geq 0}\} \cup \{(0,\varphi) : \varphi \in \mathbb{R}_{\geq 0}\} \cup \{(-\varphi,-\varphi) : \varphi \in \mathbb{R}_{\geq 0}\}.$$

Let $\mathbf{g} : \mathbb{R}^1 \to \mathbb{R}^2$ be the tropicalization of the linear map $\mathbf{f} : \mathbb{R}^1 \to \mathbb{R}^2$, and let $\mathbf{g}'$ be the tropicalization of $\mathbf{f}'$. These piecewise-linear maps are given by

$$\mathbf{g}(w) = \big(\text{trop}(f_1)(w), \text{trop}(f_2)(w)\big) = (w \oplus 0, w \oplus 0) = (\min(w,0), \min(w,0))$$

and $\quad \mathbf{g}'(w) = \big(\text{trop}(f_1')(w), \text{trop}(f_2')(w)\big) = (w \oplus 0, w) = (\min(w,0), w).$

These map $\mathbb{R}^1$ onto one or two of the three halfrays of the tropical line $\mathcal{T}(I_{\mathbf{f}})$:

$$\begin{aligned}
\text{image}(\mathbf{g}) &= \{(-\varphi,-\varphi) : \varphi \in \mathbb{R}_{\geq 0}\}, \\
\text{image}(\mathbf{g}') &= \{(0,\varphi) : \varphi \in \mathbb{R}_{\geq 0}\} \cup \{(-\varphi,-\varphi) : \varphi \in \mathbb{R}_{\geq 0}\} = \mathcal{T}^+(I_{\mathbf{f}}).
\end{aligned}$$

The tropicalization $\mathbf{g}'$ of the surjectively positive map $\mathbf{f}'$ maps onto the positive tropical variety. This is an example for the result in Theorem 3.42 below. □

Returning to our general discussion, consider an arbitrary polynomial map $\mathbf{f} : \mathbb{C}^d \to \mathbb{C}^m$. The *tropicalization* of $\mathbf{f}$ is the piecewise-linear map

$$\mathbf{g} : \mathbb{R}^d \to \mathbb{R}^m, \quad \varphi \mapsto \big(g_1(\varphi), g_2(\varphi), \ldots, g_m(\varphi)\big), \tag{3.37}$$

where $g_i = \text{trop}(f_i)$ is the tropicalization of the ith coordinate polynomial f_i of $\mathbf{f}$. To describe the geometry of the tropical polynomial map $\mathbf{g}$, we consider the Newton polytopes $\text{NP}(f_1), \text{NP}(f_2), \ldots, \text{NP}(f_m)$ of the coordinates of $\mathbf{f}$. Recall from Section 2.3 that the *Newton polytope* $\text{NP}(f_i)$ of the polynomial f_i is the convex hull of the vectors $(u_1, u_2, \ldots, u_d)$ such that $\theta_1^{u_1}\theta_2^{u_2}\cdots\theta_d^{u_d}$ appears with non-zero coefficient in f_i. The cones in the normal fan of the Newton polytope $\text{NP}(f_i)$ are the domains of linearity of the piecewise-linear map $g_i : \mathbb{R}^d \to \mathbb{R}^m$. The *Newton polytope of the map* $\mathbf{f}$ is defined as the Newton polytope of the product of the coordinate polynomials:

$$\text{NP}(\mathbf{f}) := \text{NP}(f_1 \cdot f_2 \cdots \cdot f_m) = \text{NP}(f_1) \odot \text{NP}(f_2) \odot \cdots \odot \text{NP}(f_m). \tag{3.38}$$

Here the operation $\odot$ is the Minkowski sum of polytopes (Theorem 2.26). The following theorem describes the geometry of tropicalizing polynomial maps.

Theorem 3.42 *The tropical polynomial map* $\mathbf{g} : \mathbb{R}^d \to \mathbb{R}^m$ *is linear on each cone in the normal fan of the Newton polytope* $\text{NP}(\mathbf{f})$. *Its image lies inside the tropical variety* $\mathcal{T}(I_{\mathbf{f}})$. *If* $\mathbf{f}$ *is positive then* $\text{image}(\mathbf{g})$ *lies in the positive tropical variety* $\mathcal{T}^+(I_{\mathbf{f}})$. *If* $\mathbf{f}$ *is surjectively positive then* $\text{image}(\mathbf{g}) = \mathcal{T}^+(I_{\mathbf{f}})$.

Proof See [Pachter and Sturmfels, 2004b] and [Speyer and Williams, 2004]. □

This theorem is fundamental for the study of inference functions of statistical models in Chapter 9. Imagine that $w \in \mathbb{R}^d$ is a vector of weights for a dynamic programming problem that is derived from a statistical model. The relationship between the weights and the model parameters is $w_i \sim \log(\theta_i)$.

Evaluating the tropical map **g** at w means solving the dynamic programs for all possible observations. As we vary the weights w, the vector of outcomes is piecewise constant. Whenever w crosses a boundary, the system undergoes a "phase transition", meaning that the outcome changes for some observation. Theorem 3.42 offers a geometric characterization of these phase transitions, taking into account all possible weights and all possible observations.

If the model has only one parameter, then the Newton polytope NP(**f**) is a line segment. There are only two "phases", corresponding to the two vertices of that segment. In Example 3.41 the "phase transition" occurs at $w = 0$.

One important application of this circle of ideas is sequence alignment (Section 2.2). The statistical model for alignment is the pair HMM **f** in (2.15). The tropicalization **g** of the polynomial map **f** is the tropicalized pair HMM, whose coordinates g_i are featured in (2.16). Parametric alignment is discussed in Chapters 5, 7, 8 and 9. We conclude this section with two other examples.

Example 3.43 DiaNA's model in Example 3.40 has $d = 16$ parameters

$$\theta = \left(\beta_{\mathtt{A}}^1, \beta_{\mathtt{C}}^1, \beta_{\mathtt{G}}^1, \beta_{\mathtt{T}}^1, \ \beta_{\mathtt{A}}^2, \beta_{\mathtt{C}}^2, \beta_{\mathtt{G}}^2, \beta_{\mathtt{T}}^2, \ \gamma_{\mathtt{A}}^1, \gamma_{\mathtt{C}}^1, \gamma_{\mathtt{G}}^1, \gamma_{\mathtt{T}}^1, \ \gamma_{\mathtt{A}}^2, \gamma_{\mathtt{C}}^2, \gamma_{\mathtt{G}}^2, \gamma_{\mathtt{T}}^2 \right),$$

and is specified by the homogeneous polynomial map $\mathbf{f} : \mathbb{C}^{16} \mapsto \mathbb{C}^{4 \times 4}$ with

$$p_{ij} = \beta_i^1 \beta_j^2 + \gamma_i^1 \gamma_j^2, \qquad \text{where } i, j \in \{\mathtt{A}, \mathtt{C}, \mathtt{G}, \mathtt{T}\}.$$

We know from the linear algebra literature [Cohen and Rothblum, 1993] that every positive 4×4-matrix of rank ≤ 2 is the sum of two positive 4×4-matrices of rank ≤ 1. This means that DiaNA's model **f** is a surjectively positive map. The tropicalization of **f** is the piecewise-linear map $\mathbf{g} : \mathbb{R}^{16} \mapsto \mathbb{R}^{4 \times 4}$ given by

$$q_{ij} = \beta_i^1 \odot \beta_j^2 \oplus \gamma_i^1 \odot \gamma_j^2 = \min\left(\beta_i^1 + \beta_j^2, \gamma_i^1 + \gamma_j^2\right) \quad \text{for } i, j \in \{\mathtt{A}, \mathtt{C}, \mathtt{G}, \mathtt{T}\}.$$

Theorem 3.42 says that the image of **g** equals the positive tropical variety $\mathcal{T}^+(I_\mathbf{g})$. The space $\mathcal{T}^+(I_\mathbf{g})$ consists of all 4×4-matrices of *Barvinok rank* ≤ 2 and was studied in [Develin *et al.*, 2003] and [Develin and Sturmfels, 2004]. The Newton polytope NP(**f**) of the map **f** is a *zonotope*, i.e., it is a Minkowski sum of line segments. The map **g** is piecewise linear with respect to the hyperplane arrangement dual to that zonotope. For a detailed combinatorial study of the map **g** and the associated hyperplane arrangement see [Ardila, 2005]. $\square$

Example 3.44 We consider the hidden Markov model of length $n = 3$ with binary states ($l = l' = 2$) but, in contrast to Example 3.19, we suppose that all eight parameters $\theta_{00}, \theta_{01}, \theta_{10}, \theta_{11}, \theta'_{00}, \theta'_{01}, \theta'_{10}, \theta'_{11}$ are independent unknowns. Thus our model is the homogeneous map $\mathbf{f} : \mathbb{C}^8 \to \mathbb{C}^8$ with coordinates

$$
\begin{aligned}
f_{\sigma_1\sigma_2\sigma_3} = \ & \theta_{00}\theta_{00}\theta'_{0\sigma_1}\theta'_{0\sigma_2}\theta'_{0\sigma_3} + \theta_{00}\theta_{01}\theta'_{0\sigma_1}\theta'_{0\sigma_2}\theta'_{1\sigma_3} + \theta_{01}\theta_{10}\theta'_{0\sigma_1}\theta'_{1\sigma_2}\theta'_{0\sigma_3} \\
& + \theta_{01}\theta_{11}\theta'_{0\sigma_1}\theta'_{1\sigma_2}\theta'_{1\sigma_3} + \theta_{10}\theta_{00}\theta'_{1\sigma_1}\theta'_{0\sigma_2}\theta'_{0\sigma_3} + \theta_{10}\theta_{01}\theta'_{1\sigma_1}\theta'_{0\sigma_2}\theta'_{1\sigma_3} \\
& + \theta_{11}\theta_{10}\theta'_{1\sigma_1}\theta'_{1\sigma_2}\theta'_{0\sigma_3} + \theta_{11}\theta_{11}\theta'_{1\sigma_1}\theta'_{1\sigma_2}\theta'_{1\sigma_3}.
\end{aligned}
$$

The implicitization techniques of Section 3.2 reveal that $I_{\mathbf{f}}$ is generated by

$$p_{011}^2 p_{100}^2 - p_{001}^2 p_{110}^2 + p_{000} p_{011} p_{101}^2 - p_{000} p_{101}^2 p_{110} + p_{000} p_{011} p_{110}^2$$

$$-p_{001} p_{010}^2 p_{111} + p_{001}^2 p_{100} p_{111} + p_{010}^2 p_{100} p_{111} - p_{001} p_{100}^2 p_{111} - p_{000} p_{011}^2 p_{110}$$

$$-p_{001} p_{011} p_{100} p_{101} - p_{010} p_{011} p_{100} p_{101} + p_{001} p_{010} p_{011} p_{110} - p_{010} p_{011} p_{100} p_{110}$$

$$+p_{001} p_{010} p_{101} p_{110} + p_{001} p_{100} p_{101} p_{110} + p_{000} p_{010} p_{011} p_{111} - p_{000} p_{011} p_{100} p_{111}$$

$$-p_{000} p_{001} p_{101} p_{111} + p_{000} p_{100} p_{101} p_{111} + p_{000} p_{001} p_{110} p_{111} - p_{000} p_{010} p_{110} p_{111}.$$

Thus $\mathcal{T}(I_{\mathbf{f}})$ is the tropical hypersurface defined by this degree-four polynomial. The tropicalized HMM is the map $\mathbf{g} : \mathbb{R}^8 \to \mathbb{R}^8$, $(w, w') \mapsto q$ with coordinates

$$q_{\sigma_1 \sigma_2 \sigma_3} = \min \left\{ w_{h_1 h_2} + w_{h_2 h_3} + w'_{h_1 \sigma_1} + w'_{h_2 \sigma_2} + w'_{h_3 \sigma_3} : (h_1, h_2, h_3) \in \{0, 1\}^3 \right\}.$$

This minimum is attained by the most likely explanation $(\hat{h}_1, \hat{h}_2, \hat{h}_3)$ of the observation $(\sigma_1, \sigma_2, \sigma_3)$. Hence, evaluating the tropical polynomial $q_{\sigma_1 \sigma_2 \sigma_3}$ solves the inference task in MAP estimation. See Section 4.4 for the general definition of *MAP inference*.

Let us see an example. Consider the parameters $w = \begin{pmatrix} 6 & 5 \\ 8 & 1 \end{pmatrix}$ and $w' = \begin{pmatrix} 0 & 8 \\ 8 & 8 \end{pmatrix}$.

The observation $\quad \sigma_1 \sigma_2 \sigma_3 = \quad 000 \quad 001 \quad 010 \quad 011 \quad 100 \quad 101 \quad 110 \quad 111$
has the explanation $\quad \hat{h}_1 \hat{h}_2 \hat{h}_3 = \quad 000 \quad 001 \quad 000 \quad 011 \quad 000 \quad 111 \quad 110 \quad 111$

We call $\{0, 1\}^3 \to \{0, 1\}^3$, $\sigma_1 \sigma_2 \sigma_3 \mapsto \hat{h}_1 \hat{h}_2 \hat{h}_3$ the *inference function* for the parameters (w, w'). There are $8^8 = 16,777,216$ functions from $\{0, 1\}^3$ to itself, but only 398 of them are inference functions. In Chapter 9 it is shown that the number of inference functions is polynomial in the sequence length n, while the number of functions from $\{0, 1\}^n$ to itself is doubly-exponential in n.

The inference functions are indexed by the vertices of the Newton polytope $\mathrm{NP}(\mathbf{f})$. In our example, polymake reveals that $\mathrm{NP}(\mathbf{f})$ is 5-dimensional and has 398 vertices, 1136 edges, 1150 two-faces, 478 ridges and 68 facets. Thus there are 398 inference functions, and we understand their phase transitions.

However, there are still many questions we do not yet know how to answer. What is the most practical method for listing all maximal cones in the image of $\mathbf{g}$? How does the number of these cones compare to the number of vertices of $\mathrm{NP}(\mathbf{f})$? Is the hidden Markov model $\mathbf{f}$ surjectively positive? Which points of the positive tropical variety $\mathcal{T}^+(I_{\mathbf{f}})$ lie in $\mathrm{image}(\mathbf{g})$? $\qquad \square$

3.5 The tree of life and other tropical varieties

At the 1998 International Congress of Mathematicians in Berlin, Andreas Dress presented an invited lecture titled "The tree of life and other affine buildings" [Dress and Terhalle, 1998]. Our section title is meant as a reference to that paper, which highlighted the importance and utility of understanding the geometry and structure of phylogenetic trees and networks. In Section 4.5 we return to this topic by explaining how the space of trees and its generalizations are relevant for modeling and reconstructing the tree of life.

Here we begin with a reminder of the definition of metrics and tree metrics, which were introduced in Section 2.4. A *metric* on $[n] = \{1, 2, \ldots, n\}$ is a dissimilarity map for which the triangle inequality holds. Most metrics D are not tree metrics. The set of tree metrics is the *space of trees* $\mathcal{T}_n$. This is a $(2n-3)$-dimensional polyhedral fan inside $\binom{n}{2}$-dimensional cone of all metrics. Membership in $\mathcal{T}_n$ is characterized by the *four-point condition* (Theorem 2.36). Our goal here is to derive an interpretation of $\mathcal{T}_n$ in tropical geometry.

Let $Q = (q_{ij})$ be a symmetric matrix with zeros on the diagonal whose $\binom{n}{2}$ distinct off-diagonal entries are unknowns. For each quadruple $\{i, j, k, l\} \subset \{1, 2, \ldots, n\}$ we consider the quadratic tropical polynomial

$$g_{ijkl}(Q) \quad = \quad q_{ij} \odot q_{kl} \ \oplus \ q_{ik} \odot q_{jl} \ \oplus \ q_{il} \odot q_{jk}. \tag{3.39}$$

This tropical polynomial is the tropicalization of the Plücker relation (3.14)

$$g_{ijkl}(Q) \quad = \quad \text{trop}(p_{ik}p_{jl} - p_{ij}p_{kl} - p_{il}p_{jk})$$

which defines a tropical hypersurface $\mathcal{T}(g_{ijkl})$ in the space $\mathbb{R}^{\binom{n}{2}}$. In fact, the Plücker relations (3.14) form a tropical basis for the Plücker ideal $I_{2,n}$ [Speyer and Sturmfels, 2004]. This implies that the *tropical Grassmannian* $\mathcal{T}(I_{2,n})$ equals the intersection of these $\binom{n}{4}$ tropical hypersurfaces, i.e.,

$$\mathcal{T}(I_{2,n}) \quad = \bigcap_{1 \le i < j < k < l \le n} \mathcal{T}(g_{ijkl}) \quad \subset \quad \mathbb{R}^{\binom{n}{2}}. \tag{3.40}$$

Theorem 3.45 *The space of trees $\mathcal{T}_n$ is (up to sign) the tropical Grassmannian $\mathcal{T}(I_{2,n})$.*

Proof A metric $D = (d_{ij})$ is a point in the space of trees $\mathcal{T}_n$ if and only if thef our point condition holds. This condition states that, for all $1 \le i < j < k < l \le n$, the maximum of $\{d_{ij} + d_{kl}, d_{ik} + d_{jl}, d_{il} + d_{jk}\}$ is attained at least twice. If $Q = (q_{ij}) = -D = (-d_{ij})$ then this is equivalent to saying that the minimum of $\{q_{ij} + q_{kl}, q_{ik} + q_{jl}, q_{il} + q_{jk}\}$ is attained at least twice. This is precisely the condition for Q to be in the tropical hypersurface $\mathcal{T}(g_{ijkl})$. $\square$

Example 3.46 The space of trees on five taxa $[5] = \{1, 2, 3, 4, 5\}$ is (up to sign) a tropical variety of codimension three in $\mathbb{R}^{10}$. It is not the intersection of three tropical hypersurfaces, but it is the intersection of five hypersurfaces:

$$\begin{aligned}
\mathcal{T}(I_{2,5}) \quad = \quad -\mathcal{T}_5 \quad = \quad & \mathcal{T}(q_{12} \odot q_{34} \ \oplus \ q_{13} \odot q_{24} \ \oplus \ q_{14} \odot q_{23}) \\
\cap \ & \mathcal{T}(q_{12} \odot q_{35} \ \oplus \ q_{13} \odot q_{25} \ \oplus \ q_{15} \odot q_{23}) \\
\cap \ & \mathcal{T}(q_{12} \odot q_{45} \ \oplus \ q_{14} \odot q_{25} \ \oplus \ q_{15} \odot q_{24}) \\
\cap \ & \mathcal{T}(q_{13} \odot q_{45} \ \oplus \ q_{14} \odot q_{35} \ \oplus \ q_{15} \odot q_{34}) \\
\cap \ & \mathcal{T}(q_{23} \odot q_{45} \ \oplus \ q_{24} \odot q_{35} \ \oplus \ q_{25} \odot q_{34}).
\end{aligned}$$

The space of trees $\mathcal{T}_5$ is the union of 15 seven-dimensional cones in $\mathbb{R}^{10}_{\ge 0}$. Each

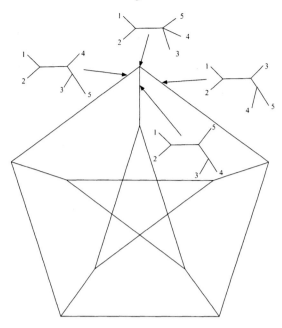

Fig. 3.6. A tropical Grassmannian of lines: the space of trees $\mathcal{T}_5$.

cone is the solution set to a system of linear inequalities such as

$$
\begin{aligned}
q_{12} + q_{34} &\geq q_{13} + q_{24} = q_{14} + q_{23}, \\
q_{12} + q_{35} &\geq q_{13} + q_{25} = q_{15} + q_{23}, \\
q_{12} + q_{45} &\geq q_{14} + q_{25} = q_{15} + q_{24}, \\
q_{13} + q_{45} &\geq q_{14} + q_{35} = q_{15} + q_{34}, \\
\text{and} \quad q_{23} + q_{45} &\geq q_{24} + q_{35} = q_{25} + q_{34}.
\end{aligned}
$$

The seven-dimensional cone specified by this linear system is isomorphic to $\mathbb{R}^7_{\geq 0}$, and corresponds to the tree with splits $(12, 345)$ and $(123, 45)$.

The combinatorial structure of the space $\mathcal{T}_5$ is that of the *Petersen graph*, shown in Figure 3.6. Vertices of the Petersen graph correspond to trees with *polytomy*, i.e trees with internal vertices of degree at least 4. The edges correspond to the seven-dimensional cones in $\mathcal{T}_5$. Two cones share a six-dimensional facet if and only if the two edges share a node in the Petersen graph. $\square$

The interpretation of the space of trees as a tropical Grassmannian opens up the possibility of modeling a wide range of problems in phylogenetics using tropical geometry. We shall demonstrate this by tropicalizing the *higher Grassmannian* $G_{d,n} = V(I_{d,n})$ and the *Pfaffian varieties* $V(I_{2,n,k})$ which we encountered towards the end of Section 3.2. We begin with a discussion of tropical linear spaces. These are relevant for evolutionary biology because:

- trees are tropical lines;
- higher-dimensional trees are tropical linear spaces.

The second statement will be made precise in Theorem 3.47.

A *tropical hyperplane* in $\mathbb{R}^n$ is any subset of $\mathbb{R}^n$ which has the form $\mathcal{T}(\ell)$, where ℓ is a tropical linear form in n unknowns q_i:

$$\ell(q) \quad = \quad a_1 \odot q_1 \ \oplus \ a_2 \odot q_2 \ \oplus \ \cdots \ \oplus \ a_n \odot q_n.$$

Here $a_1, \ldots a_n$ are arbitrary constants in $\mathbb{R} \cup \{\infty\}$. Solving linear equations in tropical mathematics means computing the intersection of finitely many hyperplanes $\mathcal{H}(\ell)$. It is tempting to define tropical linear spaces simply as intersections of tropical hyperplanes. However, this would not be a good definition because such arbitrary intersections are not always pure dimensional, and they do not behave the way linear spaces do in classical geometry. A better notion of tropical linear space is derived by allowing only those intersections of hyperplanes which are "sufficiently complete". In what follows we offer a definition which generalizes the geometric relationship between tree metrics and the Grassmannian $G_{2,n}$ that underlies Theorem 3.45. The idea is that phylogenetic trees are lines in tropical projective space, and the negated pairwise distances d_{ij} are the Plücker coordinates q_{ij} of these tropical lines.

We consider the $\binom{n}{d}$-dimensional space $\mathbb{R}^{\binom{n}{d}}$ whose coordinates $q_{i_1 \cdots i_d}$ are indexed by d-element subsets $\{i_1, \ldots, i_d\}$ of $\{1, 2, \ldots, n\}$. Let S be any $(d-2)$-element subset of $\{1, 2, \ldots, n\}$ and let i, j, k and l be any four distinct indices in $\{1, \ldots, n\} \backslash S$. The corresponding *three-term Grassmann–Plücker relation* $g_{S,ijkl}$ is the following tropical polynomial of degree two:

$$g_{S,ijkl} \quad = \quad q_{Sij} \odot q_{Skl} \ \oplus \ q_{Sik} \odot q_{Sjl} \ \oplus \ q_{Sil} \odot q_{Sjk}. \tag{3.41}$$

We define the *space of d-trees* to be the intersection of these hypersurfaces,

$$\mathcal{T}_{d,n} \quad := \quad \bigcap_{S,i,j,k,l} \mathcal{T}(g_{S,ijkl}) \quad \subset \quad \mathbb{R}^{\binom{n}{d}}, \tag{3.42}$$

where the intersection is over all S, i, j, k, l as above. If $d = 2$ then $S = \emptyset$, the polynomial (3.41) is the four-point condition (3.39), and $\mathcal{T}_{2,n}$ is the space of trees $\mathcal{T}_n = \mathcal{T}(I_{2,n})$. For $d \geq 3$, the tropical Grassmannian $\mathcal{T}(I_{d,n})$ is contained in the space of d-trees $\mathcal{T}_{d,n}$, and this containment is proper for $n \geq d + 4$. However, $\mathcal{T}_{d,n}$ is a good combinatorial approximation for $\mathcal{T}(I_{d,n})$.

The points $Q = (q_{i_1 \cdots i_d})$ in $\mathcal{T}_{d,n} \subset \mathbb{R}^{\binom{n}{d}}$ are called *d-trees*. Fix a d-tree Q. For any $(d+1)$-subset $\{j_0, j_1, \ldots, j_d\}$ of $\{1, 2, \ldots, n\}$ we consider the hyperplane specified by the following tropical linear form in the unknowns $x_1, \ldots, x_n$:

$$\ell^Q_{j_0 j_1 \cdots j_d} \quad = \quad \bigoplus_{r=0}^{d} q_{j_0 \cdots \hat{j_r} \cdots j_d} \odot x_r. \tag{3.43}$$

The *tropical linear space* associated with the d-tree Q is the intersection

$$L_Q \quad = \quad \bigcap \mathcal{T}(\ell^Q_{j_0 j_1 \cdots j_n}) \quad \subset \quad \mathbb{R}^n. \tag{3.44}$$

Here the intersection is over all $(d+1)$-subsets $\{j_0, j_1, \ldots, j_d\}$ of $\{1, 2, \ldots, n\}$. The "sufficient completeness" referred to above means that we need to solve linear equations using *Cramer's rule*, in all possible ways, in order for the intersection of hyperplanes to be a linear space. The definition of linear space given here is more inclusive than the notion one would get by tropicalizing linear spaces over the field $\mathbb{Q}(\epsilon)$. The latter are the tropical linear spaces L_Q where Q is any point in the subset $\mathcal{T}(I_{d,n})$ of $TG_{d,n}$. [Speyer, 2004] proved that all tropical linear spaces L_Q are pure-dimensional polyhedral fans.

Theorem 3.47 (Speyer's Theorem) *Let $Q \in \mathcal{T}_{d,n}$ be a d-tree. Then every maximal cone of the tropical linear space L_Q is d-dimensional.*

Tropical linear spaces have many of the properties of ordinary linear spaces. First, they have the correct dimension d. Second, every tropical linear space L_Q determines its vector of tropical Plücker coordinates Q uniquely up to tropical multiplication (= classical addition) by a common scalar. If L and L' are tropical linear spaces of dimensions d and d' with $d + d' \geq n$, then L and L' meet. It is not quite true that two tropical linear spaces intersect in a tropical linear space but it is almost so. If L and L' are tropical linear spaces of dimensions d and d' with $d + d' \geq n$ and $v \in \mathbb{R}^n$ is generic then $L \cap (L' + v)$ is a tropical linear space of dimension $d + d' - n$. One then defines the *stable intersection* of L and L' by taking the limit of $L \cap (L' + v)$ as v goes to zero.

Not every d-dimensional tropical linear space in $\mathbb{R}^n$ is the intersection of $n - d$ tropical hyperplanes. It is an open problem to determine the minimum number of tropical hyperplanes that are sufficient to cut out any tropical linear space of dimension d in $\mathbb{R}^n$. From (3.44) we see that $\binom{n}{d+1}$ hyperplanes suffice.

Theorem 3.47 is relevant for phylogenetics because tropical linear spaces can be regarded as "higher-dimensional phylogenetic trees". Indeed, suppose we have n taxa and we are given a dissimilarity measurement $-q_{i_1 i_2 \cdots i_d}$ for any d-tuple $\{i_1, i_2, \ldots, i_d\}$ of taxa in $[n]$. These $\binom{n}{d}$ real numbers form a d-dimensional dissimilarity matrix $Q = (q_{i_1 i_2 \cdots i_d}) \in \mathbb{R}^{\binom{n}{d}}$. Such a dissimilarity matrix Q is the input for the generalized neighbor-joining algorithm which is derived from Theorem 2.41. See Chapter 18.

The tropical linear space L_Q is a geometric model which plays the role of the tree for the data Q. Indeed, passing from $\mathbb{R}^n$ to the $(n-1)$-dimensional *tropical projective space* $\mathbb{R}^n / \mathbb{R}(1, 1, \ldots, 1)$, the tropical linear space L_Q is a contractible polyhedral complex of pure dimension $d - 1$. In the classical case $d = 2$, the linear space L_Q is a pure-dimensional contractible polyhedral complex of dimension 1, namely, it is precisely the tree with tree metric $-Q$.

Example 3.48 Fix $d = 3$ and $n = 6$. The *dissimilarity* of any triple $\{i, j, k\}$ of taxa in $[6] = \{1, 2, 3, 4, 5, 6\}$ is denoted by d_{ijk}, and we set $q_{ijk} = -d_{ijk}$. A point $Q = (q_{ijk}) \in \mathbb{R}^{20}$ is a 3-tree if and only if the map $(i, j) \mapsto d_{ijk}$ is a tree metric on $[6] \backslash \{k\}$ for all $k \in [6]$. Suppose this holds. Then the intersection of the $\binom{6}{4} = 15$ tropical hyperplanes $\mathcal{T}(\ell^Q_{j_0 j_1 j_2 j_3})$, is a 3-dimensional tropical linear subspace $L_Q \subset \mathbb{R}^6$. Each of the 15 defining linear forms has four terms:

$$\ell^Q_{j_0 j_1 j_2 j_3} = q_{j_0 j_1 j_2} \odot x_{j_3} \oplus q_{j_0 j_1 j_3} \odot x_{j_2} \oplus q_{j_0 j_2 j_3} \odot x_{j_1} \oplus q_{j_1 j_2 j_3} \odot x_{j_0}.$$

If we work in tropical projective 5-space, i.e. modulo the equivalence relation

$$(x_1, x_2, x_3, x_4, x_5, x_6) \quad \equiv \quad \lambda \odot (x_1, x_2, x_3, x_4, x_5, x_6),$$

then L_Q is a union of planar polygons. We call L_Q a *phylogenetic surface*.

A phylogenetic surface is a two-dimensional geometric representation of the dissimilarities among triples of taxa, just like a phylogenetic tree is a two-dimensional geometric representation of the dissimilarities among pairs of taxa. Embedded in the unbounded part of the phylogenetic surface L_Q, we find the six phylogenetic trees representing the tree metrics $(i, j) \mapsto d_{ijk}$ for fixed k.

There are 1035 combinatorial types of phylogenetic surfaces on six taxa [Speyer and Sturmfels, 2004]. They correspond to the maximal cones of the tropical Grassmannian $\mathcal{T}(I_{3,6}) = \mathcal{T}_{3,6}$, just like the 15 binary trees on five taxa correspond to the edges of the Petersen graph $\mathcal{T}_{2,5}$. If we replace $\mathbb{R}^{20}$ by its quotient modulo the subspace of dissimilarity maps of the particular form $d_{ijk} = \omega_i + \omega_j + \omega_k$, then $\mathcal{T}_{3,6}$ is a three-dimensional simplicial complex consisting of 65 vertices, 550 edges, 1395 triangles and 1035 tetrahedra. For the topologically inclined, we note that the space $\mathcal{T}_{3,6}$ of phylogenetic surfaces (i.e., the space of 3-trees on 6 taxa) is a bouquet of 126 three-dimensional spheres, just like the Petersen graph is a bouquet of 6 one-dimensional spheres.   □

We next discuss the tropical variety of the Pfaffian ideal $I_{2,n,k}$ and we offer a phylogenetic interpretation which generalizes the space of trees $\mathcal{T}_n$ (the special case $k = 2$). Suppose that $D^{(1)}, \ldots, D^{(r)}$ are $n \times n$-matrices which represent metrics. We define a new metric, denoted $D^{(1)} \vee \cdots \vee D^{(r)}$ and called the *mixture* of the given metrics, by taking the maximum distance for each pair:

$$(D^{(1)} \vee \cdots \vee D^{(r)})_{ij} \quad := \quad \max(D_{ij}^{(1)}, \ldots, D_{ij}^{(r)}).$$

Equivalently, using tropical matrix addition, the mixture of the r metrics is

$$D^{(1)} \vee \cdots \vee D^{(r)} \quad := \quad -\big((-D^{(1)}) \oplus \cdots \oplus (-D^{(r)})\big) \qquad (3.45)$$

The term "mixture" conveys the idea that each metric $D^{(\nu)}$ corresponds to a random variable $X^{(\nu)}$ on the pairs of taxa with probability distribution

$$\mathrm{Prob}(X^{(\nu)} = \{i, j\}) \quad \sim \quad \exp(-\tau D_{ij}^{(\nu)}).$$

Consider a mixture of these r random variables where $\tau \gg 0$ and the mixing probabilities $p_1, \ldots, p_r$ are positive. Then the mixed distribution satisfies

$$\mathrm{Prob}(X^{(\nu)} = \{i, j\}) \quad \sim \quad \sum_{\nu=1}^{r} p_\nu \cdot \exp(-\tau D_{ij}^{(\nu)}) \quad \sim \quad \exp\big(-\tau (D^{(1)} \oplus \cdots \oplus D^{(r)})_{ij}\big),$$

Thus defining the mixture of metrics as their sum in *max-plus-algebra* is a natural thing to do in the context of tropical geometry of statistical models.

We say that a metric D has *tree rank* $\leq r$ if there exist tree metrics $D^{(1)}$, $\ldots, D^{(r)}$ such that $D = D^{(1)} \vee \cdots \vee D^{(r)}$. Let $\mathcal{T}_n^r$ denote the subset of $\mathbb{R}^{\binom{n}{2}}$ consisting of all metrics of tree rank $\leq r$. This is a polyhedral fan, generalizing the space of trees (the case $r = 1 = k - 1$). We propose the following problem: *characterize membership in* T_n^r *and study the structure of this space.*

This problem may be relevant for the following issue in comparative genomics. If we are given an alignment of genomes then different regions (e.g. different genes) may give rise to different trees. It is desirable to create some consensus among the conflicting tree metrics. The resulting consensus object may no longer be tree-like, for instance, if we apply the refined techniques of Chapter 17. Mixtures of tree metrics may be useful models for such situations.

Fix a metric D and consider an even subset $\{i_1, \ldots, i_{2m}\}$ of $\{1, 2, \ldots, n\}$. This subset defines a complete graph K_{2m} with edge weights $d_{i_j i_k}$. A *matching* is a 1-regular subgraph of K_{2m}. The *weight* of a matching is the sum of the weights of its m edges. We are interested in the condition on D that each complete subgroup K_{2m} has more than one matching of maximum weight.

Proposition 3.49 *If a metric D has tree rank $\leq r$ then for every subset of $2r + 2$ taxa, the maximum matching among these taxa is not unique.*

For $r = 1$ this is precisely the four-point condition. In view of (3.16), we can rephrase Proposition 3.49 by tropicalizing the Pfaffians of order $2r + 2$. For instance, for $r = 2$, tropicalizing (3.17) yields the tropical 6×6-Pfaffian

$$q_{14} \odot q_{25} \odot q_{36} \ \oplus \ q_{15} \odot q_{24} \odot q_{36} \ \oplus \ q_{14} \odot q_{26} \odot q_{35} \ \oplus \ q_{15} \odot q_{26} \odot q_{34}$$
$$\oplus \ q_{16} \odot q_{24} \odot q_{35} \ \oplus \ q_{16} \odot q_{25} \odot q_{34} \ \oplus \ q_{13} \odot q_{26} \odot q_{45} \ \oplus \ q_{12} \odot q_{36} \odot q_{45}$$
$$\oplus \ q_{16} \odot q_{23} \odot q_{45} \ \oplus \ q_{13} \odot q_{25} \odot q_{46} \ \oplus \ q_{12} \odot q_{35} \odot q_{46} \ \oplus \ q_{15} \odot q_{23} \odot q_{46}$$
$$\oplus \ q_{13} \odot q_{24} \odot q_{56} \ \oplus \ q_{12} \odot q_{34} \odot q_{56} \ \oplus \ q_{14} \odot q_{23} \odot q_{56}.$$

Evaluating this tropical polynomial means finding the minimum weight matching in the complete graph K_6. We see that Proposition 3.49 is equivalent to

Proposition 3.50 *If a metric D has tree rank $\leq r$ then $-D$ lies in the intersection of the tropical hypersurfaces defined by the subpfaffians of order $2r + 2$.*

Proof Theorem 3.45 implies that, for each $i \in \{1, 2, \ldots, r\}$, there exists a skew-symmetric $n \times n$-matrix $P^{(i)}$ over the field $\mathbb{Q}(\epsilon)$ such that $P^{(i)}$ has rank 2 and $\text{order}(P^{(i)}) = -D^{(i)}$. These matrices can be chosen so that there is no cancellation of leading terms when forming the sum $P := P^{(1)} + \cdots + P^{(r)}$. Then $\text{order}(P) = -D$ and $\text{rank}(P) \leq 2r$. Every Pfaffian of order $2r + 2$ vanishes for P. Hence $-D$ lies on these tropical Pfaffian hypersurfaces. □

This proof shows how algebraic geometry in conjunction with tropicalization can suggest combinatorial constructions which may be useful for phylogenetics. Note that we are not claiming that algebraic geometry is needed for the proof; indeed, it is easy to prove Proposition 3.49 without algebra.

Unfortunately, our necessary condition for membership in T_n^r is not sufficient. The following counterexample for $n = 6, r = 2$ is due to David Bryant.

Example 3.51 Consider the path metric of K_6 with a 6-cycle removed:

$$D \quad = \quad \begin{pmatrix} 0 & 2 & 1 & 1 & 1 & 2 \\ 2 & 0 & 2 & 1 & 1 & 1 \\ 1 & 2 & 0 & 2 & 1 & 1 \\ 1 & 1 & 2 & 0 & 2 & 1 \\ 1 & 1 & 1 & 2 & 0 & 2 \\ 2 & 1 & 1 & 1 & 2 & 0 \end{pmatrix}$$

The maximum matching is attained twice; that is, $-D$ lies in the tropical hypersurface of the 6×6-Pfaffian. By examining all possible cases, one sees that D cannot be written as the mixture $D^{(1)} \vee D^{(2)}$ of two tree metrics. The tree rank of D is 3. Thus the converse to Proposition 3.50 does not hold. □

At this point, it is natural to make the following conjecture: *If every restriction of a matrix D to six points is the mixture of two trees then D is a mixture of two trees.* Or, more generally: *If every restriction of D to $2r + 2$ points is a mixture of r trees then D is a mixture of r trees.*

4

Biology

Lior Pachter

Bernd Sturmfels

This chapter describes genome sequence data and explains the relevance of the statistics, computation and algebra that we have discussed in Chapters 1–3 to understanding the function of genomes and their evolution. It sets the stage for the studies in biological sequence analysis in some of the later chapters.

Given that quantitative methods play an increasingly important role in many different aspects of biology, the question arises: why the emphasis on genome sequences? The most significant answer is that genomes are fundamental objects that carry instructions for the self-assembly of living organisms. Ultimately, our understanding of human biology will be based on an understanding of the organization and function of our genome. Another reason to focus on genomes is the abundance of high fidelity data. Current finished genome sequences have less than one error in 10,000 bases. Statistical methods can therefore be directly applied to modeling the random evolution of genomes and to making inferences about the structure and organization of functional elements; there is no need to worry about extracting signal from noisy data. Furthermore, it is possible to validate findings with laboratory experiments.

The rate of accumulation of genome sequence data has been extraordinary, far outpacing Moore's law for the increasing density of transistors on circuit chips. This is due to breakthroughs in sequencing technologies and radical advances in automation. Since the first completion of the genome of a free living organism in 1995 (*Haemophilus Influenza* [Fleischmann *et al.*, 1995]), biologists have completely sequenced over 200 microbial genomes, and dozens of complete invertebrate and vertebrate genomes. The highlight of the sequencing projects, from our *Homo sapiens* perspective, is the completion of the sequencing of the human genome, which was formally announced at the end of 2004 [Human Genome Sequencing Consortium, 2004]. Our discussion of online resources in Section 4.2 explains how to read the human genome.

In Section 4.4 we revisit hidden Markov models, and we show how these models can be used for identifying genes. Section 4.5 is concerned with statistical models for the evolution of DNA sequences. At the very end, in Example 4.29, we return to DiaNA, our fictional character on the book cover, and we explain the meaning of the cartoon in terms of comparative genomics.

4.1 Genomes

Every living organism has a genome, made up of deoxyribonucleic acids (DNA) arranged in a double helix [Watson and Crick, 1953], which encodes (in a way to be made precise) the fundamental ingredients of life. Organisms are divided into two major classes: *eukaryotes* (organisms whose cells contain nuclei, such as animals and plants) and *prokaryotes* (organisms whose cells don't contain nuclei, such as bacteria). In this book we focus on eukaryotic genomes, and, in particular, on the genomes of vertebrates. The primary example is the human genome [Human Genome Sequencing Consortium, 2004, Venter *et al.*, 2001]. This allows for the description of ongoing genome projects at the forefront of current research interests, while limiting the scope so that some detail can be provided on how to obtain and utilize the data.

Eukaryotic genomes are divided into separate molecules of DNA called *chromosomes*. In *diploid* cells, there are two copies of each chromosome. Humans have 23 pairs of chromosomes: 22 *autosomes* (two copies each in both men and women) and two *sex chromosomes*, which are denoted X and Y. Women have two X chromosomes, while men have one X and one Y chromosome. This means that there are four parental chromosomes that contribute to each chromosome in a child: two from each parent. Each chromosome in the child is a patchwork of stretches of DNA coming from the four different parent chromosomes. Theoretical aspects of genetic inheritance are studied in the well-established field of *statistical genetics* [Balding *et al.*, 2003, Hartwell *et al.*, 2003]. A connection between genetics and algebraic statistics was recently explored in [Hallgrímsdóttir and Sturmfels, 2005].

The DNA molecules in a genome are typically represented as a number of sequences (one for each chromosome) of letters from the four letter alphabet $\Sigma = \{A, C, G, T\}$. These letters correspond to the bases in the double helix, that is, the *nucleotides* Adenine, Cytosine, Guanine and Thymine. The four nucleotides fall into two pairs: *purines* (A and G) and *pyrimidines* (C and T). This grouping reflects the chemistry of the nucleotides: the purines have two rings in their structure while pyrimidines have only one. In addition to this grouping, every DNA base in the double helix is paired with a *complementary* base on the opposite helix: A is paired with T, and C with G, with hydrogen bonding serving as the main force holding the two separate chains together. Figure 4.1 illustrates these structural features of the nucleotides.

Remark 4.1 There is a natural labeling of the bases $\{A, C, G, T\}$ by the elements of the group $\mathbb{Z}_2 \times \mathbb{Z}_2$ that reflects the pyrimidine/purine dichotomy and base complementarity. This structure is the rationale behind *group-based evolutionary models* which are discussed in Section 4.4 and Chapters 15–17.

One consequence of DNA complementarity is *Chargaff's rule*, which states that every DNA sample contains the same number of A's as T's, and the same number of G's as C's [Chargaff, 1950]. A further consequence is that it suffices to list the bases in only one strand of the double helix. It is important to note

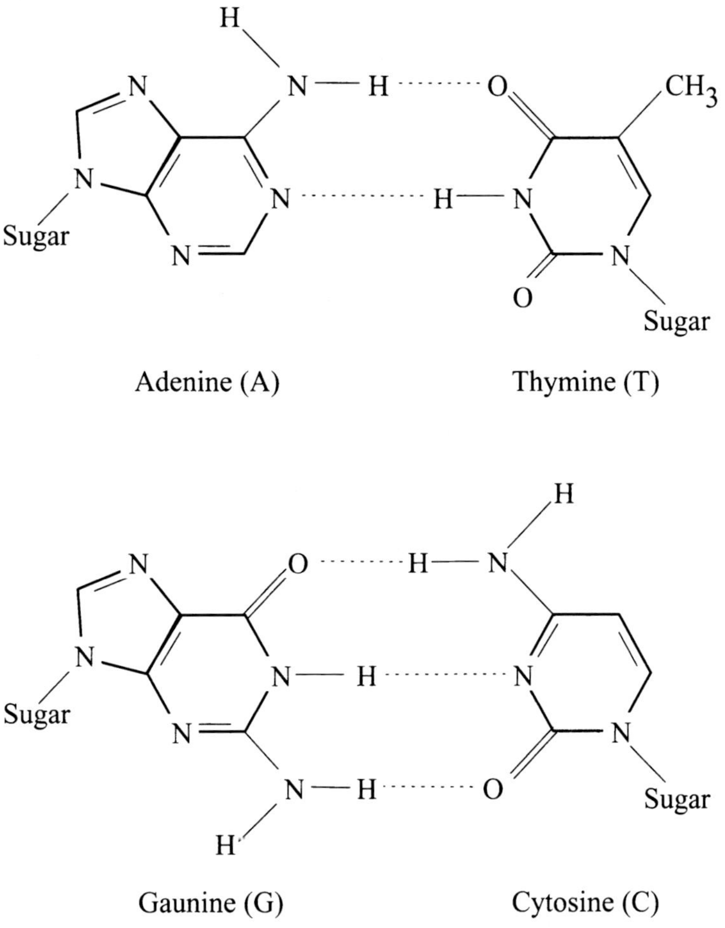

Adenine (A)   Thymine (T)

Gaunine (G)   Cytosine (C)

Fig. 4.1. The four DNA bases shown with complementary bases paired.

that each strand has a directionality that is determined by an asymmetrical arrangement of carbon atoms within the helix backbone. The asymmetry is indicated by writing the numbers $5'$ and $3'$ at the ends of the sequence. The convention is to write a single strand of DNA bases in the $5' \to 3'$ direction.

Example 4.2 The DNA sequence `GATATAGAGCGGATTACAG` of length 20 is shorthand for the double stranded sequence consisting of twenty *base pairs*

$$5' \text{ GATATCAGAGCGGATTACAG } 3'$$
$$3' \text{ CTATAGTCTCGCCTAATGTC } 5'$$

which, in turn, is shorthand for the bases along the DNA double helix. □

The human genome consists of approximately 2.8 billion base pairs. Its sequence has been obtained using high throughput sequencing technologies which allow for reading short fragments only hundreds of bases long. *Sequence assembly algorithms* are then necessary for piecing together the fragments [Myers, 1999].

Although we tend to abstract genomes as strings over the alphabet Σ, one

	T	C	A	G
T	TTT $\mapsto$ Phe TTC $\mapsto$ Phe TTA $\mapsto$ Leu TTG $\mapsto$ Leu	TCT $\mapsto$ Ser TCC $\mapsto$ Ser TCA $\mapsto$ Ser TCG $\mapsto$ Ser	TAT $\mapsto$ Tyr TAC $\mapsto$ Tyr TAA $\mapsto$ *stop* TAG $\mapsto$ *stop*	TGT $\mapsto$ Cys TGC $\mapsto$ Cys TGA $\mapsto$ *stop* TGG $\mapsto$ Trp
C	CTT $\mapsto$ Leu CTC $\mapsto$ Leu CTA $\mapsto$ Leu CTG $\mapsto$ Leu	CCT $\mapsto$ Pro CCC $\mapsto$ Pro CCA $\mapsto$ Pro CCG $\mapsto$ Pro	CAT $\mapsto$ His CAC $\mapsto$ His CAA $\mapsto$ Gln CAG $\mapsto$ Gln	CGT $\mapsto$ Arg CGC $\mapsto$ Arg CGA $\mapsto$ Arg CGG $\mapsto$ Arg
A	ATT $\mapsto$ Ile ATC $\mapsto$ Ile ATA $\mapsto$ Ile ATG $\mapsto$ Met	ACT $\mapsto$ Thr ACC $\mapsto$ Thr ACA $\mapsto$ Thr ACG $\mapsto$ Thr	AAT $\mapsto$ Asn AAC $\mapsto$ Asn AAA $\mapsto$ Lys AAG $\mapsto$ Lys	AGT $\mapsto$ Ser AGC $\mapsto$ Ser AGA $\mapsto$ Arg AGG $\mapsto$ Arg
G	GTT $\mapsto$ Val GTC $\mapsto$ Val GTA $\mapsto$ Val GTG $\mapsto$ Val	GCT $\mapsto$ Ala GCC $\mapsto$ Ala GCA $\mapsto$ Ala GCG $\mapsto$ Ala	GAT $\mapsto$ Asp GAC $\mapsto$ Asp GAA $\mapsto$ Glu GAG $\mapsto$ Glu	GGT $\mapsto$ Gly GGC $\mapsto$ Gly GGA $\mapsto$ Gly GGG $\mapsto$ Gly

Table 4.1. *The genetic code.*

must not forget that they are highly structured. Certain substrings within a genome correspond to *genes*. These play the important role of encoding *proteins*. Proteins are polymers made of twenty different types of amino acids, which are encoded by triplets of bases known as *codons*. Thus, there are 64 codons: AAA, AAC, AAG, . . . , TTT. Each triplet codes for one amino acid, so that a DNA substring of length $3k$ can code for a protein with k amino acids.

The code mapping DNA triplets to amino acids is known as the *genetic code*. Table 4.1 displays the genetic code, which maps the 64 possible codons to the twenty amino acids they code for. Each amino acid is represented by a three-letter identifier ("Phe" = Phenylalanine, "Leu" = Leucin, . . .). The genetic code is translated by ribosomes (which are partially made of proteins) that build a protein from the linear sequence of a gene. The three codons TAA, TAG and TGA are special: instead of coding for an amino acid, they are used to signal that translation should end. They are called *stop codons*.

Example 4.3 (Codon usage, GC content and genome signatures)
Codon usage refers to the relative abundances of the different codons in a genome. Codon usage varies widely between genomes, and can be used to distinguish prokaryotic genomes from each other [Campbell *et al.*, 1999], and similarly for eukaryotic genomes [Gentles and Karlin, 2001]. This is in contrast to the genetic code, which is almost universal (a list of exceptions is maintained at the National Center for Biotechnology Information (NCBI) website; see Section 4.2). Part of the difference in codon usage stems from different GC content in genomes (the term GC content means the fraction of the genome that consists of G or C). The G and C nucleotides are known to be involved in a number of genome regulation mechanisms. For example, CpG *sites* are locations in DNA sequences where a C is adjacent and upstream of a G. DNA methyltransferase recognizes CpG sites and converts the cytosine into 5-methylcytosine. Spontaneous deamination causes the 5-methylcytosine to be converted into thymine,

and the mutation is not mended by DNA repair mechanisms. This results in a gradual erosion of CpG sites in the genome. However, at sites within promoter regions of genes (Section 4.3), the mutations are repaired, leading to restored, unmethylated CpG sites. Regions with many of these sites are known as CpG islands, but they alone do not explain the vast differences in GC content seen between genomes.

A model for distinguishing organisms based on their genome composition has been proposed by Samuel Karlin and his collaborators in the Stanford mathematics department. The data consists of 16 numbers u_{ij}, $i, j \in \Sigma = \{A, C, G, T\}$, where u_{ij} counts the number of times that the nucleotides i and j appear consecutively in that order in a genome. Following [Campbell *et al.*, 1999], the *genome signature* of these data is the 4×4-matrix whose entry in row i and column j equals $u_{ij} u_{++} / u_{i+} u_{+j}$. Thus the genome signature is the coordinate-wise ratio of the empirical distribution and the maximum likelihood estimate in the independence model (Proposition 1.13).

A similar study undertaken at IHES in France, has led to *the mystery of the two straight lines* [Gorban and Zinovyev, 2004]. Given a coding DNA sequence of length $3k$, the data consists of the 64 numbers u_{ijk}, which count the number of occurrences of each codon $ijk \in \Sigma^3$. In an analysis of 143 fully sequenced bacterial genomes, it was found that the resulting 143 points in the 63-dimensional simplex are distributed along two cubic curves, one for eubacterial genomes and one for archaeal genomes. Replacing each data point by its maximum likelihood estimate in the independence model $p_{ijk} = \alpha_i \beta_j \gamma_k$, the cubic curves become straight lines in the 9-dimensional parameter space with coordinates (α, β, γ). $\square$

In order to make protein, DNA is first copied into a similar molecule called RNA. This process is called *transcription*. It is the RNA that is *translated* into protein. The link between DNA, RNA, and protein is the basis of molecular biology, and is sometimes referred to as the *central dogma*.

When protein is created from RNA, the gene that has been translated is said to have been *expressed*. Proteins can form structures, or perform complex tasks (such as regulation of expression) by interacting with the many molecules and complexes in cells. Thus, the genome is a blueprint for life. A major goal in biology, to be discussed in Section 4.3, is a complete understanding of the genes, the function of their proteins, and their expression patterns.

The human genome contains approximately 25,000 genes, although the exact number is still not known [Human Genome Sequencing Consortium, 2004]. While there are experimental methods for discovering and validating genes, and even high throughput technologies for finding gene fragments, there is still no technology for experimentally detecting all the genes in the genome. The computational problem of identifying genes, the *gene-finding problem*, is an active area of research [Dewey *et al.*, 2004, Korf *et al.*, 2001]. One of the main difficulties lies in the fact that only a small portion of the genome is genic; in fact, less than 5% of the genome is known to be coding. In Section 4.4 we discuss this problem, and the role of statistical models in formulating sound methods for distinguishing genes from non-genic sequences. The models of choice are the *hidden Markov models* whose mathematical characterizations were discussed in Section 1.4. Hidden Markov models allow for the integration of diverse biological information (such as the genetic code and the structure of genes) and are suitable for designing efficient algorithms. In spite of much progress, the current understanding of genes is not sufficient to allow for the *ab initio* identification of all the genes in a genome [Guigó *et al.*, 2004].

A key idea in biology has been that the comparison of multiple genome sequences can assist in identifying genes and other functional elements. The underlying premise of the *comparative genomics* approach is that although DNA sequences change over time, functional elements, such as genes, will tend to be conserved due to their critical role in coding for proteins or other important elements. The comparative genomics approach therefore seeks to utilize consequences of Darwin's principle of *natural selection* to sift through genome sequences for functional elements. The principle has been applied to collections of similar genomes [Boffelli *et al.*, 2003], as well as more divergent sequences [Waterston *et al.*, 2002, Dermitzakis *et al.*, 2003, Hillier *et al.*, 2004].

The different types of comparisons require an understanding of the underlying biology. For example, differences between the genomes of individuals in a population are small and are primarily due to recombination events (the process by which two copies of parental chromosomes are merged in the offspring). On the other hand, the genomes of different species tend to be much more diverse. Genome differences between species result from numerous transformations:

- *Genome rearrangement* – Comparing chromosomes of related species reveals large segments that have been reversed and flipped (*inversions*), segments that have been moved (*transpositions*), *fusions* of chromosomes, and other large scale events. Methods of discrete mathematics have led to significant progress in this field [Hannenhalli and Pevzner, 1999, Tesler, 2002], but the biological mechanisms are poorly understood [Sankoff and Nadeau, 2003].

- *Duplications and loss* – Some genomes have undergone whole genome duplications. This was recently demonstrated for yeast [Kellis *et al.*, 2004]. Individual chromosomes or genes may also be duplicated. Duplication events are often accompanied by *gene loss*, as redundant genes slowly lose or adapt their function [Eichler and Sankoff, 2003].

- *Parasitic expansion* – Large sections of genomes are repetitive, consisting of elements which have duplicated and re-integrated themselves [Brown, 2002].

- *Point mutation, insertion and deletion* – DNA sequences mutate, and, in non-functional regions, these mutations accumulate over time. Such regions are also likely to exhibit deletions; for example, strand slippage during replication can lead to an incorrect number of copies of repeated bases.

Biological questions about how these mechanisms operate lead directly to mathematical problems.

Example 4.4 (Sorting by reversals) Comparison of the X chromosome in humans, mice and rats reveals large segments within which the order and orientation of genes is conserved. For example, the human X chromosome can be divided into 16 segments labeled consecutively $1, \ldots, 16$, which appear in different orders and orientations in the mouse and rat genomes, but within which order and orientation are preserved (not counting rearrangements less than 300kb in size) [Gibbs *et al.*, 2004]. The changes in the mouse and rat can be recorded by *signed permutations*, i.e., elements of the Weyl group B_{16}. From [Bourque *et al.*, 2004] we have:

```
Human   1    2 3   4   5    6  7  8  9   10   11 12 13   14   15   16
Mouse  -5   -6 4  13  14  -15 16  1 -3   9  -10 11 12 -7    8   -2
Rat   -13  -4 5  -6 -12  -8 -7  2  1 -3   9  10 11   14 -15   16
```

Inversions correspond to *reversals* of the signed permutations. By this we mean selecting a subsequence of a signed permutation and reversing the order of the numbers and their sign. For example, a reversal in the mouse could be

```
Mouse -5 -6 4 13 14 -15 16  1 -3   9 -10 11 12 -7 8 -2
                     <------->
      -5 -6 4 13 14 -15  3 -1 -16   9 -10 11 12 -7 8 -2
```

An important genomics problem is to identify the order of genes in the ancestral chromosome, so that the number of rearrangements that have occurred over time can be counted. In [Tesler, 2002], it is shown that the distance between two multichromosomal genomes, defined as the minimum number of reversals, translocations, fissions and fusions required to transform one genome into the other, can be computed in time polynomial in the number of features being considered. Genome rearrangements are important to study because they shed light on genome evolution, and also because many diseases are known to be associated with genome rearrangement (e.g., [Strachan and Read, 2004, Raphael and Pevzner, 2004]). □

The problem of untangling the evolutionary history of genomes is complicated, and statistical methods are well-suited for modeling the different events, many of which are inherently random. Some of the connections between statistics and evolutionary models are discussed in Section 4.5.

Two distinct DNA bases that share a common ancestor are called *homologous*. Homologous bases can be related via speciation and duplication events, and are therefore divided into two classes: *paralogous* and *orthologous*. Orthologous bases are descended from a single base in an ancestral genome that underwent a speciation event, whereas paralogous bases are related via duplication. Because we cannot sequence ancestral genomes, it is never possible to prove formally that two DNA bases are homologous. However, statistical arguments can show that it is extremely likely that two bases are homologous, or even orthologous (see Chapter 22). The problem of identifying homologous bases in genomes of related species is known as the *alignment problem*. The statistical model of choice for alignment is the pair hidden Markov model. The

algebraic representation of this model and its tropicalization, which underlies
the Needleman–Wunsch algorithm, were discussed in Section 2.2.

4.2 The data

Biology is a data-driven science. This means that progress in the field is the
result of analyzing data obtained from experiments. The experiments are performed in individual laboratories, or via large-scale collaborations utilizing high
throughput technologies. In most subjects data is not usually distributed before publication, but one of the attractive aspects of genomics is the widespread
availability of large amounts of high quality genome sequence data. In fact,
many publicly funded projects are required to distribute their data through
publicly available websites within hours of sequencing. The *Fort Lauderdale
agreement*, a result of extensive discussions in 2003 among sequence providers
and sequence analyzers, provides guidance from the NIH on how and when
to publish results from genome analysis. The document can be viewed at
`http://www.genome.gov/10506537`. Researchers are generally free to publish
results derived from publicly posted genomes, and, in return, publication rights
for whole genome analyses on a newly sequenced genome are reserved for those
who sequenced the genome.

   In this section we describe some of the data that is available for analysis,
and explain how to download it from publicly accessible websites.

4.2.1 Sequence data

Most genomes today are sequenced using the *whole genome shotgun* strategy.
This strategy is based on two high throughput technologies. The first is re-
combinant DNA technology which allows for the construction of *libraries*, so
called because they consist of large pieces of DNA from a genome and can be
stored in a freezer. A library is made by shearing multiple copies of a genome
and inserting the pieces into simple replicating molecules (or organisms) called
vectors. The inserted pieces are called *inserts*. The inserts can vary in size
depending on the vector: libraries may consist of inserts ranging in size from
2 kilobases up to hundreds of kilobases long. The second important technology is high throughput sequencing, which allows for the rapid sequencing of
DNA at the ends of the inserts. The pieces that can be sequenced accurately
are typically about 500–700 base pairs long, and are called *reads*. Note that
each read comes with a mate, namely the sequence that was obtained from the
opposite end of the same insert.

Example 4.5 (Hedgehog) The following is a read from the genome of the
lesser Madagascar hedgehog (*Echinops telfairi*), sequenced in February 2005:

```
>gnl|ti|643153582 name:G753P82FG11.T0 mate:643161057
 mate_name:G753P82RG11.T0 template:G753P82G11 end:R
  TAATGAGTGGGGCGAAAGAATCGGCTCCGGTGATTCATCACTTGGCTGACCCAGGCCTGA
```

```
CCCAACCCATGGAATTGTCAAGTGCCTCGTATGCATGTGGAAGTTGGACATTGATTAAGA
AGACCAAAGAAGAATCTATGTGTTTTATTTGTGGTGCTAGAGAAGTACCTTGGACTGATA
AAAAGACAAACCAAACTGTATTGGACGAAGTAAGGCTTCTTGGAGGCAAGGATAGGAAGA
CTTTGTCTCACATACTTTGGACATATTGTCAGGACAGACCAGTCCCTGGCGAAGGACATC
ATGCTTGGTCAAGTGGAGGGGCAGTGGAAAAGAGGAAGGCGCTTAATGAGATGGATGGAT
ACAATTGCTACAATAATGGACCCAGGCATGGAAAAAAATTAAGTTTGTCACAGGACTGGG
CAGTGTTTCCTTTTGTTGTGCACAGGGTTGCTATGGGTCGGCACAGACTCAATGGCTTCA
AACAACAATAACAACAATCTAGTGATCCCAATAGTCAGCCTTTTATTTTTTCTCCCCCAA
GAAGAAAATATAATGGAGAAATTACATTCTGCTTTCATATTGAGGAAGAGAATTATGTTC
CTAATTGACCTATCATTGGCCCAGGATCCTGGATCTTCAACCCTAGTTTTTAGTGAAAGC
GTATGCTGAACTATTGTCTCCTGCATGGCATCTTCCACCCAGTTAGCTCTTGAAATGTTG
GGTTCTCTACATGACCTGATTCCTTCTTCTTCACACCCTAAGTCAAATATACATTGAGTC
CCATCAGTACCATCTCCAAAATACATTACAAATAAGACCATTTATTACCAATGCATTGCT
ATGACTCTAGACCATCTCTTCTCGTACTTGAACAATTGCAACAGCCAGTTCAATGCACCC
AGTACCCCTGTCCTCCACCTCTTCACAGGTCTCTCTATTTACACAATGGCCAAGAAGAGG
AAGAACACTTTTAATATATTGTGTGTCAAACAGCAAAAAACCACACAAC
```

This read was obtained by going to the NCBI trace archive (the raw output of a sequencing machine is called a *trace*), which is located at

`http://www.ncbi.nlm.nih.gov/Traces/trace.cgi`

This website allows visitors to browse recently deposited traces, or to perform advanced searches. Click on the `Obtaining Data` tab to learn more. Examining the read, we see that it is in `FASTA` format (see the discussion of `MAVID` in Section 2.5). The name of the read specifies which one is its mate. □

In whole genome shotgun projects, enough reads are sequenced so that there is considerable overlap among them, allowing them to be merged in order to reconstruct the genome. The problem of reconstructing the genome from the reads is called the *sequence assembly problem*. Its difficulty depends on the amount of sequencing, and also on the repetitive nature of the genome. Reads also contain errors, which complicate matters.

A few definitions are helpful in understanding sequence data, and the quality of assemblies. Reads come equipped with *quality scores*. These quality scores are estimates of the reliability of the bases in a read. The reliability improves with assembly because of the redundancy in the libraries, and the fact that every base in the genome appears in many reads. Therefore, even though sequencing machines may be only 98% accurate, it is possible to correct errors during assembly, and to estimate the uncertainty in bases in assembled genomes. Quality scores are reported on a logarithmic scale, so that if a base has a $1/10^k$ chance of being incorrect, then its quality score is $10k$. Sequencing standards have progressed to the point where quality scores of 40 are the norm. The *coverage* of a whole genome shotgun project is defined to be the average, taken over all bases in the genome, of the number of reads containing each base. For example, 5.1x coverage means that every base in the genome was contained, on average, in 5.1 reads (see Example 4.9). Thus, the coverage of a project is a measure of how much the reads overlap.

Overlapping reads are assembled into *contigs*, and contigs may be linked (by mate pairs) into super-contigs. Contigs are therefore made up of chains of overlapping reads, however the contigs within super-contigs do not overlap.

Example 4.6 (Lander–Waterman model) Making a few simplifying assumptions about sequencing procedures, [Lander and Waterman, 1988] were able to derive formulas for the expected lengths of contigs in an assembly (the original paper relates to clone fingerprinting for physical mapping, but the results apply to whole genome shotgun projects).

Let G be the length (in base pairs) of the genome being sequenced, L the length of a read, and N the number of sequenced reads. Let T be the amount of overlap in base pairs between two reads needed to detect overlap. Set $\sigma = 1 - \frac{T}{L}$ and let c be the coverage $c = \frac{LN}{G}$. The Lander–Waterman model is a Poisson model for the number of times a base is sequenced. Standard results on the Poisson distribution imply the following proposition.

Proposition 4.7 *Assume that the reads are randomly located in the genome.*

(i) *The expected number of contigs is $Ne^{-c\sigma}$.*

(ii) *The expected number of contigs consisting of j reads ($j \geq 1$) is*

$$Ne^{-2c\sigma}(1 - e^{-c\sigma})^{j-1}.$$

(iii) *The expected number of reads in a contig is $e^{c\sigma}$.*

(iv) *The expected length of a contig is*

$$L\left(\frac{e^{c\sigma} - 1}{c} + (1 - \sigma)\right).$$

These formulas can be used to calculate the amount of sequencing that is necessary for different qualities of assembly. □

The quality of an assembly is measured in terms of $N50$ sizes. Let c_k denote the fraction of bases in an assembly that lie in contigs of size at least k. Note that $c_k \geq c_{k+1}$ for all k. The $N50$ *size* of an assembly is the largest k such that $c_k \geq \frac{1}{2}$. $N50$ sizes can also be calculated for super-contigs.

Example 4.8 (Rice genome) The Beijing Institute of Genomics sequenced a cultivar of the *indica* subspecies of rice (*Oryza sativa*) using the whole genome shotgun strategy [Yu *et al.*, 2002]. The original publication describes an assembly from 4.2x coverage, built from $3,565,386$ reads, with reads of length 546 having quality score 20. The N50 contig size was 6.69 kb, and the N50 super-contig size was 362 Mb. Updates to the original assembly and comparisons with other subspecies are reported in [Yu *et al.*, 2005]. □

The whole genome shotgun strategy has certain limitations, one of which is that it is not possible to sequence long, highly repetitive portions of a genome. It is therefore not possible to sequence the *heterochromatin*, which consists of highly condensed, transcriptionally inactive regions of chromosomes that are extremely repetitive. In fact, there is no existing technology for sequencing this DNA, and it is therefore impossible to completely finish sequencing any vertebrate genome at the current time. Nevertheless, the term *finished* has come to mean a genome whose *euchromatin* (which consists of DNA that is not part of the heterochromatic portions of chromosomes) has been sequenced and assembled into high quality long contigs.

Finished genomes are useful for a number of reasons. For example, the absence in one species of a subsequence that exists in other organisms can be certified and investigated for biological relevance, something that is not possible with a poor assembly. Furthermore, continuity of the sequence allows for positional information of the sequence to be used, something which is not always possible with draft genomes consisting of many contigs.

Example 4.9 (Human genome) The human genome was finished in 2004 [Human Genome Sequencing Consortium, 2004]. The assembly as of January 2005 consists of 2.85 billion nucleotides interrupted by 341 gaps. It covers almost all (about 99%) of the euchromatic part of the genome and has only about one error in every 100, 000 bases. The latest build of the human genome can be downloaded from NCBI at
`ftp://ftp.ncbi.nih.gov/genomes/H_sapiens/`.

Another useful website is the UCSC genome browser. The following steps describe how to download a piece of the human genome from the UCSC browser:

  (i) Open a browser and load the URL `http://genome.ucsc.edu`
 (ii) Click on the `Genome Browser` tab on the left hand side.
(iii) There are three pull-down menus for selecting a clade, a genome from that clade, and a specific version. Select `Vertebrate`, `Human`, `May 2004`.
 (iv) The specific position to browse is entered in the `position` box. Enter the coordinates `chr17:38,451,220-38,530,912` and press `submit`.
  (v) You will see a `GIF` image which depicts a region of the human genome (shown in Figure 4.3). Click on the `DNA` tab on the top of the page.
 (vi) Click on the `Get DNA` button. You should see almost 100, 000 DNA bases on the screen.

This particular region of the human genome contains a gene which is linked to breast cancer. More discussion of this example follows in Section 4.3. □

Although some assembly programs are freely distributed, they are fairly complicated software tools that require large amounts of computer memory, and until recently most assembly has been done by the sequencing centers. Thus, the sources for genome assemblies are mostly the large sequencing centers, which we summarize in the list below:

• Broad Institute at M.I.T., Cambridge, Massachusetts, USA:
  `http://www.broad.mit.edu/resources.html`
• DOE Joint Genome Institute, Walnut Creek, California, USA:
  `http://genome.jgi-psf.org`

- Human Genome Sequencing Center at the Baylor College of Medicine, Houston, Texas, USA: `http://www.hgsc.bcm.tmc.edu/projects`
- Wellcome Trust Sanger Institute, Cambridge, England: `http://www.sanger.ac.uk/Projects`
- Genome Sequencing Center at Washington University, St. Louis, USA: `http://www.genome.wustl.edu`
- Genoscope – the French National Sequencing Center, Evry, France: `http://www.genoscope.cns.fr/externe/English/Projets`
- Agencourt Bioscience Coorporation, Beverly, Massachusetts, USA: `http://www.agencourt.com`
- The Institute for Genomic Research, Rockville, Maryland, USA: `http://www.tigr.org/tdb`
- Beijing Genome Institute, Beijing, China: `http://www.genomics.org.cn`
- Genome Sequencing Centre, Jena, Germany: `http://genome.imb-jena.de`

We have already seen that the UCSC Genome Browser is a useful site for browsing genomes (although it is not a sequencing center). Another similar site is Project ENSEMBL at `http://www.ensembl.org`

The most comprehensive online resource for genomic sequences is the National Center for Biotechnology Information (NCBI) `http://www.ncbi.nlm.nih.gov/`. In addition to serving as a worldwide repository for all genome related data (maintained in a database called GENBANK at `http://ncbi.nlm.nih.gov/Genbank/`), NCBI also hosts the trace archive we have mentioned. We note that the reads for some genomes are not available because the sequencing centers have not released them. Another popular trace archive is housed at `http://trace.ensembl.org/`.

4.2.2 Alignments

Finding homologous components among genomes is the first step in identifying highly conserved sequences that point to the small fraction of the genome that is being conserved by natural selection, and therefore likely to be functional. The recognition of homologous components requires two separate steps:

(i) *Sequence matching.* The identification of similar sequence elements between genomes.

(ii) *Homology mapping.* The separation of matches into homologous components and sequences that match by chance.

The combined problem of finding matching sequence elements and then sorting them into homologous components is called the *alignment problem.* Both these steps are difficult, and are active areas of research in genomics. The homology mapping problem is the topic of Chapter 13.

type	program	site
pairwise	AVID	http://pipeline.lbl.gov
pairwise	BLASTZ	http://ecrbrowser.dcode.org
multiple	MAVID	http://hanuman.math.berkeley.edu/genomes/
multiple	MAUVE	http://asap.ahabs.wisc.edu/mauve/index.php
multiple	MULTIZ	http://hgdownload.cse.ucsc.edu/downloads.html

Table 4.2. *Sites for downloading genome alignments.*

Alignments of genomes are available for download from a number of websites (Table 4.2). These sites maintain up-to-date versions of genomes, perform regular updates of the alignments, and provide tools for visualizing and retrieving alignments. Popular alignment programs include pairwise aligners such as AVID [Bray *et al.*, 2003] and BLASTZ [Schwartz *et al.*, 2003], as well as tools for multiple alignment such as MAUVE [Darling *et al.*, 2004], MAVID [Bray and Pachter, 2004], and MLAGAN [Brudno *et al.*, 2003a]. There are also programs such as MULITZ [Blanchette *et al.*, 2004] which perform multiple pairwise alignments with respect to a reference sequence. We have restricted our list in Table 4.2 to those programs and sites for which whole genome eukaryotic alignments are available for download.

Genome alignments should be used with caution. Results are very dependent on choices of parameters in the programs, and the multiple genome alignment problem is particularly difficult due to the combinatorial explosion of the possible number of alignments. The dependence of alignments on parameters is the topic of Chapter 7. See also Sections 2.2 and 3.4, and Chapters 5 and 8.

4.3 The problems

Biology is the study of living organisms. Although living organisms are complex and there are no known simple principles that completely explain their function, there are fundamental components whose organization and interaction form the basis of life. These components are distinguished by scale. At the macroscopic end of the spectrum are populations of organisms, whose interactions may be governed by certain ecological constraints. Organs within individual organisms are composed of tissues and cells. At the microscopic level there is DNA, which is itself composed of organic precursors and organized into genomes. Genomes and cells are related by a series of intermediary biomolecules: RNA and proteins, coded for by DNA. Together they form metabolites and organelles which make up cells, and, in turn, cells are the structures that house DNA and allow for its replication.

Mathematical biology is a general term that, in principle, encapsulates the parts of mathematics relevant to the study of biology. Typically, this has referred to the mathematical analysis of biological models at a macroscopic scale. This is because molecular biology has only relatively recently become an integral part of biological investigation. Even more recent is the emergence of *genomics*, or the study of genomes, as a discipline in its own right. Even though genomics is only a tiny piece of the complex puzzle of biology, a complete understanding of genomes is an essential step for learning more about the cell, which in turn is the stepping stone to higher level systems. What this chapter deals with is therefore a new branch of mathematical biology, which could be termed "mathematical genomics".

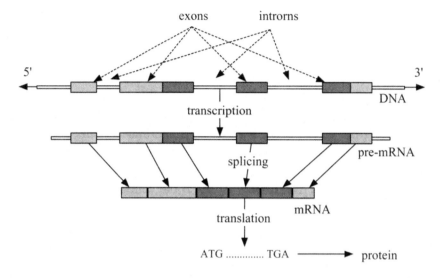

Fig. 4.2. Structure of a gene.

There are two important aspects to genome analysis. On the one hand, a key problem is to understand the organization and function of individual genomes. On the other hand, there is the equally interesting problem of understanding the evolution of genomes and the mechanisms of natural selection. The relationship between these problems is the central theme of comparative genomics, and is illustrated pictorially in Figure 4.4. Our aim in this section is to explain this figure, and survey some of the key problems in comparative genomics.

We begin by elaborating on the structure of a (eukaryotic) gene, which is represented in Figure 4.2 by the boxes on the top horizontal line. Note that the horizontal line is a cartoon for a genome, and represents a sequence of DNA. In order to understand the meaning of the boxes it is necessary to know a bit about the structure of genes.

First, we recall that a gene is a segment of the genome that codes for a protein. Figure 4.2 is a depiction of the processes by which the encoding sequence is extracted from the genome and prepared for translation into protein. In Figure 4.2, the top line is DNA, and the sequence represented by the double line and the shaded boxes is the transcribed genic sequence. Transcription results in the DNA being "copied" into pre-mRNA. The pre-mRNA differs from the DNA in that Thymine (T) is replaced by Uracil (U), and a sequence of multiple Adenines (As) (called the polyA tail) is added at the end of the transcript to keep the pre-mRNA stable. The intergenic regions (indicated by flanking arrows) at the far left and right of the top line are not transcribed.

Only some parts of the pre-mRNA encode the protein, and these parts are isolated by two refinements. First, light and dark boxes, known as *exons*, are cut out and glued or spliced together to form mRNA, which is the substrate used for translation. The thin lines connecting the exons, which are called *introns*, are discarded. Furthermore, only part of the mRNA makes up the dark coding region, which is translated into a protein. Those parts of the DNA that are spliced into the mRNA, but that do not get translated are known as *untranslated regions* (UTRs). Typically, both the 5′ and the 3′ ends of the gene contain UTRs. Transcription is regulated via a *promoter region* which lies immediately upstream (i.e., to the 5′ end of the site where transcription begins. The promoter region contains sequence elements that are identified by proteins and complexes which intitiate transcription.

Fig. 4.3. Breast cancer type I early-onset gene: snapshot from the UCSC browser.

Figure 4.3 is a depiction from the UCSC Genome Browser of a real human gene. The top line is a schematic of chromosome 17, and the lower panel is a magnified view of the red region from the first line. This region is quite long (almost 100kb) but contains only one gene, BRCA1, the breast cancer type 1 early-onset gene. The dark vertical bars in the lower panel are exons of BRCA1. Mutations in these exons are known to lead to truncated proteins, and studies have confirmed that patients with early-onset breast cancer are far more likely than the general population to have mutations in this gene.

One of the main features of eukaryotic genes is that a single gene may code for a number of different proteins. It is the splicing of exons that is unique to the eukaryotes and allows for this flexibility. The 5′ ends of introns are known as 5′ splice sites, or *donor sites*, and (almost) always begin with the nucleotides GT. Similarly, the 3′ ends of introns are known as 3′ splice sites, or *acceptor sites*, and (almost) always end in AG. The alternating sequence of introns and exons or, equivalently, the alternating sequence of 5′ and 3′ splice sites, determine how the mRNA is spliced together; this information is what we will call a gene structure. A gene may have many different possible structures or alternative splicings, and may therefore code for many different proteins. In Example 4.9, one can view some of the alternative splicings for BRCA1 by selecting "full" from the pull-down menu under the RefSeq Genes link towards the bottom of the page, and clicking the "refresh" button under the figure. We summarize our discussion of genes by proposing a definition, although it should be pointed out that any given definition will likely have to be modified as the understanding of transcription, translation and splicing improves.

Definition 4.10 (Gene) A *coding gene* is a segment of DNA thatis transcribed, and which has the property that a subset of the transcribed sequence is translated. A *gene structure* is an alternating sequence of exons and introns within a gene; thus, a gene may contain multiple gene structures. These are also known as *alternative splicings*.

Mathematically, the statement that a gene can have many alternative splicings can be formalized as follows. We represent a coding gene by a sequence of parentheses, such as ())(()(())(()()((). An opening parenthesis "(" represents a donor splice site, and a closing parenthesis ")" represents an acceptor splice site. A gene structure is a subsequence of alternating opening and closing parentheses, such as ()()(). One occurrence of the gene structure ()()() in the coding gene ())(()(())(()()((() is given by the six underlined parentheses in ())(()(())(()()((().

Proposition 4.11 *Let F_i denote the Fibonacci numbers which are defined by*

$$F_1 = 1, F_2 = 1, F_3 = 2, F_4 = 3, F_5 = 5, \ldots, F_i = F_{i-1} + F_{i-2}.$$

A coding gene that contains n donor splice sites and m acceptor splice sites has at most F_{n+m+1} gene structures. This upper bound is attained when $n = m$, and is realized by a coding gene of the form $()()()() \cdots ()$.

For example, the coding gene ()()() has $F_7 = 13$ gene structures: the empty string, the entire structure ()()(), six structures () and five structures ()().

There are many outstanding biology questions related to genes. For example, it is unknown whether there are functional roles for all intronic sequences (sometimes called "junk" DNA). Furthermore, it is still unclear if there are organizing principles that explain in simple terms the regulation of genes. There is also a need for better statistical models of gene structures. The hidden Markov model approach to gene-finding is explained in Section 4.4.

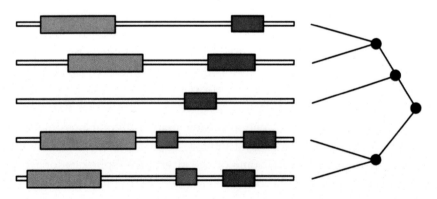

Fig. 4.4. Comparative genomics via annotation and phylogeny.

Returning to Figure 4.4, we see that the tree on the right hand side shows the evolutionary relationships between the sequences. This leads us to the alignment and evolutionary modeling components ofcomparative genomics.

In order to identify functional elements in the sequences, it is useful to find conserved regions in the alignments. Conversely, the alignment problem is easier if one knows ahead of time the functional elements in each sequence. Statistical models for alignment and evolutionary models are based on these biological considerations, and we discuss them in more detail in Section 4.5.

It is important to note that comparative genomics is not only a computational endeavor. There are many experimental techniques being developed that can be used to identify functional elements in genomes, and that also shed light on genome evolution. In this regard, we wish to mention the ENCyclopedia of DNA Elements (ENCODE) Project [Consortium, 2004]. This is an international consortium organized by the National Human Genome Reesearch Institute, working towards the goal of identifying all functional elements in the human genome sequence. The ENCODE consortium sequence and analysis repository is housed at `http://genome.ucsc.edu/encode`. The pilot phase of the project is focused on 1% of the human genome sequence. Initial efforts include the development of high throughput technologies for detecting functional elements, as well as the sequencing of orthologous regions from multiple primates, mammals and other vertebrates. The available sequence from multiple organisms complements additional sequences extracted from whole genome sequencing projects, and serves as a testbed for comparative genomics approaches to detecting functional elements. Thus, the ENCODE project is aimed at fostering interaction between computational and experimental scientists, and at identifying promising research avenues and scalable technologies. Preliminary analysis of ENCODE regions is discussed in Chapters 21 and 22.

4.4 Statistical models for a biological sequence

In Chapter 1 we introduced DiaNA, a strange fictional character who flips coins and generates words over the alphabet $\{A, C, G, T\}$. Although DiaNA does not seem to have anything to do with real biological DNA sequences, the principle of imagining DNA to have been generated by fictional entities like DiaNA, who throw dice, has proved to be extremely useful for biological sequence analysis. In order to see this, suppose that we would like to analyze 1 million bases of DNA from the human genome and identify CpG islands among them. One approach is to count, for each contiguous subsequence of length 100, the number of Cs and Gs, and to call a 100bp segment a CpG island if there are more than 70 Cs and Gs. There are a number of problems with such an approach. First, the segment size 100 is arbitrary; perhaps some biologists prefer working with segments of length 50, or 200. Secondly, for such different segment sizes, what should be the cutoff for deciding when the number of Cs and Gs indicates a CpG island? Again, there may be different intuitive guesses as to what constitutes "random looking sequence." A statistical approach to the problem helps to resolve such issues by carefully and precisely specifying the parameters and the model, thus allowing for a mathematically rigorous description of "random." This leads to sensible approaches for deciding when a region is a CpG island.

Example 4.12 DiaNA serves as the statistical surrogate for our biological intuition and understanding of CpG islands. In searching for CpG islands, we begin by specifying what non-CpG random DNA should look like (DiaNA's fair die). When she chooses to toss this die, she makes a "non-CpG DNA base". Next, our biological knowledge suggests that CpG islands should have an excess of C's and G's. The CpG island die therefore has higher probabilities for those bases. Finally, a third die may represent DNA sequences that are poor in C's and G's. Returning to Example 1.1, we recall that the probabilities were:

$$
\begin{array}{lcccc}
 & \text{A} & \text{C} & \text{G} & \text{T} \\
\text{first die} & 0.15 & 0.33 & 0.36 & 0.16 \\
\text{second die} & 0.27 & 0.24 & 0.23 & 0.26 \\
\text{third die} & 0.25 & 0.25 & 0.25 & 0.25
\end{array}
\tag{4.1}
$$

These probabilities reflect actual properties of CpG islands; they were computed from the table in [Durbin *et al.*, 1998, page 50]. Once a model is specified, statistical inference can be applied to DNA for finding CpG islands. □

One of the original applications that highlighted the use of discrete statistical models for biological sequence analysis is the gene finding problem. Hidden Markov models (HMMs) have been successfully applied to this problem. They have also been used for finding other functional elements.

Maximum a posteriori (MAP) inference with such models has become the method of choice for *ab initio* gene-finding. To give a precise definition of MAP inference, let us recall the set-up of Section 1.3. The hidden model is the map $F : \mathbb{R}^d \to \mathbb{R}^{m \times n}$ specified by a matrix of polynomials $F = \left(f_{ij}(\theta)\right)$, while the observed model is the map $\mathbf{f} : \mathbb{R}^d \to \mathbb{R}^m$ whose coordinates are the row sums of the matrix F: that is, $f_i(\theta) = \sum_{j=1}^n f_{ij}(\theta)$. In MAP inference we assume that one particular observation $i \in [m]$ has been made. The problem is to identify an index $j \in [n]$ that maximizes $f_{ij}(\theta)$. In other words, we wish to find the best explanation j for the given observation i. Traditionally, the parameters θ are assumed to be known and fixed, but here we also consider the parametric version where some or all of the parameters are unknowns.

For many models used in computational biology, including the Markov models discussed in Section 1.4, the hidden model F will be a toric model (or very close to one). This means that the entries of the matrix F are monomials in the parameters, say $f_{ij}(\theta) = \theta^{a_{ij}}$ for some $a_{ij} = (a_{ij1}, \ldots, a_{ijd}) \in \mathbb{N}^d$. Then the probability of observing state $i \in [m]$ in the model $\mathbf{f}$ equals $f_i(\theta) = \sum_{j=1}^n \theta^{a_{ij}}$. The tropicalization of this polynomial is

$$
g_i(w) = \bigoplus_{j=1}^n w^{\odot a_{ij}} = \min\{a_{ij1}w_1 + a_{ij2}w_2 + \cdots + a_{ijd}w_d\}.
$$

If we introduce logarithmic parameters $w_i = -\log(\theta_i)$ then our problem is to evaluate the tropical polynomial $g_i(w)$. We summarize this as a remark.

Remark 4.13 MAP inference is the tropical evaluation of one coordinate polynomial of an algebraic statistical model.

Implications of the connection between MAP inference and tropical arithmetic are discussed in Chapter 2. We remark that the probability of any "best explanation" derived using MAP inference will usually be very low, so that it makes sense to find explanations that are suboptimal but become optimal with a small change of the parameters. This issue is discussed in Chapters 5–9.

Example 4.14 (Google) A useful example to keep in mind when thinking of MAP inference is the Google "did you mean..." feature. A web search for the words `topicaal geom try` leads Google to respond `Did you mean: tropical geometry`. In this case, the observed sequence (or the index i in the discussion above) is `topicaal geom try`, and the alphabets for the hidden and observed sequences are the same. The MAP inference problem is to find the string (index j) that maximizes $f_{ij}(\theta)$. The model can be specified in many ways, perhaps taking advantage of patterns in the English language or among common queries from other users. Below we replace the English language by DNA, and patterns of usage in the English language by features of genes. □

In the context of biological sequence analysis, hidden Markov models can be used to model splice sites of eukaryotic genes. The underlying biology was explained in the previous section. Our model incorporates two fixed sizes for the $5'$ and $3'$ splice sites (k and k' respectively), and distinguishes exons from introns. We use the notation of Section 1.4. Our HMM has length n where n is the length of the DNA sequences we wish to model. The alphabet of hidden states is $\Sigma = \{E, 1, \ldots, k, I, 1', \ldots, k'\}$, where E is a state for "exon" sequence preceding the first $5'$ splice site which terminates this exon, and I is a state for "intron" after the first $5'$ splice site and before the first $3'$ splice site, which begins a new exon. The alphabet of observed states is $\Sigma' = \{A, C, G, T\}$.

The parameters of this model consist of a pair of matrices θ, θ' where

$$
\theta =
\begin{array}{c}
 \\
E \\
1 \\
2 \\
\vdots \\
k-1 \\
k \\
I \\
1' \\
\vdots \\
k'-1 \\
k'
\end{array}
\begin{pmatrix}
E & 1 & 2 & 3 & \cdots & k & I & 1' & 2' & \cdots & k' \\
\theta_1 & 1-\theta_1 & 0 & 0 & \cdots & 0 & 0 & 0 & 0 & \cdots & 0 \\
0 & 0 & 1 & 0 & \cdots & 0 & 0 & 0 & 0 & \cdots & 0 \\
0 & 0 & 0 & 1 & \cdots & 0 & 0 & 0 & 0 & \cdots & 0 \\
 & & & & \ddots & & & & & & \\
0 & 0 & 0 & 0 & \cdots & 1 & 0 & 0 & 0 & \cdots & 0 \\
0 & 0 & 0 & 0 & \cdots & 0 & 1 & 0 & 0 & \cdots & 0 \\
0 & 0 & 0 & 0 & \cdots & 0 & \theta_2 & 1-\theta_2 & 0 & \cdots & 0 \\
0 & 0 & 0 & 0 & \cdots & 0 & 0 & 0 & 1 & \cdots & 0 \\
 & & & & & & & & & \ddots & \\
0 & 0 & 0 & 0 & \cdots & 0 & 0 & 0 & 0 & \cdots & 1 \\
1 & 0 & 0 & 0 & \cdots & 0 & 0 & 0 & 0 & \cdots & 0
\end{pmatrix}
$$

and θ' is a $(k + k' + 2) \times 4$ matrix specifying the output probabilities. The latter matrix is known as a *position specific scoring matrix* (PSSM) or a *weight matrix.* When describing a PSSM, the output probabilities for the states I and E are typically not represented, as they are assumed to either be 0.25 for all observed possibilities, or else easily obtainable for the problem at hand.

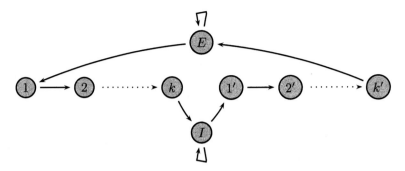

Fig. 4.5. State space diagram of a simple splice site model.

Example 4.15 Using sequence data from 139 splice site junction sequences, [Mount, 1982] estimated the parameters for a $k = 12$ donor site PSSM:

	1	2	3	4	5	6	7	8	9	10	11	12
G	0.2	0.09	0.11	0.74	1	0	0.29	0.12	0.84	0.09	0.18	0.2
A	0.3	0.4	0.64	0.09	0	0	0.61	0.67	0.09	0.16	0.39	0.24
T	0.2	0.07	0.13	0.12	0	1	0.07	0.11	0.05	0.63	0.22	0.27
C	0.3	0.44	0.11	0.06	0	0	0.02	0.09	0.02	0.12	0.2	0.28

This particular PSSM played a key role in helping to find splice sites in genomes, although the availability of much more data has revealed additional structure in splice sites that can be modeled and used to improve their identification [Abril *et al.*, 2005]. □

The transition matrix θ of an HMM is typically sparse, so it is convenient to represent the sparsity pattern with a directed graph. This graph is known as the *state space diagram.* In our example for two splice sites, that graph is a directed cycle of length of $k + k' + 2$ with two special nodes (namely, "E" and "I") which have self-loops (Figure 4.5). We shall explain MAP inference for this model. For simplicity, we assume that numerical values (perhaps those in Example 4.15) have been fixed for all entries in the PSSM θ', and that the initial distribution on Σ is the distribution that is uniform on the two states "E" and "I".

Suppose we are considering DNA sequences of length n. Then our HMM is a polynomial map $\mathbf{f} : \mathbb{R}^2 \to \mathbb{R}^{4^n}$. The coordinates of the map $\mathbf{f}$ are indexed by DNA sequences $\sigma \in (\Sigma')^n$. Omitting initial probabilities for simplicity, we see that each coordinate f_σ is a polynomial in the two model parameters θ_1 and θ_2, and is naturally written in the form

$$f_\sigma(\theta_1, \theta_2) = \sum_{i,j,k,l} \alpha_{ijkl} \cdot \theta_1^i \cdot (1 - \theta_1)^j \cdot \theta_2^k \cdot (1 - \theta_2)^l,$$

where $\alpha_{ijkl} \in \mathbb{R}_{\geq 0}$ depends polynomially on the entries in the PSSM θ'. Each sequence in Σ^n that is a walk in the directed cycle described above will contribute to one of the summands of $f_\sigma(\theta_1, \theta_2)$. In that summand, i is the number of adjacent pairs "EE" in the sequence, j is the number of pairs "$E1$", k is the number of pairs "II", and l is the number of pairs "$I1'$".

For instance, if $n = 10, k = k' = 2$ and the observation is $\sigma = $ ACGTGGTAGA, then the sequence of hidden states $EE12III1'2'E$ contributes the term

$$\left((\theta'_{EA})^2 \theta'_{EC} (\theta'_{IG})^2 \theta'_{IT} \theta'_{1G} \theta'_{2T} \theta'_{1'A} \theta'_{2'G} \right) \cdot \theta_1 \cdot (1 - \theta_1) \cdot \theta_2^2 \cdot (1 - \theta_2),$$

where the parenthesized product is a term in α_{ijkl}.

For MAP inference in this model it is convenient to think of θ_i and $1 - \theta_i$ as independent parameters. We thus introduce four different logarithmic weights:

$$w_{11} = -\log(\theta_1), \ w_{12} = -\log(1 - \theta_1), \ w_{21} = -\log(\theta_2), \ w_{22} = -\log(1 - \theta_2).$$

Then the tropicalization of $f_\sigma(\theta_1, \theta_2)$ has the form

$$g_\sigma(w) \quad = \quad \min_{i,j,k,l} \{ \beta_{ijkl} + iw_{11} + jw_{12} + kw_{21} + lw_{22} \},$$

where β_{ijkl} is the tropicalization of α_{ijkl}.

MAP inference for this model means evaluating this piecewise-linear function for fixed w_{ij}. Parametric inference means precomputing $g_\sigma(w)$ by polyhedral geometry. Using the techniques discussed in Section 3.3, this can be done for any given PSSM θ' and any observation σ. The output of that computation is a complete list of all the *Viterbi sequences*, by which we mean sequences in Σ^n whose corresponding linear forms $\beta_{ijkl} + iw_{11} + jw_{12} + kw_{21} + lw_{22}$ attain the unique minimum for some choice of weights w_{ij}. Each Viterbi sequence represents the optimal splice site locations for a range of numerical values of θ_1 and θ_2. The list of Viterbi sequences produced by parametric inference also includes a characterization of all boundaries between these ranges in the (θ_1, θ_2)-plane. Such an output may provide valuable information about the robustness of a specific Viterbi sequence to changes in the parameters.

In order to predict genes in a genome, a more sophisticated HMM than the splice site model given above needs to be used. Indeed, more recent approaches to gene-finding make use of variants of HMMs that not only model splice sites, but also ensure that predicted structures contain an *open reading frame* (ORF). This means that the translated part of the mRNA must be of length 0 mod 3, and must contain a stop codon only at the very end. In addition, exon lengths, which are not geometrically distributed, are modeled explicitly using a modification of hidden Markov models known as *semi-hidden Markov models* or *generalized HMMs*. In Example 4.16 we describe a model that ensures the correct length for open reading frames, but that does not explicitly model splice sites or exon lengths. For a description of more complete models see [Burge and Karlin, 1997, Kulp *et al.*, 1996, Alexandersson *et al.*, 2003, Pachter *et al.*, 2002].

Example 4.16 (Gene-finding HMM) The model is specified by a pair of

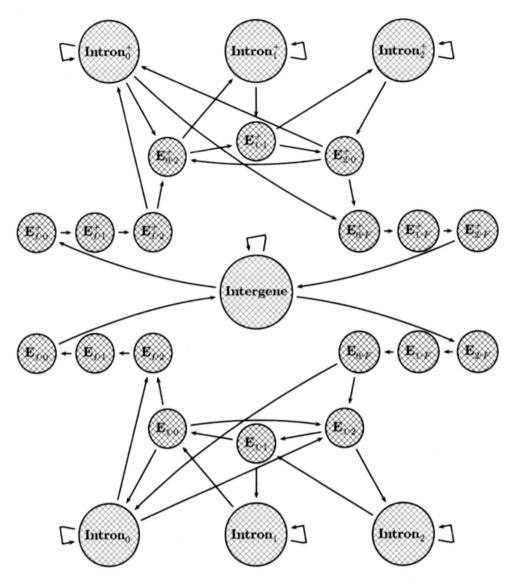

Fig. 4.6. State space diagram of a simple gene-finding HMM.

matrices, θ and θ', one of which is a 25×25 matrix θ of transition probabilities, and the other is a 25×4 matrix θ' of output probabilities. In total, there can therefore be up to 725 parameters (ignoring the requirement that probabilities should sum to 1). In practice, biological considerations simplify matters, leading to matrices θ that are very sparse. In fact, the specific model we have in mind has only 41 non-zero entries for the matrix θ. There is also additional structure in θ, for example many entries are set to 1. Models with so many parameters are summarized with a *state transition diagram* (see Figure 4.6). The state transition diagram is a graph with one node for every hidden state. There is an edge for every non-zero entry in the matrix θ. Notice that the bottom half of the state transition diagram is a mirror image of the top half (with the directions of the arrows reversed). This reflects the fact that genes can be found on either strand of the DNA sequence. □

Our gene-finding HMM is designed to respect the fact that codons are triplets. There are 18 exon states $E^{\pm}$ in the state space diagram of Figure 4.5. Consider any directed cycle that starts and ends at the intergene state. The number of exon states visited in any such a cycle must be a multiple of three. This can be seen by inspecting Figure 4.6.

Remark 4.17 A sequence of hidden states has probability zero in the fully observed Markov model underlying Example 4.16 unless the number of exon states in that sequence is divisible by three.

The parameters θ' in a gene-finding HMM are derived from observed frequencies of codons in known genes, and from the overall frequencies of the bases in intergenic and intronic DNA. In other words, the maximum likelihood estimates for these parameters can be obtained from the fully observed model using Proposition 1.18. The parameters in θ relate biologically to lengths of introns, exons, and the distance between genes. These are also derived from known genes. In principle, one could estimate the parameters using MLE with the hidden model; however this is typically not done in practice.

The model described in Example 4.16 has a number of limitations. As we have discussed, it does not model splice sites at all and there is no explicit modeling of exon lengths. There are also other gene elements that are simply not possible to model at all with current methods. For example, different cell types in an organism have different genes transcribed (and therefore translated) at different times. This process is largely regulated through enhancement or suppression of transcription, and is referred to as *regulation*. *Transcription factor binding sites* (TFBSs), also called *cis-regulatory elements*, are small sequences, typically in the neighborhood of genes, that are bound to by proteins that mediate transcription. Such proteins are said to be *trans-acting*. A complete solution to the gene-finding problem therefore requires annotation of TFBSs. Despite some encouraging results in this line with hidden Markov models, the TFBS identification problem is harder than gene-finding and has traditionally been tackled separately [Tompa *et al.*, 2005].

4.5 Statistical models of mutation

Point mutations in DNA sequences can be modeled by continuous time Markov processes on trees. This point of view has been extensively explored and developed during the past thirty years [Felsenstein, 2003]. The relevant algebraic statistics involves the hidden tree models of Subsection 1.4.4. In what follows we offer a derivation of biologically relevant hidden tree models. We also return to our discussion of pair hidden Markov models for sequence alignment, with a biological discussion of insertions and deletions, and explain what DiaNA is doing in the diagram on the book cover.

4.5.1 Evolutionary models

Although the biology of point mutation is complicated, the use of Markov processes on trees is motivated by underlying principles that capture, to some degree, the complexities of mutation. These are:

- Mutations occur at random, although possibly with different probabilities at different places in the genome.
- Mutations occur independently in different species.
- In genome locations where mutations can occur, there is, at any instant in time, a non-zero probability that mutation occurs.

The first two requirements lead to hidden tree models. The tree T corresponds to a species tree, with different species labeling the leaves of the tree. Thus, we have the structure of a phylogenetic tree appearing naturally, and associating a hidden tree model with the point mutation process is equivalent to specifying that bases observed at the leaves of the tree are the result of a stochastic process of mutation between "hidden" interior vertices of the tree. The interior vertices correspond to ancestral species.

The third requirement leads to a further restriction of hidden tree models, specifically, to *evolutionary models*. The precise characterization of what we mean by an evolutionary model will be given in Definition 4.24. It is based on the idea that the tree represents a continuous time Markov process. We begin by defining the mathematical ingredients of such a Markov process.

Definition 4.18 A *rate matrix* (or *Q-matrix*) is a square matrix $Q = (q_{ij})$, with rows and columns indexed by $\Sigma = \{A, C, G, T\}$. We could also use the binary alphabet, or the twenty-letter alphabet of amino acids, instead of Σ. Rate matrices must satisfy the following requirements:

$$q_{ij} \geq 0 \quad \text{for} \quad i \neq j,$$

$$\sum_{j \in \Sigma} q_{ij} = 0 \quad \text{for all} \quad i \in \Sigma,$$

$$q_{ii} < 0 \quad \text{for all} \quad i \in \Sigma.$$

Rate matrices capture the notion of *instantaneous rate of mutation*. From a given rate matrix Q, one computes the *substitution matrices* $\theta(t)$ by exponentiation. The entry of $\theta(t)$ in row i and column j equals the probability that the substitution $i \rightarrow \cdots \rightarrow j$ occurs in a time interval of length t.

Theorem 4.19 *Let Q be any rate matrix and $\theta(t) = e^{Qt} = \sum_{i=0}^{\infty} \frac{1}{i!} Q^i t^i$. Then*

- (i) $\theta(s + t) = \theta(s) \cdot \theta(t)$ *(Chapman–Kolmogorov equations)*,
- (ii) $\theta(t)$ *is the unique solution to the forward differential equation*
  $\theta'(t) = \theta(t) \cdot Q$, $\theta(0) = \mathbf{1}$ *for $t \geq 0$ (here $\mathbf{1}$ is the identity matrix)*,
- (iii) $\theta(t)$ *is the unique solution to the backward differential equation*
  $\theta'(t) = Q \cdot \theta(t)$, $\theta(0) = \mathbf{1}$ *for $t \geq 0$,*

(iv) $\theta^{(k)}(0) = Q^k$.

Furthermore, a matrix Q is a rate matrix if and only if the matrix $\theta(t) = e^{Qt}$ is a stochastic matrix (non-negative with row sums equal to one) for every $t \geq 0$.

Proof For any square matrix A, the matrix exponential e^A is defined by

$$e^A \quad := \quad \sum_{k=0}^{\infty} \frac{A^k}{k!}.$$

The matrix exponential is well defined because the series on the right hand side converges componentwise for any A. A standard identity that can be derived directly from the definition is that $e^{A+B} = e^A e^B$ provided that A and B are matrices that commute. Since sQ and tQ commute for any s, t, it follows that $\theta(s+t) = \theta(s) \cdot \theta(t)$. In order to derive (ii) and (iii), we need to differentiate $\theta(t)$ term-by-term, which is possible because the power series $\theta(t)$ has infinite radius of convergence. We find that

$$\theta'(t) \quad = \quad \sum_{k=1}^{\infty} \frac{t^{k-1} Q^k}{(k-1)!} \quad = \quad \theta(t) \cdot Q \quad = \quad Q \cdot \theta(t).$$

Iterated differentiation leads to the identity (iv), which says that the kth derivative of $\theta(t)$ evaluated at 0 is just the matrix Q^k. The uniqueness in parts (ii) and (iii) is a standard result on systems of ordinary linear differential equations.

The last part of the theorem provides the crucial connection between rate matrices and substitution matrices. One direction of the theorem is easy: if $\theta(t)$ is a substitution matrix for every $t \geq 0$, then $\sum_j \theta_{ij}(t) = 1$ for all $t \geq 0$. Using identity (iv) with $k = 1$, we have that

$$\sum_j q_{ij}(t) \quad = \quad \sum_j \theta'_{ij}(0) \quad = \quad 0.$$

This says that the row sums of Q are 0. The sign conditions on q_{ij} follow since $\theta(0)$ is the identity matrix and θ must remain stochastic, so Q is a rate matrix. To prove the other direction, we note that as $t \to 0$, the Taylor series expansion gives $\theta(t) = I + tQ + O(t^2)$. This immediately implies that, when t is sufficiently small, for all $i \neq j$, $q_{ij}(t) \geq 0$ if and only if $\theta_{ij}(t) \geq 0$. But, by (i), $\theta(t) = \theta(t/m)^m$ for all m, so in fact $q_{ij} \geq 0$ if and only if $\theta_{ij}(t) \geq 0$ for all $t \geq 0$. Finally, it is easy to check that since Q has row sums equal to zero, so does Q^m for all m, and so the result follows directly from the definition of $\theta(t)$ in terms of Q. □

A standard example is the *Jukes–Cantor rate matrix*

$$Q \quad = \quad \begin{pmatrix} -3\alpha & \alpha & \alpha & \alpha \\ \alpha & -3\alpha & \alpha & \alpha \\ \alpha & \alpha & -3\alpha & \alpha \\ \alpha & \alpha & \alpha & -3\alpha \end{pmatrix},$$

where $\alpha \geq 0$ is a parameter. The corresponding substitution matrix equals

$$
\theta(t) \;=\; \frac{1}{4}
\begin{pmatrix}
1+3e^{-4\alpha t} & 1-e^{-4\alpha t} & 1-e^{-4\alpha t} & 1-e^{-4\alpha t} \\
1-e^{-4\alpha t} & 1+3e^{-4\alpha t} & 1-e^{-4\alpha t} & 1-e^{-4\alpha t} \\
1-e^{-4\alpha t} & 1-e^{-4\alpha t} & 1+3e^{-4\alpha t} & 1-e^{-4\alpha t} \\
1-e^{-4\alpha t} & 1-e^{-4\alpha t} & 1-e^{-4\alpha t} & 1+3e^{-4\alpha t}
\end{pmatrix}.
$$

The expected number of mutations over time t is the quantity

$$
3\alpha t \;=\; -\frac{1}{4}\cdot \mathrm{trace}(Q)\cdot t \;=\; -\frac{1}{4}\cdot \log \det\bigl(\theta(t)\bigr). \tag{4.2}
$$

This number is called the *branch length*. It can be computed from the substitution matrix $\theta(t)$ and is used to label edges in a phylogenetic tree.

One way to specify an evolutionary model is to give a phylogenetic tree T together with the above rate matrix Q and an initial distribution for the root of T (which we here assume to be the uniform distribution on Σ). The branch lengths of the edges are unknown parameters, and the objective is to estimate these branch lengths from data. Thus, if the tree T has r edges, then such a model has r free parameters, and, according to the philosophy of algebraic statistics, we would like to regard it as an r-dimensional algebraic variety.

Such an algebraic representation does indeed exist, which is not entirely obvious since the probabilities in the substitution matrix $\theta(t)$ do not depend polynomially on the parameters α and t (see Remark 4.25). We shall explain the (algebraic representation of) the Jukes–Cantor DNA model on an arbitrary finite rooted tree T. Suppose that T has r edges and the leaves are indexed by $[n] = \{1, 2, \dots, n\}$. Let $\theta^i = \theta^i(t_i)$ denote the substitution matrix associated with the ith edge of T.

We make the following change of variables in the space of parameters. Instead of using α_i and t_i as in (4.2), we introduce the new two parameters

$$
\pi_i \;=\; \frac{1}{4}(1 - e^{-4\alpha_i t_i}) \quad \text{and} \quad \mu_i \;=\; \frac{1}{4}(1 + 3e^{-4\alpha_i t_i}).
$$

These parameters satisfy the linear constraint

$$
\mu_i + 3\pi_i \;=\; 1,
$$

and the branch length of the ith edge can be recovered as follows:

$$
3\alpha_i t_i \;=\; -\frac{1}{4}\cdot \log \det\bigl(\theta^i\bigr) \;=\; -\frac{3}{4}\cdot \log(1 - 4\pi_i).
$$

The parameters are simply the entries in the substitution matrix

$$
\theta^i \;=\;
\begin{pmatrix}
\mu_i & \pi_i & \pi_i & \pi_i \\
\pi_i & \mu_i & \pi_i & \pi_i \\
\pi_i & \pi_i & \mu_i & \pi_i \\
\pi_i & \pi_i & \pi_i & \mu_i
\end{pmatrix}.
$$

The *Jukes–Cantor model* is a submodel of the general hidden tree model in Subsection 1.4.4. Namely, the Jukes–Cantor model on the tree T with r edges and n leaves is the polynomial map

$$\mathbf{f} : \mathbb{R}^r \rightarrow \mathbb{R}^{4^n}$$

which is obtained by specializing the transition matrices along the edges to the specific 4×4 matrices θ^i above. As in Proposition 1.22 we see:

Remark 4.20 Each coordinate polynomial $f_{u_1 \cdots u_n}$ of the Jukes–Cantor model is a multilinear polynomial in the model parameters $(\mu_1, \pi_1), \ldots, (\mu_r, \pi_r)$, i.e., $f_{u_1 u_2 \cdots u_n}$ is linear in (μ_i, π_i) when the other parameters are fixed.

As an illustration we derive the model which was featured in Example 1.7.

Example 4.21 Let $n = r = 3$, and let T be the tree with three leaves, labeled by $\{1, 2, 3\}$, directly branching off the root of T. We consider the Jukes–Cantor DNA model with uniform root distribution on T. This model is a three-dimensional algebraic variety, given as the image of a trilinear map

$$\mathbf{f} : \mathbb{R}^3 \rightarrow \mathbb{R}^{64}.$$

The number of states in $\{\mathsf{A}, \mathsf{C}, \mathsf{G}, \mathsf{T}\}^3$ is $4^3 = 64$ but there are only five distinct polynomials occurring among the coordinates of the map $\mathbf{f}$. Let p_{123} be the probability of observing the same letter at all three leaves, p_{ij} the probability of observing the same letter at the leaves i, j and a different one at the third leaf, and p_{dis} the probability of seeing three distinct letters. Then

$$
\begin{aligned}
p_{123} &= \mu_1 \mu_2 \mu_3 + 3\pi_1 \pi_2 \pi_3, \\
p_{dis} &= 6\mu_1 \pi_2 \pi_3 + 6\pi_1 \mu_2 \pi_3 + 6\pi_1 \pi_2 \mu_3 + 6\pi_1 \pi_2 \pi_3, \\
p_{12} &= 3\mu_1 \mu_2 \pi_3 + 3\pi_1 \pi_2 \mu_3 + 6\pi_1 \pi_2 \pi_3, \\
p_{13} &= 3\mu_1 \pi_2 \mu_3 + 3\pi_1 \mu_2 \pi_3 + 6\pi_1 \pi_2 \pi_3, \\
p_{23} &= 3\pi_1 \mu_2 \mu_3 + 3\mu_1 \pi_2 \pi_3 + 6\pi_1 \pi_2 \pi_3.
\end{aligned}
$$

All 64 coordinates of $\mathbf{f}$ are given by these five trilinear polynomials, namely,

$$
\begin{aligned}
f_{\mathsf{AAA}} = f_{\mathsf{CCC}} = f_{\mathsf{GGG}} = f_{\mathsf{TTT}} &= \frac{1}{4} \cdot p_{123}, \\
f_{\mathsf{ACG}} = f_{\mathsf{ACT}} = \cdots = f_{\mathsf{GTC}} &= \frac{1}{24} \cdot p_{dis}, \\
f_{\mathsf{AAC}} = f_{\mathsf{AAT}} = \cdots = f_{\mathsf{TTG}} &= \frac{1}{12} \cdot p_{12}, \\
f_{\mathsf{ACA}} = f_{\mathsf{ATA}} = \cdots = f_{\mathsf{TGT}} &= \frac{1}{12} \cdot p_{13}, \\
f_{\mathsf{CAA}} = f_{\mathsf{TAA}} = \cdots = f_{\mathsf{GTT}} &= \frac{1}{12} \cdot p_{23}.
\end{aligned}
$$

This means that our Jukes–Cantor model is the image of the simplified map

$$\mathbf{f}' : \mathbb{R}^3 \rightarrow \mathbb{R}^5, \; \big((\mu_1, \pi_1), (\mu_2, \pi_2), (\mu_3, \pi_3)\big) \mapsto (p_{123}, p_{dis}, p_{12}, p_{13}, p_{23}).$$

There are only three parameters since $\mu_i + 3\pi_i = 1$. Algebraists prefer the above representation with $(\mu_i : \pi_i)$ as homogeneous coordinates on the projective line.

To characterize the image of $\mathbf{f}'$ algebraically, we perform the following linear change of coordinates:

$$
\begin{aligned}
q_{111} &= p_{123} + \tfrac{1}{3}p_{dis} - \tfrac{1}{3}p_{12} - \tfrac{1}{3}p_{13} - \tfrac{1}{3}p_{23} = (\mu_1 - \pi_1)(\mu_2 - \pi_2)(\mu_3 - \pi_3) \\
q_{110} &= p_{123} - \tfrac{1}{3}p_{dis} + p_{12} - \tfrac{1}{3}p_{13} - \tfrac{1}{3}p_{23} = (\mu_1 - \pi_1)(\mu_2 - \pi_2)(\mu_3 + 3\pi_3) \\
q_{101} &= p_{123} - \tfrac{1}{3}p_{dis} - \tfrac{1}{3}p_{12} + p_{13} - \tfrac{1}{3}p_{23} = (\mu_1 - \pi_1)(\mu_2 + 3\pi_2)(\mu_3 - \pi_3) \\
q_{011} &= p_{123} - \tfrac{1}{3}p_{dis} - \tfrac{1}{3}p_{12} - \tfrac{1}{3}p_{13} + p_{23} = (\mu_1 + 3\pi_1)(\mu_2 - \pi_2)(\mu_3 - \pi_3) \\
q_{000} &= p_{123} + p_{dis} + p_{12} + p_{13} + p_{23} = (\mu_1 + 3\pi_1)(\mu_2 + 3\pi_2)(\mu_3 + 3\pi_3).
\end{aligned}
$$

Following Section 3.3, we introduce the homogeneous prime ideal $P_{\mathbf{f}'}$ which consists of all polynomials that vanish on the image of $\mathbf{f}'$.

The change of coordinates reveals that our model is the hypersurface in Δ_4 whose ideal equals

$$
P_{\mathbf{f}'} \;=\; \langle\, q_{000}q_{111}^2 - q_{011}q_{101}q_{110} \,\rangle.
$$

If we set $\mu_i = 1 - 3\pi_i$ then we get the additional constraint $q_{000} = 1$. $\qquad\square$

Remark 4.22 A standardized notation for probabilities and parameters of small tree models will be introduced in Chapter 15. In that notation, the probabilities $p_{123}, p_{12}, p_{13}, p_{23}, p_{dis}$ will be simply p_1, p_2, p_3, p_4, p_5, and the Fourier coordinates $q_{111}, q_{110}, q_{101}, q_{011}, q_{000}$ become q_1, q_2, q_3, q_4, q_5.

The construction in this example generalizes to arbitrary trees T. There exists a linear change of coordinates, simultaneously on the *parameter space* $\mathbb{R}^r$ and on the *probability space* $\mathbb{R}^{4^n}$, such that the map $\mathbf{f}$ becomes a monomial map in the new coordinates. This change of coordinates is known in the phylogenetics literature as the *Fourier transform* or as the *Hadamard conjugation* (see [Evans and Speed, 1993, Hendy and Penny, 1993, Semple and Steel, 2003]).

We can regard the Jukes–Cantor DNA model on a tree T with n leaves and r edges as an algebraic variety of dimension r in the probability simplex Δ of dimension $4^n - 1$. That variety is the image of the map $\mathbf{f}$, and it can be computed (in principle) using the methods of Section 3.2. The ideal $P_{\mathbf{f}}$ is generated by differences of monomials $q^a - q^b$ in the Fourier coordinates. In the phylogenetics literature (including the books [Felsenstein, 2003, Semple and Steel, 2003]), the polynomials in the ideal $P_{\mathbf{f}}$ are known as *phylogenetic invariants* of the model. The model in Example 4.21 was characterized by the following phylogenetic invariant:

$$
q^a - q^b \;=\; q_{000}q_{111}^2 - q_{011}q_{101}q_{110}.
$$

The following general result was shown in [Sturmfels and Sullivant, 2005].

Theorem 4.23 *The prime ideal $P_{\mathbf{f}}$ of the Jukes–Cantor model on a binary tree T is generated by monomial differences $q^a - q^b$ of degree at most three.*

If we consider the hidden tree model where there are no constraints on the substitution matrices θ^i, then we get what is known in phylogenetics as the *general Markov model.* A beautiful recent paper [Allman and Rhodes, 2004a] determines a system of phylogenetic invariants for the general Markov model on a tree T. For binary states, the invariants are polynomials of degree three. which have a determinantal presentation as in Chapter 19.

We wish to emphasize that the general Markov model is much bigger than the models typically used for phylogenetic analysis. We shall now define those models, keeping in mind the idea of a continuous time Markov process.

Definition 4.24 An *evolutionary model* is specified by a phylogenetic tree T together with a rate matrix Q and an initial distribution π for the root of T. The branch lengths t_i of the edges are unknown parameters. The substitution matrix on the ith edge of the tree T is the matrix exponential

$$\theta^i(t_i) \quad = \quad e^{Qt_i}.$$

The evolutionary model $\mathbb{R}^r_{\geq 0} \to \mathbb{R}^{4^n}$ is the specialization of the hidden tree model to these matrices. Note that this map is not a polynomial map, since we are taking the branch lengths t_i to be the coordinates on $\mathbb{R}^r_{\geq 0}$.

Remark 4.25 The Jukes–Cantor model (JC69) can be re-parameterized so that it is algebraic, as shown in the previous discussion. This is not possible for most other evolutionary models, unless we allow a different rate matrix on every edge of the phylogenetic tree. For instance, the strand symmetric model (CS05) is presented in the algebraic incarnation in Chapter 16, as opposed to its formulation in the Felsenstein hierarchy below where there is a common rate matrix for all the edges in the phylogenetic tree.

An important problem in phylogenomics is to identify the maximum likelihood branch lengths given a phylogenetic tree T, a rate matrix Q and an alignment of sequences. For the Jukes–Cantor DNA model on three taxa, described in Example 4.21, the exact "analytic" solution of this optimization problem leads to an algebraic equation of degree 23 (Chapter 18). The specialization in Example 3.26 led to an algebraic equation of degree 16.

The *Felsenstein hierarchy* is a nested family of evolutionary models. It is the cumulative result of experimentation and development of many special continuous time Markov models with rate matrices that incorporate biologically meaningful parameters. The models are summarized in Figure 4.7, with arrows indicating the nesting of the models, and the more general models on top. Each matrix shown is a rate matrix Q, and it is assumed that $\pi_A + \pi_C + \pi_G + \pi_T = 1$. The diagonal entries (marked by a dot) are forced, by the definition of a rate matrix, to equal the negative of the sum of the other entries in their row.

A short description of each model is provided below:

- JC69 [Jukes and Cantor, 1969]: The simplest model. Highly structured (equal transition probabilities among the bases) with uniform root distribution. Although this model does not capture the biology very well, it is easy to work with and is often used for quick calculations. Phylogenetic invariants for this model were characterized in [Sturmfels and Sullivant, 2005].

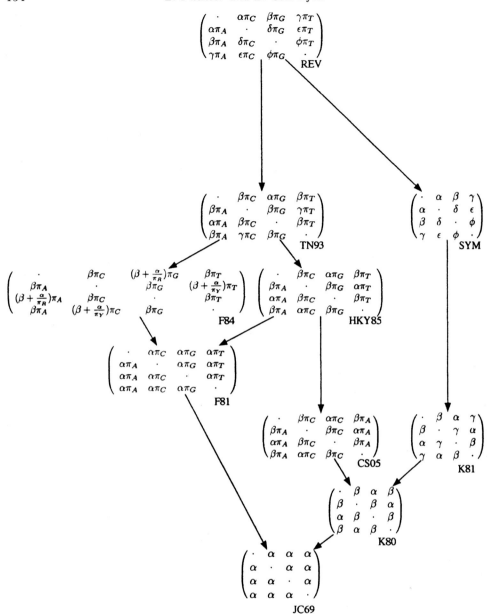

Fig. 4.7. The Felsenstein hierarchy of evolutionary models

- K80 [Kimura, 1980]: An extension of the Jukes–Cantor model that has different rate parameters for transitions (between purines, or between pyrimidines) and transversions (between a purine and a pyrimidine).
- K81 [Kimura, 1981]: A modification of the K80 model that includes an extra parameter for the two different types of transversions.
- CS05 [Yap and Pachter, 2004]: Strand symmetric models are time reversible models in which it is assumed that in the root distribution $\pi_A = \pi_T$, and $\pi_C = \pi_G$. In [Yap and Pachter, 2004] it is shown that REV rate matrices estimated from human, mouse and rat alignments indicate that strand symmetry is a reasonable assumption. This is the basis for the study of the

strand symmetric model in Chapter 16 by Casanellas and Sullivant, hence the name CS05.

- F81 [Felsenstein, 1981]: An extension to the Jukes–Cantor model that allows for non-uniform root distribution.

- HKY85 [Hasegawa *et al.*, 1985]: A widely used "compromise model" that allows for different transition and transversion rates as dictated by the chemistry of DNA (Figure 4.1). Also allows for non-uniform root distribution.

- F84 [Felsenstein, 1989]: Similar to the HKY85 model. This model is implemented in the PHYLIP package [Felsenstein, 2004].

- TN93 [Tamura and Nei, 1993]: An extension of the HKY and F84 models that includes an extra parameter for the two different types of transversions.

- SYM [Zharkikh, 1994]: Assumes uniform root distribution and time reversibility.

- REV [Lanave *et al.*, 1984, Tavaré, 1986]: This is the most general time reversible Markov model. That is, the only restriction on the rate matrix is that it is symmetric. It lacks the group-based structure of the simpler models, and the larger number of parameters makes it harder to estimate models from data using maximum likelihood.

We conclude with two examples: one that shows how mutation models are used in distance-based tree reconstruction, and another which illustrates the use of evolutionary models for identifying conserved positions in genomes.

Example 4.26 (Jukes–Cantor correction) Suppose that we are given a multiple alignment from which we would like to infer a tree; for example,

Human:	ACAATGTCATTAGCGAT...
Mouse:	ACGTTGTCAATAGAGAT...
Rat:	ACGTAGTCATTACACAT...
Chicken:	GCACAGTCAGTAGAGCT...

If there are many more than four taxa, it is not feasible to inspect all trees (Lemma 2.32), so instead a metric is constructed and mapped to a tree metric. The neighbor-joining algorithm 2.41 is the most widely used projection.

In order to obtain a metric from an alignment, evolutionary models are used to compute the maximum likelihood distance between each pair of taxa. In the case of the Jukes–Cantor model, this is known as the *Jukes–Cantor correction*. More generally, such estimates are called *pairwise distance estimates*. Here the tree T has only two leaves, labeled by $\{1, 2\}$, directly branching off the root of T. The Jukes–Cantor model for two taxa is given by a bilinear map

$$\mathbf{f} : \mathbb{R}^1 \times \mathbb{R}^1 \rightarrow \mathbb{R}^2, \quad ((\mu_1, \pi_1), (\mu_2, \pi_2)) \mapsto (p_{12}, p_{dis}). \qquad (4.3)$$

Here $\mu_i = 1 - 3\pi_i$ for $i = 1, 2$, as before. The coordinates of the map $\mathbf{f}$ are

$$
\begin{aligned}
p_{12} &= \mu_1\mu_2 + 3\pi_1\pi_2, \\
p_{dis} &= 3\mu_1\pi_2 + 3\mu_2\pi_1 + 6\pi_1\pi_2.
\end{aligned}
$$

One crucial difference between the model (4.3) and Example 4.21 is that the parameters in (4.3) are *not identifiable*. Indeed, the inverse image of any point in the line $\{p_{12} + p_{dis} = 1\}$ under the map $\mathbf{f}$ is a curve in $\mathbb{R}^1 \times \mathbb{R}^1$. Suppose we are given data consisting of two aligned DNA sequences of length n where k of the bases are different. The corresponding point in $\mathbb{R}^2$ is $u = (n - k, k)$. The inverse image of $\frac{1}{n}u$ under the map $\mathbf{f}$ is the plane curve with the equation

$$
12n\pi_1\pi_2 - 3n\pi_1 - 3n\pi_2 + k = 0.
$$

Every point (π_1, π_2) on this curve is an *exact fit* for the data $u = (n - k, k)$. Hence this curve equals the set of all maximum likelihood parameters for this model and the given data. We rewrite the equation of the curve as follows:

$$
(1 - 4\pi_1)(1 - 4\pi_2) = 1 - \frac{4k}{3n}. \tag{4.4}
$$

Recall from (4.2) that the branch length from the root to leaf i equals

$$
3\alpha_i t_i = -\frac{1}{4} \cdot \log\det\big(\theta^i(t_i)\big) = -\frac{3}{4} \cdot \log(1 - 4\pi_i).
$$

By taking logarithms on both sides of (4.4), we see that the curve of all maximum likelihood parameters becomes a line in the branch length coordinates:

$$
3\alpha_1 t_1 + 3\alpha_2 t_2 = -\frac{3}{4} \cdot \log\left(1 - \frac{4k}{3n}\right). \tag{4.5}
$$

The sum on the left hand side equals the distance from leaf 1 to leaf 2 in the tree T. We summarize our discussion of the two-taxa model as follows:

Proposition 4.27 *Given an alignment of two sequences of length n, with k differences between the bases, the ML estimate of the branch length equals*

$$
\delta_{12} = -\frac{3}{4} \cdot \log\left(1 - \frac{4k}{3n}\right).
$$

Similar results exist for all other models in the Felsenstein hierarchy. $\qquad\square$

Example 4.28 (Phylogenetic shadowing) Suppose we are given a multiple alignment of n sequences and we wish to decide which columns of the alignment are conserved. To address this question, we fix a model in the Felsenstein hierarchy, but we consider two different rate matrices Q_S (slow model) and Q_F (fast model). This gives rise to two different points on the evolutionary model, with coordinates p_σ^F and p_σ^S for each $\sigma = \sigma_1\sigma_2\cdots\sigma_n \in \{\mathtt{A},\mathtt{C},\mathtt{G},\mathtt{T}\}^n$. To determine whether a particular column σ in our alignment is conserved or not, we perform a *likelihood ratio test*. Namely, we compute the ratio of logarithms,

$$
\log\frac{p_\sigma^F}{p_\sigma^S}, \tag{4.6}
$$

where p_σ^F is the probability of σ with rate matrix Q_F, and p_σ^S is the probability of σ with rate matrix Q_S. The ratio in (4.6) can be used to test the hypothesis that a column is conserved by computing the distribution of the test statistic (4.6) under the model corresponding to the rate matrix Q_F.

Such a test is especially sensible if the species being compared to each other are closely related, so that the multiple alignment is reliable without many insertions and deletions [Boffelli *et al.*, 2003]. The term *phylogenetic shadowing* originates with the proposal that multiple sequences that are all similar to each other can nonetheless be used to distinguish between conserved and non-conserved regions in an alignment. The method formalizes the idea that in an alignment of multiple similar sequences a column casts a "shadow" whenever there is at least one sequence that differs from the others, and that the search for conserved regions in an alignment is therefore akin to identifying regions that are not shadowed. Closely related primate sequences suitable for phylogenetic shadowing can be found at `http://pga.lbl.gov/seq/`.

Phylogenetic hidden Markov models are extensions of hidden Markov models to more general tree models that take advantage of this idea. More details can be found in [Siepel and Haussler, 2004, McAuliffe *et al.*, 2004]. □

4.5.2 *Insertion and deletion*

Two of the main mechanisms by which DNA sequences change are insertion and deletion. Insertions and deletions can happen during DNA replication. Repetitive sequences are particularly prone to a phenomenon known as *strand slippage*, during which non-pairing of the complementary strand results in small insertions or deletions [Levinson and Gutman, 1987]. In order to align sequences correctly, it is therefore necessary to accurately model insertions and deletion, as well as point mutation. The pair hidden Markov models of Section 2.2 are supposed to address this problem. The parameters for these models are chosen based on biological considerations and it makes sense to vary these parameters.

Inference with the pair HMM takes us back to DiaNA, and her picture on the cover of the book. DiaNA is again our surrogate for biological intuition. We close this chapter with a precise description of what DiaNA is doing.

Example 4.29 (DiaNA hopping on the alignment graph) The graph on which DiaNA hops is the alignment graph $\mathcal{G}_{n,m}$ (Section 2.2). This is a square grid which includes diagonal edges. DiaNA begins at the upper left corner and will hop along edges of the graph until she reaches the opposite corner (Figure 2.1). She must always move towards the far corner; so in particular she cannot backtrack her steps. She hops randomly, which means that at every vertex, she decides at random which of the three directions to take. When she is on the right, or lower boundary, she is constrained to hop in only one direction since she must always progress towards the lower right corner.

When DiaNA is at a node of the alignment graph, she chooses randomly

which of the three outgoing edges to use next. Her random choice depends
on which incoming edge she used to reach that node. Each time DiaNA hops
across an edge, she tosses her two tetrahedral dice. Each of these dice has the
letters A,C,G,T written on the four sides. The dice land (at random) and the
result is recorded for us, the observers. Unfortunately, we are blindfolded, so
we cannot see DiaNA as she hops along the graph. Fortunately, her tetrahedral
dice tosses are recorded for us and we read them after she is done hopping.
Our goal is to guess which path DiaNA took on the graph.   □

Recall from (2.14) that the pair hidden Markov model has 33 parameters
which are given by two matrices, θ and θ'. DiaNA's path on the alignment
graph is determined by the nine entries of the matrix θ'. For instance, $\theta'_{I,H}$
is the probability that DiaNA makes a diagonal hop after a vertical hop, and
$\theta'_{D,D}$ is the probability that DiaNA makes two consecutive horizontal hops.

The biological role of the 3×3 matrix θ' is to model insertions and deletions,
with the probabilities associated to the length distributions. Suppose that DiaNA makes an insertion move. The probability that she repeats an insertion
is $\theta'_{I,I}$. It follows that the probability that she remains walking in the same
direction (i.e., keeps making insertions) for k steps is therefore $(\theta'_{I,I})^k$. The insertions generated by DiaNA are therefore geometrically distributed in length,
with expected length $\frac{1}{1-(\theta'_{I,I})^k}$. The average length of observed insertions and
deletions can therefore be used to set the parameters $\theta'_{I,I}$ and $\theta'_{D,D}$. Similarly,
the frequency of insertions and deletions determines the parameters $\theta'_{H,I}$ and
$\theta'_{H,D}$. Unfortunately, to date there has not been enough data to accurately
measure these quantities. However, it should be possible in the near future to
make such measurements, thanks to the extraordinary amount of new sequence
data, much of it from closely related species pairs.

DiaNA's pair hidden Markov model also consists of a matrix θ, which contains the probabilities for pairs of nucleotides. For instance, $\theta_{G,C}$ is the probability that DiaNA's first die lands on G and that her second die lands on C as
she hops across a diagonal edge, and $\theta_{A,-}$ is the probability that her first die
lands on A as she hops across a vertical edge.

Thus, DiaNA's pair hidden Markov model contains an evolutionary model as
an ingredient (albeit with two taxa in the case of pairwise sequence alignment).
Indeed, the sixteen output probabilities for her dice roles along the diagonal
edges represent a point on an evolutionary model. Thus, building a pair HMM
may involve choosing a model from the Felsenstein hierarchy.

In Chapter 7, we explore the question of whether pair hidden Markov models
are sufficient for modeling insertion and deletion. In particular, it is shown
that there are genomic sequences for which no choice of parameters yields the
correct alignment, thus indicating that there is lots of room for improvement
in the modeling of sequences and their mutations. In other words, we do not
yet understand exactly what it is that DiaNA is doing.

Our objective, of which we must not lose sight, is to find the best alignment
with respect to a particular choice of model parameters. This is a problem

of MAP inference, which means (by Remark 4.13) that we must evaluate one coordinate polynomial of our statistical model tropically. Thus, guessing which path DiaNA took on the alignment graph is really an algebraic computation: it is the tropical evaluation of one coordinate of the polynomial map (2.13).

Part II
Studies on the four themes

The contributions in this part of the book were all written by students, postdocs and visitors who in some way were involved in the graduate course *Algebraic Statistics for Computational Biology* that we taught in the mathematics department at UC Berkeley during the fall of 2004. The eighteen chapters offer a more in-depth study of some of the themes which were introduced in Part I. Most of the chapters contain original research that has not been published elsewhere. Highlights among new research results include:

- New results about polytope propagation and parametric inference (Chapters 5, 6 and 8).
- An example of a biologically correct alignment which is not the optimal alignment for any choice of parameters in the pair HMM (Chapter 7).
- Theorem 9.3 which states that the number of inference functions of a graphical model grows polynomially for fixed number of parameters.
- Theorem 10.5 which states that, for alphabets with four or more letters, every toric Viterbi sequence is a Viterbi sequence.
- Explicit calculations of phylogenetic invariants for the strand symmetric model which interpolate between the general reversible model and group based models (Chapter 16).
- Tree reconstruction based on singular value decomposition (Chapter 19).

The other chapters also include new mathematical results or methodological advances in computational biology. Chapter 15 introduces a standardized framework for working with small trees. Even results on the smallest nontrivial tree (with three leaves) are interesting, and are discussed in Chapter 18. Similarly, Chapter 14 presents a unified algebraic statistical view of mutagenetic tree models. In terms of tools, Chapters 11 and 20 describe computational approaches to solving long-standing problems (fast implicitization and exact maximum likelihood for phylogeny). Chapter 12 is a thorough exposition of the Baum–Welch algorithm, and addresses some of the pitfalls to beware of when applying it. Chapter 13 shows how Markov random fields can be applied to the problem of homology mapping between genomes, thus providing an elegant probabilistic framework for a problem that has until now been solved by ad hoc means. Chapter 21 investigates the placement of the rodents within the vertebrate lineage using evolutionary models. The analysis of the "rodent problem" provides useful insights into evolutionary models and a practical introduction to sequence data and phylogeny software packages. Finally, Chapter 22 is about the mysteries of ultra-conserved elements found in multiple alignments of vertebrate and fly genomes.

We present a brief biography for each of our twenty-seven contributors that summarizes their backgrounds, and explains their connection to our class.

- **Jameel Al-Aidroos** is a PhD student in mathematics at UC Berkeley. He has worked on algebraic geometry (supervised by Tom Graber), and was first introduced to computational biology during our course in fall 2004.

- **Niko Beerenwinkel** received his PhD in computer science in 2004 from the University of Saarbrücken, Germany (supervised by Thomas Lengauer). His thesis was on computational biology using machine learning methods. Upon graduation, he was awarded the prestigious Emmy Noether fellowship which he is using to pursue postdoctoral research at UC Berkeley.

- **Nicolas Bray** was a mathematics undergraduate at UC Berkeley and continues now as a PhD student in the same department (supervised by Lior Pachter). He is the developer of the `MAVID` multiple alignment program, and continues to work on comparative genomics.

- **David Bryant** is an assistant professor of mathematics and computer science at McGill University in Montreal, Canada, and also holds a faculty position at the University of Auckland, New Zealand. An expert on phylogenetic analysis, he is a co-developer of the software package `SplitsTree`. David Bryant was invited to lecture in our course during the fall of 2004.

- **Marta Casanellas** is a Ramón y Cajal researcher at Universitat Politècnica de Catalunya in Barcelona, Spain. She is an expert in algebraic geometry, and became interested in computational biology in 2004 while visiting Roderic Guigó and his computational biology group at the Institut Municipal d'Investigació Mèdica in Barcelona. Marta came to Berkeley in the fall of 2004 for an extended visit, and now works on phylogenetics.

- **Anat Caspi** is a PhD student in the joint UC San Francisco/UC Berkeley graduate group in bioengineering (supervised by Lior Pachter). She has a Masters degree in computer science from Stanford University, and is interested in machine learning applications to computational biology.

- **Mark Contois** is an undergraduate student in mathematics at San Francisco State University (supervised by Serkan Hoşten). He has also worked on computational biology in the laboratory of Eric Routman.

- **Colin Dewey** studied biology and computer science as an undergraduate at UC Berkeley, and is now a PhD student in computer science (supervised by Lior Pachter). He has worked on homology mapping and parametric sequence alignment, and is the developer of the `Mercator` mapping program.

- **Mathias Drton** received his PhD in statistics from the University of Washington in 2004 (supervised by Michael Perlman and Thomas Richardson). He has worked on maximum likelihood estimation in graphical models, and is a postdoctoral researcher at UC Berkeley. In Fall 2005 he starts a tenure track position in the Department of Statistics at the University of Chicago.

- **Sergi Elizalde** received his PhD in applied mathematics from MIT in 2004 (supervised by Richard Stanley) where he worked on combinatorics and enumeration. He is currently a postdoctoral fellow at the Mathematical Sciences

Research Institute at Berkeley (MSRI). Starting in Fall 2005, he will be a John Wesley Instructor in Mathematics at Dartmouth College.

- **Nicholas Eriksson** is a PhD student in mathematics at UC Berkeley (supervised by Bernd Sturmfels). He has worked on algebraic statistics and is the first mathematics graduate student at UC Berkeley to enroll in the designated emphasis in computational biology interdisciplinary program.
- **Luis David Garcia** received his PhD in mathematics from Virginia Polytechnic Institute in 2004 (supervised by Reinhard Laubenbacher), where he worked at the Virginia Bioinformatics Institute. After spending a postdoctoral semester in the fall 2004 program on *Hyperplane Arrangements* at MSRI, he is now a Visiting Assistant Professor at Texas A & M University. He works in computational algebra and algebraic statistics.
- **Ingileif B. Hallgrímsdóttir** will receive her PhD in statistics from UC Berkeley in May 2005 (supervised by Terence Speed). She works in statistical genetics, a topic which she had already pursued while working at DeCODE Genetics in Iceland, and for her Masters in Gothenburg, Sweden.
- **Michael Joswig** is an expert in mathematical software and polyhedral geometry. He co-developed the software `polymake`. Michael holds a professorship in Mathematics at the Technische Universität Darmstadt, Germany.
- **Eric Kuo** will receive his PhD in computer science from UC Berkeley in May 2005 (supervised by Lior Pachter). His interests range from theoretical computer science to convex polytopes and discrete mathematics. Starting in Fall 2005, he will be a postdoc at George Mason University.
- **Fumei Lam** was an undergraduate at UC Berkeley. She will receive her PhD in applied mathematics from MIT in the summer of 2005 (supervised by Michel Goemans). She has worked on graph theory, approximation algorithms and computational biology.
- **Garmay Leung** is a PhD student in the joint UC San Francisco/UC Berkeley graduate group in bioengineering and is doing rotations (currently with Michael B. Eisen). She did research on computational biology and cell biology while an undergraduate at Cornell University.
- **Dan Levy** is receiving his PhD in mathematics during the summer of 2005 (supervised by Lior Pachter and Rainer Sachs), and will stay in Berkeley for one more postdoctoral year. His research is in mathematical biology.
- **Radu Mihaescu** was an undergraduate at Princeton University and is now a PhD student in mathematics at UC Berkeley (supervised by Lior Pachter and Satish Rao). He is interested in theoretical computer science.
- **Alex Milowski** received his MA in mathematics from San Francisco State University in May 2004 (supervised by Serkan Hoşten). Before that he was a developer of the Extensible Markup Language (XML) and related standards. Starting in the Fall 2005, he will be a mathematics PhD student at UC Davis.
- **Jason Morton** is a second-year PhD student in mathematics at UC Berkeley (supervised by Bernd Sturmfels). He is interested in algebraic statistics.

- **Raazesh Sainudiin** will receive his PhD in statistics from Cornell University in May 2005 (supervised by Rick Durrett). His research is on statistical inference in population genetics, phylogenetics and molecular evolution.

- **Sagi Snir** received his PhD in computer science in May 2004 from the Technion - Israel Institute of Technology (supervised by Benny Chor). He has worked on analytic maximum likelihood solutions for phylogenetic reconstruction, and on convex recoloring. He is a postdoc at UC Berkeley.

- **Seth Sullivant** received his MA from San Francisco State University in 2002 (supervised by Serkan Hoşten) and will receive his PhD from UC Berkeley in May 2005 (supervised by Bernd Sturmfels). Both degrees are in mathematics. He works on algebraic statistics. Starting in July 2005, Seth will be a junior fellow with the Society of Fellows at Harvard University.

- **Kevin Woods** received his PhD in mathematics from the University of Michigan in May 2004 (supervised by Alexander Barvinok). He is interested in combinatorics, specifically topics in discrete and computational geometry, and is an NSF postdoc in the Mathematics Department at UC Berkeley.

- **Ruriko Yoshida** received her PhD in mathematics from UC Davis in May 2004 (supervised by Jesus DeLoera). She has worked on phylogeny reconstruction, combinatorics, contingency tables and integer programming. After visiting the Berkeley math department during the summer of 2004, she is now an Assistant Research Professor of Mathematics at Duke University.

- **Josephine Yu** was an undergraduate at UC Davis and is now a PhD student in mathematics at UC Berkeley (supervised by Bernd Sturmfels). She has worked on matrix integrals, finite metric spaces and tropical geometry.

The results in this part of the book only begin to hint at the vast number of mathematical, computational, statistical and biological challenges that will need to be overcome in order to understand the function and organization of genomes. In a future version of our graduate course we imagine numerous new class projects on a wide range of topics including Bayesian statistics, graphical models with cycles and loopy propagation, aspects of real algebraic geometry, information geometry, multiple sequence alignment, motif finding, RNA structure, whole genome phylogeny, and protein sequence analysis, to name a few.

5

Parametric Inference

Radu Mihaescu

Graphical models are powerful statistical tools that have been applied to a wide variety of problems in computational biology: sequence alignment, ancestral genome reconstruction, etc. A graphical model consists of a graph whose vertices have associated random variables representing biological objects, such as entries in a DNA sequence, and whose edges have associated parameters that model transition or dependence relations between the random variables at the nodes. In many cases we will know the contents of only a subset of the model vertices, the *observed random variables*, and nothing about the contents of the remaining ones, the *hidden random variables*. A common example is a phylogenetic tree on a set of current species with given DNA sequences, but with no information about the DNA of their extinct ancestors. The task of finding the most likely set of values of the hidden random variables (also known as the *explanation*) given the set of observed random variables and the model parameters, is known as *inference in graphical models*.

Clearly, inference drawn about the hidden data is highly dependent on the topology and parameters (transition probabilities) of the graphical model. The topology of the model will be determined by the biological process being modeled, while the assumptions one can make about the nature of evolution, site mutation and other biological phenomena, allow us to restrict the space of possible transition probabilities to certain parameterized families. This raises several questions. If our choice of parameters is slightly off, will the explanation change? What other choices of parameters will give the same explanation? Can we find all possible explanations and the sets of parameters which yield them? These are the sorts of questions we will answer in this chapter, using the tools of *parametric inference*, which solves the inference problem for all possible sets of parameters simultaneously.

We present the *polytope propagation algorithm* for parametric inference, which was first introduced in [Pachter and Sturmfels, 2004a]. This algorithm is the polytope algebra version (see Section 2.3) of a classical method in the theory of graphical models, known as *sum-product decomposition*. We examine the polytope propagation algorithm in Section 5.2, and, in particular, we describe the details of the algorithm in the context of two very important problems in computational biology: the hidden Markov model for gene annotation and the pair hidden Markov model for genome alignment. The analysis relies heavily on the theory developed in Sections 1.4, 2.2 and 2.3.

The running time of polytope propagation is exponential in the number of parameters. Therefore, in applications where the number of parameters is very large, it is of practical interest to specialize most of them to fixed values and study the dependence of the explanation upon variations of the remaining few parameters. In Section 5.4 we give an explicit presentation of an algorithm that does this efficiently, together with an analysis of its running time. As we will see, the complexity of the algorithm is polynomial in the length of the sequences for a fixed number of unspecialized parameters.

5.1 Tropical sum-product decompositions

We begin by showing that the problem of inference for fixed parameters can be regarded as the tropical version of computing the marginal probability of the observed data. For a graphical model with hidden and observed random variables denoted by σ and τ respectively, the probability of the observed sequence τ is

$$\text{Prob}(\tau) = \sum_{\sigma} p_{\sigma,\tau}, \tag{5.1}$$

where $p_{\sigma,\tau}$ is the probability of having states σ at the hidden nodes and states τ at the observed nodes of the model. This is the probability of the observed data marginalized over all possible values for the hidden data. The task of finding an explanation corresponds to identifying the set of hidden states $\bar{\sigma}$ with maximum a posteriori probability of generating the observed data τ. In other words:

$$\bar{\sigma} = \text{argmax}_{\sigma}\{p_{\sigma,\tau}\}.$$

Now following the notation of Chapter 2, let $w_* = -\ln(p_*)$. Then the above equation becomes

$$\bar{\sigma} = \text{argmin}_{\sigma}\{w_{\sigma,\tau}\},$$

which is exactly the marginalization in (5.1), performed in tropical algebra:

$$w_{\bar{\sigma}} = \bigoplus_{\sigma} w_{\sigma,\tau}. \tag{5.2}$$

The reader is referred to Section 2.1 for more details on tropical algebra.

In general, marginal probabilities for acyclic graphical models can be computed in time which is polynomial in the size of the model using the *sum-product decomposition*, which is a recursive representation of a polynomial in terms of smaller polynomials. Such a decomposition is very useful for computing values of polynomial expressions with a large number of monomials, where a direct symbolic computation would be very costly. In the literature on hidden Markov models this is known as the *forward algorithm*.

As we can see from the above analysis, evaluating the marginal probability tropically is equivalent to solving the inference problem. See also Remarks 2.17, 2.18 and 4.13. Therefore, the sum-product decomposition of marginal probabilities, when it exists, naturally yields efficient algorithms for inference with fixed parameters. In the following subsections we exemplify this with the *Viterbi algorithm* for hidden Markov models and the *Needleman–Wunsch algorithm* for sequence alignment (see Section 2.2).

5.1.1 The sum-product algorithm for HMMs

The hidden Markov model is one of the simplest and most popular models used in computational biology. In this subsection we will describe the sum-product algorithm for the HMM. We use the notation of Section 1.4, to which we also refer the reader unfamiliar with the model. Suppose that we have an HMM of length n, with hidden states σ_i, $i \in [n]$, taking values in an alphabet Σ with l letters, and observed variables τ_i, $i \in [n]$, taking values in the alphabet Σ' of size l'. The model parameters are the "transition" probability matrix $\theta \in \mathbb{R}^{l \times l}$ and the "emission" probability matrix $\theta' \in \mathbb{R}^{l \times l'}$. If one assumes a uniform initial distribution on the states of the first hidden node, the probability of occurrence of the sequence (σ, τ) is therefore

$$p_{\sigma,\tau} = \frac{1}{l}\theta'_{\sigma_1,\tau_1}\theta_{\sigma_1,\sigma_2}\theta'_{\sigma_2,\tau_2}\theta_{\sigma_2,\sigma_3}\cdots\theta'_{\sigma_n,\tau_n}.$$

The marginal probability of the observed sequence $\tau = \tau_1\tau_2\ldots\tau_n$ is

$$p_\tau = \sum_\sigma p_{\sigma,\tau}. \tag{5.3}$$

By tropicalizing and maintaining the notation from the beginning of the section we see that, as before, the explanation for the sequence of observations τ is given by

$$\begin{aligned}
\bar{\sigma} &= \operatorname{argmin}_\sigma\{w_{\sigma,\tau}\}, \\
w_{\bar{\sigma}} &= \bigoplus_\sigma w_{\sigma,\tau}.
\end{aligned} \tag{5.4}$$

The problem of computing (5.3) can easily be solved by noticing that the probability p_τ has the following decomposition:

$$p_\tau = \sum_{\sigma_n=1}^{l}\theta'_{\sigma_n,\tau_n}\left(\sum_{\sigma_{n-1}=1}^{l}\theta_{\sigma_{n-1},\sigma_n}\theta'_{\sigma_{n-1},\tau_{n-1}}\left(\cdots\left(\sum_{\sigma_1=1}^{l}\theta_{\sigma_1,\sigma_2}\theta'_{\sigma_1,\tau_1}\right)\cdots\right)\right). \tag{5.5}$$

Computing p_τ using this decomposition is known as the *forward algorithm* for HMMs. Its time complexity is $O(l^2 n)$, as can easily be checked.

Observe that tropicalizing this algorithm gives us a way of efficiently solving equation (5.4). By taking $u_{i,j} = -\log(\theta_{i,j})$ and $v_{i,j} = -\log(\theta'_{i,j})$, we obtain

$$\bigoplus_{\sigma} \left(v_{\sigma_1\tau_1} \odot u_{\sigma_1\sigma_2} \odot v_{\sigma_2\tau_2} \ldots \odot u_{\sigma_{n-1}\sigma_n} \odot v_{\sigma_n\tau_n}\right) = \qquad (5.6)$$

$$\bigoplus_{\sigma_n}(v_{\sigma_n\tau_n} \odot (\bigoplus_{\sigma_{n-1}} (v_{\sigma_{n-1}\tau_{n-1}} \odot u_{\sigma_{n-1}\sigma_n} \cdots \odot (\bigoplus_{\sigma_1}(v_{\sigma_1\tau_1} \odot u_{\sigma_1\sigma_2}))\ldots))).$$

Evaluating this quantity by recursively computing the parentheses in the above formula is known as the *Viterbi algorithm*, and has the same time complexity as its non-tropical version, the forward algorithm.

5.1.2 The sum-product algorithm for sequence alignment

The sequence alignment problem asks for the best possible alignment between two words $\sigma^1 = \sigma^1_1\sigma^1_2\ldots\sigma^1_n$ and $\sigma^2 = \sigma^2_1\sigma^2_2\ldots\sigma^2_n$ over the alphabet $\Sigma = \{\mathtt{A}, \mathtt{C}, \mathtt{G}, \mathtt{T}\}$ that have evolved from a common ancestor via insertions, deletions or mutations of sites in the genetic sequence. A full description of the problem can be found in Section 2.2. As in Section 2.2, we represent an alignment by an edit string h over the alphabet $\{H, I, D\}$ such that $\#H + \#D = n$ and $\#H + \#I = m$. Let $\mathcal{A}_{n,m}$ be the set of all edit strings.

Each element $h \in \mathcal{A}_{n,m}$ corresponds naturally to a pair of words (μ^1, μ^2) over the alphabet $\Sigma \cup \{-\}$ such that μ^1 consists of a copy of σ^1 together with inserted "$-$" characters, and similarly μ^2 is a copy of σ^2 with inserted "$-$" characters, see (2.8).

Now consider the pair hidden Markov model for sequence alignment presented in Section 2.2. Equation (2.15) gives us the marginal probability f_{σ^1,σ^2} of observing the pair of sequences σ^1 and σ^2:

$$f_{\sigma^1,\sigma^2} \quad = \quad \sum_{h\in\mathcal{A}_{n.m}} \prod_{i=1}^{|h|}\theta_{\mu^1_i,\mu^2_i} \cdot \prod_{i=2}^{|h|}\theta'_{h_{i-1},h_i}. \qquad (5.7)$$

Here the parameters θ and θ' are as in Section 2.2.

Just as before, we will be interested in the tropical version of the above formula, which gives the alignment with the largest *a posteriori* probability, given the parameters of the model and the observed sequences. Letting $w_{i,j} = -\ln\theta_{i,j}$ and $w'_{i,j} = -\ln\theta'_{i,j}$, equation (5.7) yields

$$\mathrm{trop}(f_{\sigma^1,\sigma^2}) \quad = \quad \bigoplus_{h\in\mathcal{A}_{n.m}} \bigodot_{i=1}^{|h|} w_{\mu^1_i,\mu^2_i} \cdot \bigodot_{i=2}^{|h|} w'_{h_{i-1},h_i}. \qquad (5.8)$$

The above relation computes the negative logarithm of the *maximum a posteriori* (MAP) probability over the set of possible alignments. This is equivalent to finding a minimum path in the alignment graph of Section 2.2, which can be solved through the Needleman–Wunsch algorithm, a version of the sum-product algorithm, based on the recursive decomposition of (5.8) described below.

Let $\sigma^1_{\leq i}$ denote the sequence $\sigma^1_1\sigma^1_2\cdots\sigma^1_i$. Let $\sigma^2_{\leq j}$ be defined in the same way. Also define $\Phi^X(i,j)$ to be the maximum negative log probability among alignments of $\sigma^1_{\leq i}$ and $\sigma^2_{\leq j}$ such that the last character in the corresponding edit string is X. Equation (5.8) then gives us the following recursive formula(s):

$$\text{trop}(f_{\sigma^1,\sigma^2}) = \bigoplus_X \Phi^X(n,m), \tag{5.9}$$

where

$$
\begin{aligned}
\Phi^I(i,j) &= w_{-,\sigma^2_j} \odot \bigoplus_X (\Phi^X(i,j-1) \odot w'_{X,I}), \\
\Phi^D(i,j) &= w_{\sigma^1_i,-} \odot \bigoplus_X (\Phi^X(i-1,j) \odot w'_{X,D}), \\
\Phi^H(i,j) &= w_{\sigma^1_i,\sigma^2_j} \odot \bigoplus_X (\Phi^X(i-1,j-1) \odot w'_{X,H}),
\end{aligned}
\tag{5.10}
$$

and

$$
\begin{aligned}
\Phi^X(0,0) &= 0 \text{ for all } X, \\
\Phi^X(0,j) &= 0 \text{ for all } X \neq I, \\
\Phi^X(i,0) &= 0 \text{ for all } X \neq D, \\
\Phi^I(0,j) &= w_{-,\sigma^2_1} \odot \bigodot_{k=2}^{j} (w'_{I,I} \odot w_{-,\sigma^2_k}), \\
\Phi^D(i,0) &= w_{\sigma^1_1,-} \odot \bigodot_{k=2}^{i} (w'_{D,D} \odot w_{\sigma^1_k,-}).
\end{aligned}
$$

The running time of the Needleman–Wunsch algorithm is $O(nm)$ as we perform a constant number of $\oplus$ and $\odot$ operations for each pair of indices (i,j).

5.2 The polytope propagation algorithm

In this section we will describe parametric *maximum a posteriori probability* estimation for probabilistic models. Our goal is to explain how parametric MAP estimation is related to linear programming and polyhedral geometry. The parametric MAP estimation problem comes in two different versions. First there is the local one: given a particular choice of parameters determine the set of all parameters which have the same MAP estimate. This version is an important problem because it can be used to decide how sensitive the MAP estimate is to perturbations in the parameters. The global version of parametric MAP estimation problem asks for a partition of the space of parameters such that any two parameters lie in the same part if and only if they yield the same MAP estimate. We will show that for arbitrary statistical models, the local problem is solved by computing a certain polyhedral cone (the normal cone at a vertex of the Newton polytope) and the global problem is solved by computing a certain polyhedral fan (the normal fan of the Newton polytope). If the underlying statistical model has a sum-product decomposition, there is a natural extension of the tropical sum-product algorithm which replaces numbers with polyhedra and solves the parametric MAP estimation problem.

We will now show how to perform the tropical sum-product algorithm in a general fashion, finding an explanation for all possible sets of parameters. Let us consider the polynomial

$$f(p) = \sum_{i=1}^{d} p_1^{e_{i1}} \cdots p_k^{e_{ik}}$$

and suppose that f is the density function associated to a statistical model where $p = (p_1, \ldots, p_k)$ is the vector of parameters, and each possible sequence of hidden states corresponds to some monomial of f. See (5.3) as an example. We maintain this assumption throughout the rest of this chapter.

For a fixed value of p, finding an explanation is equivalent to finding the index j of the monomial of f with maximum value

$$j = \mathrm{argmax}_i \{ p_1^{e_{i1}} \cdots p_k^{e_{ik}} \}.$$

Letting $w_i = -\log p_i$, this amounts to finding the index j of the monomial of f which minimizes the linear expression $e_j \cdot w = \sum_{i=1}^{k} w_i e_{j,i}$. We observe that e_j can be an explanation for some choice of parameters if and only if the point $P_j = (e_{j1}, \ldots, e_{jk})$ is on the convex hull of the set $\{(e_{i1}, \ldots, e_{ik}) : i \in [d]\}$, i.e., it is a vertex of the Newton polytope of f, $\mathrm{NP}(f)$.

The optimization problem of finding an explanation for fixed parameters w can therefore be interpreted geometrically as a linear programming problem on the Newton polytope $\mathrm{NP}(f)$. In the notation of Section 2.3, the optimization problem described above means finding $(e_{j,1}, e_{j,2}, \ldots, e_{j,k}) \in \mathrm{face}_w(\mathrm{NP}(f))$. Conversely, the parametric version of this problem asks for the set of parameter vectors w for which a vertex P_j gives the explanation. In Section 2.3 it is shown that this is the cone in the normal fan of the polytope $\mathrm{NP}(f)$ which corresponds to the vertex P_j, namely $N_{\mathrm{NP}(f)}(P_j)$. Constructing the normal fan $\mathcal{N}_{\mathrm{NP}(f)}$ therefore amounts to partitioning the parameter space into regions such that the explanation(s) for all sets of parameters in a given region is given by the polytope vertex(face) associated to that region. We can obtain $\mathrm{NP}(f)$ and $\mathcal{N}_{\mathrm{NP}(f)}$ through the *polytope propagation algorithm*, which is the polytope algebra version of the sum-product decomposition. We refer the reader to Section 2.3 for details on Newton polytopes, normal fans and the polytope algebra.

For example, solving the parametric version of (5.4) for hidden Markov models or (5.8) for sequence alignment amounts to finding the normal fan of the Newton polytopes $\mathrm{NP}(p_\tau)$ and $\mathrm{NP}(f_{\sigma^1, \sigma^2})$. As can easily be observed, in both examples our polynomials will have an exponential number of monomials. It is thus not feasible to compute the Newton polytope by first computing the polynomial explicitly. We will therefore make use of the recursive representations given by (5.6) and (5.10). Theorem 2.26 immediately gives us a recursive representation of the needed Newton polytopes: simply translate (5.6) and (5.10) into the polytope algebra of Section 2.3.

For the hidden Markov model we obtain the following:

$$\mathrm{NP}(p_\tau) \;=\; \bigoplus_{\sigma_n}(\mathrm{NP}(\theta'_{\sigma_n\tau_n}) \odot \bigoplus_{\sigma_{n-1}}(\mathrm{NP}(\theta'_{\sigma_{n-1}\tau_{n-1}}\theta_{\sigma_{n-1}\sigma_n})$$

$$\odot \ldots \odot \bigoplus_{\sigma_1}(\mathrm{NP}(\theta'_{\sigma_1\tau_1}\theta_{\sigma_1\sigma_2}))\ldots)).$$

In the sequence alignment example, take $\mathcal{P}^I(i,j)$ to be the Newton polytope of the sum of the scores of all alignments of the two partial sequences $\sigma^1_{\leq i}$ and $\sigma^2_{\leq j}$ which end with an insertion. This corresponds to the sum of the weights of all paths from the origin to the insertion vertex of the $K_{3,3}$ corresponding to position (i,j) in the alignment graph of Figure 2.2. Define $\mathcal{P}^D(i,j)$ and $\mathcal{P}^H(i,j)$ similarly and (5.10) gives us

$$\mathrm{NP}(f_{\sigma^1,\sigma^2}) = \bigoplus_X \mathcal{P}^X(n,m), \tag{5.11}$$

where

$$\mathcal{P}^I(i,j) \;=\; \mathrm{NP}(\theta_{-,\sigma^2_j}) \odot \bigoplus_X (\mathcal{P}^X(i,j-1) \odot \mathrm{NP}(\theta'_{X,I})),$$

$$\mathcal{P}^D(i,j) \;=\; \mathrm{NP}(\theta_{\sigma^1_i,-}) \odot \bigoplus_X (\mathcal{P}^X(i-1,j) \odot \mathrm{NP}(\theta'_{X,D})),$$

$$\mathcal{P}^H(i,j) \;=\; \mathrm{NP}(\theta_{\sigma^1_i,\sigma^2_j}) \odot \bigoplus_X (\mathcal{P}^X(i-1,j-1) \odot \mathrm{NP}(\theta'_{X,H})),$$

and

$$\mathcal{P}^X(0,0) \;=\; \{(0,\ldots,0)\} \text{ for all } X,$$

$$\mathcal{P}^X(0,j) \;=\; \{(0,\ldots,0)\} \text{ for all } X \neq I,$$

$$\mathcal{P}^X(i,0) \;=\; \{(0,\ldots,0)\} \text{ for all } X \neq D,$$

$$\mathcal{P}^I(0,j) \;=\; \mathrm{NP}(\theta_{-,\sigma^2_1} \prod_{k=2}^{j}(\theta'_{I,I}\theta_{-,\sigma^2_k})),$$

$$\mathcal{P}^D(i,0) \;=\; \mathrm{NP}(\theta_{\sigma^1_1},- \prod_{k=2}^{i}(\theta'_{D,D}\theta_{\sigma^1_k,-})).$$

The above decompositions naturally yield straightforward algorithms for computing the Newton polytopes $\mathrm{NP}(p_\tau)$ and $\mathrm{NP}(f_{\sigma^1,\sigma^2})$, and one can easily extend this method to any polynomial f with a sum-product decomposition. Once the polytope $\mathrm{NP}(f)$ has been computed, the final step of our algorithm is to compute the normal fan $\mathcal{N}_{\mathrm{NP}(f)}$.

To conclude our presentation of polytope propagation, we have to make two very important observations. First, we note that several sequences of hidden states may yield equivalent monomials. In general the coefficient of the monomial is equal to the number of different sequences of hidden states that map to it. The Newton polytope of f, and thus the polytope propagation algorithm, is completely oblivious to this aspect. One can however recover the full set of hidden sequences corresponding to a given vertex by backtracking along the steps of the polytope propagation algorithm, a common technique in dynamic programming. In sequence alignment for example, finding the edit strings corresponding to an extremal vertex is the parametric version of finding all optimal paths in the alignment graph of Section 2.2.

The second observation is that our assumption that each set of hidden states maps to a monomial of f implicitly insures that the coefficient of that monomial is not zero: in other words that no cancellation occurs in the sum-product decomposition. In general, for polynomials representing marginal probabilities in a graphical model, all the coefficients in their sum-product representations are positive. This is clearly the case for our two running examples. In fact, as Newton polytopes are oblivious to coefficients, running polytope propagation instead of evaluating f and $\mathrm{NP}(f)$ directly automatically insures that no vertices are omitted because of cancellation. Therefore, all sequences of values for the hidden variables which are optimal for some set of parameters are represented in the final polytope, which will contain $\mathrm{NP}(f)$ if cancellation occurs.

5.2.1 A small alignment example

To illustrate our algorithm, we give below a small example of parametric sequence alignment under a highly simplified version of the scoring scheme of Section 2.2. The parameters of our model are a "reward" assigned to all matches and a "penalty" assigned to all mismatches and gaps. We disregard completely the scores assigned to transitions between hidden nodes. In the language of Section 2.2, this is equivalent to $w'_{X,Y} = 0$ for all $X, Y \in \{H, I, D\}$, $w_{a,a} = x$ for all $a \in \{\mathtt{A}, \mathtt{C}, \mathtt{G}, \mathtt{T}\}$ and $w_{a,b} = y$ for all $a, b \in \{\mathtt{A}, \mathtt{C}, \mathtt{G}, \mathtt{T}, -\}$, $a \neq b$. This model is commonly known as the *2-parameter model for sequence alignment*.

Notice that the absence of transition probabilities between hidden nodes eliminates the need for the triple recurrence present in the sum-product decomposition of the generalized scoring scheme. Letting $\Phi(i, j)$ denote the score of the best alignment of the sequences $\sigma^1_{\leq i}$ and $\sigma^2_{\leq j}$, we have:

$$\Phi(i,j) = (\Phi(i-1, j-1) \odot w_{\sigma^1_i, \sigma^2_j}) \oplus (\Phi(i-1, j) \odot y) \oplus (\Phi(i, j-1) \odot y). \quad (5.12)$$

In polytope algebra, let $\mathcal{P}(i, j) = \mathrm{NP}(\Phi(i, j))$. The above relation becomes:

$$\mathcal{P}(i,j) = (\mathcal{P}(i-1, j-1) \odot \mathrm{NP}(w_{\sigma^1_i, \sigma^2_j})) \oplus (\mathcal{P}(i-1, j) \odot (y)) \oplus (\mathcal{P}(i-1, j-1) \odot (y)),$$

where (x) denotes the Newton polytope with the single vertex $(1, 0)$ and (y) denotes the Newton polytope with the single vertex $(0, 1)$.

Figure 5.1 illustrates the polytope propagation algorithm for the alignment of the two sequences $\sigma^1 = \mathtt{ATCG}$ and $\sigma^2 = \mathtt{TCGG}$. At each cell in the matrix, the displayed polytope is the convex hull of the points given by alignments of the corresponding partial sequences. Each is obtained by taking the convex hull of the $(0, 1)$ translation of the polytope in the cell above (alignments ending with a deletion) , the $(0, 1)$ translation of the polytope in the left cell (alignments ending with an insertion), and the $(1,0)$ or $(0,1)$ translation of the polytope in the above-left cell (alignments ending with a match or mismatch, respectively).

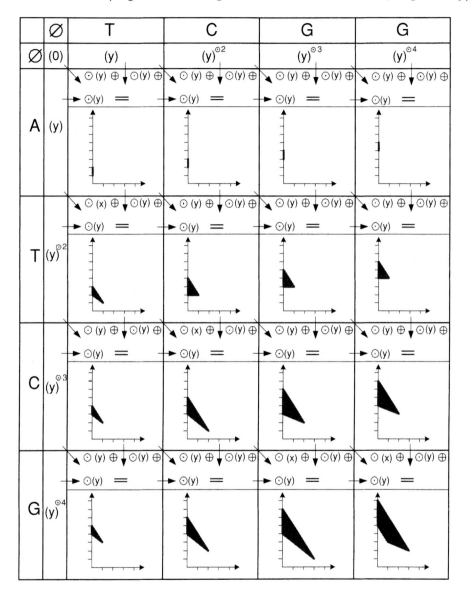

Fig. 5.1. Polytope propagation for sequence alignment.

5.3 Algorithm complexity

In this section we analyze the time complexity of the polytope propagation algorithm. Given a polynomial f together with a sum-product decomposition, we want to compute $\mathrm{NP}(f)$ and $\mathcal{N}_{\mathrm{NP}(f)}$. The questions we want to answer are:

(i) How many polytope algebra operations do we have to perform?
(ii) What is the individual time complexity of these operations?
(iii) What is the complexity of computing the normal fan of $\mathcal{N}_{\mathrm{NP}(f)}$?

Let us consider the first question. In the examples of the previous section, the only multiplicative (Minkowski sum) operations we needed to perform were multiplication by a single point, i.e., shifting a polytope by a given vector. For a D-dimensional polytope with N vertices, this takes $O(ND)$ time and is strongly dominated by the additive operations. In this section we will limit our analysis to models where the only multiplicative operations are the trivial ones, as this turns out to be the case with most acyclic graphical models.

In general, the number of polytope algebra operations we need to perform will be the product of the number of levels in the sum-product decomposition of f and the number of operations per level. This is exactly the time complexity of computing the explanation for a *given* set of parameters, using the sum-product decomposition. For instance, in the case of the hidden Markov model, the total number of polytope algebra operations will be $O(l^2n)$. In the sequence alignment example, at each step we compute three sums of exactly three polynomials, therefore the total number of operations is $O(nm)$.

In order to answer question (ii), we need to choose an efficient representation for our Newton polytopes. Section 2.3 provides us with two options: the V-representation and the H-representation. In general, the two are roughly equivalent in terms of computational versatility, due to the principle of duality. In the context of parametric inference, the V-representation is more natural, as we are able to prove upper bounds on the number of vertices of the Newton polytopes we encounter.

To facilitate our subsequent discussion, let $\nu_D(K)$ denote the computational complexity of finding the V-representation of the convex hull of K points in D dimensions. Also let N be the maximum number of vertices among all the polytopes encountered by the algorithm. It is clear that the additive polytope algebra operation will have a time complexity of at most $O(\nu_D(2N))$.

The next step is providing upper bounds for the number N. Since we are dealing with Newton polytopes of polynomials, all vertices will have integer coordinates. Now suppose that the degree of f is n. Then at every intermediate step of the algorithm, the degree of any variable will always be at most n. We can therefore assert that all polytopes encountered by the algorithm will lie inside the D-dimensional hypercube of side-length n. We will make use of the following theorem of [Andrews, 1963]:

Theorem 5.1 *For every fixed integer D there exists a constant C_D such that the number of vertices of any convex lattice polytope $\mathcal{P}$ in $\mathbb{R}^D$ is bounded above by $C_D \cdot \mathrm{vol}(P)^{(D-1)/(D+1)}$.*

Unfortunately, Theorem 5.1 only applies to full dimensional polytopes: the polytope must not be contained in a lower-dimensional affine subspace of $\mathbb{R}^D$. As it turns out, this will almost always be the case. Now suppose that the final polytope we compute has dimension $d < D$. It is easy to see that the polytope algebra operations can only result in an increase of the dimension of the polytopes, i.e., $\dim(\mathcal{P} \oplus \mathcal{Q}) \geq \max\{\dim(\mathcal{P}), \dim(\mathcal{Q})\}$ and $\dim(\mathcal{P} \odot \mathcal{Q}) \geq \max\{\dim(\mathcal{P}), \dim(\mathcal{Q})\}$ for any two polytopes $\mathcal{P}$ and $\mathcal{Q}$. Then all of the intermediate polytopes will have dimension at most d. We call d the *true dimension* of the model or polynomial.

Let us illustrate. In the HMM example, each monomial $p_{\sigma,\tau}$ is a product of n variables θ'_{ij} and $n-1$ variables θ_{ij}. Thus, for each point in $\mathrm{NP}(p_\tau)$, the sum of the ll' coordinates corresponding to the θ'_{ij} variables is n and the sum of the l^2 coordinates corresponding to the θ_{ij} variables is $n-1$. These two inherent constraints of the model mean that $d \leq D - 2$.

The following easy lemma will be needed in the derivation of our running time bounds.

Lemma 5.2 *Let S be a d-dimensional linear subspace of $\mathbb{R}^D$. Then, among the set of D coordinate axes of $\mathbb{R}^D$, there exists a subset $\{i_1, i_2, ..., i_d\}$ such that the projection $\phi : S \to \mathbb{R}^d$ given by $\phi((x_1, x_2, \ldots, x_D)) = (x_{i_1}, x_{i_2}, \ldots, x_{i_D})$ is injective.*

Proof Let $v^1, \ldots, v^d \in \mathbb{R}^D$ be a basis for the subspace S. Let A be the $D \times d$ matrix with columns $v^1, \ldots, v^d$. Then the rank of the matrix A is exactly d, the dimension of S. Now suppose that for any choice of indices $\{i_1, i_2, ..., i_d\}$ the projection $\phi((x_1, x_2, \ldots, x_D)) = (x_{i_1}, x_{i_2}, \ldots, x_{i_D})$ is not injective on S. This means that the $d \times d$ minor of A given by the rows indexed by the i_j's has rank strictly less than d. Since the choice of rows was arbitrary, this contradicts the fact that the rank of A is d. □

Let S be the d-dimensional affine span of a polytope $\mathcal{P}$. By Lemma 5.2, there is a set of d coordinate axes such that the projection ϕ of the linear subspace S onto the space determined by those d axes is injective. Then $\phi(\mathcal{P})$ is also a d-dimensional polytope whose vertices have integer coordinates and are in 1-1 correspondence with the vertices of $\mathcal{P}$. Moreover, $\phi(\mathcal{P})$ lies inside a d-dimensional hypercube with edge length n (all the exponents are at most n). Therefore the volume of $\phi(\mathcal{P})$ is at most n^d. Applying Theorem 5.1, we infer that $\mathcal{P}$ has at most $N = C_d \cdot n^{d(d-1)/(d+1)}$ vertices. Since all intermediate polytopes have dimension less than or equal to that of $\mathrm{NP}(f)$, the above upper bound on the number of vertices will hold for all intermediate polytopes if we let d be the true dimension of our model. We obtain the following theorem:

Theorem 5.3 *Let f be a polynomial of degree n in D variables. If the dimension of $\mathrm{NP}(f)$ is d, then all the polytopes computed by the polytope propagation algorithm will have at most $C_d \cdot n^{d(d-1)/(d+1)}$ vertices.*

We have not yet elucidated the mystery surrounding the function ν_D. It turns out that for $D = 2$ and $D = 3$, things are much simpler than in higher dimensions. The dominant operation in our algorithm is that of taking the convex hull of the union of two convex polytopes. For $D = 2$, this can be done in $O(N)$ time if the input polytopes have a total of N vertices, although computing the extremal vertices of a general set of N points takes $O(N \log N)$ time. For $D = 3$, computing the union of two disjoint convex polytopes with

N vertices takes $O(N)$ time, but when the polytopes are not disjoint it is asymptotically no easier than computing the convex hull of the union of their sets of vertices. This can be done in $O(N \log N)$ time by the algorithm of Preparata and Hong. The results for both 2 and 3 dimensions can be found in [Preparata and Shamos, 1985].

Now consider the case $D \geq 4$. A point $v_j = (x_{j1}, \ldots, x_{jD}) \in \mathbb{R}^D$ is a vertex of the convex hull of the set $\mathcal{A} = \{v_1, \ldots, v_N\}$ if and only if there is a hyperplane of $\mathbb{R}^D$ separating v_j from the rest of the points in $\mathcal{A}$. In turn, the existence of such a hyperplane is equivalent to the feasibility of the following linear program:

$$\text{Maximize} \quad \epsilon \text{ subject to}$$
$$\lambda_0, \ldots, \lambda_D \in \mathbb{R}, \, \epsilon \geq 0,$$
$$\sum_{k=1}^{N} \lambda_k x_{ik} \leq \lambda_0 - \epsilon \text{ for all } i \neq j$$
$$\sum_{k=1}^{N} \lambda_k x_{jk} \geq \lambda_0 + \epsilon. \tag{5.13}$$

We can therefore solve the problem of computing the extremal points of a set of N points in $\mathbb{R}^D$ by solving N linear programs, each with D variables and N constraints. The time complexity of linear programming however is no simple matter. It is often the case that the algorithms which are optimal in some theoretical sense do not perform well in practice, whereas algorithms with no theoretical guarantees are very practical. See [Megiddo, 1984] and [Grötschel *et al.*, 1993] for more details.

If one regards the dimension D as fixed, solving a linear program with N constraints can be done $O(N)$ time by using the algorithms of [Megiddo, 1984]. However, the constant of proportionality is $O(2^{D \log D})$, which is exponential in D. On the other hand, Khachiyan's algorithm [Khachiyan, 1980] solves a linear program in time polynomial in the length of the bit representation of the program parameters. For our purposes, this implies a strongly polynomial algorithm, since all the points of our polytopes have integer coordinates of size at most n, therefore the linear programs will have integer parameters of size linear in n. The length of the representation of the linear programs will therefore be polynomial in n and d.

Theorem 5.4 *Let f be a polynomial of degree n in D variables, and suppose that f has a sum-product decomposition into k steps with at most l additions and l multiplications by a monomial per step. Let $d = \dim(\mathrm{NP}(f))$ and let $N = C_d n^{d(d-1)/(d+1)}$. Then for $D = 2$ and $D = 3$, the V-representation of $\mathrm{NP}(f)$ can be computed in time $O(klN)$ and $O(klN \log N)$, respectively. For $D \geq 4$, the V-representation of $\mathrm{NP}(f)$ can be computed in time $O(kl\nu_D(2N))$, where $\nu_D(2N)$ is either $O(2^{O(D \log D)} N^2)$ or polynomial in both N and D.*

It is worth mentioning that in practice the well-known simplex method is often the fastest way to solve linear programs. The method relies on heuristic ways of moving along the boundary of a convex polytope called *pivoting strategies*. Although pivoting strategies which run in polynomial time have been found for special classes of linear programs, no such strategy is known for the general instance of linear programming [Megiddo, 1984].

Finally, we need to address our third question, the complexity of computing the normal fan of $\mathrm{NP}(f)$. Note that if one is only interested in the set of extremal vertices of $\mathrm{NP}(f)$, together with a single parameter vector for which each such vertex is optimal, then the V-representation of $\mathrm{NP}(f)$ suffices and the computation of the normal fan is not needed. Linear programming provides us with a certificate for each vertex, i.e. a direction in which that vertex is optimal. If one is interested in the full set of parameter vectors associated to each vertex, then one needs to compute the normal fan $\mathcal{N}_{\mathrm{NP}(f)}$. This requires the computation of the full convex hull of the polytope $\mathrm{NP}(f)$ and its running time is in fact dominated by this computation. For $D \leq 3$, maintaining a full description of the convex hulls of our polytopes is no harder than maintaining the V-representation, however for fixed $D > 3$, an asymptotically optimal algorithm for computing the convex hull of a D-dimensional polytope with N vertices is given in [Chazelle, 1993] and has a time complexity of $O(N^{\lfloor D/2 \rfloor})$. Unfortunately, the constant of proportionality is again exponential in D.

Theorem 5.5 *Let f be a polynomial of degree n in D variables, and suppose that f has a sum-product decomposition into k steps with at most l additions and l multiplications by a monomial per step. Let $d = \dim(\mathrm{NP}(f))$ and let $N = C_d n^{d(d-1)/(d+1)}$. Then for $D = 2$ and $D = 3$, $\mathcal{N}_{\mathrm{NP}(f)}$ can be computed in time $O(klN)$ and $O(klN \log N)$, respectively. For $D \geq 4$, $\mathcal{N}_{\mathrm{NP}(f)}$ can be computed in time $O(klN^2 + N^{\lfloor D/2 \rfloor})$, where the constant of proportionality is exponential in D.*

5.4 Specialization of parameters

5.4.1 Polytope propagation with specialized parameters

In this section we present a variation of the polytope propagation algorithm in which we may specialize the values of some of the parameters. Let us return to the generic example

$$f(p) = \sum_{j=1}^{n} p_1^{e_{j1}} \cdots p_k^{e_{jk}}.$$

Now assume that we assign the values $p_i = a_i$ for $h < i \leq k$. Our polynomial becomes

$$f_a(p) = \sum_{j=1}^{n} p_1^{e_{1j}} \cdots p_h^{e_{hj}} a_{h+1}^{e_{(h+1)j}} \cdots a_k^{e_{kj}},$$

which we can write as

$$f_a(p) = \sum_{j=1}^{n} p_1^{e_{1j}} \ldots p_h^{e_{hj}} \exp(e_{(h+1)j} \ln a_{h+1} + \cdots e_{kj} \ln a_k). \qquad (5.14)$$

We can treat the number e as a general free parameter and this new representation of f_a as a polynomial with only $h + 1$ free parameters, with the only generalization that the exponent of the last parameter need not be an integer, but an integer combination of the logarithms of the specialized parameters.

Now suppose that the polynomial f has a sum-product decomposition. It is clear that such a decomposition automatically translates into a decomposition for f_a by setting p_i^x to $e^{x \cdot \ln p_i}$. Furthermore, a monomial $p_1^{e_{j1}} \ldots p_k^{e_{jk}}$ gives an explanation for some parameter specialization $p = b$ such that $b_i = a_i$ for $i > h$ if and only if $b \cdot e_j = \max_i \{b \cdot e_i\}$, so only if the corresponding vertex $(e_1, \ldots, e_h, \sum_{i=h+1}^{k} e_i \ln a_i)$ is on the Newton polytope $\mathrm{NP}(f_a)$ of f_a. We have reduced the problem to that of computing $\mathrm{NP}(f_a)$ given the sum-product decomposition of f_a induced by that of f.

Finally, we have to give an association of each set of parameters $(e_1, \ldots, e_h)$ with a vertex of $\mathrm{NP}(f_a)$. We can do this in two ways, both of which have comparable time complexity. The first method involves computing the normal fan of $\mathrm{NP}(f_a)$, just as before. This gives a decomposition of the space into cones, each of which corresponds to a vertex of $\mathrm{NP}(f_a)$. For all vectors of parameters with negative logarithms lying inside a certain cone, the corresponding explanation vertex is the one associated with that cone. However, the last parameter in the expression of f_a is the number e, therefore the only relevant part of the parameter hyperspace $\mathbb{R}^{h+1}$ is the hyperplane given by $p_{h+1} = e$, so $-\ln p_{h+1} = -1$. The decomposition G_{f_a} of this hyperplane induced by the normal fan of $\mathrm{NP}(f_a)$ in $\mathbb{R}^{h+1}$ is what we are interested in. Every region in this decomposition is associated with a unique vertex on the *upper half* of $\mathrm{NP}(f_a)$ (with respect to the $(h + 1)$th coordinate).

Alternatively we can compute the decomposition G_{f_a} directly. First we project the upper side of the polytope $\mathrm{NP}(f_a)$ onto the first h coordinates. This gives a regular subdivision R_{f_a} of an h-dimensional polytope. The reason for projecting the upper side alone is that we are looking for minima of linear functionals given by the negative logarithms of the parameter vectors. However, the last parameter is e, so the corresponding linear coefficient is -1. Taking the real line in $\mathbb{R}^{h+1}$ corresponding to a fixed set of values for the first h coordinates, we can see that the point on the intersection of this line with the polytope $\mathrm{NP}(f_a)$ which minimizes the linear functional is the one with the highest $(h + 1)$th coordinate. Thus when projecting on the first h coordinates we will only be interested in the upper half of $\mathrm{NP}(f_a)$.

In order to partition the hyperplane $\mathbb{R}^h$ into regions corresponding to vertices of R_{f_a}, we need to identify the dividing hyperplanes in $\mathbb{R}^h$. Each such hyperplane is given by the intersection of the hyperplanes in the normal fan of $\mathrm{NP}(f_a)$ with the h-dimensional space given by setting the last coordinate in $\mathbb{R}^{h+1}$ to -1. Therefore, each dividing hyperplane corresponds uniquely to an edge (v_i, v_j) in R_{f_a}, and is given by the set of solutions $(x_1, \ldots x_h)$ to the following linear equation:

$$x_1 e_{i1} + \cdots + x_h e_{ih} - [e_{(h+1)i} \ln a_{h+1} + \cdots e_{ki} \ln a_k] =$$
$$x_1 e_{j1} + \cdots + x_h e_{jh} - [e_{(h+1)j} \ln a_{h+1} + \cdots e_{kj} \ln a_k]. \qquad (5.15)$$

The subdivision of $\mathbb{R}^h$ induced by these hyperplanes will be geometrically dual to R_{f_a}, with each region uniquely associated to a vertex of R_{f_a}, so of $\mathrm{NP}(f_a)$. This is the object which we are interested in, as it gives a unique monomial of f for every set of values for the first h parameters, given the specialization of the last $k - h$ parameters.

5.4.2 Complexity of polytope propagation with parameter specialization

Let us now compute the running time of the algorithm described above. Much of the discussion in the previous section will carry through and we will only stress the main differences. As before, the key operation is taking convex hulls of unions of polytopes. Let N be the maximum number of vertices among all the intermediate polytopes we encounter. We again define the function $\nu_{h+1}(N)$ to be the complexity of finding the vertices of the convex hull of a set of N points in $\mathbb{R}^{h+1}$. Note that in the case of parameter specialization, the last coordinate of the vertices of our polytopes is not necessarily an integer.

This last observation implies that we will not be able to use Khachiyan's algorithm for linear programming to get a strongly polynomial algorithm for finding the extremal points. However, in practice this will not be a drawback, as one is nevertheless forced to settle for a certain floating point precision. Assuming that the values a_i we assign to the parameters p_i for $h < i \leq k$ are bounded, we may still assume that the binary representation of the linear programs needed to find the extremal points is polynomial in N. On the other hand, Megiddo's algorithm for linear programming is strongly polynomial in N for fixed h. It will run in time $O(2^{O(h \log h)} N)$. Of course, for the special case when our algorithms run in 2 or 3 dimensions, so $h = 1$ or $h = 2$, the analysis and running time bounds of the previous section still apply. We do not include them here for the sake of brevity.

Finally, what is an upper bound on N? Our polytopes have vertices with co-ordinate vectors $(e_1, \ldots, e_h, \sum_{i=h+1}^{k} e_i \ln a_i)$. By projecting on the first h coordinates, we see that each vertex must project onto a lattice point $(e_1, \ldots, e_h) \in \mathbb{Z}_{\geq 0}^h$ such that $e_1, \ldots, e_h \leq n$, where n is the degree of f. There are n^h such points. Moreover, at most two vertices of the polytope can project to the same point in $\mathbb{R}^h$. Therefore, $N \leq 2n^h$ and we have the following theorem.

Theorem 5.6 *Let f be a polynomial of degree n in D variables, and suppose that f has a sum-product decomposition into k steps with at most l additions and l multiplications by a monomial per step. Also suppose that all but h of* *f*'s variables are specialized. Then the running time required to compute a V-representation of the Newton polytope of *f* with all but *h* parameters specialized is $O(kl\nu_{h+1}(N))$, where $N = 2n^h$ and $\nu_{h+1}(N) = O(2^{O(h \log h)} N^2)$.

For the last part of our task, computing the normal fan $\mathcal{N}_{NP(f_a)}$ and the regular subdivision it induces on the hyperplane $-\ln e_{h+1} = -1$, we remark that the running time is dominated by the computation of the full description of the convex hull of $NP(f_a)$. Again, if we consider *h* fixed, Chazelle's algorithm solves this in $O(N^{\lfloor (h+1)/2 \rfloor})$ time, where the constant of proportionality is exponential in *h*.

Theorem 5.7 *Let f be a polynomial of degree n in D variables, and suppose that f has a sum-product decomposition into k steps with at most l additions and l multiplications by a monomial per step. Also suppose that all but h of f's variables are specialized. Then the running time required to compute all extremal vertices of* $NP(f_a)$, *together with their associated sets of parameter vectors, is* $O(kl\nu_{h+1}(N) + N^{\lfloor (h+1)/2 \rfloor})$, *where* $N = 2n^h$, $\nu_{h+1}(N) = O(N^2)$ *and all constants of proportionality are exponentials in h.*

We therefore observe that if one disregards the preprocessing step of retrieving a sum-product decomposition of f_a from a decomposition of *f*, the running time of our algorithm will not depend on the total number of parameters, but only on the number of unspecialized parameters. Also, the complexity of the pre-processing step is only linear in the size of the decomposition of *f*. This may prove a very useful feature when one is interested in the dependence of explanations on only a small subset of the parameters.

Conclusions: In this chapter we have presented the polytope propagation algorithm as a general method for performing parametric inference in graphical models. For models with a sum-product decomposition, the running time of polytope propagation is the product between the size of the decomposition and the time required to perform a convex hull computation. If the number of parameters *D* is fixed, this is polynomial in the size of the graphical model. However, as *D* increases the asymptotic running time is exponential in *D*. For large numbers of parameters, we present a variation of the algorithm in which one allows only *h* of the parameters to vary. This variation is computationally equivalent to running the initial algorithm on a model of the same size with only *h* + 1 parameters.

6

Polytope Propagation on Graphs

Michael Joswig

Polytope propagation associated with hidden Markov models or arbitrary tree models, as introduced in [Pachter and Sturmfels, 2004a], can be carried to a further level of abstraction. Generalizing polytope propagation allows for a clearer view of the algorithmic complexity issues involved.

We begin with the simple observation that a graphical model associated with a model graph G, which may or may not be a tree, defines another directed graph $\Gamma(G)$ which can roughly be seen as a product of G with the state space of the model (considered as a graph with an isolated node for each state). Polytope propagation actually takes place on this product graph $\Gamma(G)$. The nodes represent the polytopes propagated while each arc carries a vector which represents the multiplication with a monomial in the parameters of the model.

The purpose of this chapter is to collect some information about what happens if $\Gamma(G)$ is replaced by an arbitrary directed acyclic graph and to explain how this more general form of polytope propagation is implemented in `polymake`.

6.1 Polytopes from directed acyclic graphs

Let $\Gamma = (V, A)$ be a finite directed graph with node set V and arc set A. Also, let $\alpha : A \to \mathbb{R}^d$ be some function. We assume that Γ does not have any directed cycles. As such, Γ has at least one source and one sink where a *source* is a node of in-degree zero and a *sink* is a node of out-degree zero.

Such a pair (Γ, α) inductively defines a convex polytope $P_v \subset \mathbb{R}^d$ at each node $v \in V$ as follows.

Definition 6.1 For each *source* $q \in V$ let $P_q = 0 \in \mathbb{R}^d$. For all non-source nodes v let P_v be the joint convex hull of suitably translated polytopes: more precisely,

$$P_v = \bigoplus_{i=1}^{k} \left(P_{u_i} \odot \alpha(u_i, v) \right),$$

where $u_1, \ldots, u_k$ are the *predecessors* of v, that is, $(u_1, v), \ldots, (u_k, v) \in A$ are the arcs of Γ pointing to v. This polytope P_v is the polytope *propagated* by

(Γ, α) at v. Often we will be concerned with graphs which have only one sink s, in which case we will write $P(\Gamma, \alpha) = P_s$. The operators $\oplus$ and $\odot$ are the polytope algebra's arithmetic operations "joint convex hull" and "Minkowski sum", respectively; see Section 2.3.

It is a key feature of polytope propagation that each vertex of a propagated polytope corresponds to a (not necessarily unique) directed path from one of the sinks. In the context of statistical models such paths are known as *Viterbi sequences*; see Section 4.4 and Chapter 10.

Example 6.2 Let $\Gamma = (V, A)$ have a unique source q, and assume that there is a function $\nu : V \to \mathbb{R}^d$ with $\nu(q) = 0$ such that $\alpha(u, v) = \nu(v) - \nu(u)$. Then the polytope P_n at each node n is the point $\nu(n) - \nu(q) = \nu(n)$.   □

If Γ has exactly one source and exactly one sink we call it *standard*. The following observation is immediate.

Proposition 6.3 *Let Γ be standard. Then the propagated polytope $P(\Gamma, \alpha)$ is a single point if and only if there is a function $\nu : V \to \mathbb{R}^d$ with $\alpha(u, v) = \nu(v) - \nu(u)$ for all arcs (u, v) of Γ.*

While these definitions may seem restricting, the following example shows that the propagated polytope is not restricted in any way.

Example 6.4 Let $x_1, \ldots, x_n \in \mathbb{R}^d$ be a finite set of points. We define a directed graph with the $n + 2$ nodes $0, 1, \ldots, n + 1$ such that there are arcs from 0 to the nodes $1, 2, \ldots, n$ and arcs from each of the nodes $1, 2, \ldots, n$ to $n + 1$. Further we define $\alpha(0, k) = x_k$ and $\alpha(k, n + 1) = 0$, for $1 \leq k \leq n$. As defined, 0 is the unique source, and the propagated polytope at the unique sink $n + 1$ is $\mathrm{conv}(x_1, \ldots, x_n)$.   □

Example 6.5 Let Γ be a (standard) directed graph on the nodes $0, 1, \ldots, 7$ with arcs as shown in Figure 6.1. The node 0 is the unique source, and the node 7 is the unique sink in Γ. The black arrows indicate the lengths and the directions of the vectors in $\mathbb{R}^2$ associated with each arc: $\alpha(0, 1) = \alpha(1, 3) = \alpha(3, 5) = (1, 0)$, $\alpha(1, 4) = \alpha(3, 6) = (0, 1)$, $\alpha(2, 4) = \alpha(4, 6) = (0, 2)$, and the vectors on the remaining arcs are zero. The propagated polytope is the pentagon

$$P(\Gamma, \alpha) = \mathrm{conv}((0, 1), (0, 4), (1, 0), (1, 3), (3, 0)).$$

□

Remark 6.6 Suppose that $p \in K[x_1^{\pm}, \ldots, x_d^{\pm}]$ is a Laurent polynomial over some field K, such that p is a sequence of sums and products with a monomial. Then, in view of Theorem 2.26, the computation of the Newton polytope of p can be paralleled to the computation of p itself by means of polytope propagation. The nodes of the corresponding graph correspond to the Newton

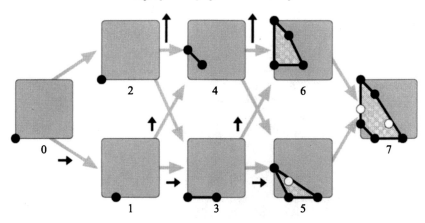

Fig. 6.1. Propagated pentagon in $\mathbb{R}^2$. The white points correspond to translated vertices of predecessor polytopes which are redundant for the propagated polytopes.

polytopes of Laurent polynomials which arise as sub-expressions. For instance, Figure 6.1 can be interpreted as the computation of the Newton polytope of $p = x + x^3 + xy + xy^3 + x^2y + y + y^2 + y^4$. Denoting by $p(i)$ the polynomial associated with node i, we have $p(0) = p(2) = 1$, $p(1) = x$, $p(3) = 1 + x^2$, $p(4) = xy + y^2$, $p(5) = x + x^3 + xy + y^2$, $p(6) = x^2y + xy^3 + y + y^4$, and $p(7) = p$. Note that since we keep adding polynomials with non-negative coefficients cancellation does not occur.

Example 6.7 A *zonotope* Z is the Minkowski sum of k line segments $[p_i, q_i] \subset \mathbb{R}^d$ or, equivalently, an affine projection of the regular k-cube. Each zonotope can be obtained by polytope propagation as follows. Let Γ_k be a graph with node set $\{0, 1, \ldots, 2k + 1\}$ and $4k$ arcs $(2i, 2i + 2)$, $(2i, 2i + 1)$, $(2i - 1, 2i + 2)$, $(2i-1, 2i+1)$ for all relevant integers i. Also, let $\alpha(2i, 2i+2) = \alpha(2i-1, 2i+2) = q_i - p_i$ and $\alpha(a) = 0$ for all other arcs a. Up to a translation by the vector $p_1 + \cdots + p_k$ the propagated polytope at the unique sink $2k+1$ is the zonotope Z.

The directed paths in Γ_k from the source to the sink can be identified with the bit-strings of length k and hence also with the vertices of the regular k-cube. This defines a projection of the k-cube onto the propagated polytope. Figure 6.2 shows the construction of a centrally symmetric hexagon. See [Ziegler, 1995] for more information on zonotopes. □

Polytope propagation leads to an efficient algorithm in the sense that it is output-sensitive in the following way. Consider polytope propagation defined by the standard directed acyclic graph $\Gamma = (V, A)$ and the function $\alpha : A \to \mathbb{R}^d$. We define the *size* of the polytope propagation with respect to (Γ, α) as the sum of the sizes of all propagated polytopes plus the size of the function α. Here the *size* of a polytope P is the number of bits required to store its V-representation (that is, as a matrix of row vectors corresponding to the vertices of P); likewise the *size* of α is the number of bits required to store α as a matrix with d columns and as many rows as there are arcs in A.

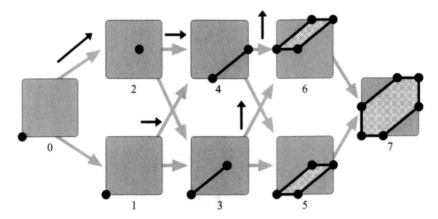

Fig. 6.2. The graph Γ_3 and its propagated zonotope $P(\Gamma_3) \subset \mathbb{R}^2$; see Example 6.7.

Theorem 6.8 *The vertices of the polytope propagated by Γ can be computed in time which is polynomially bounded in the size of the polytope propagation with respect to (Γ, α).*

Proof We only have to prove that the computation at each node of Γ takes polynomial time.

Let w be some node of Γ. By induction we can assume that the vertices of the polytopes propagated to the nodes which have an arc to w can be computed in polynomial time. Our task is to compute the vertices of the polytope propagated at w given as the convex hull of finitely many points. Note that by definition the number of these points does not exceed the total size of the polytope propagation.

If $P = \text{conv}\{x_1, \ldots, x_n\}$ is a polytope in $\mathbb{R}^d$ then a point x_k is a vertex if and only if it can be strictly separated from $x_1, \ldots, x_{k-1}, x_{k+1}, \ldots, x_n$ by an affine hyperplane. The linear optimization problem:

$$\text{maximize } \epsilon \text{ subject to}$$
$$\lambda_0, \ldots, \lambda_d \in \mathbb{R}, \ \epsilon \geq 0,$$
$$\sum_{j=1}^{d} \lambda_j x_{ij} \leq \lambda_0 - \epsilon, \text{ for each } i \neq k, \tag{6.1}$$
$$\sum_{j=1}^{d} \lambda_j x_{kj} \geq \lambda_0 + \epsilon$$

has a solution with $\epsilon > 0$ if and only if x_k is a vertex of P. Since linear optimization is solvable in polynomial time [Khachiyan, 1980, Grötschel *et al.*, 1993], the claim follows. $\square$

The above rough estimation of the complexity of polytope propagation is slightly incomplete. The linear optimization problem 6.1 can be solved in linear time if d is fixed, see [Megiddo, 1984], and so finding all the vertices of $P =$

conv$\{x_1, \ldots, x_n\}$ takes $O(n^2)$ time in fixed dimension. Moreover, computing convex hulls in $\mathbb{R}^2$ or $\mathbb{R}^3$ only takes at most $O(n \log n)$ time; see [Seidel, 2004]. This means that for the special case of $d \leq 3$, computing convex hulls directly is superior to the approach described above. For $d = 2$ the complexity can further be reduced to linear time by an inductive construction of the convex hulls involved.

Note that Theorem 6.8 does not claim that polytope propagation is polynomial-time in the size of the pair (Γ, α). In fact, the regular d-dimensional cube can be propagated by a graph Γ_d with $2(d+1)$ nodes and $4(d+1)$ arcs as a trivial zonotope; see Example 6.7. However, since the d-cube has 2^d vertices, its size is not polynomially bounded by the size of (Γ_d, α). In certain situations better bounds can be obtained for the sizes of the propagated polytopes; see, for example, Section 5.3.

6.2 Specialization to hidden Markov models

We consider a finite Markov process with l states and transition matrix $\theta = (\theta_{ij}) \in \mathbb{R}^{l \times l}$ satisfying the conditions $\theta_{ij} \geq 0$ and $\sum_j \theta_{ij} = 1$. The value θ_{ij} is the transition probability of going from state i into state j. In the hidden Markov model we additionally take into account that the observation itself may be a probabilistic function of the state. That is to say, there are l' possible observations (where l' may even be different from l) and a non-negative matrix $\theta' = (\theta'_{ij}) \in \mathbb{R}^{l \times l'}$ satisfying $\sum_j \theta'_{ij} = 1$. The entry θ'_{ij} expresses the probability that j has been observed provided that the actual state was i. In the following we will be concerned with parameterized hidden Markov models as in Chapter 1.

Example 6.9 The simplest possible (non-trivial) HMM occurs for $l = l' = 2$ and $\theta = \begin{pmatrix} \theta_{00} & \theta_{01} \\ \theta_{10} & \theta_{11} \end{pmatrix}$, $\theta' = \begin{pmatrix} \theta'_{00} & \theta'_{01} \\ \theta'_{10} & \theta'_{11} \end{pmatrix}$. If, for instance, we observe our primary process (defined by θ) for three steps, the sequence of observations is some bit-string $\beta_1 \beta_2 \beta_3 \in \{0, 1\}^3$. There are eight possible sequences of states which may have led to this observation. As in Chapter 1 we assume that the initial state of the primary process attains both states with equal probability $\frac{1}{2}$. Then we have that

$$\text{Prob}(Y_1 = \beta_1, Y_2 = \beta_2, Y_3 = \beta_3) = \frac{1}{2} \sum_{\sigma \in \{0,1\}^3} \theta_{\sigma_1 \sigma_2} \theta_{\sigma_2 \sigma_3} \theta'_{\sigma_1 \beta_1} \theta'_{\sigma_2 \beta_2} \theta'_{\sigma_3 \beta_3}.$$

Now let us have a look at the parameterized model, where, for the sake of simplicity, we assume that $\theta = \theta' = \begin{pmatrix} u & v \\ v & w \end{pmatrix}$ is symmetric. Since otherwise our primary process would be stationary, we can additionally assume $v \neq 0$. Neglecting the probabilistic constraints and re-scaling then allows us to study the parameterized HMM of

$$\hat{\theta} = \hat{\theta}' = \begin{pmatrix} x & 1 \\ 1 & y \end{pmatrix} \in \mathbb{R}[x, y]^{2 \times 2}.$$

The probability to make the specific observation 011 in the parameterized model yields

$$\text{Prob}(Y_1 = 0, Y_2 = 1, Y_3 = 1) = \frac{1}{2}(x^3 + x^2y + xy + xy^3 + x + y + y^2 + y^4),$$

which happens to be the polynomial p in Remark 6.6 (up to the constant factor $\frac{1}{2}$), and whose Newton polytope is the pentagon in Figure 6.1.   □

The fixed observation $\beta_1 \cdots \beta_n$ of a hidden Markov model (θ, θ') for n steps gives rise to the standard acyclic directed graph $\Gamma^n_{(\theta,\theta')}$ with node set $[n] \times [l] \cup \{-\infty, +\infty\}$ as follows: There is an arc from $-\infty$ to all nodes $(1, j)$, and there is an arc from (n, j) to $+\infty$ for all $j \in [l]$, and there is an arc between $(i - 1, j)$ and (i, k) for any i, j, k. The directed paths from the unique source $-\infty$ to the unique sink $+\infty$ directly correspond to the l^n possible sequences of states of the primary process that may or may not have led to the given observation $\beta_1 \cdots \beta_n$. Such a path passes through the node (i, j) if and only if the ith primary state X_i has been j.

The probability that $X_{i-1} = j$ and $X_i = k$ under the condition that $Y_i = \beta_i$ is the monomial $\theta_{jk}\theta'_{k,\beta_i} \in \mathbb{R}[\theta_{11}, \dots, \theta_{ll}, \theta'_{11}, \dots, \theta'_{ll'}]$. We associate its (integral) exponent vector with the arc $(i - 1, j) \to (i, k)$. Likewise, the exponent vector of θ'_{k,β_0} is associated with the arc $-\infty \to (1, k)$ and 0 to all arcs $(n, k) \to +\infty$.

The graph $\Gamma^n_{(\theta,\theta')}$ gives rise to a propagated polytope whose vertices correspond to most likely explanations (in the a posteriori maximum likelihood model) for the observation made for some choice of parameters; see Chapter 1, Section 4.4, and Chapter 10.

This scheme can be modified by specializing some of the parameters as in Example 6.9. The associated graph is given as Example 6.5, and its propagated polytope is the Newton polytope of the polynomial $p(7)$ in Remark 6.6, a pentagon. Section 6.4 shows how polymake can be used to investigate this particular example further.

In our version of polytope propagation we chose to restrict the information on the arcs to be single points (which are 0-dimensional polytopes). For this reason we chose our parametric HMM to have monomials only for the transition probabilities. It is conceivable (but not implemented in polymake) to allow for arbitrary polynomials for the transition probabilities. This would then require to iteratively compute Minkowski sums of higher dimensional polytopes. For Minkowski sum algorithms for general polytopes see [Fukuda, 2004].

6.3 An implementation in polymake

A software system called polymake is available for the study of convex polytopes [Gawrilow and Joswig, 2000, Gawrilow and Joswig, 2001]. This software implements many existing algorithms for computations on polytopes. Additionally, there is a large array of interfaces to other software packages. This integration via interfaces is entirely transparent to the user. This way it does not make much of a difference whether a function is built into the system's core or whether it is actually realized by calling an external package. A key feature of the **polymake** system is that polytopes (and a few other things) are seen by the system as objects with a set of opaque functions which implement the polytope's properties known to the system. Calling such a function automatically triggers the construction of a sequence of steps to produce information about the desired property from the data which defines the polytope object. In this way **polymake** somewhat resembles an expert system for polytopes.

As far as the implementation is concerned, **polymake** is a Perl/C++ hybrid, and both languages can be used to extend the system's functionality. This modular design allows for extensions by the user which are technically indistinguishable from the built-in functions. Internally, the polytope propagation algorithm is implemented in C++, and it is part of **polymake**'s current version 2.1 as the client **sum_product**. The fully annotated source code of this program is listed below. Since it combines many of the system's features, while it is still a fairly short program, it should give a good idea how **polymake** can be extended in other ways, too.

We give a complete listing of the program **sum_product**, which implements the algorithm described. In order to increase the legibility the code has been re-arranged and shortened slightly.

6.3.1 *Main program and the* polymake *template library*

The main program sets up the communication with the **polymake** server for the polytope object **p**, which contains the propagation graph Γ and the function α. The graph Γ is assumed to be standard, and the polytope object **p** corresponds to the propagated polytope $P(\Gamma, \alpha)$. The class **Poly** is derived from **iostream**. The function which computes the polytope propagation is **sum_product** in the namespace **polymake::polytope**.

```
using namespace polymake;

int main(int argc, char *argv[]) {
   if (argc != 2) {
      cerr << "usage: " << argv[0] << " <file>" << endl;
      return 1;
   }
   try {
      Poly p(argv[1], ios::in | ios::out);
      polytope::sum_product(p);
   }
   catch (const std::exception& e) {
      cerr << e.what() << endl;
      return 1;
   }
   return 0;
}
```

The **polymake** system comprises a rich template library which complies with the Standard Template Library (STL). This includes a variety of container template classes, such as **Array** (which is a variation of STL's **vector** class) and template classes for doing linear algebra, such as `Vector` and `Matrix`. Additionally, there are several classes which special functionality useful in algorithmic polytope theory, for example, `Graph` and `IncidenceMatrix`.

A common feature of `polymake`'s container classes is a memory management based on reference counting (with copy-on-write). As a result, copying a non-altered `Array` costs next to nothing. Furthermore, all these classes provide a range of features useful for debugging. The reader is referred to the `polymake`'s documentation for a detailed description.

The arithmetic operations use exact representations of rational numbers. Our class `Rational` wraps the corresponding implementation from the GNU Multiprecision Library [Swox, 2004].

The rest of this section lists the function `sum_product` in the namespace `polymake::polytope`.

```
namespace polymake { namespace polytope { ... } }
```

6.3.2 The core of the implementation

The graph that the client reads already comes with a (translation) vector specified for each edge. This is the second template parameter. The first template parameter is set to an artificial type `nothing` which means that there is no data at the nodes. The third template parameter (again some artificial type) indicates that our graph is directed. We assume that the graph is acyclic with a unique sink. Arbitrarily many sources are allowed.

```
typedef Graph< nothing, Vector<Rational>, directed > graph;
```

We now start to describe the actual polytope propagation algorithm. It operates on a single `Poly` object from which the `SUM_PRODUCT_GRAPH` is read. It writes the `VERTICES` and `VERTEX_NORMALS` of the polytope corresponding to the unique sink in the graph.

```
void sum_product(Poly& p) { ... }
```

The function starts by reading its input, which is the graph and the dimension of the ambient space. The member function `give()` of the `Poly` class induces the polymake server to deliver the named property of the object `p`. If necessary, the server triggers further computations to answer the request as defined by an extensible set of rules. There is a reciprocal function `take()` to be used further below.

```
const graph G=p.give("SUM_PRODUCT_GRAPH");
const int n=G.nodes();
if (n==0)
   throw std::runtime_error("SUM_PRODUCT_GRAPH must be non-empty");

const int d=p.give("AMBIENT_DIM");
```

We continue to step through the code line by line. On the way we explain the relevant concepts.

A description of the origin as a 0-dimensional polytope in d-space is given below. It has one vertex and an empty vertex-facet incidence matrix which is used to define the initial objects of the type `Poly` for the sources of the graph.

```
const Matrix<Rational> single_point_vertices(vector2row(
                           unit_vector<Rational>(d+1,0)));
const IncidenceMatrix<> single_point_vif;
```

The following defines an array where all the intermediate polytopes are stored. The nodes in the graph are consecutively numbered, starting with 0. The corresponding polytope can be accessed by indexing the array `pa` with the node number. In the beginning the `Poly` objects are undefined. The list `next_nodes` stores those nodes of the graph whose propagated polytopes are not yet known but whose predecessors are.

```
Array<Poly> pa(n);
std::list<int> next_nodes;
```

Next, initialization is accomplished by assigning a single point (origin) to each source in the graph. The function `add_next_generation` is listed in Section 6.3.3.

```
for (int v=0; v<n; ++v) {
    if (G.in_degree(v)==0) {
        pa[v].init(0, ios::in | ios::out | ios::trunc,
                               "RationalPolytope");
        pa[v].take("VERTICES") << single_point_vertices;
        pa[v].take("VERTICES_IN_FACETS") << single_point_vif;
        add_next_generation(next_nodes,v,G,pa);
    }
}
```

At each node of the graph we recursively define a polytope as the convex hull of the translated predecessors. We also try to find the sink on the way.

```
int sink=-1;
```

The number -1 does not correspond to any valid node number, and so it indicates that no sink has been found yet.

```
while(!next_nodes.empty()) {
```

Inside the loop we get some node `w` for which we already know all its predecessors, and initialize the client-server communication for that node's polytope. Until now it was undefined.

```
const int w=next_nodes.front(); next_nodes.pop_front();
pa[w].init(0, ios::in | ios::out, "RationalPolytope");
```

The polytope will be specified as the convex hull of points, which will be collected from other polytopes. The special data type `ListMatrix` is efficient in terms of concatenating rows (which correspond to points) but it is not efficient in terms of matrix operations (although all operations are defined). This is acceptable, since we do not compute anything in this step.

```
ListMatrix< Vector<Rational> > points(0,d+1);
for (Entire<graph::in_edge_list>::const_iterator
            e=entire(G.in_edges(w)); !e.at_end(); ++e) {
```

The node v is the current predecessor to process. The arc from v to w is e. The operator * extracts the associated vector from an arc.

```
const int v=e.from_node();
const Vector<Rational> vec=*e;
```

Next we read the vertices of the predecessor polytope at the node v. The system's standard set of rules by default uses cdd's implementation to check for redundant (= non-vertex) points among the input by solving linear programs [Fukuda, 2003]. Note that this implementation is based on the Simplex Method so in the worst case we actually do not achieve the complexity promised in Theorem 6.8. No convex hull computation is necessary. What is going on behind the scene is hidden from the user. The polymake server decides how to produce the VERTICES of the polytope object pa[v]. Note that this is the same behavior as if polymake would be asked for these vertices via the command line interface.

```
const Matrix<Rational> these_vertices=pa[v].give("VERTICES");
```

Now we concatenate the translated matrix (where the rows correspond to the vertices of the predecessor) to what we already have. The C++ operator /= takes care of the concatenation of compatible matrix blocks on top of one another. Note that the final } closes the for-loop through all the predecessors of w:

```
    points /= these_vertices*translation_by(vec);
}
```

As the last step in the while-loop we define the polytope object as the convex hull of all those points collected and proceed.

```
pa[w].take("POINTS") << points;
if (G.out_degree(w)==0) sink=w;
else add_next_generation(next_nodes,w,G,pa);
}
```

Afterwards, this will be just any sink. It is only indeterministic if the graph does have several sinks.

```
if (sink<0)
    throw std::runtime_error("no sink found in digraph");
```

Finally, the sink defines the polytope we are after and this is going to be defined as the polytope object p. A set of vertex normals serves as a certificate that the claimed points are indeed vertices.

```
const Matrix<Rational>
    sink_vertices = pa[sink].give("VERTICES"),
    sink_normals  = pa[sink].give("VERTEX_NORMALS");

p.take("VERTICES") << sink_vertices;
```

```
        p.take("VERTEX_NORMALS") << sink_normals;
}
```

6.3.3 Two auxiliary functions

The first function `add_next_generation` gathers the next generation of graph nodes which can be defined at that stage, since all predecessors known. It is one step in a common breadth first search. The graph nodes which can be processed (because all their predecessors are known) are stored in a doubly linked list that is STL's type `list`.

```
void add_next_generation(std::list<int>& next_nodes, const int v,
                 const graph& G, const Array<Poly>& pa)
{
   for (Entire<graph::out_edge_list>::const_iterator
            e=entire(G.out_edges(v)); !e.at_end(); ++e) {
      const int x=e.to_node();
      Entire<graph::in_edge_list>::const_iterator
            f=entire(G.in_edges(x));
      for( ; !f.at_end() && pa[f.from_node()].get_mode(); ++f);
      if (f.at_end())
         next_nodes.push_back(x);
   }
}
```

The following function `translation_by` returns a translation matrix (to be applied to row vectors from the right) for given vector. The C++ operators | and / are overloaded for the `Matrix` class: They define the concatenation of two matrix blocks side by side and one on top of the other, respectively.

```
Matrix<Rational> translation_by(const Vector<Rational>& vec)
{
   const int d=vec.dim();
   return unit_vector<Rational>(d+1,0) |
          (vec / unit_matrix<Rational>(d));
}
```

6.4 Returning to our example

There is a standard client program `binary-markov-graph` which produces the polytope propagation graphs for the special HMM discussed in Example 6.9. It can be called from the command line as follows:

```
> binary-markov-graph b011.poly 011
```

Here the argument `011` specifies the observation. The command yields a file `b011.poly` which contains a description of the polytope propagation defining pair (Γ, α) suitable as input for the `sum_product` client. This is what it looks like:

```
> polymake b011.poly SUM_PRODUCT_GRAPH
SUM_PRODUCT_GRAPH
{(1 1 0) (2 0 0)}
```

```
{(3 1 0) (4 0 1)}
{(3 0 0) (4 0 2)}
{(5 1 0) (6 0 1)}
{(5 0 0) (6 0 2)}
{(7 0 0)}
{(7 0 0)}
{}
```

The rows below the keyword `SUM_PRODUCT_GRAPH` are implicitly numbered $0, 1, \ldots, 7$. Each row corresponds to one node in the graph Γ, and the triplet $(j\ x\ y)$ in row i says that there is an arc from i to j with $\alpha(i,j) = (x,y) \in \mathbb{R}^2$.

```
> sum_product b011.poly
```

The result of this is to define a polytope object which is accessible to all of **polymake**'s functions. For example, this can be used to list all the vertices and corresponding vertex normals of the propagated polytope. In the output, each row corresponds to a vertex of our pentagon. In the numbered output each line i starts with "i:". Note that the first coordinate column (following the colon) is used for homogenization.

```
> polymake b011.poly "numbered(VERTICES)" \
                     "numbered(VERTEX_NORMALS)"
numbered(VERTICES)
0:1 3 0
1:1 1 0
2:1 0 1
3:1 1 3
4:1 0 4

numbered(VERTEX_NORMALS)
0:0 1/2 0
1:0 -1/2 -3/2
2:0 -2 -1
3:0 4 3
4:0 -1/2 1/2
```

For example, the vertex with coordinates $(1,3)$, numbered as 3, yields the most likely explanation for our observation 011 if we are up to maximizing the linear function $4x + 3y$ on our parameters x and y (this is a vertex normal of $(1,3)$, also numbered as 3). The vertex $(1,3)$ corresponds to the unique path $0 \to 1 \to 4 \to 6 \to 7$ in the graph depicted in Figure 6.1 and this in turn corresponds to the state sequence 011 in the primary process. Phrased differently, if our goal is to maximize $4x + 3y$ in the parameterized model, then we can trust our observation.

7

Parametric Sequence Alignment

Colin Dewey

Kevin Woods

Biological sequence alignment is a fundamental problem in computational biology. Section 2.2 introduced some scoring schemes and algorithms used for global pairwise alignment. Each scoring scheme is dependent on a set of parameters. As shown in Example 2.14, the optimal alignment can change significantly as these parameters are varied. We would like to know how the results obtained from alignment programs are affected by the values of the parameters and how confident we can be in a given optimal alignment. Such questions can be answered using *parametric sequence alignment* methods. Here we introduce the techniques of parametric sequence alignment and apply them to characterize different scoring schemes. Parametric alignment algorithms can be implemented almost as efficiently as algorithms for conventional sequence alignment. Hence, parametric algorithms may become powerful and important components of the computational biologist's toolbox.

7.1 Few alignments are optimal

The primary goal of biological sequence alignment is to match up positions in the input sequences that are *homologous*. Two sequence positions are homologous if the characters at those positions are derived from the same position in some ancestral sequence. It is important to note that two positions can be homologous even though the states of the positions are different. For example, there may exist sequences σ^1 and σ^2 such that position 5 in σ^1 and position 9 in σ^2 are homologous despite the fact that $\sigma_5^1 = $ A and $\sigma_9^2 = $ C. Alignments indicate that positions in two sequences are homologous by matching up the characters at those positions. An alignment is *biologically correct* if it matches up all positions that are truly homologous and no others. Because this chapter focuses on the *global* alignment problem for two sequences, we will ignore cases in which a duplication event causes one position to be homologous to multiple positions in the other sequence.

What does it mean for a global alignment to be "optimal"? If our goal is to discover the biological truth, we should say that an "optimal" global alignment is one that is biologically correct. Having biologically correct alignments is critical, because other analyses, such as phylogenetic tree reconstruction, are heavily dependent on sequence alignments. Unfortunately, in most cases, we do not know what the biological truth is. To make a guess at the truth, we treat sequence alignment as an optimization problem, where we choose the alignment that is optimal with respect to some objective function. Although many types of objective functions are imaginable, the most common is to assign a score to each alignment, according to a scoring scheme such as the one described in Section 2.2. We then seek to find the alignment that maximizes this score. Which scoring scheme should we use to guess at the biologically correct alignment? Once we have chosen the parameter values for a given scoring scheme, how confident can we be that the resulting alignment is a good guess? It could be the case that, if we vary our chosen parameter values just slightly, we will obtain very different alignments. Even worse, it could be that no values for the parameters of our chosen scoring scheme will give the biologically correct alignment as being optimal.

The methods of parametric sequence alignment help to answer these questions, for specific input sequences, by analyzing a scoring scheme over all possible parameter values. Specifically, parametric sequence alignment subdivides the parameter space into regions such that parameter values in the same region give rise to the same optimal alignments. Such a subdivision tells us which of the exponentially many possible alignments can be optimal for some choice of parameter values and how many classes of optimal alignments there can be. In addition, if we find that the point in the parameter space corresponding to our current choice of values for the parameters is close to the boundary of a region in the subdivision, we may be less confident in our results. Lastly, we may wish to take a Bayesian approach and place a prior distribution on the parameters. In this case, we can determine what the most likely optimal alignments are by integrating over the different regions in the subdivision (note that this is different from finding the *most likely alignment*) [Pachter and Sturmfels, 2004a].

Parametric sequence alignment is feasible, because, although two given sequences have exponentially many possible alignments, there are only a few subsets of these alignments (corresponding to regions of the parameter space subdivision) that can be optimal for some choice of parameters. For two sequences of length at most n, it has been shown that, for a simple scoring scheme with 2 parameters allowed to vary, the number of regions in the subdivision is $O(n^{\frac{2}{3}})$ [Gusfield *et al.*, 1994]. This bound corresponds exactly with the bound given in Section 5.3 for the number of vertices of a sequence alignment Newton polytope with $d = 2$. In fact, for a scoring scheme with any number of parameters, the number of regions is bounded by a polynomial in n (see Section 5.3). The bound on the number of regions in the subdivision allows for the subdivision to be determined in just slightly more time than it takes to simply align the sequences in question. For the simple scoring scheme just mentioned, the subdivision can be found in $O(n^{\frac{8}{3}})$ time, as opposed to $O(n^2)$ time for aligning the sequences with a fixed set of parameter values.

The concept of parametric sequence alignment is not a new one, and there are several programs available to perform parametric analysis [Gusfield, 1997].

Existing methods are restricted to the analysis of scoring schemes with at most 2 parameters allowed to vary at one time, whereas we may also like to analyze scoring schemes that have many parameters, such as the general 33-parameter scoring scheme described in Section 2.2. In this chapter, we apply the techniques of parametric inference described in Chapter 5 to the problem of sequence alignment and thus give a general method for parametric sequence analysis with scoring schemes involving any number of parameters.

7.2 Polytope propagation for alignments

In this section, we will describe how to efficiently compute a parametric sequence alignment of two sequences, σ^1 and σ^2, with lengths n and m, respectively. While the method we describe can be used to analyze a fully-parameterized scoring scheme (with the 33 parameters comprising the matrices shown in the equations (2.11) and (2.12)), we will concentrate on a simple 4-parameter scoring scheme. In this scoring scheme we have parameters M, X, S, and G, corresponding to the weights for matches, mismatches, spaces (- symbols in the alignment), and gaps (contiguous sets of spaces), respectively. This scoring scheme is just a special case of the general scoring scheme with

$$
\begin{aligned}
w_{\pi,\pi} &= M \quad \forall \pi \in \Sigma \\
w_{\pi_1,\pi_2} &= X \quad \forall \pi_1, \pi_2 \in \Sigma, \pi_1 \neq \pi_2 \\
w_{\pi,-} = w_{-,\pi} &= S \quad \forall \pi \in \Sigma \\
w'_{H,I} = w'_{H,D} = w'_{I,D} = w'_{D,I} &= G \\
w'_{H,H} = w'_{I,H} = w'_{D,H} = w'_{I,I} = w'_{D,D} &= 0
\end{aligned}
$$

Because we expect biologically correct alignments to have a large number of matches and a limited number of gaps, the parameters of this scoring scheme are normally set to reward matches and penalize gaps. We will refer to *biologically reasonable* parameter values as those that satisfy

$$ M > S, \qquad M > X, \qquad \text{and} \qquad G < 0. \tag{7.1} $$

An even simpler scoring scheme, which we will refer to as the 3-parameter scoring scheme, lacks a gap parameter (that is, it is the 4-parameter scoring scheme with $G = 0$). We shall present a method for parametric alignment with the 3-parameter scoring scheme, but it should be noted that this method easily generalizes to the 4-parameter and 33-parameter scoring schemes.

With the 3-parameter scoring scheme, the weight of an alignment, h, is

$$ W_{\sigma^1,\sigma^2}(h) = M m_h + X x_h + S s_h, $$

where m_h, x_h, and s_h denote the number of matches, mismatches, and spaces in h, respectively. We define the monomial

$$ f_{\sigma^1,\sigma^2,h}(\theta_M, \theta_X, \theta_S) = \theta_M^{m_h} \theta_X^{x_h} \theta_S^{s_h} $$

and the polynomial

$$f_{\sigma^1,\sigma^2} = \sum_{h \in \mathcal{A}_{n,m}} f_{\sigma^1,\sigma^2,h}, \qquad (7.2)$$

where $\mathcal{A}_{n,m}$ is the set of all possible alignments of σ^1 and σ^2. The weight, $W(h)$, is simply $\log f_{\sigma^1,\sigma^2,h}(e^M, e^X, e^S)$. As we saw in Section 2.2, finding the alignment, h, that minimizes $W(h)$ is equivalent to evaluating f_{σ^1,σ^2} tropically. Following with tradition, we will choose to maximize $W(h)$ in this chapter, but these are equivalent problems.

The parametric alignment problem for this scoring scheme, then, is to compute which values of M, X, and S result in which optimal alignments. The key object in our parametric alignment of σ^1 and σ^2 is the Newton polytope of the polynomial f_{σ^1,σ^2}. The Newton polytope of f_{σ^1,σ^2}, denoted by $\text{NP}(f_{\sigma^1,\sigma^2})$, is the convex hull of all points (m_h, x_h, s_h), for $h \in \mathcal{A}_{n,m}$. Recall from Section 2.3 that each vertex of $\text{NP}(f_{\sigma^1,\sigma^2})$ corresponds to an alignment (or set of alignments, all having the same number of matches, mismatches, and spaces) that will be optimal for a certain set of values for the parameters, M, X, and S. For a vertex, v, of $\text{NP}(f_{\sigma^1,\sigma^2})$, the set of parameters for which the alignments corresponding to v are all optimal is given by its normal cone $N(v)$. The normal fan of the Newton polytope, $\mathcal{N}(\text{NP}(f_{\sigma^1,\sigma^2}))$ is thus a subdivision of the parameter space with the property that, for all parameter values in the same region (normal cone), the same alignments are optimal. This subdivision is exactly the desired output of parametric sequence alignment.

Having shown that the Newton polytope of f_{σ^1,σ^2} and its corresponding normal fan solve the parametric alignment problem, we now turn to how to compute these objects efficiently using the polytope propagation algorithm of Chapter 5. First, however, we make two remarks that will make our presentation cleaner.

Remark 7.1 For an alignment, h, of σ^1 and σ^2, we must have that $2m_h + 2x_h + s_h = n + m$.

Proof Recall from Section 2.2 that h is a string over the alphabet $\{H, I, D\}$. Matches and mismatches correspond to H characters in the alignment, while spaces correspond to I and D characters. The remark follows from combining the equalities in (2.7).   □

Remark 7.2 Any 3-parameter scoring scheme (M, X, S) is equivalent to another scoring scheme (M', X', S'), where $M' = 0$, $X' = X - M$, and $S' = S - \frac{M}{2}$.

Proof Using Remark 7.1, we have that

$$W'(h) = M'm_h + X'x_h + S's_h = W(h) - \frac{M}{2}(n + m) = W(h) - C,$$

where $C = \frac{M}{2}(n + m)$ is a constant with respect to the alignment h. Since all

	A	A	C
0	-3	-6	-9
-3	-2	-5	-6
-6	-3	-2	-5
-9	-6	-3	-4

Fig. 7.1. $A[i,j]$ with $\sigma^1 = $ CAA, $\sigma^2 = $ AAC, $M = 0$, $X = -2$, and $S = -3$.

scores are shifted by the same constant, the rankings of possible alignments under the two scoring schemes are the same. □

Having made this remark, we now assume that $M = 0$ for the remainder of this section. With this constraint (and setting $\theta_M = e^M = 1$), we have

$$f_{\sigma^1,\sigma^2} = \sum_{h \in \mathcal{A}_{n,m}} \theta_X^{x_h} \theta_S^{s_h},$$

and our Newton polytope and normal fan will be two-dimensional.

Let us recall the Needleman–Wunsch algorithm, which, given specific parameter values X and S and sequences σ^1 and σ^2, computes the optimal global alignment. Let $\sigma^1 = \sigma_1^1 \sigma_2^1 \cdots \sigma_n^1$ and $\sigma^2 = \sigma_1^2 \sigma_2^2 \cdots \sigma_m^2$. For $0 \le i \le n$, define

$$\sigma_{\le i}^1 = \sigma_1^1 \sigma_2^1 \cdots \sigma_i^1,$$

and similarly define $\sigma_{\le j}^2$ to be the first j characters in σ^2. Define $A[i,j]$ to be the score of the optimal alignment of $\sigma_{\le i}^1$ and $\sigma_{\le j}^2$. We would like to find $A[n,m]$, and we do this recursively as follows. For characters π_1 and π_2 in Σ, define $w(\pi_1, \pi_2)$ to be 0 if $\pi_1 = \pi_2$ and X if $\pi_1 \ne \pi_2$. For the base cases of the recursion, we have that $A[i,0] = i \cdot S$ and $A[0,j] = j \cdot S$ (since the only alignments have i spaces and j spaces, respectively). Then we recursively apply the formula

$$A[i,j] = \max \begin{cases} A[i-1,j-1] + w(\sigma_i^1, \sigma_j^2) \\ A[i-1,j] + S \\ A[i,j-1] + S \end{cases} \tag{7.3}$$

for $1 \le i \le n$ and $1 \le j \le m$.

Example 7.3 Suppose that $\sigma^1 = $ CAA and $\sigma^2 = $ AAC, and we have parameter values $M = 0$, $X = -2$, and $S = -3$. Then Figure 7.1 shows $A[i,j]$ for $0 \le i, j \le 3$. In particular, the optimal alignment for σ^1 and σ^2 is $\begin{smallmatrix} \text{CAA} \\ \text{AAC} \end{smallmatrix}$ with score $A[3,3] = -4$. □

In the recursive formula for the Needleman–Wunsch algorithm, we use the two operations, max and $+$. In other words, the recursion takes place in the $(\max, +)$ algebra (which is equivalent to the tropical algebra from Section 2.1).

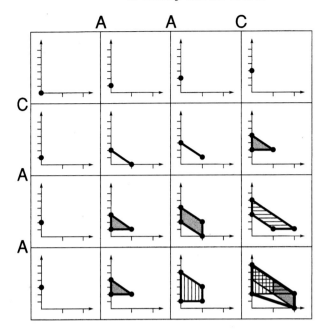

Fig. 7.2. $P[i,j]$ with $\sigma^1 = $ CAA and $\sigma^2 = $ AAC.

The polytope propagation algorithm is the exact same recursion, but in the polytope algebra (Section 2.3), where addition is the convex hull of the union of two polytopes and multiplication is the Minkowski sum of two polytopes. To be precise, let $P[i,j]$ be the Newton polytope for aligning the two strings $\sigma^1_{\leq i}$ and $\sigma^2_{\leq j}$, and define $v(\pi_1, \pi_2)$ to be $\{(1,0)\}$ (the Newton polytope of the monomial θ_X^{Γ}) if $\pi_1 \neq \pi_2$ and $\{(0,0)\}$ if $\pi_1 = \pi_2$. Then, if $\odot$ denotes the Minkowski sum operation and $\oplus$ is the convex hull of the union of two polytopes, we have the recursion

$$\begin{aligned} P[i,j] \quad = \quad & P[i-1, j-1] \odot v(\sigma^1_i, \sigma^2_j) \\ & \oplus P[i-1, j] \odot \{(0,1)\} \\ & \oplus P[i, j-1] \odot \{(0,1)\} \end{aligned}$$

for $1 \leq i \leq n$ and $1 \leq j \leq m$. Compare this to (7.3). We must also describe the base cases for the recursion: $P[i,0] = \{(0,i)\}$, because the only possible alignment has 0 mismatches and i spaces, and similarly $P[0,j] = \{(0,j)\}$. This recursion is exactly the polytope propagation algorithm run on a directed acyclic graph (Chapter 6), namely, the alignment graph $\mathcal{G}_{n,m}$ (Section 2.2).

Example 7.4 Using the same sequences as in Example 7.3, Figure 7.2 shows $P[i,j]$ for $0 \leq i, j \leq 3$, including an illustration of how to determine $P[3,3]$ from $P[2,2]$, $P[2,3]$, and $P[3,2]$. Table 7.1 lists, for each vertex of the polytope

vertex	alignment	parameter values
(2,0)	CAA AAC	$S \leq 0$ and $S \leq X$
(0,2)	CAA- -AAC	$X \leq S \leq 0$
(0,6)	CAA--- ---AAA	$S \geq 0$ and $S \geq \frac{1}{2}X$
(2,2)	-CAA AAC-	$0 \leq S \leq \frac{1}{2}X$

Table 7.1. *Vertices of $P[3,3]$, a corresponding optimal alignment, and the parameter values for which it is optimal.*

$P[3,3]$, an optimal alignment corresponding to that vertex, and the parameter values for which this alignment is optimal (these parameters are obtained by taking the normal fan of $P[3,3]$). Note that $(0,6)$ and $(2,2)$ do not correspond to biologically reasonable parameter values (as given by (7.1)), because their corresponding alignments are optimal for $S \geq 0 = M$. ☐

As we noted before, the parametric alignment method we have described generalizes to scoring schemes of any number of parameters by translating the extended Needleman–Wunsch algorithm to the polytope algebra. The algorithm presented runs efficiently for small numbers of parameters. It runs in $O(n^{\frac{8}{3}})$ and $O(n^{\frac{7}{2}})$ time for the 3- and 4-parameter scoring schemes, respectively (Section 5.3). For a large number of parameters, however, the algorithm is computationally expensive, as the convex hull operation is much more costly in higher dimensions. In practice, for scoring schemes with many parameters, one would fix all but a few of the parameters using the methods outlined in Section 5.4. If one is only concerned with computing the normal cone containing given parameter values, this can also be done efficiently by simply keeping track of the relevant vertices of the Newton polytope.

7.3 Retrieving alignments from polytope vertices

Parametric alignment is made feasible by the fact that the number of vertices of the Newton polytope is fairly small. In fact, for the 3-parameter scoring scheme, the number of vertices is bounded by

$$\text{constant} \cdot n^{\frac{2}{3}},$$

where n is the length of the longer of the two sequences (see Section 5.3). In contrast, the total number of possible alignments is exponential in n (see Proposition 2.9), which becomes unmanageable for large n.

However, the bound on the number of vertices of the Newton polytope does not imply a bound on the number of actual optimal alignments that each vertex may correspond to. Is this number also small, i.e., is it bounded by a polynomial in n? Unfortunately, in our 3-parameter scoring scheme, this is easily seen not to be the case.

Example 7.5 Assuming that $M > S$ (a biologically reasonable constraint, as defined in (7.1)), the number of optimal alignments of

$$\underbrace{\texttt{CC} \cdots \texttt{CC}}_{2n} \text{ and } \underbrace{\texttt{C} \cdots \texttt{C}}_{n}$$

is $\binom{2n}{n}$, which is exponential in n. In these optimal alignments, the $2n$ C's of the first string are aligned with any ordering of n C's and n -'s. For example, when $n = 4$, one of the optimal alignments is

$$\texttt{CCCCCCCC}$$
$$\texttt{-CC--C-C.}$$ □

One would hope that a more robust scoring scheme would not have this problem. For example, let us look at the 4-parameter scoring scheme discussed at the beginning of Section 7.2. Note that if we allow parameter values that are not biologically reasonable (that is, not satisfying (7.1)), then it is easy to get exponentially many optimal alignments. For example, if $G = 0$, then we are reduced to the 3-parameter scoring scheme, and Example 7.5 shows sequences with $\binom{2n}{n}$ optimal alignments.

If the parameter values are biologically reasonable, is it true that few alignments are optimal? Unfortunately, for each such choice of parameter values, there are sequences with exponentially many optimal alignments.

Proposition 7.6 *Given parameters M, X, G, and S such that $M > X$, $M > S$, and $G < 0$, choose any $k \in \mathbb{Z}_+$ such that*

$$k > \frac{-G}{M - \max\{X, S\}}.$$

For a given $r \in \mathbb{Z}_+$, define the sequences σ^1 and σ^2 of lengths $4kr$ and $3kr$, respectively, as follows: σ^1 is

$$\underbrace{\texttt{AA} \cdots \texttt{AA}}_{2k} \underbrace{\texttt{CC} \cdots \texttt{CC}}_{2k}$$

repeated r times, and σ^2 is

$$\underbrace{\texttt{AA} \cdots \texttt{AA}}_{2k} \underbrace{\texttt{C} \cdots \texttt{C}}_{k}$$

also repeated r times.

Then there are exactly $(k + 1)^r$ optimal alignments, which is exponential in the lengths of σ^1 and σ^2.

Proof The obvious choice for optimal alignments are ones like

$$\texttt{AAAACCCCAAAACCCCAAAACCCC}$$
$$\texttt{AAAACC--AAAAC--CAAAA--CC}$$

(in this example, $k = 2$ and $r = 3$) with all of the A's aligned and with one gap in each block of C's in σ^2. Since there are r blocks of C's and $k + 1$ choices of where to place the gap in each block, there are $(k+1)^r$ of these alignments. Let $\mathcal{O}$ denote the set of such alignments. We must prove that there are no alignments better than those in $\mathcal{O}$.

Note that the alignments in $\mathcal{O}$ have the greatest possible number of matches $(3kr)$, the least possible number of mismatches (zero), and the least possible number of spaces (kr). Therefore, the only way to improve the alignment is to have fewer gaps. Suppose we have a new alignment, h, with a better score than those in $\mathcal{O}$. We will divide the alignment scores into 2 parts:

(i) scores from gaps and spaces in σ^2, and

(ii) scores from gaps and spaces in σ^1, and from matches and mismatches in the alignment.

Let g be the number of gaps in h appearing in σ^2 (there may also be gaps in σ^1). Then $g < r$, because having fewer gaps is the only way to improve on the score of the alignments in $\mathcal{O}$. Part (i) of the score is increased by at most

$$(r - g)(-G),$$

since the alignments in $\mathcal{O}$ have r gaps and kr spaces, and h has g gaps and at least kr spaces in σ^2.

To see how much part (ii) of the score is decreased by changing from an alignment in $\mathcal{O}$ to the alignment h, we partition σ^2 into $r+1$ blocks

$$
\overbrace{\underbrace{\text{A}\cdots\text{A}}_{k} \mid \underbrace{\text{A}\cdots\text{A}}_{k}\underbrace{\text{C}\cdots\text{C}}_{k}\underbrace{\text{A}\cdots\text{A}}_{k} \mid \cdots \mid \underbrace{\text{A}\cdots\text{A}}_{k}\underbrace{\text{C}\cdots\text{C}}_{k}\underbrace{\text{A}\cdots\text{A}}_{k} \mid \underbrace{\text{A}\cdots\text{A}}_{k}\underbrace{\text{C}\cdots\text{C}}_{k}}^{r-1}.
$$

Ignoring the first block, we concentrate on the last r blocks. In the alignment h, at least $r - g$ of these blocks have no gaps inside them. No matter what part of σ^1 is aligned to one of these blocks, there must be at least k total of mismatches or spaces (each placed at the expense of a match); for example,

```
AACCCC          AACC--          AA--AA
AACCAA          AACCAA          AACCAA
```

are possibilities for $k = 2$. Then part (ii) of the score must be decreased by at least

$$(r - g) \cdot k \cdot \left(M - \max\{X, S\}\right)$$

by changing from an alignment in $\mathcal{O}$ to the alignment h.

Since we have assumed that the alignment h has a better score, we combine parts (i) and (ii) of the score and have that

$$(r - g)(-G) - (r - g)k\left(M - \max\{X, S\}\right) \geq 0.$$

But then

$$k \leq \frac{-G}{M - \max\{X, S\}},$$

contradicting our choice of k. Therefore, the alignments in $\mathcal{O}$ must have been optimal (and, in fact, must have been the only optimal alignments), and the proof follows.   □

A few comments are in order. Sequences with exponentially many optimal alignments might often appear in reality. The key condition yielding this exponential behavior is that there are large regions that are well aligned (the blocks of A's) interspersed with regions with more than one optimal alignment (the blocks of C's). In situations like this, however, there is some consolation: consensus alignment would work very well. A *consensus alignment* is a partial alignment of two sequences σ^1 and σ^2 that aligns only those segments of σ^1 and σ^2 that all of the optimal alignments agree on. In the example, the consensus alignment would have all of the A's aligned perfectly.

7.4 Biologically correct alignments

The scoring schemes that we have discussed thus far attempt to produce biologically correct alignments by modeling the evolutionary events undergone by biological sequences. The 4-parameter scoring scheme introduced in Section 7.2 models evolution by simply assigning weights to the events of mutation, deletion, and insertion. The hope is that a suitable choice of values for the parameters of this scoring scheme will lead to biologically correct alignments. In this section, we explore the capabilities of the 3 and 4-parameter scoring schemes to produce correct alignments. Through the use of the parametric inference methods, we may characterize the optimal alignments determined by these scoring schemes over their entire parameter spaces.

As in Section 7.3, we restrict ourselves to the biologically reasonable parameter values defined by (7.1). Note that some biologically unreasonable parameter values can lead to strange results, as the following remark shows.

Remark 7.7 For the 4-parameter scoring scheme with parameter values $(M, X, S, G) = (2\alpha, 2\alpha, \alpha, 0)$, for $\alpha \in \mathbb{R}$, all possible alignments of two sequences, σ^1 and σ^2, are optimal.

Proof With this scoring scheme, we have that

$$
\begin{aligned}
W_{\sigma^1, \sigma^2}(h) &= 2\alpha(m_h + x_h) + \alpha s_h \\
&= \alpha(2m_h + 2x_h + s_h) \\
&= \alpha(n + m),
\end{aligned}
$$

where in the second step we have used Remark 7.1. This weight is independent of the specific alignment h. Under this scoring scheme all alignments receive the same weight and are thus all considered to be optimal.   □

In the first use of parametric sequence alignment [Fitch and Smith, 1983], an example is given of two biological sequences that cannot be correctly aligned by scoring schemes lacking a gap parameter. Although their methods differ from those presented in this chapter, the result is the same: a subdivision of the parameter space into regions giving rise to the same optimal alignments. Fitch and Smith show that, for a 4-parameter scoring scheme with match and mismatch scores fixed, the region of the (S, G) parameter space that gives rise to the biologically correct alignment does not include a setting of parameters where either S or G is zero. From this result, they conclude that a scoring scheme must have non-zero values for both S and G to give the correct alignment as optimal. However, they do not mention the effect of having the match and mismatch scores fixed, and their methods for subdividing the parameter space are not rigorously shown to be correct. In this section, we analyze their sequences using the parametric alignment methods presented in this chapter and give a more complete picture of how the parameters affect the alignment of these sequences.

The sequences in question are portions of the mRNAs for chicken α and β hemoglobin. As claimed in [Fitch and Smith, 1983], the biologically correct alignment of the amino acid sequences coded for by these mRNA segments is

$$\beta \quad \text{FASFGNLSSPTAILGNPMV}$$
$$\alpha \quad \text{FPHF-DLSH-----GSAQI.}$$

The mRNA alignment corresponding to this protein alignment is

β UUUGCGUCCUUUGGGAACCUCUCCAGCCCCACUGCCAUCCUUGGCAACCCCAUGGUC
α UUUCCCCACUUC---GAUCUGUCACAC--------------GGCUCCGCUCAAAUC.

However, as is noted by Fitch and Smith, this mRNA alignment is not necessarily the biologically correct one. Another possible alignment of the mRNA sequences is

β UUUGCGUCCUUUGGGAACCUCUCCAGCCCCACUGCCAUCCUUGGCAACCCCAUGGUC
α UUUCCCCACUUCG---AUCUGUCACAC--------------GGCUCCGCUCAAAUC,

where the leftmost gap was not the result of an exact deletion or insertion at codon boundaries. We will consider both of these mRNA alignments as potentially biologically correct and will refer to them as h_1 and h_2, respectively.

To determine how these two alignments might be obtained through the use of the 4-parameter scoring scheme, we use the parametric alignment methods described in Section 7.2, with the mRNA sequences as input. Using Remark 7.2, we run our analysis with $M = 0$ without loss of generality. The resulting alignment polytope is 3-dimensional and has 64 vertices (Figure 7.3).

Alignments h_1 and h_2 are represented by the points $p_1 = (19, 18, 2)$ and $p_2 = (18, 18, 2)$, where the coordinates represent the number of mismatches, spaces, and gaps in the alignments, respectively. Point p_2 is vertex 1 of the alignment polytope, while p_1 is in the interior the upper-right facet of the polytope, a facet that also includes vertex 1 and corresponds to all possible alignments having 18 spaces (the minimum number of spaces possible). The normal cone of vertex 1 is given by the inequalities

$$2X \geq G \geq 3X \qquad \text{and} \qquad X \geq \frac{5}{3}S. \qquad (7.4)$$

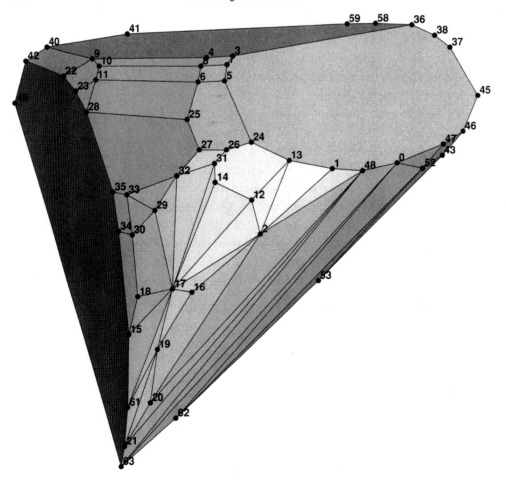

Fig. 7.3. Alignment polytope for two chicken hemoglobin sequences.

Thus, for h_2 to be optimal, the parameter values of the scoring scheme must satisfy the inequalities in (7.4). An example of a set of values for the 4-parameter scoring scheme that gives h_2 as optimal is $(M, X, S, G) = (0, -2, -2, -5)$. Using an argument similar to that used in proving Remark 7.2, it can be shown that $(M, X, S, G) = (4, 2, 0, -5)$ is an equivalent set of values. Thus, unlike in [Fitch and Smith, 1983], if we are allowed to change the match and mismatch scores, it is not the case that we need a non-zero value for S to obtain the alignment h_2.

We now turn to alignment h_1 and its corresponding point, p_1. Because p_1 lies in the interior of a facet of the polytope, h_1 is optimal only for a very restricted set of parameter values. Alignment h_1 is optimal for

$$X = 0, \qquad G = 0, \qquad \text{and} \qquad S \le 0. \tag{7.5}$$

Noting the sets of parameter values that make these two alignments optimal, we get the following theorem.

Theorem 7.8 *There exist two biological sequences, σ^1 and σ^2, such that the* *biologically correct alignment of these sequences is not optimal, under the 3-parameter scoring scheme, for any choice of parameter values satisfying the biologically reasonable constraints $M > S$ and $M > X$.*

Proof Let σ^1 and σ^2 be the mRNA sequences presented in this section. Figure 7.3 shows the alignment polytope for these sequences under the 4-parameter scoring scheme. Under the 3-parameter scoring scheme, we have that $G = 0$. Therefore, we will consider the intersection of the normal fan of this polytope with the hyperplane $G = 0$. Alignments h_1 and h_2 are the only candidates for the biologically correct alignment. From (7.4), we see that h_2 is optimal if and only if $X = M = 0$ when $G = 0$. Therefore, h_2 requires biologically unreasonable parameter values to be optimal. Similarly, from (7.5), h_1 requires that $X = M = 0$ for it to be optimal, and thus neither potentially correct alignment is optimal for any choice of biologically reasonable parameter values. ☐

Theorem 7.8 shows that the 3-parameter scoring scheme is inadequate for correctly aligning some biological sequences. However, as the inequalities in (7.4) show, alignment h_2 *can* be obtained with biologically reasonable values for the 4-parameter scoring scheme. Alignment h_1, on the other hand, cannot be obtained with biologically reasonable values for the 4-parameter scoring scheme, since G must be zero for h_1 to be optimal. If it is the case that h_1 is the true biological alignment of these two sequences, then we would have proved that the 4-parameter model is also inadequate for some biological sequences. Due to the uncertainty of the correct alignment, we leave this as a conjecture.

Conjecture 7.9 *There exist two biological sequences, σ^1 and σ^2, such that the biologically correct alignment of these sequences is not optimal, under the 4-parameter scoring scheme, for any choice of parameter values satisfying the constraints (7.1).*

As in the 3-parameter case, parametric sequence alignment should be an invaluable tool in proving this conjecture.

8

Bounds for Optimal Sequence Alignment

Sergi Elizalde

Fumei Lam

One of the most frequently used techniques in determining the similarity between biological sequences is optimal sequence alignment. In the standard instance of the sequence alignment problem, we are given two sequences (usually DNA or protein sequences) that have evolved from a common ancestor via a series of mutations, insertions and deletions. The goal is to find the best alignment between the two sequences. The definition of "best" here depends on the choice of scoring scheme, and there is often disagreement about the correct choice. In *parametric sequence alignment*, this problem is circumvented by instead computing the optimal alignment as a function of *variable* scores. In this chapter, we address one such scheme, in which all matches are equally rewarded, all mismatches are equally penalized and all spaces are equally penalized. An efficient parametric sequence alignment algorithm is described in Chapter 7. Here we will address the *structure* of the set of different alignments, and in particular the number of different alignments of two given sequences which can be optimal. For a detailed treatment on the subject of sequence alignment, we refer the reader to [Gusfield, 1997].

8.1 Alignments and optimality

We first review some notation from Section 2.2. In this chapter, all alignments will be *global alignments* between two sequences σ^1 and σ^2 of the same length, denoted by n. An *alignment* between σ^1 and σ^2 is specified either by a string over $\{H, I, D\}$ satisfying $\#H + \#D = \#H + \#I = n$, or by a pair (μ^1, μ^2) obtained from σ^1, σ^2 by possibly inserting "$-$" characters, or by a path in the alignment graph $\mathcal{G}_{n,n}$. A *match* is a position in which μ^1 and μ^2 have the same character, a *mismatch* is a position in which μ^1 and μ^2 have different characters, and a *space* (or *indel*) is a position in which exactly one of μ^1 or μ^2 has a space. Observe that in any alignment of two sequences of equal length, the number of spaces in each sequence is the same. This number equals $\#I$ and we call it the number of *insertions* of the alignment. It is half the total number of spaces for sequences of equal length.

In the general sequence alignment problem for DNA, the number of parameters in a scoring scheme is 33 (see Section 2.2). In this chapter, we consider the special case in which the score of any mismatch is $-\alpha$ and the score of any space is $-\beta/2$ (equivalently, the score of any insertion is $-\beta$). Without loss of generality, we fix the reward for a match to be 1. We then have only two parameters and the following pair (w, w') represents our scoring scheme:

$$
w = \begin{pmatrix}
1 & -\alpha & -\alpha & -\alpha & -\beta/2 \\
-\alpha & 1 & -\alpha & -\alpha & -\beta/2 \\
-\alpha & -\alpha & 1 & -\alpha & -\beta/2 \\
-\alpha & -\alpha & -\alpha & 1 & -\beta/2 \\
-\beta/2 & -\beta/2 & -\beta/2 & -\beta/2 &
\end{pmatrix}, \quad
w' = \begin{pmatrix}
0 & 0 & 0 \\
0 & 0 & 0 \\
0 & 0 & 0
\end{pmatrix}.
$$

The score of an alignment $\mathcal{A}$ with z matches, x mismatches, and y insertions (i.e., $2y$ spaces) is then

$$
score(\mathcal{A}) = z - x\alpha - y\beta.
$$

The vector (x, y, z) will be called the *type* of the alignment. Note that for sequences of length n, the type always satisfies $x + y + z = n$.

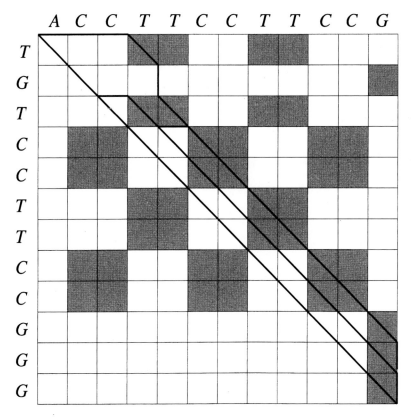

Fig. 8.1. Shaded squares denote the positions in which σ^1 and σ^2 agree. The four alignments shown and corresponding scores are

```
TGTCCTTCCGGG    TG-TCCTTCCGGG    TG-T-CCTTCCGGG    ---TGTCCTTCCGGG
ACCTTCCTTCCG    ACCTTCCTTCCG-    ACCTTCCTTCCG--    ACCT-TCCTTCCG--
   1 - 11α         5 - 6α - β        8 - 2α - 2β          9 - 3β
```

8.2 Geometric interpretation

For two fixed sequences σ^1 and σ^2, different choices of parameters α and β may yield different optimal alignments. Observe that there may be alignments that are not optimal for any choice of α and β. When α, β are not fixed, an alignment will be called *optimal* if there is some choice of the parameters that makes it optimal for those values of the parameters. Given two sequences, it is an interesting problem to determine how many different types of alignments can be optimal for some value of α and β.

For each of these types, consider the region in the $\alpha\beta$-plane corresponding to the values of the parameters for which the alignments of the given type are optimal. This gives a decomposition of the $\alpha\beta$-plane into *optimality regions*. Such regions are convex polyhedra; more precisely, they are translates of cones. To see this, note that the score of an alignment with z matches, x mismatches, and y insertions is $z - x\alpha - y\beta = n - x(\alpha+1) - y(\beta+1)$ (since $x+y+z = n$). To each such alignment, we can associate a point $(x, y) \in \mathbb{R}^2$. The convex hull of all such points is a polygon, which we denote P_{xy}. Then an alignment is optimal for those values of α and β such that the corresponding point (x, y) minimizes the product $(x, y) \cdot (\alpha + 1, \beta + 1)$ among all the points (x, y) associated with alignments of the two given sequences. Thus, an alignment with x mismatches and y insertions is optimal for some α and β if and only if (x, y) is a vertex of P_{xy}. From this, we see that the decomposition of the $\alpha\beta$-plane into optimality regions is the translation of the normal fan of P_{xy} by $(-1, -1)$.

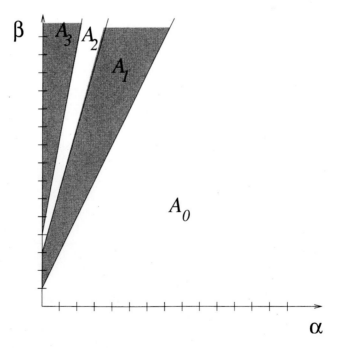

Fig. 8.2. Decomposition of the parameter space for sequences TGTCCTTCCGGG ACCTTCCTTCCG from Figure 8.1. The boundaries of the regions are given by the coordinate axes and the lines $\beta = 1 + 2\alpha$, $\beta = 3 + 4\alpha$ and $\beta = 4 + 5\alpha$.

We are interested in the number of optimality regions, or equivalently, the number of vertices of P_{xy}. The parameters are only biologically meaningful for $\alpha, \beta \geq 0$, the case in which spaces and mismatches are penalized. Thus, we will only consider the optimality regions intersecting this quadrant. Equivalently, we are only concerned with the vertices of P_{xy} along the lower-left border (i.e., those for which there is no other vertex below and to the left).

Note that for $\alpha = \beta = -1$, the score of *any* alignment is $z - (-1)x - (-1)y = z + x + y = n$; therefore, the lines bounding each optimality region are either coordinate axes or lines passing through the point $(-1, -1)$. This shows that all boundary lines of optimality regions are either coordinate axes or lines of the form $\beta = c + (c+1)\alpha$ for some constant c (see [Gusfield *et al.*, 1994]).

The polygon P_{xy} is a projection of the convex hull of the points (x, y, z) giving the number of mismatches, insertions and matches of each alignment. All these points lie on the plane $x + y + z = n$ and their convex hull is a polygon which we denote P. We call P the *alignment polygon*. Note that P is in fact the Newton polytope of the polynomial f_{σ^1, σ^2} in equation (7.2), which is a coordinate of the pair hidden Markov model after specializing several parameters in equation (2.15). One obtains polygons combinatorially equivalent to P and P_{xy} by projecting onto the xz-plane or the yz-plane instead. It will be convenient for us to consider the projection onto the yz-plane, which we denote P_{yz}.

8.2.1 The structure of the alignment polygon

The polygon P_{yz} is obtained by taking the convex hull of the points (y, z) whose coordinates are the number of insertions and matches of each alignment of the two given sequences. Note that any mismatch of an alignment can be replaced with two spaces, one in each sequence, without changing the rest of the alignment. If we perform such a replacement in an alignment of type (x, y, z), we obtain an alignment of type $(x - 1, y + 1, z)$. By replacing all mismatches with spaces we obtain an alignment of type $(0, x+y, z) = (0, n-z, z)$. Similarly, replacing a match with two spaces in an alignment of type (x, y, z) yields one of type $(x, y + 1, z - 1)$, and performing all such replacements results in an alignment of type $(x, y+z, 0) = (x, n-x, 0)$. Note however that the replacement of a match with spaces never gives an optimal alignment for non-negative values of α and β.

From this, we see that if a point (y, z) inside P_{yz} comes from an alignment and $y + z < n$ (resp. $z > 0$), then the point $(y + 1, z)$ (resp. $(y + 1, z - 1)$) must also come from an alignment. A natural question is whether all lattice points (i.e., points with integral coordinates) inside P_{yz} come from alignments. We will see in the construction of Proposition 9.4 that this is not the case in general. This means that there are instances in which a lattice point (y', z') lies inside the convex hull of points corresponding to alignments, but there is no alignment with y' insertions and z' matches.

From the above observation, it follows that P has an edge joining the vertices

$(0, n, 0)$ and $(0, n - z_{max}, z_{max})$, where z_{max} is the maximum number of matches in any alignment of the two given sequences. Similarly, P has an edge joining $(0, n, 0)$ and $(x_{max}, n - x_{max}, 0)$, where x_{max} is the maximum number of mismatches in any alignment. The alignment with no spaces also gives a vertex of P, which we denote $v_0 = (x_0, 0, z_0)$. An example of an alignment polytope appears in Figure 8.3. Going around P starting from v_0 in the positive direction of the z axis, we label the vertices $v_1, v_2, \ldots, v_r = (0, n - z_{max}, z_{max})$, $v_{r+1} = (0, n, 0)$, $v_{r+2} = (x_{max}, n - x_{max}, 0), \ldots$

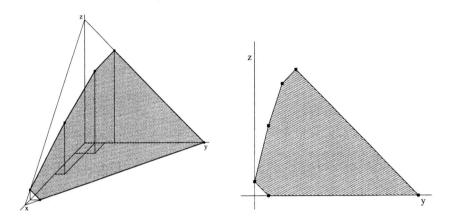

Fig. 8.3. The alignment polygon P for the sequences in Figure 8.1 and its projection P_{yz}.

For the meaningful values of the parameters, optimal alignments correspond to the vertices $v_0, v_1, \ldots, v_r$. We call them *relevant vertices* of P. For $v_i = (x_i, y_i, z_i)$, we have

$$z_0 < z_1 < \cdots < z_{r-1} \leq z_r = z_{max},$$

$$0 = y_0 < y_1 < \cdots < y_{r-1} < y_r = n - z_{max},$$

$$x_0 > x_1 > \cdots > x_{r-1} > x_r = 0.$$

Each vertex corresponds to an optimality region in the first quadrant of the $\alpha\beta$-plane. For sequences σ^1, σ^2, we define $g(\sigma^1, \sigma^2) = r + 1$ to be the number of such optimality regions. Note that $g(\sigma^1, \sigma^2)$ also equals the number of relevant vertices of P, and the number of different types of optimal alignments of σ^1 and σ^2 for non-negative values of α and β.

Let Σ be a fixed alphabet (which can be finite or infinite). We define

$$f_\Sigma(n) = \max_{\sigma^1, \sigma^2 \in \Sigma^n} g(\sigma^1, \sigma^2).$$

In other words, $f_\Sigma(n)$ is the maximum number of optimality regions in the decomposition of the first quadrant induced by a pair of sequences of length n in the alphabet Σ. We are interested in bounds on $f_\Sigma(n)$.

8.3 Known bounds

The first non-trivial upper bound on $f_\Sigma(n)$ was given in [Gusfield *et al.*, 1994], where it was shown that $f_\Sigma(n) = O(n^{2/3})$ for any alphabet Σ (finite or infinite). The more precise bound $f_\Sigma(n) = 3(n/2\pi)^{2/3} + O(n^{1/3} \log n)$ is shown in [Fernández-Baca *et al.*, 2002], and an example is given for which the bound is tight if Σ is an infinite alphabet.

To establish the upper bound, [Fernández-Baca *et al.*, 2002] uses the fact that the slopes of the segments connecting pairs of consecutive relevant vertices of P_{yz} must be all different. The bound is obtained by calculating the maximum number of different rational numbers such that the sum of all the numerators and denominators is at most n. To show that this bound is tight for an infinite alphabet, for every n, they construct a pair of sequences of length n for which the above bound on the number of different slopes between consecutive vertices of P_{yz} is attained. In their construction, the number of different symbols that appear in the sequences of length n grows linearly in n. It is an interesting question whether a similar $\Omega(n^{2/3})$ bound can be achieved using fewer symbols, even if the number of symbols tends to infinity as n increases.

A lower bound example which works only when the alphabet Σ is infinite is not very practical. This is because biological sequences encountered in practice are over a finite alphabet, usually the 4-letter alphabet $\{A, C, G, T\}$ or the 20-letter amino acid alphabet. For finite alphabets Σ, the asymptotic behavior of $f_\Sigma(n)$ is not known. The upper bound $f_\Sigma(n) \leq 3(n/2\pi)^{2/3} + O(n^{1/3} \log n)$ seems to be far from the actual value for a finite alphabet Σ. We discuss this in the next section.

Questions analogous to the ones addressed in this chapter can be asked for the 4-parameter scoring scheme described at the beginning of Section 7.2, which has an extra parameter giving the penalty of a gap (i.e., a contiguous set of spaces). After restricting the reward for a match to 1, this model has three free parameters. Again, for two given sequences of length n, the corresponding 3-dimensional parameter space is decomposed into optimality regions, according to the values of the parameters for which the same class of alignments is optimal (with the new scoring scheme). As in the 2-dimensional case, we can now associate to each alignment a point (y, z, w) whose coordinates are the number of insertions, matches, and gaps, respectively. The convex hull of these points is a 3-dimensional lattice polytope contained in the box $[0, n] \times [0, n] \times [0, 2n]$. Therefore, it follows from Theorem 5.1 that the number of vertices of this polytope (equivalently, the number of optimality regions in which the parameter space is divided) is at most $O(n^{3/2})$. Note that this improves the $O(n^{5/3})$ bound given in [Fernández-Baca *et al.*, 2004]. It is an open question whether $O(n^{3/2})$ is a tight bound. We believe that this is not the case, i.e., there is no pair of sequences of length n (even over an infinite alphabet) whose corresponding 3-dimensional alignment polytope has $\Theta(n^{3/2})$ vertices. We will not discuss this model in more detail here. In the rest of the chapter, we focus on the scoring scheme with two free parameters α and β.

8.4 The square root conjecture

In the case of a binary alphabet $\Sigma = \{0,1\}$, for any n, there is a pair of sequences of length n over Σ such that the number of different types of optimal alignments is $\Omega(\sqrt{n})$ [Fernández-Baca *et al.*, 2002]. More precisely, if we let $s(n) = s$ be the largest integer such that $s(s-1)/2 \leq n$, then one can construct sequences of length n such that the number of relevant vertices of the alignment polygon is $s(n)$. It follows that $f_{\{0,1\}}(n) \geq s(n) = \Omega(\sqrt{n})$. Clearly, for any alphabet Σ of size $|\Sigma| \geq 2$, the same construction using only two symbols also shows $f_\Sigma(n) \geq s(n) = \Omega(\sqrt{n})$. For a finite alphabet, the best known bounds are $f_\Sigma(n) = \Omega(\sqrt{n})$ and $f_\Sigma(n) = O(n^{2/3})$. It is an open problem to close this gap. We believe that the actual asymptotic behavior of $f_{\{0,1\}}(n)$ is given by the lower bound.

Conjecture 8.1 $f_{\{0,1\}}(n) = \Theta(\sqrt{n})$.

Experimental results from [Fernández-Baca *et al.*, 2002] suggest that the expected number of regions generated by random pairs of sequences over a finite alphabet is also $\Theta(\sqrt{n})$. In these experiments, the probability distribution of the number of regions seems to be concentrated sharply around $\sqrt{n}$, and very few sequences are found to exhibit a significantly worse behavior.

It is clear that increasing the number of symbols in the alphabet Σ creates a larger number of possible pairs of sequences. In particular, we have that $f_{\Sigma'}(n) \geq f_\Sigma(n)$ whenever $|\Sigma'| \geq |\Sigma|$, since the sequences over Σ whose alignment polygon has $f_\Sigma(n)$ relevant vertices can be interpreted as sequences over Σ' as well. Intuitively, a larger alphabet gives more freedom on the different alignment polygons that arise, which potentially increases the upper bound on the number of relevant vertices.

This is indeed the case in practice, as the following example shows. Let $\Sigma = \{0,1\}$ be the binary alphabet, and let $\Sigma' = \{w_1, w_2, \ldots, w_6\}$. Consider the pair of sequences $\sigma^1 = w_6 w_1 w_2 w_6 w_3 w_6 w_6 w_4 w_6$ and $\sigma^2 = w_1 w_2 w_3 w_4 w_5 w_5 w_5 w_5 w_5$ of length $n = 9$ over Σ' (note that this construction is similar to the one in Figure 8.4). Then, the alignment polygon has 5 relevant vertices, namely $v_0 = (9,0,0)$, $v_1 = (6,1,2)$, $v_2 = (4,2,3)$, $v_3 = (1,4,4)$ and $v_4 = (0,5,4)$. It is not hard to see that in fact $f_{\Sigma'}(n) = 5$. However, one can check by exhaustive computer search that there is no pair of binary sequences of length 9 such that their alignment polytope has 5 relevant vertices. Thus, $f_{\{0,1\}}(n) = 4 < f_{\Sigma'}(n)$.

In contrast to this result, the construction that gives the best known lower bound $\Omega(\sqrt{n})$ for finite alphabets is in fact over the binary alphabet. Thus, one interesting question is whether the bounds on $f_\Sigma(n)$ are asymptotically the same for all finite alphabets. In particular, it is an open question whether an improved upper bound in the case of the binary alphabet would imply an improved upper bound in the case of a finite alphabet.

One possible approach to reduce from the finite to the binary alphabet case is to consider the sequences σ^1 and σ^2 under all maps $\pi : \Sigma \to \{0,1\}$. If we let $k = |\Sigma|$, there are 2^k such maps, which we denote by π_j, $j = 1, \ldots, 2^k$. For each j, let P_{xy}^j be the convex hull of the points (x, y) giving the number of mismatches and insertions of the alignments of sequences $\pi_j(\sigma^1)$ and $\pi_j(\sigma^2)$. We would like to relate the vertices of P_{xy} to the vertices of P_{xy}^j.

Conjecture 8.2 *For each relevant vertex (x_i, y_i) of P_{xy}, there exists a map $\pi_j : \Sigma \to \{0, 1\}$ such that P_{xy}^j has a relevant vertex whose second coordinate is y_i.*

Let Σ be a finite alphabet on at least two symbols and let $k = |\Sigma|$. If this conjecture is true, then $f_\Sigma(n) \leq 2^k f_{\{0,1\}}(n)$ for every n, implying the following stronger version of Conjecture 8.1.

Conjecture 8.3 *For any finite alphabet Σ, $f_\Sigma(n) = \Theta(\sqrt{n})$.*

Note that the case of $\Sigma = \{A, C, G, T\}$ is particularly interesting since most biological sequences are over this 4-letter alphabet.

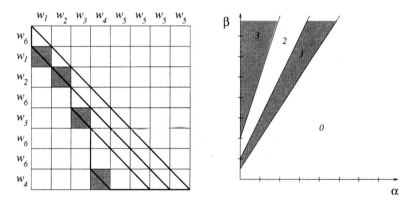

Fig. 8.4. Optimal alignments for $\sigma^1 = w_6 w_1 w_2 w_6 w_3 w_6 w_6 w_4$ and $\sigma^2 = w_1 w_2 w_3 w_4 w_5 w_5 w_5 w_5$, and corresponding optimality regions A_0, A_1, A_2 and A_3 given by lines $\beta = \frac{1}{2} + \frac{3}{2}\alpha, \beta = 1 + 2\alpha$ and $\beta = 2 + 3\alpha$. . In this example, for any mapping $\pi : \{w_1, w_2, \ldots, w_6\} \to \{0, 1\}$ and any region B_j^π in the resulting decomposition, $B_j^\pi \not\subseteq A_1$ and $B_j^\pi \not\subseteq A_2$.

Instead of finding a relationship between the vertices of P_{xy} and the vertices of P_{xy}^j, another approach would be to try to find a relationship between the optimality regions in the decomposition of the $\alpha\beta$ parameter space under all mappings to the binary alphabet. For sequences σ^1 and σ^2, let $A_0, A_1, \ldots A_m$ denote the optimality regions of the decomposition of the parameter space, and for any map $\pi : \Sigma \to \{0, 1\}$, let $B_0^\pi, B_1^\pi, \ldots B_{i_\pi}^\pi$ denote the optimality regions for alignments of $\pi(\sigma^1)$ and $\pi(\sigma^2)$. If for every A_i, $0 \leq i \leq m$, there exists a mapping π such that $B_j^\pi \subseteq A_i$ for some j, $0 \leq j \leq i_\pi$, then this would imply $f_\Sigma(n) \leq 2^k f_{\{0,1\}}(n)$ for every n, proving that Conjecture 8.3 follows from Conjecture 8.1. However, this is not true for the example in Figure 8.4.

The construction of two binary sequences of length n with $s = s(n)$ optimal alignments used to obtain the lower bound in [Fernández-Baca *et al.*, 2002] has the following peculiarity. The number of insertions in the alignments giving relevant vertices of the alignment polygon P are $y_0 = 0$, $y_1 = 1$, $y_2 = 2$, ..., $y_{s-2} = s - 2$ (see Figure 8.5).

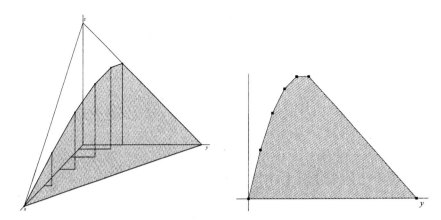

Fig. 8.5. The alignment polygon P for binary sequences of length $n = 15$ with $s(n) = 6$ relevant vertices.

The slopes between consecutive relevant vertices in the projected alignment polygon P_{yz} are then $s, s - 1, s - 2, \ldots, 1, 0$. In particular, all the slopes are integral. If this was true of all alignment polytopes coming from binary sequences, then Conjecture 8.1 would follow, because the maximum number of different integral slopes in P_{yz} is $s(n)$. However, it is not the case in general that the slopes obtained from binary sequences are always integral.

The smallest counterexample is given by the pair of sequences $\sigma^1 = 001011$ and $\sigma^2 = 111000$. The vertices of the corresponding polytope P_{yz} are $(0, 2)$, $(3, 3)$, $(6, 0)$ and $(1, 0)$. The slope between the first two vertices is $1/3$, which shows that not all slopes are integral. In fact, it follows from the proof of Proposition 9.5 that the situation is quite the opposite. We will see in the next chapter that for any positive integers u, v with $u < v$, one can construct a pair of binary sequences of length at most $6v - 2u$ such that the corresponding (projected) alignment polytope P_{yz} has a slope u/v between two consecutive relevant vertices.

9

Inference Functions

Sergi Elizalde

Some of the statistical models introduced in Chapter 1 have the feature that, aside from the observed data, there is hidden information that cannot be determined from an observation. In this chapter we consider graphical models with hidden variables, such as the hidden Markov model and the hidden tree model. A natural problem in such models is to determine, given a particular observation, what is the most likely hidden data (which is called the *explanation*) for that observation. This problem is called MAP inference (Remark 4.13). Any fixed values of the parameters determine a way to assign an explanation to each possible observation. A map obtained in this way is called an *inference function*.

Examples of inference functions include *gene-finding functions* which were discussed in [Pachter and Sturmfels, 2005, Section 5]. These inference functions of a hidden Markov model are used to identify gene structures in DNA sequences (see Section 4.4. An observation in such a model is a sequence over the alphabet $\Sigma' = \{A, C, G, T\}$.

After a short introduction to inference functions, we present the main result of this chapter in Section 9.2. We call it the *Few Inference Functions Theorem*, and it states that in any graphical model the number of inference functions grows polynomially if the number of parameters is fixed. This theorem shows that most functions from the set of observations to possible values of the hidden data cannot be inference functions for any choice of the model parameters. In Section 9.3 we apply this theorem to the model for sequence alignment described in Chapter 8. We obtain a quadratic bound on the number of inference functions of this model, and we show that it is asymptotically tight.

9.1 What is an inference function?

Let us introduce some notation and make the definition of an inference function more formal. Consider a graphical model (as defined in Section 1.5) with n observed random variables $Y_1, Y_2, \ldots, Y_n$, and q hidden random variables $X_1, X_2, \ldots, X_q$. To simplify the notation, we make the assumption, which is often the case in practice, that all the observed variables take their values in the same finite alphabet Σ', and that all the hidden variables are on the finite alphabet Σ. The state space is then $(\Sigma')^n$. Let $l = |\Sigma|$ and $l' = |\Sigma'|$. Let d be the number of parameters of the model, which we denote by $\theta_1, \theta_2, \ldots, \theta_d$. They represent transition probabilities between states. The model is represented by a positive polynomial map $\mathbf{f} : \mathbb{R}^d \longrightarrow \mathbb{R}^{(l')^n}$. For each observation $\tau \in (\Sigma')^n$, the corresponding coordinate f_τ is a polynomial in $\theta_1, \theta_2, \ldots, \theta_d$ that gives the probability of making the particular observation τ. Thus, we have $f_\tau = \mathrm{Prob}(\mathbf{Y} = \tau) = \sum_{\mathbf{h} \in \Sigma^q} \mathrm{Prob}(\mathbf{X} = \mathbf{h}, \mathbf{Y} = \tau)$, where each summand $\mathrm{Prob}(\mathbf{X} = \mathbf{h}, \mathbf{Y} = \tau)$ is a monomial in the parameters $\theta_1, \theta_2, \ldots, \theta_d$.

For fixed values of the parameters, the basic inference problem is to determine, for each given observation τ, the value $\mathbf{h} \in \Sigma^q$ of the hidden data that maximizes $\mathrm{Prob}(\mathbf{X} = \mathbf{h} | \mathbf{Y} = \tau)$. A solution to this optimization problem is denoted $\widehat{\mathbf{h}}$ and is called an *explanation* of the observation τ. Each choice of parameter values $(\theta_1, \theta_2, \ldots, \theta_d)$ defines an *inference function* $\tau \mapsto \widehat{\mathbf{h}}$ from the set of observations $(\Sigma')^n$ to the set of explanations Σ^q. A brief discussion of inference functions and their geometric interpretation in terms of the tropicalization of the polynomial map $\mathbf{f}$ was given at the end of Section 3.4.

It is possible that there is more than one value of $\widehat{\mathbf{h}}$ attaining the maximum of $\mathrm{Prob}(\mathbf{X} = \mathbf{h} | \mathbf{Y} = \tau)$. In this case, for simplicity, we will pick only one such explanation, according to some consistent tie-breaking rule decided ahead of time. For example, we can pick the least such $\widehat{\mathbf{h}}$ in some given total order of the set Σ^q of hidden states. Another alternative would be to define inference functions as maps from $(\Sigma')^n$ to subsets of Σ^q. This would not affect the results of this chapter, so for the sake of simplicity, we consider only inference functions as defined above.

It is important to observe that the total number of functions $(\Sigma')^n \longrightarrow \Sigma^q$ is $(l^q)^{(l')^n} = l^{q(l')^n}$, which is doubly-exponential in the length n of the observations. However, most of these are not inference functions for any values of the parameters. In this chapter we give an upper bound on the number of inference functions of a graphical model.

We conclude this section with some more notation and preliminaries that will be needed in the chapter. We denote by E the number of edges of the underlying graph of the graphical model. The logarithms of the model parameters will be denoted by $v_i = \log \theta_i$.

The coordinates of our model are polynomials of the form $f_\tau(\theta_1, \theta_2, \ldots, \theta_d) = \sum_i \theta_1^{a_{1,i}} \theta_2^{a_{2,i}} \cdots \theta_d^{a_{d,i}}$, where each monomial represents $\mathrm{Prob}(\mathbf{X} = \mathbf{h}, \mathbf{Y} = \tau)$ for some $\mathbf{h}$. Recall from Section 2.3 that the *Newton polytope* of such a polynomial is defined as the convex hull in $\mathbb{R}^d$ of the exponent vectors $(a_{1,i}, a_{2,i}, \ldots, a_{d,i})$. We denote the Newton polytope of f_τ by $\mathrm{NP}(f_\tau)$, and its *normal fan* by $\mathcal{N}(\mathrm{NP}(f_\tau))$.

Recall also that the *Minkowski sum* of two polytopes P and P' is $P \odot P' := \{\mathbf{x} + \mathbf{x}' : \mathbf{x} \in P, \mathbf{x}' \in P'\}$. The Newton polytope of the map $\mathbf{f} : \mathbb{R}^d \longrightarrow \mathbb{R}^{(l')^n}$ is defined as the Minkowski sum of the individual Newton polytopes of its coordinates, namely $\mathrm{NP}(\mathbf{f}) := \bigodot_{\tau \in (\Sigma')^n} \mathrm{NP}(f_\tau)$.

The *common refinement* of two or more normal fans is the collection of cones obtained as the intersection of a cone from each of the individual fans.

For polytopes $P_1, P_2, \ldots, P_k$, the common refinement of their normal fans is denoted $\mathcal{N}(P_1) \wedge \cdots \wedge \mathcal{N}(P_k)$. The following lemma states the well-known fact that the normal fan of a Minkowski sum of polytopes is the common refinement of their individual fans (see [Ziegler, 1995, Proposition 7.12] or [Gritzmann and Sturmfels, 1993, Lemma 2.1.5]):

Lemma 9.1 $\mathcal{N}(P_1 \odot \cdots \odot P_k) = \mathcal{N}(P_1) \wedge \cdots \wedge \mathcal{N}(P_k)$.

We finish with a result of Gritzmann and Sturmfels that will be useful later. It gives a bound on the number of vertices of a Minkowski sum of polytopes.

Theorem 9.2 ([Gritzmann and Sturmfels, 1993]) *Let $P_1, P_2, \ldots, P_k$ be polytopes in $\mathbb{R}^d$, and let m denote the number of non-parallel edges of $P_1, \ldots, P_k$. Then the number of vertices of $P_1 \odot \cdots \odot P_k$ is at most*

$$2 \sum_{j=0}^{d-1} \binom{m-1}{j}.$$

Note that this bound is independent of the number k of polytopes.

9.2 The few inference functions theorem

For fixed parameters, the inference problem of finding the explanation $\widehat{\mathbf{h}}$ that maximizes $\mathrm{Prob}(\mathbf{X} = \mathbf{h} | \mathbf{Y} = \tau)$ is equivalent to identifying the monomial $\theta_1^{a_{1,i}} \theta_2^{a_{2,i}} \cdots \theta_d^{a_{d,i}}$ of f_τ with maximum value. Since the logarithm is a monotonically increasing function, the desired monomial also maximizes the quantity

$$
\begin{aligned}
\log(\theta_1^{a_{1,i}} \theta_2^{a_{2,i}} \cdots \theta_d^{a_{d,i}}) &= a_{1,i} \log(\theta_1) + a_{2,i} \log(\theta_2) + \cdots + a_{d,i} \log(\theta_d) \\
&= a_{1,i} v_1 + a_{2,i} v_2 + \cdots + a_{d,i} v_d.
\end{aligned}
$$

This is equivalent to the fact that the corresponding vertex $(a_{1,i}, a_{2,i}, \ldots, a_{d,i})$ of the Newton polytope $\mathrm{NP}(f_\tau)$ maximizes the linear expression $v_1 x_1 + \cdots + v_d x_d$. Thus, the inference problem for fixed parameters becomes a linear programming problem.

Each choice of the parameters $\theta = (\theta_1, \theta_2, \ldots, \theta_d)$ determines an inference function. If $\mathbf{v} = (v_1, v_2, \ldots, v_d)$ is the vector in $\mathbb{R}^d$ with coordinates $v_i = \log(\theta_i)$, then we denote the corresponding inference function by

$$\Phi_{\mathbf{v}} : (\Sigma')^n \longrightarrow \Sigma^q.$$

For each observation $\tau \in (\Sigma')^n$, its explanation $\Phi_{\mathbf{v}}(\tau)$ is given by the vertex of $\mathrm{NP}(f_\tau)$ that is maximal in the direction of the vector $\mathbf{v}$. Note that for certain values of the parameters (if $\mathbf{v}$ is perpendicular to a positive-dimensional face of $\mathrm{NP}(f_\tau)$) there may be more than one vertex attaining the maximum. It is also possible that a single point $(a_{1,i}, a_{2,i}, \ldots, a_{d,i})$ in the polytope corresponds to several different values of the hidden data. In both cases, we pick the explanation according to the tie-braking rule determined ahead of time. This simplification does not affect the asymptotic number of inference functions.

Different values of θ yield different directions $\mathbf{v}$, which can result in distinct inference functions. We are interested in bounding the number of different inference functions that a graphical model can have. The next theorem gives an upper bound which is polynomial in the size of the graphical model. In fact, very few of the $l^{q(l')^n}$ functions $(\Sigma')^n \longrightarrow \Sigma^q$ are inference functions.

Theorem 9.3 (The few inference functions theorem) *Let d be a fixed positive integer. Consider a graphical model with d parameters, and let E be the number of edges of the underlying graph. Then, the number of inference functions of the model is at most $O(E^{d(d-1)})$.*

Before proving this theorem, observe that the number E of edges depends on the number n of observed random variables. In the graphical models discussed in this book, E is a linear function of n, so the bound becomes $O(n^{d(d-1)})$. For example, the hidden Markov model has $E = 2n - 1$ edges.

Proof In the first part of the proof we will reduce the problem of counting inference functions to the enumeration of the vertices of a certain polytope. We have seen that an inference function is specified by a choice of the parameters, which is equivalent to choosing a vector $\mathbf{v} \in \mathbb{R}^d$. The function is denoted $\Phi_{\mathbf{v}} : (\Sigma')^n \longrightarrow \Sigma^q$, and the explanation $\Phi_{\mathbf{v}}(\tau)$ of a given observation τ is determined by the vertex of $\mathrm{NP}(f_\tau)$ that is maximal in the direction of $\mathbf{v}$. Thus, cones of the normal fan $\mathcal{N}(\mathrm{NP}(f_\tau))$ correspond to sets of vectors $\mathbf{v}$ that give rise to the same explanation for the observation τ. Non-maximal cones (i.e., those contained in another cone of higher dimension) correspond to directions $\mathbf{v}$ for which more than one vertex is maximal. Since ties are broken using a consistent rule, we disregard this case for simplicity. Thus, in what follows we consider only maximal cones of the normal fan.

Let $\mathbf{v}' = (v_1', v_2', \ldots, v_d')$ be another vector corresponding to a different choice of parameters (see Figure 9.1). By the above reasoning, $\Phi_{\mathbf{v}}(\tau) = \Phi_{\mathbf{v}'}(\tau)$ if and only if $\mathbf{v}$ and $\mathbf{v}'$ belong to the same cone of $\mathcal{N}(\mathrm{NP}(f_\tau))$. Thus, $\Phi_{\mathbf{v}}$ and $\Phi_{\mathbf{v}'}$ are the same inference function if and only if $\mathbf{v}$ and $\mathbf{v}'$ belong to the same cone of $\mathcal{N}(\mathrm{NP}(f_\tau))$ for all observations $\tau \in (\Sigma')^n$. Consider the common refinement of all these normal fans, $\bigwedge_{\tau \in (\Sigma')^n} \mathcal{N}(\mathrm{NP}(f_\tau))$. Then, $\Phi_{\mathbf{v}}$ and $\Phi_{\mathbf{v}'}$ are the same function exactly when $\mathbf{v}$ and $\mathbf{v}'$ lie in the same cone of this common refinement.

This implies that the number of inference functions equals the number of cones in $\bigwedge_{\tau \in (\Sigma')^n} \mathcal{N}(\mathrm{NP}(f_\tau))$. By Lemma 9.1, this common refinement is the normal fan of $\mathrm{NP}(\mathbf{f}) = \bigodot_{\tau \in (\Sigma')^n} \mathrm{NP}(f_\tau)$, the Minkowski sum of the polytopes $\mathrm{NP}(f_\tau)$ for all observations τ. It follows that enumerating inference functions is equivalent to counting vertices of $\mathrm{NP}(\mathbf{f})$. In the remaining part of the proof we give an upper bound on the number of vertices of $\mathrm{NP}(\mathbf{f})$.

Note that for each τ, the polytope $\mathrm{NP}(f_\tau)$ is contained in the hypercube $[0, E]^d$, since each parameter θ_i can appear as a factor of a monomial of f_τ at most E times. Also, the vertices of $\mathrm{NP}(f_\tau)$ have integral coordinates, because they are exponent vectors. Polytopes whose vertices have integral coordinates are called *lattice polytopes*. It follows that the edges of $\mathrm{NP}(f_\tau)$ are given by

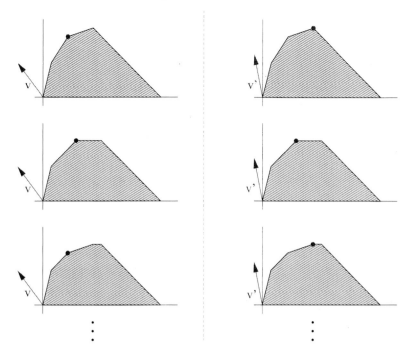

Fig. 9.1. Two different inference functions, $\Phi_{\mathbf{v}}$ (left column) and $\Phi_{\mathbf{v}'}$ (right column). Each row corresponds to a different observation. The respective explanations are given by the marked vertices in each Newton polytope.

vectors where each coordinate is an integer between $-E$ and E. There are only $(2E+1)^d$ such vectors, so this is an upper bound on the number of different directions that the edges of the polytopes $NP(f_\tau)$ can have.

This property of the Newton polytopes of the coordinates of the model will allow us to give an upper bound on the number of vertices of their Minkowski sum $NP(\mathbf{f})$. The last ingredient that we need is Theorem 9.2. In our case we have a sum of polytopes $NP(f_\tau)$, one for each observation $\tau \in (\Sigma')^n$, having at most $(2E+1)^d$ non-parallel edges in total. Hence, by Theorem 9.2, the number of vertices of $NP(\mathbf{f})$ is at most

$$2 \sum_{j=0}^{d-1} \binom{(2E+1)^d - 1}{j}.$$

As E goes to infinity, the dominant term of this expression is

$$\frac{2^{d^2-d+1}}{(d-1)!} E^{d(d-1)}.$$

Thus, we get an $O(E^{d(d-1)})$ upper bound on the number of inference functions of the graphical model. □

9.3 Inference functions for sequence alignment

We now show how Theorem 9.3 can be applied to give a tight bound on the number of inference functions of the model for sequence alignment used in Chapter 8. Recall the sequence alignment problem from Section 2.2, which consists in finding the best alignment between two sequences. Given two strings σ^1 and σ^2 of lengths n_1 and n_2 respectively, an *alignment* is a pair of equal length strings (μ^1, μ^2) obtained from σ^1, σ^2 by inserting dashes "$-$" in such a way that there is no position in which both μ^1 and μ^2 have a dash. A *match* is a position where μ^1 and μ^2 have the same character, a *mismatch* is a position where μ^1 and μ^2 have different characters, and a *space* is a position in which one of μ^1 and μ^2 has a dash. A simple scoring scheme consists of two parameters α and $\widetilde{\beta}$ denoting mismatch and space penalties respectively. The reward of a match is set to 1. The score of an alignment with z matches, x mismatches, and $\widetilde{y}$ spaces is then $z - x\alpha - \widetilde{y}\widetilde{\beta}$. Observe that these numbers always satisfy $2z + 2x + \widetilde{y} = n_1 + n_2$. For pairs of sequences of equal length, this is the same scoring scheme used in Chapter 8, with $\widetilde{y} = 2y$ and $\widetilde{\beta} = \beta/2$. In this case, y is called the number of insertions, which is half the number of spaces, and β is the penalty for an insertion.

This model for sequence alignment is a particular case of the pair hidden Markov model discussed in Section 2.2. The problem of determining the highest scoring alignment for given values of α and $\widetilde{\beta}$ is equivalent to the inference problem in the pair hidden Markov model, with some parameters set to functions of α and $\widetilde{\beta}$, or to 0 or 1. In this setting, an observation is a pair of sequences $\tau = (\sigma^1, \sigma^2)$, and the number of observed variables is $n = n_1 + n_2$. An explanation is then an optimal alignment, since the values of the hidden variables indicate the positions of the spaces.

In the rest of this chapter we will refer to this as the 2-*parameter model for sequence alignment*. Note that it actually comes from a 3-parameter model where the reward for a match has, without loss of generality, been set to 1. The Newton polytopes of the coordinates of the model are defined in a 3-dimensional space, but in fact they lie on a plane, as we will see next. Thus, the parameter space has only two degrees of freedom.

For each pair of sequences τ, the Newton polytope of the polynomial f_τ is the convex hull of the points $(x, \widetilde{y}, z)$ whose coordinates are the number of mismatches, spaces, and matches, respectively, of each possible alignment of the pair. This polytope lies on the plane $2z + 2x + \widetilde{y} = n_1 + n_2$, so no information is lost by considering its projection onto the $x\widetilde{y}$-plane instead. This projection is just the convex hull of the points $(x, \widetilde{y})$ giving the number of mismatches and spaces of each alignment. For any alignment of sequences of lengths n_1 and n_2, the corresponding point $(x, \widetilde{y})$ lies inside the square $[0, n]^2$, where $n = n_1 + n_2$. Therefore, since we are dealing with lattice polygons inside $[0, n]^2$, it follows from Theorem 9.3 that the number of inference functions of this model is $O(n^2)$.

Next we show that this quadratic bound is tight. We first consider the case in which the alphabet Σ' of the observed sequences is allowed to be sufficiently large.

Proposition 9.4 *Consider the 2-parameter model for sequence alignment described above. Assume for simplicity that the two observed sequences have the same length n. Let $\Sigma' = \{\omega_0, \omega_1, \ldots, \omega_n\}$ be the alphabet of the observed sequences. Then, the number of inference functions of this model is $\Theta(n^2)$.*

Proof The above argument shows that $O(n^2)$ is an upper bound on the number of inference functions of the model. To prove the proposition, we will argue that there are at least $\Omega(n^2)$ such functions.

Since the two sequences have the same length, we will use y to denote the number of insertions, where $y = \widetilde{y}/2$, and $\beta = 2\widetilde{\beta}$ to denote the insertion penalty. For fixed values of α and β, the explanation of an observation $\tau = (\sigma^1, \sigma^2)$ is given by the vertex of $\mathrm{NP}(f_\tau)$ that is maximal in the direction of the vector $(-\alpha, -\beta, 1)$. In this model, $\mathrm{NP}(f_\tau)$ is the convex hull of the points (x, y, z) whose coordinates are the number of mismatches, spaces and matches of the alignments of σ^1 and σ^2.

The argument in the proof of Theorem 9.3 shows that the number of inference functions of this model is the number of cones in the common refinement of the normal fans of $\mathrm{NP}(f_\tau)$, where τ runs over all pairs of sequences of length n in the alphabet Σ'. Since the polytopes $\mathrm{NP}(f_\tau)$ lie on the plane $x + y + z = n$, it is equivalent to consider the normal fans of their projections onto the yz-plane. These projections are lattice polygons contained in the square $[0, n]^2$. We denote by P_τ the projection of $\mathrm{NP}(f_\tau)$ onto the yz-plane.

We claim that for any two integers a and b such that $a, b \geq 1$ and $a + b \leq n$, there is a pair $\tau = (\sigma^1, \sigma^2)$ of sequences of length n in the alphabet Σ' so that the polygon P_τ has an edge of slope b/a.

Before proving the claim, let us show that it implies the proposition. First note that the number of different slopes b/a obtained by numbers a and b satisfying the above conditions is $\Theta(n^2)$. Indeed, this follows from the fact that the proportion of relative prime pairs of numbers in $\{1, 2, \ldots, m\}$ tends to a constant (which is known to be $6/\pi^2$) as m goes to infinity (see for example [Apostol, 1976]). For each edge of P_τ, the normal fan of the polygon has a 1-dimensional ray (the border between two maximal cones) perpendicular to it. Thus, different slopes give different rays in $\bigwedge_{\tau \in (\Sigma')^n \times (\Sigma')^n} \mathcal{N}(P_\tau)$, the common refinement of fans. In two dimensions, the number of maximal cones equals the number of rays. Thus, $\bigwedge_\tau \mathcal{N}(P_\tau)$ has at least $\Omega(n^2)$ cones. Equivalently, the model has $\Omega(n^2)$ inference functions.

Let us now prove the claim. Given a and b as above, construct the sequences σ^1 and σ^2 as follows:

$$\sigma^1 = \underbrace{\omega_0 \cdots \omega_0}_{a \text{ times}} \omega_1 \omega_2 \cdots \omega_b \underbrace{\omega_{b+2} \cdots \omega_{b+2}}_{n-a-b \text{ times}},$$

$$\sigma^2 = \omega_1\omega_2 \cdots \omega_b \underbrace{\omega_{b+1} \cdots \omega_{b+1}}_{n-b\,\text{times}}.$$

Then, it is easy to check that for $\tau = (\sigma^1, \sigma^2)$, P_τ has an edge between the vertex $(0,0)$, corresponding to the alignment with no spaces, and the vertex (a,b), corresponding to the alignment

$$\mu^1 = \omega_0 \cdots \omega_0\, \omega_1\, \omega_2 \cdots \omega_b\, \omega_{b+2} \cdots \omega_{b+2} \quad - \quad \cdots \quad -$$
$$\mu^2 = - \quad \cdots \quad - \quad \omega_1\, \omega_2 \cdots \omega_b\, \omega_{b+1} \cdots \omega_{b+1}\, \omega_{b+1} \cdots \omega_{b+1}.$$

The slope of this edge is b/a. In fact, the four vertices of P_τ are $(0,0)$, (a,b), $(n-b, b)$ and $(n,0)$. This proves the claim. Figure 9.2 shows a pair of sequences for which P_τ has an edge of slope $4/3$. $\square$

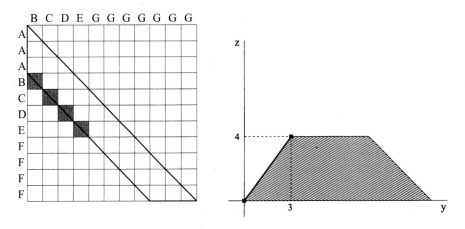

Fig. 9.2. A pair of sequences of length 11 and two alignments giving the slope $4/3$ in the alignment polygon.

Chapter 8 raised the question of whether all the lattice points inside the projected alignment polygon P_τ come from alignments of the pair of sequences τ. The construction in the proof of Proposition 9.4 gives instances for which some lattice points inside P_τ do not come from any alignment.

For example, take P_τ to be the projection of the alignment polygon corresponding to the pair of sequences in Figure 9.2. The points $(1,1)$, $(2,1)$ and $(2,2)$ lie inside P_τ. However, there is no alignment of these sequences with less than 3 insertions having at least one match, so these points do not correspond to alignments. Figure 9.3 shows exactly which lattice points in P_τ come from alignments of this pair of sequences.

Next we will show that, even in the case of the binary alphabet, our quadratic upper bound on the number of inference functions of the 2-parameter model for sequence alignment is tight as well. Thus, the large alphabet Σ' from Proposition 9.4 is not needed to obtain $\Omega(n^2)$ slopes in the alignment polytopes.

Proposition 9.5 *Consider the 2-parameter model for sequence alignment for two observed sequences of length n and let $\Sigma' = \{0,1\}$ be the binary alphabet. The number of inference functions of this model is $\Theta(n^2)$.*

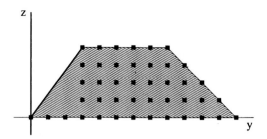

Fig. 9.3. The thick dots are the points (y, z) giving the number of insertions and matches in alignments of the sequences in Figure 9.2.

Proof We follow the same idea as in the proof of Proposition 9.4. We will construct a collection of pairs of binary sequences $\tau = (\sigma^1, \sigma^2)$ so that the total number of different slopes of the edges of the polygons $\mathrm{NP}(f_\tau)$ is $\Omega(n^2)$. This will imply that the number of cones in $\bigwedge_\tau \mathcal{N}(\mathrm{NP}(f_\tau))$ is $\Omega(n^2)$, where τ ranges over all pairs of binary sequences of length n.

Recall that P_τ denotes the projection of $\mathrm{NP}(f_\tau)$ onto the yz-plane, and that it is a lattice polygon contained in $[0, n]^2$.

We claim that for any positive integers u and v with $u < v$ and $6v - 2u \leq n$, there exists a pair τ of binary sequences of length n such that P_τ has an edge of slope u/v. This will imply that the number of different slopes created by the edges of the polygons P_τ is $\Omega(n^2)$.

Thus, it only remains to prove the claim. Given positive integers u and v as above, let $a := 2v$, $b := v - u$. Assume first that $n = 6v - 2u = 2a + 2b$. Consider the sequences

$$\sigma^1 = 0^a 1^b 0^b 1^a, \qquad \sigma^2 = 1^a 0^b 1^b 0^a,$$

where 0^a indicates that the symbol 0 is repeated a times. Let $\tau = (\sigma^1, \sigma^2)$. Then, it is not hard to see that the polygon P_τ for this pair of sequences has four vertices: $v_0 = (0, 0)$, $v_1 = (b, 3b)$, $v_2 = (a + b, a + b)$ and $v_3 = (n, 0)$. The slope of the edge between v_1 and v_2 is $\frac{a - 2b}{a} = \frac{u}{v}$.

If $n > 6v - 2u = 2a + 2b$, we just append $0^{n - 2a - 2b}$ to both sequences σ^1 and σ^2. In this case, the vertices of P_τ are $(0, n - 2a - 2b)$, $(b, n - 2a + b)$, $(a + b, n - a - b)$, $(n, 0)$ and $(n - 2a - 2b, 0)$.

Note that if $v - u$ is even, the construction can be done with sequences of length $n = 3v - u$ by taking $a := v$, $b := \frac{v - u}{2}$. Figure 9.4 shows the alignment graph and the polygon P_τ for $a = 7$, $b = 2$. $\qquad \square$

In most cases, one is interested only in those inference functions that are biologically meaningful. In our case, meaningful values of the parameters occur when $\alpha, \beta \geq 0$, which means that mismatches and spaces are penalized instead of rewarded. Sometimes one also requires that $\alpha \leq \beta$, which means that a mismatch should be penalized less than two spaces. It is interesting to observe that our constructions in the proofs of Propositions 9.4 and 9.5 not only show that the total number of inference functions is $\Omega(n^2)$, but also that the number of biologically meaningful ones is still $\Omega(n^2)$. This is because the different rays

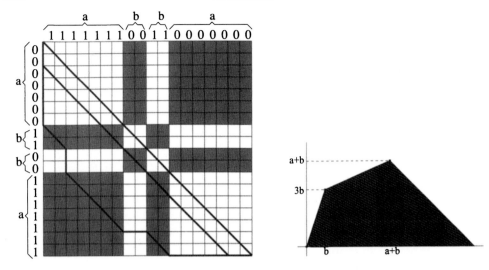

Fig. 9.4. A pair of binary sequences of length 18 giving the slope 3/7 in their alignment polytope.

created in our construction have a biologically meaningful direction in the parameter space.

Let us now relate the results from this section with the bounds given in the previous chapter. In Chapter 8 we saw that in the 2-parameter model for sequence alignment, if τ is a pair of sequences of length n in an arbitrarily large alphabet, then the polygon P_τ can have $\Theta(n^{2/3})$ vertices in the worst case. In Proposition 9.4 we have shown that the Minkowski sum of these polygons for all possible such observations τ, namely $\bigodot_\tau P_\tau$, has $\Theta(n^2)$ vertices.

In the case where the alphabet for the sequences is binary (or more generally, finite), we conjectured in Chapter 8 that the polygon P_τ can only have $\Theta(\sqrt{n})$ vertices. In Proposition 9.5 we have proved that the polygon $\bigodot_\tau P_\tau$, where τ runs over all pairs of binary sequences of length n, has $\Theta(n^2)$ vertices as well.

An interpretation of Theorem 9.3 is that the ability to change the values of the parameters of a graphical model does not give as much freedom as it may appear. There is a very large number of possible ways to assign an explanation to each observation. However, only a tiny proportion of these come from a consistent method for choosing the most probable explanation for a certain choice of parameters. Even though the parameters can vary continuously, the number of different inference functions that can be obtained is at most polynomial in the number of edges of the model, assuming that the number of parameters is fixed.

In the case of sequence alignment, the number of possible functions that associate an alignment to each pair of sequences of length n is doubly-exponential in n. However, the number of functions that pick the alignment with highest score in the 2-parameter model, for some choice of the parameters α and β, is only $\Theta(n^2)$. Thus, most ways of assigning alignments to pairs of sequences do not correspond to any consistent choice of parameters. If we use a model with more parameters, say d, the number of inference functions may be larger, but still polynomial in n, namely $O(n^{d(d-1)})$.

Having shown that the number of inference functions of a graphical model is polynomial in the size of the model, an interesting next step would be to find an efficient way to precompute all the inference functions for given models. This would allow us to give the answer (the explanation) to a query (an observation) very quickly. It follows from this chapter that it is computationally feasible to precompute the polytope $\mathrm{NP}(\mathbf{f})$, whose vertices correspond to the inference functions. However, the difficulty arises when we try to describe a particular inference function efficiently. The problem is that the characterization of an inference function involves an exponential number of observations.

10

Geometry of Markov Chains

Eric Kuo

Graphical models play an important role in solving problems in computational biology. A typical problem setup consists of n observed random variables $Y_1, \ldots, Y_n$ and m hidden random variables $X_1, \ldots, X_m$. Suppose we observe $Y_1 = \sigma_1, \ldots, Y_n = \sigma_n$. A standard inference problem is to find the hidden assignments h_i that produce the maximum a posteriori (MAP) log probability

$$\delta_{\sigma_1 \cdots \sigma_n} = \min_{h_1, \ldots, h_m} -\log(\Pr[X_1 = h_1, \ldots, X_m = h_m, Y_1 = \sigma_1, \ldots, Y_n = \sigma_n]),$$

where the h_i range over all the possible assignments for the hidden random variables. However, when the parameters of the graphical model change, the hidden assignments may also vary. The parametric inference problem is to solve this inference problem for all model parameters simultaneously. Chapter 5 presented one technique, polytope propagation, that solves this problem.

This chapter discusses a parametric inference problem for Markov chains. Certain sequences of states are uniquely maximal for some Markov chain; these sequences are called *Viterbi sequences*. In Section 10.1 we will define Viterbi sequences formally and discover some of their properties. It is easier to enumerate Viterbi sequences for *toric* Markov chains (defined in Section 1.4) than for regular Markov chains. This chapter explains the differences between Viterbi sequences of Markov chains and toric Markov chains. In Section 10.2 we find that when the chains have two or three states, there are some sequences that are Viterbi for a toric Markov chain, but not for any Markov chain. But in Section 10.3, we show that the sets of Viterbi sequences are identical for both Markov chains and toric Markov chains on at least four states. Therefore enumerating Viterbi sequences becomes the same task for both types of Markov chains on at least four states. Finally, Section 10.4 discusses maximal probability sequence pairs for fully observed Markov models. We will show that each sequence pair must include a Viterbi sequence.

10.1 Viterbi sequences

This chapter uses the same terminology and notation for Markov chains introduced in Section 1.4. In addition, we number the states of an l-state Markov chain from 0 to $l - 1$. Most of the time we will assume the initial state is always 0 (as opposed to the uniform distribution used elsewhere in this book). Given a Markov chain M, the probability of a sequence is the product of the probability of the initial state and all the transition probabilities between consecutive states. There are l^n possible sequences of length n. A *Viterbi path of length n* is a sequence of n states (containing $n - 1$ transitions) with the highest probability. Viterbi paths of Markov chains can be computed in polynomial time [Forney, 1973, Viterbi, 1967].

A Markov chain may have more than one Viterbi path of length n. For instance, if 012010 is a Viterbi path of length 6, then 010120 must also be a Viterbi path since each sequence begins with the state 0 and each transition appears with the same frequency in each sequence. The only difference between these sequences is the order of their transitions. Two sequences are *equivalent* if they start with the same state and each transition occurs with equal frequency in each sequence.

The Viterbi paths of a Markov chain might not all be equivalent. Consider the Markov chain on l states that has a uniform initial distribution and a uniform transition matrix (i.e. $\theta_{ij} = \frac{1}{l}$ for all states i, j). Since each sequence of length n has the same probability $\frac{1}{l^n}$, every sequence is a Viterbi path for this Markov chain.

If every Viterbi path of a Markov chain is equivalent to a fixed sequence S, then we say S a *Viterbi sequence*, or more simply, *Viterbi*. A simple example of a Viterbi sequence of length 4 is 0000 since it is the only Viterbi path for the two-state Markov chain with initial distribution $\pi_0 = 1$ and $\pi_1 = 0$, and transition matrix

$$\theta = \begin{pmatrix} 1 & 0 \\ 0 & 1 \end{pmatrix}.$$

An example of a sequence that is not a Viterbi sequence is 0011. If it were, then its probability must be greater than those of sequences 0001 and 0111. Since $p_{0011} > p_{0001}$,

$$\theta_{00}\theta_{01}\theta_{11} > \theta_{00}^2\theta_{01},$$

from which we conclude $\theta_{11} > \theta_{00}$. But from $p_{0011}p_{0111}$, we get

$$\theta_{00}\theta_{01}\theta_{11} > \theta_{01}\theta_{11}^2,$$

from which we get $\theta_{00} > \theta_{11}$, a contradiction.

Let us introduce a few verbal conventions when speaking of Viterbi sequences. A Viterbi sequence S represents the entire class of sequences equivalent to S. Thus, when enumerating Viterbi sequences, equivalent sequences are considered to be the same. When we say that "sequence S' can end with (subsequence) X," we may refer to a sequence equivalent to S rather than S itself. Thus 0010 can end with 00 since 0010 is equivalent to 0100, which ends with 00. When we say that S is "a Viterbi sequence on l states" or "an l-state Viterbi sequence," we mean that there is a Markov chain on l states for which S is Viterbi. Sequence S itself might not contain all l states; for instance, 0000 is a Viterbi sequence on 2 states despite the absence of state 1 in 0000.

Finally, when we speak of a *subsequence* of a sequence S, we will always refer to a contiguous subsequence.

We can similarly define Viterbi paths and sequences for toric Markov chains. Recall from Section 1.4 that the transition matrix for a toric Markov chain is not necessarily stochastic, i.e. the entries in each row need not sum to 1. Because of this fact, the set T_l of Viterbi sequences of l-state toric Markov chains may be different from the set V_l of Viterbi sequences of l-state (regular) Markov chains. Note that V_l is a subset of T_l since every Markov chain is a toric Markov chain. We call the sequences in the set difference $T_l \setminus V_l$ *pseudo-Viterbi sequences*. For the rest of this chapter, the term *Viterbi path* and *sequence* will refer to regular Markov chains; for toric Markov chains, we use the terms *toric Viterbi paths* and *sequences*.

The main result in this chapter is that although pseudo-Viterbi sequences exist for $l = 2$ and 3, there are none for higher l. When $l \geq 4$, the sets of Viterbi sequences and toric Viterbi sequences are equal.

To help prove these results, we will first need to prove some general properties about Viterbi sequences.

Proposition 10.1 *Let T_1 and T_2 be two non-equivalent sequences of length t that both begin with state q_1 and end with state q_2. If a Viterbi sequence S contains T_1 as a subsequence, then S cannot also contain T_2 as a subsequence, nor can S be rearranged so that T_2 appears as a subsequence.*

Proof Suppose that S contains both subsequences T_1 and T_2. Then let S_1 be the sequence obtained by replacing T_2 with T_1 in S so that $p_{S_1} = p_S p_{T_1} / p_{T_2}$. Similarly, let S_2 be the sequence obtained by replacing T_1 with T_2 in S so that $p_{S_2} = p_S p_{T_2} / p_{T_1}$. Since S is Viterbi and not equivalent to S_1, we must have $p_S > p_{S_1}$, which implies $p_{T_2} > p_{T_1}$. But since S is also not equivalent to S_2, we also have $p_S > p_{S_2}$, which implies $p_{T_1} > p_{T_2}$. This is a contradiction, so S cannot contain both T_1 and T_2.

Similarly, suppose that S could be rearranged so that T_2 appears as a subsequence. We could define S_1 by replacing T_2 with T_1 in a rearrangement of S, and define S_2 by replacing T_1 with T_2. The proof follows as before. □

As an example, 01020 is not a Viterbi sequence since 010 and 020 are non-equivalent subsequences of the same length beginning and ending with 0. Therefore, either p_{01010} or p_{02020} will be greater than or equal to p_{01020}.

The next proposition was illustrated with an earlier example, 0011.

Proposition 10.2 *If a transition ii exists in a Viterbi sequence S, then no other transition jj can appear in S, where $j \neq i$.*

Proof Suppose S did contain transitions ii and jj. Consider the sequence S_1 where one subsequence ii is replaced by i and another subsequence jj is replaced by jjj. Since S is Viterbi, we must have $p_S > p_{S_1}$, from which we conclude $p_{ii} > p_{jj}$. However, we can also create another sequence S_2 in which

$A*$	0^{2m+1}	$(2m,0,0,0)$	$A*$	0^{2m+2}	$(2m+1,0,0,0)$	
B	$0^{2m}1$	$(2m-1,1,0,0)$	B	$0^{2m+1}1$	$(2m,1,0,0)$	
C	$0(01)^m$	$(1,m,m-1,0)$	$C*$	$0(01)^m0$	$(1,m,m,0)$	
$D*$	$(01)^m1$	$(0,m,m-1,1)$	D	$(01)^m10$	$(0,m,m,1)$	
$E*$	$(01)^m0$	$(0,m,m,0)$	$E*$	$(01)^{m+1}$	$(0,m+1,m,0)$	
F	$01^{2m-1}0$	$(0,1,1,2m-2)$	F	$01^{2m}0$	$(0,1,1,2m-1)$	
$G*$	01^{2m}	$(0,1,0,2m-1)$	$G*$	01^{2m+1}	$(0,1,0,2m)$	

Table 10.1. *Left: Toric Viterbi sequences of length $2m+1$. Right: Toric Viterbi sequences of length $2m+2$. Starred sequences are Viterbi, unstarred are pseudo-Viterbi. The ordered 4-tuples give the number of* 00, 01, 10 *and* 11 *transitions for the sequences.*

ii is replaced with iii and jj is replaced with j. Once again, $p_S > p_{S_2}$, so $p_{jj} > p_{ii}$, giving us a contradiction. □

Note that Propositions 10.1 and 10.2 also apply to toric Viterbi sequences.

10.2 Two- and three-state Markov chains

In [Kuo, 2005], it is shown that there are seven toric Viterbi sequences on $l = 2$ states in which the initial state is 0 and the length is at least 5. They are listed in Table 10.1.

Not all toric Viterbi sequences are Viterbi sequences. In fact, for each $n > 3$, three sequences are pseudo-Viterbi because of the following proposition.

Proposition 10.3 *No Viterbi sequence on two states can end with 001 or 110, nor can it be rearranged to end with 001 or 110.*

Proof Suppose that 001 is a Viterbi sequence. Then since $p_{001} > p_{010}$, we must have $\theta_{00} > \theta_{10}$. Also, $p_{001} > p_{000}$, so $\theta_{01} > \theta_{00}$. Finally, $p_{001} > p_{011}$, so $\theta_{00} > \theta_{11}$. But then

$$1 = \theta_{00} + \theta_{01} > \theta_{10} + \theta_{00} > \theta_{10} + \theta_{11} = 1,$$

which is a contradiction. Thus no Viterbi sequence can end with 001 (or by symmetry, 110). □

In particular, sequence $0(01)^m$ is not a Viterbi sequence since the transition 00 can be moved from the beginning to the end to form the equivalent sequence $(01)^{m-1}001$, which is not Viterbi because of Proposition 10.3.

The remaining four toric Viterbi sequences are Viterbi sequences. Stochastic transition matrices are easily constructed to produce them.

We can view each two-state toric Viterbi sequence of length n that starts with state 0 as the vertex of a Newton polytope, as defined in Section 2.3. This Newton polytope is for the polynomial $\sum p_S$ where S ranges over all sequences of length n that start with 0. These polytopes are shown in Figure 10.1. The left and right polytopes are for odd and even length sequences, respectively.

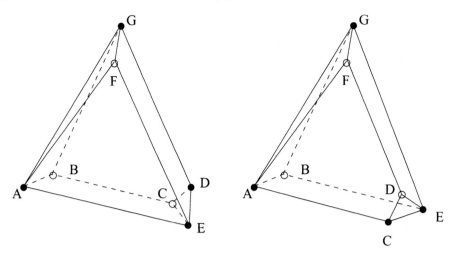

Fig. 10.1. Polytopes of two-state toric Viterbi sequences.

The vertices share the labels listed in Table 10.1. Black vertices represent Viterbi sequences, and white vertices represent pseudo-Viterbi sequences. The third and sixth columns of Table 10.1 represent the coordinates of each vertex. These coordinates represent the frequencies of the transitions 00, 01, 10, and 11. Since $(01)^m 0$ has 0 instances of 00 and 11, m instances of 01, and m instances of 10, that sequence is represented by the vertex $(0, m, m, 0)$. Although the points occupy a 4-dimensional space, the polytope itself is only 3-dimensional.

When $l = 3$ states, the number of toric Viterbi sequences that start with state 0 is 93 for even n and 95 for odd n, when $n \geq 12$. These sequences are enumerated in [Kuo, 2005]. Of these sequences, only four are pseudo-Viterbi. All four pseudo-Viterbi sequences on 3 states can end in 11210 or some symmetric variant such as 00102.

Proposition 10.4 *A Viterbi sequence (or an equivalent sequence) on three states cannot end with 11210, or equivalently, 12110.*

Proof Suppose that a Viterbi sequence could end with 11210. Since the sequence ends with 10, we must have $\theta_{10} > \theta_{11}$. Since 110 is a Viterbi subsequence with higher probability than 100, we also have $\theta_{11} > \theta_{00}$. Thus $\theta_{10} > \theta_{00}$. Moreover, since $p_{110} > p_{101}$, we must have $\theta_{11}\theta_{10} > \theta_{10}\theta_{01}$, which means $\theta_{11} > \theta_{01}$. Finally, 112 has higher probability than 102, so $\theta_{11}\theta_{12} > \theta_{10}\theta_{02}$. Then

$$\theta_{12} > \frac{\theta_{10}\theta_{02}}{\theta_{11}} > \theta_{02}$$

where we use the fact that $\theta_{10} > \theta_{11}$. Thus

$$1 = \theta_{10} + \theta_{11} + \theta_{12} > \theta_{00} + \theta_{01} + \theta_{02} = 1$$

which is a contradiction.   □

However, 0212110 is a toric Viterbi sequence for the toric Markov chain with the following transition matrix:

$$\begin{pmatrix} 0 & 0 & 0.1 \\ 0.5 & 0.3 & 0.2 \\ 0 & 0.6 & 0 \end{pmatrix}.$$

It is possible to create such a matrix since we can choose values for θ_{10}, θ_{11}, and θ_{12} that are each bigger than those for θ_{00}, θ_{01}, and θ_{02}. Such a task is impossible if the matrix were required to be stochastic: the sum of one of the rows would be unequal to 1.

10.3 Markov chains with many states

Of the seven two-state toric Viterbi sequences of length n that start with state 0, three are pseudo-Viterbi sequences. Out of the 93 or 95 (depending on the parity of n) three-state toric Viterbi sequences of length n starting with state 0, only four of them are pseudo-Viterbi sequences. So we might ask, how many pseudo-Viterbi sequences exist as the number of states increases? The answer lies in the following theorem:

Theorem 10.5 *Every toric Viterbi sequence on $l \geq 4$ states is a Viterbi sequence on l states.*

Proof In order to show that a toric Viterbi sequence S (of length n) is Viterbi, we need the following facts:

  (i) For each state i, there exists a state j (which may be the same as i) for which the transition ij does not appear in S.
  (ii) Given that (i) is true, we can find a stochastic matrix of transition probabilities whose Viterbi sequence of length n is S.

To prove the first fact, we assume there is a state q for which each transition $q0, q1, \ldots, q(l-1)$ exists at least once in S. Let's assume q is state 0 (for we could merely relabel the states in S). For each transition $0i$, select one instance of it in S. Let x_i be the (possibly empty) string of states after the selected $0i$ and before either the next 0 or the end of S. We will also assume that our selected transition $0(l-1)$ appears last in S except for possibly 00. Thus S will have subsequences $00, 01x_10, 02x_20, \ldots, 0(l-2)x_{l-2}0$, and $0(l-1)x_{l-1}$.

Since $l \geq 4$, strings x_1 and x_2 are followed by state 0 (as opposed to possibly being the end of S when $l \leq 3$). Strings x_1 and x_2 cannot both be empty, for otherwise S would contain the subsequences 010 and 020, violating Proposition 10.1. (Note that for $l \leq 3$, this proof fails since x_1 and x_2 could both be empty.) Suppose that x_1 is non-empty. The first state of x_1 (immediately following state 1) cannot be state 1, for then subsequences 00 and 11 would appear in S, violating Proposition 10.2. So it must be some state other than 0 or 1; call it state r. We could rearrange S so that transition 00 precedes transition $0r$. This rearrangement is done by replacing the 00 with a 0 and then replacing $0r$ with $00r$. Thus S could be rearranged so that either sub-sequence $00r$ or $01r$ could appear, violating Proposition 10.1. Similarly, if we suppose that x_2 is non-empty, we would conclude that S can be rearranged to have either subsequence $00r$ or $02r$. Thus there must be at least one state s for which the transition $0s$ does not appear in S.

To prove the second statement, we construct a stochastic transition matrix θ for which S is a Viterbi sequence. Because of the first fact, we can assign every transition ij in S a probability θ_{ij} greater than $1/l$. Note that this would be impossible in the cases 001 for $l = 2$ and 11210 for $l = 3$.

For the set of transition probabilities θ_{ij} for which ij appears in S, the values will range from a minimum of $\frac{1}{l} + \epsilon$ to a maximum of $\frac{1}{l} + \alpha\epsilon$. Both extreme values will have at least one transition in S with that probability. If S contains only one kind of transition (such as 00000 having only the transition 00), then $\alpha = 1$. We choose values for α and ϵ that satisfy the following inequalities:

$$\frac{1}{l} < \frac{1}{l} + \epsilon \ \leq \ \frac{1}{l} + \alpha\epsilon < \frac{1}{l-1} \tag{10.1}$$

$$\frac{1}{l}\left(\frac{1}{l} + \alpha\epsilon\right)^{n-2} \ < \ \left(\frac{1}{l} + \epsilon\right)^{n-2}\left(\frac{1}{l} + \alpha\epsilon\right). \tag{10.2}$$

First we choose an ϵ satisfying (10.1). Then if we set $\alpha = 1$, both inequalities are automatically satisfied. If necessary, we can still increment α very slightly and still satisfy the inequality. The meaning of (10.2) will be explained later.

If state i does not appear in S or only appears as the final state of S, then each transition ij will be assigned a probability $\theta_{ij} = \frac{1}{l}$. Otherwise, for each transition ij that appears in S, θ_{ij} will be assigned a value between $\frac{1}{l} + \epsilon$ and $\frac{1}{l} + \alpha\epsilon$. Since at most $l - 1$ transitions starting with i can appear in S and each $\theta_{ij} \leq \frac{1}{l-1}$, the sum of those θ_{ij} cannot reach 1. Thus for the remaining transitions ij' that are not in S, we can set $0 < \theta_{ij'} < 1/l$.

Next we explain the meaning of (10.2). The right hand side of (10.2) represents the minimum possible value for the probability of S, for it has at least one transition with the maximum probability $\frac{1}{l} + \alpha\epsilon$ while all the others have at least the minimum probability $\frac{1}{l} + \epsilon$. The left hand side of (10.2) represents the maximum probability any other sequence S' can attain in which S' contains a transition not in S. At least one transition has probability at most $\frac{1}{l}$, while all the other transitions may have the maximum probability $\frac{1}{l} + \alpha\epsilon$. Therefore S will have a probability p_S strictly greater than $p_{S'}$ for any other sequence S' that contains at least one transition not in S.

It remains to find appropriate values for θ_{ij}. Since S is a toric Viterbi sequence, there is a transition matrix ϕ for which S is the toric Viterbi sequence. We then take logarithms so that $w_{ij} = \log \phi_{ij}$ for each transition ij for which $\phi_{ij} \neq 0$. Let w_1 and w_2 be the minimum and maximum values of w_{ij}, respectively, of all the transitions ij that appear in S. We then let $L(w) = aw + b$ (where $a > 0$) be the linear transformation that maps w_1 to $-\log(\frac{1}{l} + \alpha\epsilon)$ and w_2 to $-\log(\frac{1}{l} + \epsilon)$. For each transition ij that appears in S, we set θ_{ij} such that

$$-\log\theta_{ij} = L(w_{ij}) = aw_{ij} + b. \tag{10.3}$$

If every transition in another sequence S' appears in S, we have

$$
\begin{aligned}
-\log p_S = \sum_{i=1}^{n-1} L(w_{S_i S_{i+1}}) &= b(n-1) + a\sum_{i=1}^{n-1} w_{S_i S_{i+1}} \\
&< b(n-1) + a\sum_{i=1}^{n-1} w_{S'_i S'_{i+1}} \\
&= \sum_{i=1}^{n-1} L(w_{S'_i S'_{i+1}}) = -\log p_{S'}.
\end{aligned}
\tag{10.4}
$$

The middle inequality follows since

$$\sum_{i=1}^{n-1} w_{S_i S_{i+1}} = -\log \prod_{i=1}^{n-1} \phi_{S_i S_{i+1}} < -\log \prod_{i=1}^{n-1} \phi_{S'_i S'_{i+1}} = \sum_{i=1}^{n-1} w_{S'_i S'_{i+1}}$$

and S is a toric Viterbi sequence.

We note how for each ij in S, the value of θ_{ij} fits between $\frac{1}{l} + \epsilon$ and $\frac{1}{l} + \alpha\epsilon$:

$$\frac{1}{l} + \epsilon = \exp(-L(w_2)) \le \theta_{ij} = \exp(-L(w_{ij})) \le \exp(-L(w_1)) = \frac{1}{l} + \alpha\epsilon.$$

Finally, for all other transitions that are not in S, we assign probabilities such that none of them exceed $\frac{1}{l}$ and the entries of each row in θ sums to 1.

From (10.2) we know that $p_S > p_{S'}$ when S' has a transition not in S. From (10.4) we know that $p_S > p_{S'}$ when every transition in S' also appears in S. Therefore we conclude that θ is a stochastic matrix for which S is the only Viterbi path. $\qquad\square$

10.4 Fully observed Markov models

We now turn our attention to the *fully observed Markov model* (Section 1.4). The states of a fully observed Markov model are represented by two alphabets Σ and Σ' of l and l' letters, respectively. We parameterize the fully observed Markov model by a pair of matrices (θ, θ') with dimensions $l \times l$ and $l \times l'$. The entry θ_{ij} is the transition probability from state $i \in \Sigma$ to state $j \in \Sigma$. Entry θ'_{ij} represents an emission probability from state $i \in \Sigma$ to output $j \in \Sigma'$.

The fully observed Markov model generates a pair of sequences $\sigma \in \Sigma^n$ and $\tau \in \Sigma'^n$. This model generates σ in the same way as a regular Markov chain with transition matrix θ. Each symbol in τ is generated individually from the corresponding state in σ with the emission matrix θ'. Thus if σ_i is the ith state in σ, then the ith symbol of τ would be τ_i with probability $\theta'_{\sigma_i \tau_i}$. We will call σ the *state sequence* and τ the *output sequence* of a *fully observed sequence pair* (σ, τ). The fully observed Markov model with matrices (θ, θ') generates a fully observed sequence pair (σ, τ) of length n with probability

$$p_{\sigma,\tau} = \pi_{\sigma_1} \theta'_{\sigma_1 \tau_1} \prod_{i=2}^{n} \theta_{\sigma_{i-1}\sigma_i} \theta'_{\sigma_i \tau_i}$$

where π_{σ_1} is the probability of the initial state σ_1.

In a fully observed Markov model, both the transition matrices, θ and θ', are stochastic. When we don't require either matrix to be stochastic, we call it a *fully observed toric Markov model*.

We now define some terms referring to fully observed sequences of maximal likelihood. If (σ, τ) and (σ', τ') are two fully observed sequence pairs for which each transition $(i, j) \in \Sigma \times \Sigma$ and emission $(i, j') \in \Sigma \times \Sigma'$ appears in each sequence pair an equal number of times, then we say that (σ, τ) and (σ', τ') are *equivalent*. For example, $(0010, ABCD)$ is equivalent to $(0100, DCBA)$ since each pair contains one instance of each transition 00, 01, and 10, and once instance of each emission $0A$, $0B$, $1C$, and $1D$. The pair (σ, τ) will usually represent the entire equivalence class of sequence pairs. If there exists a fully observed Markov model for which (σ, τ) and any equivalent sequence pairs are the only ones generated with the maximum probability, then (σ, τ) is a *fully observed Viterbi sequence pair*. We analogously define *fully observed toric Viterbi sequence pairs* for fully observed toric Markov models.

Lemma 10.6 *Let (σ, τ) be a fully observed (toric) Viterbi sequence pair. If σ_i and σ_j are the same state from Σ, then τ_i and τ_j are also the same output from Σ'.*

Proof Suppose that $\sigma_i = \sigma_j$, but $\tau_i \neq \tau_j$. We can create two additional output sequences τ' and τ'' in which we set $\tau'_i = \tau'_j = \tau_i$ and $\tau''_i = \tau''_j = \tau_j$. Then since $p_{\sigma,\tau} > p_{\sigma,\tau'}$, we must have

$$\theta'_{\sigma_j \tau_j} > \theta'_{\sigma \tau'_j} = \theta'_{\sigma_i \tau_i}.$$

Since we also have $p_{\sigma,\tau} > p_{\sigma,\tau''}$, we must also have

$$\theta'_{\sigma_i \tau_i} > \theta'_{\sigma_i \tau''_i} = \theta'_{\sigma_j \tau_j},$$

which contradicts the first inequality. $\qquad \square$

So if symbol $j \in \Sigma'$ appears with state $i \in \Sigma$ in a fully observed Viterbi sequence pair (σ, τ), then j appears in τ with every instance of i in σ.

We now prove some properties about the state sequence of a fully observed toric Viterbi sequence.

Lemma 10.7 *The state sequence of every fully observed toric Viterbi sequence is a toric Viterbi sequence.*

Proof Let σ be the state sequence for the fully observed toric Viterbi sequence

(σ, τ) of a fully observed Markov model M with matrices (θ, θ'). Now create a new toric Markov chain M' with transition matrix ϕ such that

$$\phi_{ij} = \max_k \{\theta_{ij} \theta'_{jk}\}.$$

Now we show that σ is also the only toric Viterbi path of M'. Its probability is greater than that of any other sequence σ' since the probability of σ' in the Markov chain is equal to the maximum probability $p_{\sigma', \tau'}$ of all fully observed sequences with state sequence σ'. The value $p_{\sigma', \tau'}$ is less than $p_{\sigma, \tau}$. □

We finally deduce a criterion for determining whether a sequence pair (σ, τ) is a fully observed Viterbi sequence pair.

Theorem 10.8 *The state sequence σ from a fully observed toric Viterbi sequence pair (σ, τ) is a Viterbi sequence if and only if (σ, τ) is also a fully observed Viterbi sequence pair.*

Proof First suppose σ is a Viterbi sequence. Then there is a stochastic matrix θ for which σ is the only Viterbi path. Let us create another stochastic matrix θ' in which $\theta'_{ij} = 1$ whenever emission ij appears in (σ, τ). (Recall from Lemma 10.6 that for each i, at most one output state j matches with i.) So for all other sequence pairs (σ', τ'),

$$p_{\sigma, \tau} = p_\sigma > p_{\sigma'} \geq p_{\sigma', \tau'}. \tag{10.5}$$

Thus (σ, τ) is a fully observed Viterbi sequence for the fully observed Markov model with matrices (θ, θ').

To show the other direction, we assume that the state sequence σ is not a Viterbi sequence. However, we know that σ is a toric Viterbi sequence by Lemma 10.7. Thus σ is a pseudo-Viterbi sequence. So there are two cases to consider:

Case I: $\Sigma = \{0, 1\}$. In [Kuo, 2005, §4] it is shown that σ (or an equivalent sequence) must end with 001 (or symmetrically, 110). Let θ'_{0A} and θ'_{1B} be emissions with the maximum probability respectively from states 0 and 1 in θ', where $A, B \in \Sigma'$. Note that A and B need not be distinct. By Lemma 10.6, τ must end with AAB. Then, since $p_{001, AAB} > p_{010, ABA}$, we must have $\theta_{00} > \theta_{10}$. Also, $p_{001, AAB} > p_{000, AAA}$, so $\theta_{01} \theta'_{1B} > \theta_{00} \theta'_{0A}$. Finally, $p_{001, AAB} > p_{011, ABB}$, so $\theta_{00} \theta'_{0A} > \theta_{11} \theta'_{1B}$. Thus $\theta_{01} \theta'_{1B} > \theta_{11} \theta'_{1B}$, so $\theta_{01} > \theta_{11}$. But then

$$1 = \theta_{00} + \theta_{01} > \theta_{10} + \theta_{11} = 1,$$

which is a contradiction. Thus σ is not the state sequence of a fully observed Viterbi sequence.

Case II: $\Sigma = \{0, 1, 2\}$. From [Kuo, 2005, §5], the only three-state toric Viterbi sequences that are pseudo-Viterbi are the ones that can end with 11210 or a symmetric variant like 00102. So σ (or an equivalent sequence) must end with 11210. Let $A, B, C \in \Sigma'$ such that $\theta'_{0A}, \theta'_{1B}$, and θ'_{2C} are the greatest probabilities for transitions in θ' from states 0, 1, and 2, respectively. Once again, A, B, and C need not be distinct. Since the sequence ends with 10, we must have

$$\theta_{10}\theta'_{0A} > \theta_{11}\theta'_{1B}. \tag{10.6}$$

We also have $p_{11210,BBCBA} > p_{12100,BCBAA}$, which implies that

$$\theta_{11}\theta'_{1B} > \theta_{00}\theta'_{0A}. \tag{10.7}$$

From inequalities (10.6) and (10.7), we conclude $\theta_{10}\theta'_{0A} > \theta_{00}\theta'_{0A}$, which means

$$\theta_{10} > \theta_{00}. \tag{10.8}$$

And since $p_{11210,BBCBA} > p_{12101,BCBAB}$, we must also have

$$\theta_{11} > \theta_{01}. \tag{10.9}$$

Finally, $p_{11210,BBCBA} > p_{10210,BABCA}$, so

$$\theta_{11}\theta_{12}\theta'_{1B} > \theta_{10}\theta_{02}\theta'_{0A}. \tag{10.10}$$

Combining inequalities (10.10) and (10.6), we get

$$\theta_{12} > \frac{\theta_{10}\theta_{02}\theta'_{0A}}{\theta_{11}\theta'_{1B}} > \theta_{02}. \tag{10.11}$$

Finally, by combining inequalities (10.8), (10.9), and (10.11), we conclude

$$1 = \theta_{10} + \theta_{11} + \theta_{12} > \theta_{00} + \theta_{01} + \theta_{02} = 1, \tag{10.12}$$

which is a contradiction. $\qquad\square$

We conclude with a remark that it would be interesting to extend our results to hidden Markov models. In the case of gene finding (Section 4.4), Viterbi sequences of both hidden Markov models [Burge and Karlin, 1997] and toric hidden Markov models [Parra *et al.*, 2000] have been used for gene prediction, and it is therefore interesting to know which toric Viterbi sequences are not Viterbi sequences.

11

Equations Defining Hidden Markov Models

Nicolas Bray

Jason Morton

In this chapter, we investigate the ideal of polynomial invariants satisfied by the probabilities of observations in a hidden Markov model. Two main techniques for computing this ideal are employed. First, we describe elimination using Gröbner bases. This technique is only feasible for small models and yields invariants that may not be easy to interpret. Second, we present a technique using linear algebra refined by two gradings of the ideal of relations. Finally, we classify some of the invariants found in this way.

11.1 The hidden Markov model

The *hidden Markov model* was described in Section 1.4.3 (see also Figure 1.5) as the algebraic statistical model defined by composing the fully observed Markov model F with the marginalization ρ, giving a map $\rho \circ F : \Theta_1 \subset \mathbb{C}^d \longrightarrow \mathbb{C}^{(l')^n}$, where Θ_1 is the subset of Θ defined by requiring row sums equal to one. Here we will write the hidden Markov model as a composition of three maps, $\rho \circ F \circ g$, beginning in a coordinate space $\Theta'' \subset \mathbb{C}^d$ which parameterizes the $d = l(l-1) + l(l'-1)$-dimensional linear subspace Θ_1 lying in the $l^2 + ll'$-dimensional space Θ, so that $\Theta_1 = g(\Theta'')$. These maps are shown in the following diagrams:

$$\mathbb{C}^{l(l-1)+l(l'-1)} \xrightarrow{\ g\ } \mathbb{C}^{l^2+ll'} \xrightarrow{\ F\ } \mathbb{C}^{l^n l'^n} \xrightarrow{\ \rho\ } \mathbb{C}^{l'n}$$

$$\mathbb{C}[\Theta_1] \xleftarrow{\ g^*\ } \mathbb{C}[\theta_{ij}, \theta'_{ij}] \xleftarrow{\ F^*\ } \mathbb{C}[p_{\sigma,\tau}] \xleftarrow{\ \rho^*\ } \mathbb{C}[p_\tau]$$

In the bottom row of the diagram, we have phrased the hidden Markov model in terms of rings by considering the ring homomorphism g^*, F^* and ρ^*. The marginalization map $\rho^* : \mathbb{C}[p_\tau] \to \mathbb{C}[p_{\sigma,\tau}]$ expands p_τ to a sum across hidden states, $p_\tau \mapsto \sum_\sigma p_{\sigma,\tau}$. The fully observed Markov model ring map $F^* : \mathbb{C}[p_{\sigma,\tau}] \to \mathbb{C}[\theta_{ij}, \theta'_{ij}]$ expands each probability in terms of the parameters, $p_{\sigma,\tau} \mapsto \theta_{\sigma_1 \sigma_2} \dots \theta_{\sigma_{n-1}\sigma_n} \theta'_{\sigma_1 \tau_1} \dots \theta'_{\sigma_n \tau_n}$. The map g^* gives the Θ'' coordinates of the Θ parameters, $g^* : \theta_{ij} \mapsto \theta'_k$; for example, in the binary case our final

237

parameter ring will be $\mathbb{C}[x, y, z, w]$ with the map g^* from $\mathbb{C}[\theta_{ij}, \theta'_{ij}]$ given by

$$(\theta_{ij}) \quad \overset{g^*}{\mapsto} \quad \begin{array}{c} \quad\;\; \sigma = 0 \quad \sigma = 1 \\ \begin{array}{c} \sigma = 0 \\ \sigma = 1 \end{array} \left(\begin{array}{cc} x & 1 - x \\ 1 - y & y \end{array} \right) \end{array} \qquad \text{and}$$

$$(\theta'_{ij}) \quad \overset{g^*}{\mapsto} \quad \begin{array}{c} \quad\;\; \tau = 0 \quad \tau = 1 \\ \begin{array}{c} \sigma = 0 \\ \sigma = 1 \end{array} \left(\begin{array}{cc} z & 1 - z \\ 1 - w & w \end{array} \right) \end{array}$$

As discussed in Section 3.2, the Zariski closure of the image of $\mathbf{f} := \rho \circ F \circ g$ is an algebraic variety in the space of probability distributions. We are interested in the ideal $I_{\mathbf{f}}$ of invariants (polynomials) in $\mathbb{C}[p_\tau]$ which vanish on this variety. By plugging observed data into these invariants (even if we don't know all of them) and observing if the result is close to zero, it can be checked whether a hidden Markov model might be an appropriate model. For details on using invariants to reconstruct tree models, see Chapters 15 and 19. In addition, since this ideal captures the geometry and the restrictions imposed by the choice of model, it may be useful for parametric inference. For more on parameter inference for the hidden Markov model, see Chapter 12.

The equations defining the hidden Markov model are precisely the elements of the kernel of the composed ring map $g^* \circ F^* \circ \rho^*$, so one way to investigate this kernel is to look at the kernels of these maps separately. In other words, if we have an invariant f in the ring $\mathbb{C}[p_\tau]$, and we trace it to its image in $\mathbb{C}[\theta''_i]$, at which point does it become zero? In particular we distinguish invariants which are in the kernel of $F^* \circ \rho^*$ as they have a helpful multi-graded structure not shared by all of the invariants of the constrained model. Mihaescu [Mihaescu, 2005] has investigated the map F^* and classified its invariants. Fully observed Markov models with binary hidden states and nine or fewer nodes were shown to have ideals of invariants I_{F^*} generated in degree 2, and this property was conjectured to hold for any number of nodes. In Section 11.4 we trace how some of the invariants of this map become invariants of the hidden Markov model map.

11.2 Gröbner bases

Elimination theory provides an algorithm for computing the implicitization of a polynomial map such as the one corresponding to the hidden Markov model. We recall the method from Section 3.2.

Let $\mathbb{C}[\theta''_i, p_\tau]$ be the ring containing both the parameter and probability variables, where the θ''_i are the variables in the final parameter ring. Now let I be the ideal $I = (p_\tau - (g^* \circ F^* \circ \rho^*)(p_\tau))$, where $(g^* \circ F^* \circ \rho^*)(p_\tau)$ is p_τ expanded as a polynomial in the final parameters, considered as an element of $\mathbb{C}[\theta''_i, p_\tau]$.

Then the ideal of the hidden Markov model is just the elimination ideal consisting of elements of I involving only the p_τ, $I_e = I \cap \mathbb{C}[p_\tau]$, and $V(I_e)$ is the smallest variety in probability space containing the image of the hidden

Markov model map. To actually compute I_e, we compute a Gröbner basis G for I under a term ordering (such as lexicographic) which makes the parameter indeterminates "expensive." Then the elements of G not involving the parameters form a basis for I_e; see Example 3.19.

The computer packages `Macaulay2` [Grayson and Stillman, 2002] as well as `Singular` [Greuel *et al.*, 2003] contain optimized routines for computing such Gröbner bases, and so can be used to find the ideal of the implicitization. These packages are discussed in Section 2.5. However, Gröbner basis computations suffer from intermediate expression swell. In the worst case, the degrees of the polynomials appearing in intermediate steps of the computation can be doubly exponential in the number of variables [Bayer and Mumford, 1993]. Thus, these methods are only feasible for small models. Perhaps more importantly, the basis obtained this way tends to involve complex expressions and many redundant elements, which makes interpretation of the generators' statistical meaning difficult.

The three node binary model is the largest model which has succumbed to a direct application of the Gröbner basis method. For the binary, unconstrained model ($F^* \circ \rho^*$, no g^*), the computation takes about seven hours on a dual 2.8GHz, 4 GB RAM machine running `Singular` and yields the single polynomial, reported in [Pachter and Sturmfels, 2004b]:

$$p_{011}^2 p_{100}^2 - p_{001}p_{011}p_{100}p_{101} - p_{010}p_{011}p_{100}p_{101} + p_{000}p_{011}p_{101}^2$$

$$+ p_{001}p_{010}p_{011}p_{110} - p_{000}p_{011}^2 p_{110} - p_{010}p_{011}p_{100}p_{110} + p_{001}p_{010}p_{101}p_{110}$$

$$+ p_{001}p_{100}p_{101}p_{110} - p_{000}p_{101}^2 p_{110} - p_{001}^2 p_{110}^2 + p_{000}p_{011}p_{110}^2$$

$$- p_{001}p_{010}^2 p_{111} + p_{000}p_{010}p_{011}p_{111} + p_{001}^2 p_{100}p_{111} + p_{010}^2 p_{100}p_{111}$$

$$- p_{000}p_{011}p_{100}p_{111} - p_{001}p_{100}^2 p_{111} - p_{000}p_{001}p_{101}p_{111} + p_{000}p_{100}p_{101}p_{111}$$

$$+ p_{000}p_{001}p_{110}p_{111} - p_{000}p_{010}p_{110}p_{111}.$$

Note that the polynomial is homogeneous. It is also homogeneous with respect to a multigrading by the total number of 1s appearing in the subscripts in a given monomial. Our implicitization program using linear algebraic techniques and a multigrading of the kernel (to be discussed in Section 11.3) computes this polynomial in 45 seconds. The relevant commands for our `Singular`-based implementation would be:

```
makemodel(4,2,2); setring R_0; find_relations(4,1);
```

The Gröbner basis method is very sensitive to the number of ring variables. Consequently, adding in the constraining map g^* makes the 3-node binary Gröbner basis computation much faster, taking only a few seconds.

The variety obtained has dimension 4, as expected, and has degree 11. The computation yields fourteen polynomials in the reduced Gröbner basis for the graded reverse lexicographic order, though the ideal is in fact generated by a subset of five generators: the partition function together with one generator of degree 2, and three of degree 3. The ideal can be computed using the `Singular` code in Example 3.19 of Section 3.2. One of the generators (homogenized using the sum of the p_τ; see Proposition 11.2) is the following:

$$
\begin{aligned}
g_4 =\ & 2p_{010}p_{100}^2 + 2p_{011}p_{100}^2 - p_{000}p_{100}p_{101} - p_{001}p_{100}p_{101} + 3p_{010}p_{100}p_{101} \\
& + 3p_{011}p_{100}p_{101} - p_{100}^2p_{101} - p_{000}p_{101}^2 - p_{001}p_{101}^2 + p_{010}p_{101}^2 + p_{011}p_{101}^2 \\
& - 2p_{100}p_{101}^2 - p_{101}^3 - p_{000}p_{010}p_{110} - p_{001}p_{010}p_{110} - p_{010}^2p_{110} \\
& - p_{000}p_{011}p_{110} - p_{001}p_{011}p_{110} - 2p_{010}p_{011}p_{110} - p_{011}^2p_{110} - p_{000}p_{100}p_{110} \\
& - p_{001}p_{100}p_{110} + 2p_{010}p_{100}p_{110} + 2p_{011}p_{100}p_{110} + p_{100}^2p_{110} - 2p_{000}p_{101}p_{11} \\
& - 2p_{001}p_{101}p_{110} + p_{010}p_{101}p_{110} + p_{011}p_{101}p_{110} - p_{101}^2p_{110} - 2p_{000}p_{110}^2 \\
& - 2p_{001}p_{110}^2 - p_{010}p_{110}^2 - p_{011}p_{110}^2 + p_{100}p_{110}^2 + p_{000}^2p_{111} + 2p_{000}p_{001}p_{111} \\
& + p_{001}^2p_{111} + p_{000}p_{010}p_{111} + p_{001}p_{010}p_{111} + p_{000}p_{011}p_{111} + p_{001}p_{011}p_{111} \\
& + 3p_{010}p_{100}p_{111} + 3p_{011}p_{100}p_{111} + p_{100}^2p_{111} - p_{000}p_{101}p_{111} - p_{001}p_{101}p_{111} \\
& + 2p_{010}p_{101}p_{111} + 2p_{011}p_{101}p_{111} - p_{101}^2p_{111} - 3p_{000}p_{110}p_{111} \\
& - 3p_{001}p_{110}p_{111} - p_{010}p_{110}p_{111} - p_{011}p_{110}p_{111} + 2p_{100}p_{110}p_{111} - p_{000}p_{111}^2 \\
& - p_{001}p_{111}^2 + p_{100}p_{111}^2
\end{aligned}
$$

Unfortunately, it is not obvious how these polynomials relate to the map $g^* \circ F^* \circ \rho^*$. We would like to be able to write down a more intuitive set, which admits a combinatorial characterization, with the aim of elucidating their statistical and geometric meaning. To this end, we turn to alternative methods of finding the invariants of the model.

11.3 Linear algebra

We may also consider the implicitization problem as a linear algebra problem by limiting our search to those generators of the ideal which have degree less than some bound. As we speculate that the ideal of a binary hidden Markov model of any length is generated by polynomials of low degree (see Conjecture 11.9) this approach seems reasonable. Our implementation is written in `Singular` and C++. It is available at `http://bio.math.berkeley.edu/ascb/chapter11/`. In the implementation, we make use of `NTL`, a number theory library, for finding exact kernels of integer matrices [Shoup, 2004], which is in turn made quicker by the `GMP` library [Swox, 2004] .

11.3.1 Finding the invariants of a statistical model

Turning for a moment to the general implicitization problem, given a polynomial map $\mathbf{f}$ with ring map $\mathbf{f}^* : \mathbb{C}[p_1, \ldots, p_m] \to \mathbb{C}[\theta_1, \ldots, \theta_d]$, we are interested in calculating the ideal $I_\mathbf{f}$ of relations among the p_i. If we denote $\mathbf{f}^*(p_i)$ by f_i, $\prod_i p_i^{a_i}$ by p^a, and $\mathbf{f}^*(p^a)$ by f^a, then $I_\mathbf{f}$ is all polynomials $\sum_a \alpha_a p^a$ such that $\sum_a \alpha_a f^a = 0$.

The Gröbner basis methods discussed in the previous section will generate a basis for $I_{\mathbf{f}}$ but as we have stated these computations quickly become intractable. However by restricting our attention to $I_{\mathbf{f},\delta}$, the ideal of relations generated by those of degree at most δ, there are linear-algebraic methods which are more practical. As $I_{\mathbf{f}}$ is finitely generated, there is some δ such that $I_{\mathbf{f}} = I_{\mathbf{f},\delta}$. Hence these problems eventually coincide. Deciding which δ will suffice is a difficult question. Some degree bounds are available in Gröbner basis theory, but they are not practically useful.

We begin with the simplest case, namely that of $\delta = 1$. Let $\mathbf{f}^*\mathcal{P} = \{f_0 := 1, f_1, \ldots, f_m\}$. A polynomial relation of degree at most $\delta = 1$ is a linear relation among the f_i.

Let $\mathcal{M} = \{m_i\}_{i=1}^k$ be the vector of all monomials in the θ_j occurring in $\mathbf{f}^*\mathcal{P}$, so we can write each f_i in the form $f_i = \sum_j \beta_{ij} m_j$. An invariant then becomes a relation

$$\sum_{i=1}^m \alpha_i \left(\sum_{j=1}^k \beta_{ij} m_j \right) = \sum_{i=1}^m \sum_{j=1}^k \alpha_i \beta_{ij} m_j = \sum_{j=1}^k \left(\sum_{i=1}^m \beta_{ij} \alpha_i \right) m_i = 0$$

This polynomial will equal the zero polynomial if and only if the coefficient of each monomial is zero. Thus all linear relations between the given polynomials are given by the common solutions to the relations $\sum_{i=1}^m \beta_{ij} \alpha_i = 0$ for $j \in \{1, \ldots, k\}$. To say that such a vector $\alpha - (\alpha_i)$ satisfies these relations is to say it belongs to the kernel of $B = (\beta_{ij})$, the matrix of coefficients, and so a set of generators for $I_{\mathbf{f},1}$ can be found by computing a linear basis for the kernel of B. A straightforward method for computing a set of generators for $I_{\mathbf{f},\delta}$ for arbitrary δ now presents itself: a polynomial relation of degree at most δ is a linear relation between the products of the p_is of degree at most δ, and so we can simply compute all the polynomials f^a and then find linear relations among them as above.

Proposition 11.1 *Let $\mathcal{P}$ be the row vector consisting of all monomials in the unknowns $p_1, \ldots, p_m$ of degree at most δ, and $\mathcal{M}$ be the column vector consisting of all monomials occurring in the images under $\mathbf{f}^*$ of entries in $\mathcal{P}$. Now let $B_{\mathbf{f},\delta}$ be the matrix whose columns are labeled by entries of $\mathcal{P}$ and whose rows are labeled by entries of $\mathcal{M}$, and whose (i, j)th entry is the coefficient of $\mathcal{M}_i$ in $\mathbf{f}^*(\mathcal{P}_j)$. If K is a matrix whose columns form a basis of the kernel of $B_{\mathbf{f},\delta}$ then the elements of $\mathcal{P}K$ generate $I_{\mathbf{f},\delta}$.*

As exactly computing the kernel of a large matrix is difficult, we will introduce some refinements to this technique. Some of these are applicable to any statistical model, while some depend on the structure of the hidden Markov model in particular.

Suppose we increment the maximum degree, compute the generators of at most this degree by taking the kernel as above, and repeat. Let $\mathcal{L}$ be the resulting list of generators, in the order in which they are found. There will be many generators in $\mathcal{L}$ which, while linearly independent of preceding generators, lie in the ideal of those generators. We can save steps and produce a shorter list of generators by eliminating those monomials p^a in higher degree that can be expressed, using previous relations in $\mathcal{L}$, in terms of monomials of lower degree. This elimination can be accomplished, after each addition of a generator g to the list $\mathcal{L}$, by deleting all columns of $B_{\mathbf{f},\delta}$ that correspond to monomials p^a which are multiples of leading terms of relations so far included in $\mathcal{L}$. In fact, we can delete all columns whose corresponding monomials lie in the initial ideal of the ideal generated by the entries of $\mathcal{L}$.

Since $\mathbf{f}$ is an algebraic statistical model, we automatically have the trivial invariant $1 - \sum_{i=1}^{m} p_i$. If the constant polynomial 1 is added to our set of polynomials then this invariant will automatically be found in the degree one step of the above procedure, and one of the p_i variables (depending on how they are ordered) will then be eliminated from all subsequent invariants. However, there is a better use for this invariant.

Proposition 11.2 *Suppose $\mathbf{f}$ is an algebraic statistical model, and $I_{\mathbf{f}}$ is its ideal of invariants. Then there exists a set $\mathcal{L}$ of homogeneous polynomials in the p_i such that $\{1 - \sum_i p_i\} \cup \mathcal{L}$ is a basis for $I_{\mathbf{f}}$.*

Proof By Hilbert's basis theorem (Theorem 3.2), $I_{\mathbf{f}}$ has a finite basis $\mathcal{B}$. For $g \in \mathcal{B}$, let δ be the smallest degree of a monomial occurring in g. If δ is also the largest degree of a monomial occurring in g, then g is homogeneous. If not, let g_δ be the degree δ part of g. Since $1 - \sum_i p_i \in I_{\mathbf{f}}$, so is $(1 - \sum_i p_i)g_\delta$. Then if we replace $g \in \mathcal{B}$ with $g - (1 - \sum_i p_i)g_\delta$, $\mathcal{B}$ still generates $I_{\mathbf{f}}$, but the minimum degree of a monomial occurring in g has increased by at least one. Repeating this finitely many times, we have the required $\mathcal{L}$. $\qquad\square$

Thus we may restrict our search for invariants to homogeneous polynomials. We summarize the method we have described in Algorithm 11.3. Let $\mathfrak{m}_\delta$ be the set of all degree δ monomials in $\mathbb{C}[p_i]$.

Algorithm 11.3
Input: An algebraic statistical model $\mathbf{f}$ and a degree bound D.
Output: Generators for the ideal $I_{\mathbf{f}}$ up to degree D.

Recursively compute $\ker(\mathbf{f}^*)$ by letting $I = \text{find-relations}(D, (0))$

- *Subroutine:* find-relations(δ, I_R):

  - If $\delta = 1$, return find-linear-relations($\mathfrak{m}_1$, (0))
  - else {

    - $I_R = \text{find-relations}(\delta - 1, I_R)$
    - Return find-linear-relations($\mathfrak{m}_\delta$, I_R)

    }

- *Subroutine:* find-linear-relations(m, J):

- Delete the monomials in m which lie in the initial ideal of J to form a list P
- Write the coefficient matrix M by mapping P to the parameter ring, using the given $\mathbf{f}^*$
- Return the kernel of M as relations among the monomials in P

11.3.2 Hidden Markov model refinements

A $\mathbb{N}^r$-*multigraded* ring is a ring S together with a direct sum decomposition $\bigoplus_{\delta \in \mathbb{N}^r} S_\delta$, such that if $s \in S_\delta$ and $t \in S_{\delta'}$ then $st \in S_{\delta+\delta'}$. Addition of vectors $\delta \in \mathbb{N}^r$ is performed coordinatewise. For example, the grading of a ring of polynomials in m variables by total degree is a $\mathbb{N}^1$-multigrading.

In the hidden Markov model, both the rings $\mathbb{C}[p_\tau]$ and $\mathbb{C}[\theta_{ij}, \theta'_{ij}]$ can be $\mathbb{N}^{l'}$-multigraded by assigning to an indeterminate a weight $(c_1, \ldots, c_{l'})$ where c_k is the number of occurrences of output state k in its subscript. So θ_{ij} will have a weight vector of all zeros while θ'_{ij} will have $c_j = 1$ and all other vector entries zero. The key fact is that this multigrading is preserved by the map $F^* \circ \rho^*$.

Proposition 11.4 *If $f \in \mathbb{C}[p_\tau]$ is homogeneous with weight $(c_1, \ldots, c_{l'})$ then so is $(F^* \circ \rho^*)(f)$. Thus the kernel of $F^* \circ \rho^*$ is also multigraded by this grading.*

Proof It will suffice to assume that f is a monomial. Each monomial in the p_τ has an image of the form $\prod \sum_\sigma \theta_{\sigma_1 \sigma_2} \ldots \theta_{\sigma_{n-1} \sigma_n} \theta'_{\sigma_1 \tau_1} \ldots \theta'_{\sigma_n \tau_n}$ and expanding the product, the same multiset of τ_i appear in the subscripts of each resulting monomial. If $f \in \ker F^* \circ \rho^*$ the images of its terms in $\mathbb{C}[\theta_{ij}, \theta'_{ij}]$ must cancel. As there are no relations among the θ'_{ij}, this means that each monomial must cancel only with others possessed of the same multiset of τ_i. Then the ideal decomposes according to the multigrading. $\qquad\square$

Note that if f is homogeneous with weight $(c_1, \ldots, c_{l'})$ then $\sum_k c_k$ is n times the degree of f and so the kernel of $F^* \circ \rho^*$ is homogeneous in the usual sense. When we move to the constrained model $g^* \circ F^* \circ \rho^*$ the above multigrading is no longer preserved, but by Proposition 11.2, the grading by degree nearly is. Moreover, any invariants of $g^* \circ F^* \circ \rho^*$ which fail to be multigraded must contain all output symbols, since otherwise there will still be no relations among the appearing θ''_{ij} and the proof of Proposition 11.4 goes through.

Using the multigrading to find the kernel of $F^* \circ \rho^*$ yields an immense advantage in efficiency. For example in the binary case, in degree δ, instead of finding the kernel of a matrix with $\binom{2^n + \delta - 1}{\delta}$ columns, we can instead use $n\delta$ matrices with at most $\binom{n\delta}{n\delta/2} = \frac{(n\delta)!}{(n\delta/2)!^2}$ columns. Computing the ideal of the four-node unconstrained $(F^* \circ \rho^*)$ binary hidden Markov model, intractable by the Gröbner basis method, takes only a few minutes to find the invariants up to degree 4 using the multigrading. The method yields 43 invariants, 9 of degree 2, 34 of degree 3, and none of degree 4; the coefficients are small, just 1s and 2s. After building the multigraded kernel for $F^* \circ \rho^*$, the full kernel of

$g^* \circ F^* \circ \rho^*$ up to degree 3 can be computed, yielding 21 additional relations in degree 2 and 10 in degree 3.

A related condition applies which concerns what we will call the count matrix of a polynomial. For a monomial $m = \prod_{k=1}^{\delta} p_{\tau_k}$, we define the count matrix $C(m)$ of m to be the $l' \times n$ matrix where the entry $C(m)_{ij}$ is the number of times symbol i appears in position j among the $\tau_1, \ldots, \tau_\delta$. This is extended to polynomials by defining the count matrix of a polynomial to be the sum of the count matrices of its monomials, weighted by coefficients. For example, the leading monomial of the invariant given at the beginning of Section 11.4.2 is $p_{0000}p_{0101}p_{1111}$ and has count matrix

$$\begin{pmatrix} 2 & 1 & 2 & 1 \\ 1 & 2 & 1 & 2 \end{pmatrix}.$$

The count matrix of the entire invariant in fact has all entries zero, as the reader may verify. We will show that this holds for any invariant of an unconstrained hidden Markov model with at least two hidden states (which we call 0 and 1), but first we will prove a technical lemma relating the count matrix of a polynomial to the images of certain graded pieces of its expansion in $\mathbb{C}[p_{\sigma,\tau}]$ (where the grade of a variable $p_{\sigma,\tau}$ is given by the multiset of hidden state transitions in σ). The lemma is in essence a counting argument: the coefficients of these images will count occurrences of a given output symbol. In an attempt to simplify notation, we define a map ϕ_s for every output variable s which maps $\mathbb{C}[\theta, \theta']$ to $\mathbb{C}[\theta'_{0s}]$ by simply sending every variable but θ'_{0s} to 1.

Lemma 11.5 *Suppose that $f \in \mathbb{C}[p_\tau]$ is a homogeneous polynomial of degree δ and that i is between 0 and $n-1$. Let T_i be the multiset of hidden state transitions consisting of one $0 \to 1$ transition, i $0 \to 0$ transitions, and $\delta(n-1) - i - 1$ $1 \to 1$ transitions. If $f_{T_i} := (\rho^* f)_{T_i}$ is the T_i-graded part of the expansion of f in $\mathbb{C}[p_{\sigma,\tau}]$ and the image of f_{T_i} in $\mathbb{C}[\theta'_{0s}]$ under $\phi_s \circ F^*$ is $\sum_{j=0}^{i+1} a_j \theta'^j_{0s}$, then $\sum_{j=1}^{i+1} j a_j = \sum_{j=1}^{i+1} C(f)_{sj}$.*

Proof It will suffice to prove the statement when f is a monomial $\prod_{j=1}^{\delta} p_{\tau_j}$ since taking count matrices is by definition linear. The image of f in $\mathbb{C}[p_{\sigma,\tau}]$ will be $\prod_{j=1}^{\delta} \sum_{\sigma} p_{\sigma,\tau_j}$ and when we expand the product, we get $\sum_{(\sigma_1, \ldots, \sigma_\delta)} \prod_{j=1}^{\delta} p_{\sigma_j, \tau_j}$ where the sum is over all δ-tuples of hidden state sequences $(\sigma_1, \ldots, \sigma_\delta) \in (\Sigma^n)^\delta$. Suppose that the term indexed by $(\sigma_1, \ldots, \sigma_\delta)$ in the sum has grade T_i. Then we must have that for some k, σ_k is $i + 1$ zeros followed by all ones, and all the other hidden state sequences $\sigma_j, j \neq k$ are all ones. In this case, the image of $\prod_{j=1}^{\delta} p_{\sigma_j, \tau_j}$ in $\mathbb{C}[\theta'_{0s}]$ is θ'^m_{0s} where m is the number of times s occurs in the first $i + 1$ positions of τ_k. Thus the image of all of f_{T_i} in $\mathbb{C}[\theta'_{0s}]$ will be $\sum_{j=0}^{i+1} a_j \theta'^j_{0s}$ where a_j is the number of times a τ appearing as a subscript in our monomial has exactly j occurrences of output symbol s in its first $i + 1$ positions. We will then have that $\sum_{j=1}^{i+1} j a_j$ is the total number of occurrences of s in the first $i + 1$ positions of any of our subscripts but this is exactly $\sum_{j=1}^{i+1} C(f)_{sj}$. $\qquad\square$

The application of this lemma to invariants rests on the fact that the ideal of invariants for the unconstrained fully observed Markov model is also graded by the multiset of hidden state transitions.

Proposition 11.6 *If f is an invariant of the unconstrained hidden Markov model, then $C(f)$ consists of all zeroes.*

Proof Since the homogeneous parts of an invariant are again invariants and taking count matrices is linear, it will suffice to prove the statement when f is homogeneous. Since f is an invariant of the unconstrained model, we have that for any s, f_{T_i} will get mapped to zero in $\mathbb{C}[\theta, \theta']$ and thus in $\mathbb{C}[\theta'_{0s}]$ as well, for every i. In the notation of the above lemma, the image in $\mathbb{C}[\theta'_{0s}]$ is $\sum_{j=0}^{i+1} a_j \theta'_{0s}{}^j$ and will then have that a_j is zero for all j. Thus, by the lemma, $\sum_{j=1}^{i+1} C(f)_{sj} = 0$. Since this holds for all i, taking consecutive differences $0 - 0 = \sum_{j=1}^{k} C(f)_{sj} - \sum_{j=1}^{k-1} C(f)_{sj} = C(f)_{sk}$ shows that $C(f)_{si} = 0$ for all i. As this holds for all s, $C(f)$ consists entirely of zeros. $\square$

To further reduce the number of invariants to be considered, we can also show that once we have found an invariant, we have found a whole class of invariants which result from permuting the output alphabet Σ'. In fact, any permutation of the output alphabet preserves invariants.

Proposition 11.7 *Let $\pi \in S_{\Sigma'}$ be a permutation of the output alphabet, and let π^* be the automorphism of $\mathbb{C}[p_\tau]$ induced by $\pi^*(p_\tau) = p_{\pi(\tau)}$. Then if f is in the kernel of $\mathbf{f}^* = g^* \circ F^* \circ \rho^*$, so is $\pi^*(f)$.*

Proof We have two maps from $\mathbb{C}[p_\tau]$ to $\mathbb{C}[\theta''_i]$, namely $\mathbf{f}^*$ and $\mathbf{f}^* \circ \pi^*$. If there exists an automorphism ϕ^* of $\mathbb{C}[p_\tau]$ such that $\phi^* \circ \mathbf{f}^* \circ \pi^* = \mathbf{f}^*$, then if $\mathbf{f}^*(f) = 0$, so does $\mathbf{f}^* \circ \pi^*(f)$ as ϕ is injective:

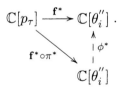

Thus we need only show that for any $\pi \in S_{\Sigma'}$, there exists such a ϕ^*. But π is equivalent to simply permuting the columns of the matrix $g^*(\theta')$, which are labeled by Σ'. Thus we define ϕ^* to be the map induced by π as a permutation of the columns of $g^*(\theta')$. Note that ϕ^* is a ring homomorphism, and in fact an automorphism of $\mathbb{C}[\theta'']$, as required. $\square$

As an example, consider the map induced by $\pi = (12345)$ as a permutation of the columns in the occasionally dishonest casino (described in Example 1.21). Then ϕ^* would be $f_1 \mapsto f_2, f_2 \mapsto f_3, \dots f_5 \mapsto 1 - \sum f_i$, which implies $1 - \sum f_j \mapsto f_1$, and similarly in the second row $l_1 \mapsto l_2, l_2 \mapsto l_3, \dots l_5 \mapsto 1 - \sum l_i$, which implies $1 - \sum l_j \mapsto l_1$. Note that in the multigraded case, we now need only look at a representative of each equivalence class (given by partitions of

$n\delta$ objects into at most l' places); the proof of Proposition 11.7 works for the unconstrained case as well. We now revise the algorithm to take into account these refinements.

Algorithm 11.8

Input: An hidden Markov model $\mathbf{f}^* = g^* \circ F^* \circ \rho^*$ and a degree bound D.

Output: Generators for the ideal $I_{\mathbf{f}}$ up to degree D.

Step 1: Compute $\ker(F^* \circ \rho^*)$ by letting $I_u = $ find-relations(D,(0),multigraded)

Step 2: Compute $\ker(g^* \circ F^* \circ \rho^*)$ by letting $I = $ find-relations(D,I_u,$\mathbb{N}$-graded)

Subroutine: find-relations(bound δ, ideal I, grading):

- If $\delta = 1$, return find-linear-relations($\mathfrak{m}_1$, (0), grading)
- else {
  - $I_R = $ find-relations($\delta - 1$, I_R, grading)
  - return find-linear-relations($\mathfrak{m}_\delta$, I_R, grading)

  }

Subroutine: find-linear-relations(m, J, grading):

- Delete the monomials in m which lie in the initial ideal of J to form a list P
- If grading=multigrading {
  - For each multigraded piece P_w of P, modulo permutations of Σ'
  - Write the coefficient matrix M_w by mapping P_w to the parameter ring by $F^* \circ \rho^*$
  - Append to a list K the kernel of M_w as relations among the monomials in P_w

  }
- else {
  - Write the coefficient matrix M by mapping P to the parameter ring by $\mathbf{f}^*$
  - Let K be the kernel of M as relations among the monomials in P

  }
- Return K

While the above suffices as an algorithm to calculate $I_{\mathbf{f},\delta}$, our goal is to calculate all of $I_{\mathbf{f}}$. One could simply calculate $I_{\mathbf{f},\delta}$ for increasing δ, but this would require a stopping criterion to yield an algorithm. One approach is to bound the degree using a conjecture based on the linear algebra. We strongly suspect that the ideals of hidden Markov models of arbitrary length are generated in low degree, perhaps as low as 3. We can observe directly that this is true of the 3-node binary model. For small n, the matrix $B_{\mathbf{f},\delta}$ of Proposition 11.1 has more rows than columns. However, the number of columns grows exponentially in the length of the model while the number of rows grows only polynomially. Hence for large n, we might expect to be able to write each monomial in $\mathcal{M}$ in terms of the monomials in $\mathcal{P}$. This, together with the results and conjecture of [Mihaescu, 2005] that the ideal of the fully observed binary hidden Markov model F^* is generated in degree 2, lead us to conjecture something similar, which may begin to hold for n as small as 5.

Conjecture 11.9 *The ideal of invariants of the binary hidden Markov model of length n is generated by linear and quadratic polynomials for large n.*

The results obtained using the linear algebra technique are summarized in Table 11.1; dashes indicate the limit of what has been computed.

Table 11.1. *Number of Generators*

Degree	1	2	3	4	time
n, l, l' Type					
3, 2, 2 Unc	0	0	0	1	45s
4, 2, 2 Unc	0	9	34	0	6m
5, 2, 2 Unc	2	127	120	-	159m
6, 2, 2 Unc	11	681	-	-	14m
7, 2, 2 Unc	40	-	-	-	17s
3, 2, 3 Con	1	87	-	-	181s
3, 2, 6 Con	20	-	-	-	29s
4, 2, 4 Con	46	-	-	-	90s
4, 2, 6 Con	330	-	-	-	

11.4 Combinatorially described invariants

We discuss two types of invariants of the hidden Markov model, found by the linear algebra technique, and which admit a statistical interpretation. They are *Permutation Invariants* and *Determinantal Invariants*

11.4.1 Permutation invariants

As discussed in [Mihaescu, 2005], the unhidden Markov model has certain simple invariants called *shuffling invariants*. These are invariants of the form $p_{\sigma,\tau} - p_{\sigma,t(\tau)}$ where t is a permutation of τ which preserves which hidden state each symbol is output from (clearly such an invariant can be non-trivial only if there are repeated hidden states in σ). To translate these invariants to the hidden Markov model, it is tempting to try to simply sum over σ. However, this is not possible unless every σ has repeated hidden states, and even then the permutation t will depend on σ and so one would end up with $p_\tau - \sum_\sigma p_{\sigma,t_\sigma(\tau)}$. However, if we also sum over certain permutations of τ then this problem can be avoided. Let S_n be the group of permutations on n letters.

Proposition 11.10 *If σ has two identical states then for any τ, the polynomial*

$$\sum_{\pi \in S_n} (-1)^\pi p_{\sigma, \pi(\tau)} \tag{11.1}$$

is an invariant of the unhidden Markov model.

Proof Suppose that $\sigma_i = \sigma_j$ and let $t_\sigma = (ij)$. We now have that

$$\sum_{\pi \in S_n} (-1)^\pi p_{\sigma, \pi(\tau)} = \sum_{\pi \in A_n} p_{\sigma, \pi(\tau)} - p_{\sigma, (t_\sigma \pi)(\tau)}$$

The result now follows from the fact that each $p_{\sigma, \pi(\tau)} - p_{\sigma, t_\sigma(\pi(\tau))}$ is a shuffling invariant.   □

Corollary 11.11 *If $l < n$ then*

$$\sum_{\pi \in S_n} (-1)^\pi p_{\pi(\tau)}$$

is an invariant of the hidden Markov model.

Proof If $l < n$ then any σ for the n-node model with l hidden states will have some repeated hidden state and so we can choose a t_σ for every σ and sum over σ to get an invariant of the hidden Markov model.   □

However, if $l' < n$ then for any τ, there will be two repeated output states. If $\tau_i = \tau_j$ and $t = (ij)$ then we have $\sum_{\pi \in S_n} (-1)^\pi p_{\pi(\tau)} = \sum_{\pi \in A_n} p_{\pi(\tau)} - p_{(\pi t)(\tau)} = \sum_{\pi \in A_n} p_{\pi(\tau)} - p_{\pi(\tau)} = 0$. So the above corollary gives non-trivial invariants only when $l < n \leq l'$ in which case there are $\binom{l'}{n}$ such invariants, each corresponding to the set of unique output letters occurring in the subscripts.

Note, however, that we did not actually have to sum over the full set of permutations of τ in the above. In Proposition 11.10, any set $B \subset S_\tau$ which is closed under multiplication by t_σ would suffice and in the corollary, we need B to be closed under multiplication by every t_σ for some choice of t_σ for every σ. In particular, if we let F be a subset of the nodes of our model and let B be all permutations fixing F then we will have permutation invariants so long as $l < n - \#(F)$, since we will then have repeated hidden states outside of F allowing us to choose t_σ to transpose two identical hidden states outside of F. This will imply that $t_\sigma B = B$. Again these permutation invariants will be non-trivial only if the output states outside of F are unique and so we will have permutation invariants for every F such that $l < n - \#(F) \leq l'$. In particular, we will have permutation invariants for every model with $l < \min(n, l')$.

Proposition 11.12 *If $F \subset \{1, \ldots, n\}$ satisfies $l < n - \#(F)$ and B is defined to be the subset of S_n which fixes F, then for any τ,*

$$\sum_{\pi \in B} (-1)^\pi p_{\pi(\tau)}$$

is an invariant of the hidden Markov model. If $n - \#(F) \leq l'$ then some of these invariants will be non-trivial.

All the linear invariants of the 3-node occasionally dishonest casino ($l = 2, l' = 6$) are of the type described in Corollary 11.11 while the four-node ODC exhibits permutation invariants with fixed nodes. For example,

$$-p_{0253} + p_{0352} + p_{2053} - p_{2350} - p_{3052} + p_{3250}$$

is the sum over permutations of $0, 2, 3$ with the third letter, 5, fixed.

11.4.2 Determinantal invariants

Among the degree three invariants of the four-node binary hidden Markov model discovered using the linear algebra method described in the previous section is the invariant

$$p_{00000}p_{01011}p_{11111} + p_{00001}p_{01111}p_{11100} + p_{00011}p_{01100}p_{11011}$$

$$-p_{00011}p_{01011}p_{11100} - p_{00000}p_{01111}p_{11011} - p_{00001}p_{01100}p_{11111}$$

Consider the length n hidden Markov model with l hidden states and l' observed states. For $i, j \in \Sigma$, $\tau_1 \in \Sigma'^k$, and $\tau_2 \in \Sigma'^{n-k}$, let $p_{\tau_1, i}$ be the total probability of outputting τ_1 and ending in hidden state i and p_{j, τ_2} be the total probability of starting in state j and outputting τ_2. Conditional independence for the hidden Markov model then implies that

$$p_{\tau_1 \tau_2} = \sum_{i=1}^{l} \sum_{j=1}^{l} p_{\tau_1, i} \theta_{ij} p_{j, \tau_2}.$$

Let P be the l^k by l^{n-k} matrix whose entries are indexed by pairs (τ_1, τ_2) with $P_{\tau_1, \tau_2} = p_{\tau_1 \tau_2}$, let F be the $l^k \times l$ matrix with $F_{\tau_1, i} = p_{\tau_1, i}$ and let G be the $l \times l^{n-k}$ matrix with $G_{j, \tau_2} = p_{j, \tau_2}$. Then the conditional independence statement says exactly that $P = F \theta G$.

Since $\operatorname{rank}(\theta) = l$, this factorization implies that $\operatorname{rank}(P) \leq l$ or, equivalently, that all of its $(l+1) \times (l+1)$ minors vanish. These minors provide another class of invariants which we call *determinantal invariants*. For example, the invariant above is the determinant of the matrix

$$\begin{pmatrix} p_{00000} & p_{00001} & p_{00011} \\ p_{00100} & p_{00101} & p_{00111} \\ p_{01100} & p_{01101} & p_{01111} \end{pmatrix}$$

which occurs as a minor when the four-node model is split after the second node. See Chapter 19 for more on invariants from flattenings along splits.

12

The EM Algorithm for Hidden Markov Models

Ingileif B. Hallgrímsdóttir

R. Alexander Milowski

Josephine Yu

As discussed in Chapter 1, the EM algorithm is an iterative procedure used to obtain maximum likelihood estimates (MLEs) for the parameters of statistical models which are induced by a hidden variable construct, such as the hidden Markov model (HMM). The tree structure underlying the HMM allows us to organize the required computations efficiently, which leads to an efficient implementation of the EM algorithm for HMMs known as the Baum–Welch algorithm. For several examples of two-state HMMs with binary output we plot the likelihood function and relate the paths taken by the EM algorithm to the gradient of the likelihood function.

12.1 The EM algorithm for hidden Markov models

The hidden Markov model is obtained from the fully observed Markov model by marginalization; see Sections 1.4.2 and 1.4.3. We will use the same notation as there, so $\sigma = \sigma_1\sigma_2\ldots\sigma_n \in \Sigma^n$ is a sequence of states and $\tau = \tau_1\tau_2\ldots\tau_n \in (\Sigma')^n$ a sequence of output variables. We assume that we observe N sequences, $\tau^1, \tau^2, \ldots, \tau^N \in (\Sigma')^n$, each of length n but that the corresponding state sequences, $\sigma^1, \sigma^2, \ldots, \sigma^N \in \Sigma^n$, are not observed (hidden).

In Section 1.4.2 it is assumed that there is a uniform distribution on the first state in each sequence, i.e., $\mathrm{Prob}(\sigma_1 = r) = 1/l$ for each $r \in \Sigma$ where $l = |\Sigma|$. We will allow an arbitrary distribution on the initial state and introduce the parameter vector $\theta_0 = (\theta_{01}, \ldots, \theta_{0l})$, where $\theta_{0r} = \mathrm{Prob}(\sigma_1 = r)$ for each $r \in \Sigma$. Note that $\sum_{r\in\Sigma} \theta_{0r} = 1$ and $\theta_{0r} > 0$ for all $r \in \Sigma$. The parameters of both the fully observed and the hidden Markov models are therefore: an $1 \times l$ vector θ_0 of initial probabilities, an $l \times l$ matrix θ of transition probabilities and an $l \times l'$ matrix θ' of emission probabilities. The entry θ_{rs} represents the probability of transitioning from state $r \in \Sigma$ to state $s \in \Sigma$, and θ'_{rt} represents the probability of emitting the symbol $t \in \Sigma'$ when in state $r \in \Sigma$. Recall that the data matrix of the fully observed model is $\check{U} = (u_{\tau,\sigma}) \in \mathbb{N}^{(l')^n \times l^n}$ where $u_{\tau,\sigma}$ is the number of times the pair (τ, σ) was observed. We also let $u_\tau = \sum_{\sigma\in\Sigma^n} u_{\tau,\sigma}$ be the number of times the output sequence τ was observed in the dataset.

The likelihood function for the hidden Markov model is

$$L_{\text{obs}}(\theta_0, \theta, \theta') = \prod_{\tau \in (\Sigma')^n} f_\tau(\theta_0, \theta, \theta')^{u_\tau},$$

where $f_\tau(\theta_0, \theta, \theta')$ is the probability of the output sequence τ. Given a set of output sequences u_τ, we wish to find the parameter values that maximize this likelihood. There is no closed form solution for the MLEs for HMMs, but in Proposition 1.18 it is shown that obtaining the MLEs for the fully observed Markov model is easy. In that case, the likelihood function is

$$L_{\text{hid}}(\theta_0, \theta, \theta') = \prod_{(\tau, \sigma) \in (\Sigma')^n \times \Sigma^n} f_{\tau, \sigma}(\theta_0, \theta, \theta')^{u_{\tau, \sigma}}.$$

where $f_{\tau, \sigma}(\theta_0, \theta, \theta')$ is the probability of going through the state path σ and observing the output sequence τ, so

$$f_{\tau, \sigma}(\theta_0, \theta, \theta') = \theta_{0\sigma_1} \cdot \theta'_{\sigma_1 \tau_1} \cdot \theta_{\sigma_1 \sigma_2} \cdot \theta'_{\sigma_2 \tau_2} \cdots \theta_{\sigma_{n-1} \sigma_n} \cdot \theta'_{\sigma_n \tau_n}.$$

Note that $f_\tau(\theta_0, \theta, \theta') = \sum_\sigma f_{\tau, \sigma}(\theta_0, \theta, \theta')$. Since we added the parameter vector θ_0 to the model, we have to adjust the linear map used in Proposition 1.18 to get

$$A : \mathbb{N}^{l^n \times (l')^n} \to \mathbb{N}^l \oplus \mathbb{N}^{l \times l} \oplus \mathbb{N}^{l \times l'}.$$

The image under A of the basis vector $e_{\sigma, \tau}$ corresponding to a single observation (σ, τ) is now the triple (w_0, w, w'), where w_0 is a $1 \times l$ vector, w is an $l \times l$ matrix and w' is an $l \times l'$ matrix. The entry w_{0r} is 1 if $\sigma_1 = r$ and 0 otherwise, and as before the entry w_{rs} is the number of indices i such that $\sigma_i \sigma_{i+1} = rs$, and the entry w'_{rt} is the number of indices i such that $\sigma_i \tau_i = rt$. The sufficient statistics v_0, v, and v' are obtained by applying the linear map A to the data matrix U, $A \cdot U = (v_0, v, v')$, where v_0, v, and v' have the same dimensions as w_0, w, and w' respectively. The entry v_{0r} is the number of observed paths that start in state r, the entry v_{rs} is the number of times a transition from state r to state s is observed in the data, and each entry v'_{rt} equals the number of times the symbol t is emitted from state r in the data. It is easy to show that the MLEs for the parameters of the fully observed Markov model are

$$\widehat{\theta}_{0r} = \frac{v_{0r}}{N}, \quad \widehat{\theta}_{rs} = \frac{v_{rs}}{\sum_{s' \in \Sigma} v_{rs'}}, \quad \text{and} \quad \widehat{\theta}'_{rt} = \frac{v'_{rt}}{\sum_{t' \in \Sigma'} v'_{rt'}} \tag{12.1}$$

for all $r, s \in \Sigma$ and $t \in \Sigma'$. Note that $\widehat{\theta}_{0r}$ is the proportion of the number of sequences that started in the state r, $\widehat{\theta}_{rs}$ is the proportion of the number of transitions from state r to state s out of all transitions starting in state r, and $\widehat{\theta}'_{rt}$ is the proportion of times the symbol t was emitted when in state r. If the denominator of one of the $\widehat{\theta}_{rs}$ or $\widehat{\theta}'_{rt}$ is 0, the numerator is also 0, and we set the MLE to 0. This situation is unlikely to arise unless there is insufficient data, or the model has been misspecified.

To calculate the above MLEs the full data matrix U is needed, but in the hidden Markov model, when only the output sequences have been observed,

U is not available. This problem is circumvented in the EM algorithm by using the expected values of the data counts, $E(u_{\tau,\sigma}) = \frac{u_\tau}{f_\tau(\theta_0,\theta,\theta')} f_{\tau,\sigma}(\theta_0,\theta,\theta')$, to obtain the parameter estimates. In the E-step the expected values of the sufficient statistics v_0, v, and v' are obtained based on the expected data counts, and in the M-step these expected values are used to obtain updated parameter values based on the solution of the maximum likelihood problem for the fully observed Markov model as in (12.1). One can write the expected values of v_0, v, and v' in a way that leads to a more efficient implementation of the EM algorithm using dynamic programming, namely the Baum–Welch algorithm. To formulate the algorithm we need to introduce the forward and backward probabilities.

Definition 12.1 The **forward probability**

$$\tilde{f}_{\tau,r}(i) = \mathrm{Prob}(\tau_1, \ldots, \tau_i, \sigma_i = r)$$

is the probability of the observed sequence τ up to and including τ_i, requiring that we are in state r at position i.

The **backward probability**

$$\tilde{b}_{\tau,r}(i) = \mathrm{Prob}(\tau_{i+1}, \ldots, \tau_n | \sigma_i = r)$$

is the probability of the observed sequence τ from τ_{i+1} to the end of the sequence, given that we are in state r at position i.

Proposition 12.2 *The expected values of the sufficient statistics v_0, v, and v' can be written in terms of the forward and backward probabilities. For all $r, s \in \Sigma$ and $t \in \Sigma'$,*

$$E(v_{0r}) = \sum_{\tau \in (\Sigma')^n} \frac{u_\tau}{f_\tau(\theta_0,\theta,\theta')} \cdot \theta_{0r} \cdot \theta'_{r\tau_1} \cdot \tilde{b}_{\tau,r}(1), \tag{12.2}$$

$$E(v_{rs}) = \sum_{\tau \in (\Sigma')^n} \frac{u_\tau}{f_\tau(\theta_0,\theta,\theta')} \sum_{i=1}^{n-1} \tilde{f}_{\tau,r}(i) \cdot \theta_{rs} \cdot \theta'_{s\tau_{i+1}} \cdot \tilde{b}_{\tau,s}(i+1), \tag{12.3}$$

$$E(v'_{rt}) = \sum_{\tau \in (\Sigma')^n} \frac{u_\tau}{f_\tau(\theta_0,\theta,\theta')} \sum_{i=1}^{n} \tilde{f}_{\tau,r}(i) \cdot \tilde{b}_{\tau,r}(i) \cdot I_{(\tau_i=t)}. \tag{12.4}$$

Proof We will only include the proof of (12.3) since the proofs of (12.2) and (12.4) are similar. Recall that for a given pair (σ, τ) the entry w_{rs} is the number of indices i such that $\sigma_i\sigma_{i+1} = rs$. We will denote it by $w_{rs}(\sigma)$. Let I_A be the indicator function. It takes the value 1 if the statement A is true and 0 otherwise. We can write $w_{rs}(\sigma) = \sum_{i=1}^{n-1} I_{(\sigma_i\sigma_{i+1}=rs)}$ and

$$v_{rs} = \sum_{\tau \in (\Sigma')^n} \sum_{\sigma \in \Sigma^n} w_{rs}(\sigma) \cdot u_{\tau,\sigma}$$

$$= \sum_{\tau \in (\Sigma')^n} \sum_{\sigma \in \Sigma^n} \sum_{i=1}^{n-1} I_{(\sigma_i\sigma_{i+1}=rs)} \cdot u_{\tau,\sigma}.$$

The expected value of a sum of random variables is the sum of their expected values, so, since $E(u_{\tau,\sigma}) = \frac{u_\tau}{f_\tau(\theta_0,\theta,\theta')} f_{\tau,\sigma}(\theta_0,\theta,\theta')$, we get

$$E(v_{rs}) = \sum_{\tau \in (\Sigma')^n} \frac{u_\tau}{f_\tau(\theta_0,\theta,\theta')} \sum_{i=1}^{n-1} \sum_{\sigma \in \Sigma^n} I_{(\sigma_i\sigma_{i+1}=rs)} \cdot f_{\tau,\sigma}(\theta_0,\theta,\theta'). \qquad (12.5)$$

The innermost sum in (12.5) is the sum of all joint probabilities of pairs (σ,τ), for an output sequence τ and all σ such that $\sigma_i\sigma_{i+1} = rs$. In other words the probability of observing the path τ and a transition from state r to state s in position i is

$$\begin{aligned}
&\sum_{\sigma \in \Sigma^n} I_{(\sigma_i\sigma_{i+1}=rs)} \cdot f_{\tau,\sigma}(\theta_0,\theta,\theta') \\
&= \mathrm{Prob}(\tau, \sigma_i = r, \sigma_{i+1} = s) \\
&= \mathrm{Prob}(\tau_1,\ldots,\tau_i, \sigma_i = r) \cdot \mathrm{Prob}(\sigma_{i+1} = s | \sigma_i = r) \\
&\qquad \cdot \mathrm{Prob}(\tau_{i+1} | \sigma_{i+1} = s) \cdot \mathrm{Prob}(\tau_{i+2},\ldots,\tau_n | \sigma_{i+1} = s) \\
&= \tilde{f}_{\tau,r}(i) \cdot \theta_{rs} \cdot \theta'_{s\tau_{i+1}} \cdot \tilde{b}_{\tau,s}(i+1).
\end{aligned}$$

We obtain the desired result by plugging the above expression into (12.5). □

We need the probability $f_\tau(\theta_0,\theta,\theta')$ of the whole sequence τ in order to evaluate the expected values of the sufficient statistics. It can be calculated based on the forward probabilities, $f_\tau(\theta_0,\theta,\theta') = \sum_{r \in \Sigma} \tilde{f}_{\tau,r}(n)$, or on the backward probabilities, $f_\tau(\theta_0,\theta,\theta') = \sum_{r \in \Sigma} \theta_{0r} \cdot \theta'_{r\tau_1} \cdot \tilde{b}_{\tau,r}(1)$.

For each τ the matrices $\tilde{f}_\tau$ and $\tilde{b}_\tau$ of forward and backward probabilities are of size $l \times n$ giving a total of $2 \cdot (l')^n \cdot l \cdot n$ values. Each entry in $\tilde{f}_\tau$ (resp. $\tilde{b}_\tau$) can be efficiently obtained based on the values in the previous (resp. subsequent) column using the following recursions [Durbin *et al.*, 1998], which are valid for all $r \in \Sigma$:

$$\begin{aligned}
\tilde{f}_{\tau,r}(1) &= \theta'_{r\tau_1} \cdot \theta_{0r}, \\
\tilde{f}_{\tau,r}(i) &= \theta'_{r\tau_i} \sum_{s \in \Sigma} \tilde{f}_{\tau,s}(i-1) \cdot \theta_{sr} \quad \text{for } i \in \{2,3,\ldots,n\}
\end{aligned}$$

and

$$\begin{aligned}
\tilde{b}_{\tau,r}(n) &= 1, \\
\tilde{b}_{\tau,r}(i) &= \sum_{s \in \Sigma} \theta_{rs} \cdot \theta'_{s\tau_{i+1}} \cdot \tilde{b}_{\tau,s}(i+1) \quad \text{for } i \in \{1,2,\ldots,n-1\}.
\end{aligned}$$

This results in a great saving of both computer memory and processor time compared to calculating and storing the $(l')^n \times l^n$ matrix $F = (f_{\tau,\sigma}(\theta_0,\theta,\theta'))$, where each entry, a monomial of degree $2n$ in $l + l^2 + l \cdot l'$ variables, has to be evaluated separately.

In each iteration of the Baum–Welch algorithm the parameters $\widehat{\theta}_0$, $\widehat{\theta}$ and $\widehat{\theta'}$ are updated in the following steps:

Algorithm 12.3 (Baum–Welch)

Input: Observed data u_τ and starting values for the parameters, θ_0, θ and θ'.
Output: Proposed MLEs $\widehat{\theta}_0, \widehat{\theta}$ and $\widehat{\theta}'$.
Initialization: Pick arbitrary model parameters θ_0, θ and θ'.
Recurrence:

>   Calculate $\tilde{f}_{\tau,r}(i)$ and $\tilde{b}_{\tau,r}(i)$ using θ_0, θ and θ'.
>   Calculate the expected values of v_0, v and v' using Proposition 12.2.
>   Calculate new parameter values, θ_0^*, θ^* and θ'^*, using (12.1).

Termination: If the change in ℓ_{obs} is less than some predefined threshold, then set $\widehat{\theta}_0 = \theta_0^*$, $\widehat{\theta} = \theta^*$ and $\widehat{\theta}' = \theta'^*$. Otherwise set $\theta_0 = \theta_0^*$, $\theta = \theta^*$ and $\theta' = \theta'^*$ and repeat the recurrence.

12.2 An implementation of the Baum–Welch algorithm

In this section we provide pseudo-code for implementing the Baum–Welch algorithm. An online application which allows the user to generate data from any HMM, and to run the Baum–Welch algorithm is also available at

`http://bio.math.berkeley.edu/ascb/chapter12/`

The Java source can also be downloaded.

We use the following five functions in the implementation of the Baum–Welch algorithm:

Baum_Welch implements the Baum–Welch algorithm.
count_transitions calculates v_{rs} using forward and backward probabilities.
count_emissions calculates v'_{rt} using forward and backward probabilities.
forward_array implements a dynamic programming algorithm to calculate the forward probabilities for each position i and each state σ_i, given an output sequence τ.
backward_array calculates the backward probabilities in a similar way.

The values $\tilde{f}_{\tau,r}(i)$ and $\tilde{b}_{\tau,r}(i)$ are in general so small that underflow problems can occur on most computer systems. Thus it is necessary to either scale the values or work with logarithms. Since it is convenient to work with logarithms, all calculations in the pseudocode below are performed in log-space. To evaluate the sum $x + y$ based on $\log x$ and $\log y$ without converting back to x and y, we use,

$$\log(x + y) = \log x + \log(1 + e^{\log y - \log x})$$

which is codified in the utility function `add_logs`. The numerical stability of the result depends on the implementation of the log and exp functions. Checks should be included in the code to ensure that the calculations remain stable. Note that these checks are not included in the pseudo-code.

In the implementation of the Baum–Welch algorithm below the matrices $\tilde{f}_\tau$ and $\tilde{b}_\tau$ are calculated at the beginning of each iteration (we only need to keep one $\tilde{f}_\tau$ and one $\tilde{b}_\tau$ matrix in memory at any time). One can also use the recursive property of the forward and backward probabilities to calculate them for each position i, state r, and output sequence τ, as they are needed in the evaluation of v_{rs} and v'_{rt}. This adds computation time but removes the need for storing the matrices $\tilde{f}_\tau$ and $\tilde{b}_\tau$. In the code we use S to denote the transition matrix θ and T to denote the emission matrix θ'.

```
add_logs(x, y):
    return x + log(1 + exp(y − x))


Baum_Welch(S, T, sequences, limit):
    lastLogLikelihood ← −∞
    repeat
        logLikelihood ← 0
        S_counts ← zero_matrix(S)
        new_S ← zero_matrix(S)
        T_counts ← zero_matrix(T)
        new_T ← zero_matrix(T)
        for s ← 0 to length[sequences] − 1
        do
            sequence ← sequences[s]
            // Get the forward/backward values for the current sequence & model parameters
            forward ← forward_array(S, T, sequence, length[sequence] − 1)
            backward ← backward_array(S, T, sequence, 0)
            // Calculate sequence probability
            seqprob ← forward[1][length[sequence]] + S[1][0]
            for state ← 2 to row_count[S] − 1
            do
                seqprob ← add_logs(seqprob, forward[state][length[sequence]] + S[state][0])
            // Add contribution to log-likelihood
            logLikelihood ← logLikelihood + seqprob
            // Calculate the "counts" for this sequence
            S_counts ← count_transitions(S_counts, S, T, forward, backward, sequence)
            T_counts ← count_emissions(T_counts, T, forward, backward, sequence)
            // Calculate contribution for this sequence to the transitions
            for from ← 0 to row_count[S] − 1
            do
                for to ← 1 to row_count[S] − 1
                do
                    if s = 0
                    then
                        new_S[from][to] ← S_counts[from][to] − seqprob
                    else
                        new_S[from][to] ← add_logs(new_S[from][to],
                                                    S_counts[from][to] − seqprob)
                // Calculate contribution for this sequence to the transitions
                for sym ← 0 to row_count[T] − 1
                do
                    if s = 0
                    then
                        new_T[from][sym] ← T_counts[from][sym] − seqprob
                    else
                        new_T[from][sym] ← add_logs(new_T[from][sym],
                                                     T_counts[from][sym] − seqprob)
        // We stop when the log-likelihood changes a small amount
        change ← logLikelihood − lastLogLikelihood
    until change < limit
    return S, T


count_transitions(S_counts, S, T, forward, backward, sequence):
    // Count initial transitions 0 → n
    for to ← 1 to row_count[S] − 1
    do
        S_counts[0][to] ← S[0][to] + (T[to][sequence[0]] + backward[to][1])
    // Count final transitions n → 0
    for from ← 1 to row_count[S] − 1
```

```
do
    S_counts[from][0] ← S[from][0] + forward[from][length[sequence] − 1]
    // Count transitions k → l where k,l!=0
for from ← 1 to row_count[S] − 1
do
    for to ← 1 to row_count[S] − 1
    do
        S_counts[from][to] ← forward[from][0]+
                        (S[from][to] + (T[to][sequence[1]] + backward[to][1]))
        for pos ← 1 to length[sequence] − 2
        do
            v ← forward[from][pos]+
                (S[from][to] + (T[to][sequence[pos]] + backward[to][pos + 1]))
            S_counts[from][to] ← add_logs(S_counts[from][to], v)
return S_counts
```

count_emissions(T_counts, T, forward, backward, sequence):
```
    // Count initial transitions 0 → n
for state ← 1 to row_count[S] − 1
do
    T_counts[state][sequence[0]] ← forward[state][0] + backward[state][0]
for state ← 1 to row_count[S] − 1
do
    for pos ← 1 to length[sequence] − 1
    do
        T_counts[state][sequence[pos]] ←
            add_logs(T_counts[state][sequence[pos]], forward[state][pos] + backward[state][pos])
return T_counts
```

forward_array(S, T, sequence, end):
```
    result ← allocate_matrix(length[sequence], row_count[S])
for state ← 1 to row_count[S]
do
    // Start with initial transitions
    result[0][state] ← S[0][state] + T[state][sequence[0]]
    // Calculate the value forward from the start of the sequence
for pos ← 0 to end
do
    // Calculate the next step in the forward chain
    for state ← 1 to row_count[S] − 1
    do
        // Traverse all the paths to the current state
        result[pos][state] ← result[pos − 1][1] + S[1][state]
        for from ← 2 to row_count[S] − 1
        do
            // log formula for summation chains
            result[pos][state] ← add_logs(result[pos][state], result[pos − 1][from] + S[from][state])
        // Add in the probability of emitting the symbol
        result[pos][state] ← result[pos][state] + T[state][sequence[pos]]
return result
```

backward_array(S, T, sequence, start):
```
    result ← allocate_matrix(length[sequence], row_count[S])
for state ← 1 to row_count[S]
do
    // Start with end transitions
    result[length[sequence]][state] ← S[state][0]
    // Calculate the value backward from end
for pos ← length[sequence] − 1 to start
do
    for state ← 1 to row_count[S] − 1
    do
        result[pos][state] ← result[pos + 1][1] + (S[state][1] + T[1][sequence[pos]])
        for to ← 2 to row_count[S] − 1
        do
            // log formula for summation chains
            result[pos][state] ←
                add_logs(result[pos][state], result[pos + 1][to] + (S[state][to] + T[to][sequence[pos]]))
return result
```

12.3 Plots of the likelihood surface

The Baum–Welch algorithm is widely used to obtain the MLEs $\widehat{\theta}_0$, $\widehat{\theta}$ and $\widehat{\theta}'$ because it is in general not possible to obtain them algebraically. The likelihood function L_{obs} for a hidden Markov model is a polynomial of degree $2nN$, whose variables are the unknown parameters in the matrices θ_0, θ and θ', and finding the roots of a polynomial in many variables and of high degree is well known to be difficult. However, for short HMMs with few hidden and output states, it is possible to obtain the MLEs algebraically using the methods described in Section 3.3. Based on our experience, this is not feasible if the number of parameters is larger than two and the maximum likelihood degree is high. We will look at a few examples of likelihood functions that arise from two state HMMs of length three with binary output ($l = l' = 2, n = 3$). If we fix the initial probabilities at $\theta_0 = (0.5, 0.5)$, the model has four free parameters, and we will reduce the number to two or three parameters by restricting some of them. For simplicity, we denote the transition and emission probabilities by

$$\theta = \begin{pmatrix} x & 1-x \\ 1-y & y \end{pmatrix} \quad \text{and} \quad \theta' = \begin{pmatrix} z & 1-z \\ 1-w & w \end{pmatrix}.$$

For a fixed set of observed data, the likelihood function L_{obs} is a polynomial in the variables x, y, z, and w. We are interested in maximizing L_{obs} over the region $0 \le x, y, z, w \le 1$. More generally, we want to study how many critical points L_{obs} typically has. Note that $L_{\text{obs}}(x, y, 0.5, 0.5)$ is constant with respect to x and y, and that $L_{\text{obs}}(x, y, z, w) = L_{\text{obs}}(y, x, 1-w, 1-z)$ for any x, y, z, w. Therefore the critical points and global maxima of L_{obs} occur in pairs for the four-parameter model.

In the rest of this section, we will look at the likelihood functions for the six examples listed in Table 12.1. In each example we either fix one or two of the emission probabilities or impose symmetry constraints on the transition probabilities, thereby reducing the number of free parameters to 2 or 3 and allowing visualization in 3 dimensions. For each model a data vector u_τ is given. The likelihood functions corresponding to Examples 2, 3, and 6 in Figure 12.1 have local maxima and saddle points. This shows that even for these very small models finding the global maxima of the likelihood function is non-trivial. However, out of several hundred simulated data sets, most of them have only one (or a pair of) global maxima as in Examples 1, 4, and 5, and very few of them have local maxima or saddle points.

In the first three examples we restrict z and w to a constant, so there are only two free parameters x and y. Using a `Singular` implementation of the algebraic method for obtaining MLEs described in Section 3.3, we find that the likelihood function in the first example has a single critical point $(0.688, 0.337)$, which is a local and global maximum.

We can plot the likelihood as a function of x and y using the software `Mathematica`, which was described in Chapter 2. For simplicity, we use 1 instead of 0.5 for the initial probabilities. This only scales the likelihood function

	000	001	010	011	100	101	110	111	N	z	w
Ex. 1	113	102	80	59	53	32	28	33	500	12/17	12/17
Ex. 2	26	31	44	4	9	16	40	35	205	12/17	12/17
Ex. 3	37	20	35	46	29	13	50	33	263	0.635	0.635
Ex. 4	73	56	49	51	70	53	67	81	500	free	0.1
Ex. 5	116	88	67	85	51	37	31	25	500	$z = w$	$z = w$
Ex. 6	37	20	35	46	29	13	50	33	263	$z = w$	$z = w$

Table 12.1. *Observed data and emission probability constraints for our examples.*

by a constant 2^N and does not change the critical points. The `Mathematica` code is:

```
s[0,0] := x; s[0,1] := 1-x; s[1,0] := 1-y; s[1,1] := y;
t[0,0] := 12/17; t[0,1] := 5/17; t[1,0] := 5/17; t[1,1] := 12/17;
f[i_,j_,k_]:=Sum[t[a,i]*Sum[s[a,b]*t[b,j]*Sum[s[b,c]*t[c,k],{c,0,1}],{b,0,1}],{a,0,1}];
L := Product[Product[Product[f[i,j,k]^u[i,j,k], {k,0,1}], {j,0,1}], {i,0,1}];
u[0,0,0] = 113; u[0,0,1] = 102; u[0,1,0] = 80; u[0,1,1] = 59;
u[1,0,0] = 53; u[1,0,1] = 32; u[1,1,0] = 28; u[1,1,1] = 33;
Plot3D[L,{x,0,1}, {y,0,1}, PlotPoints -> 60, PlotRange -> All];
```

The resulting plot can be viewed in the first panel of Figure 12.1. This is a typical likelihood function for this model. In fact we randomly generated several hundred data vectors u_τ and for each of them we examined the likelihood as a function of x and y for various values of $z = w$. There was almost always only one local (and global) maximum. We did find a handful of more interesting cases, including the ones in Example 2 and Example 3. The following code was used to generate three sets of data in `Mathematica` and visualize them:

```
<< Graphics'Animation'
s[0, 0] := x; s[0, 1] := 1-x; s[1, 0] := 1-y; s[1, 1] := y;
t[0, 0] := z; t[0, 1] := 1-z; t[1, 0] := 1-z; t[1, 1] := z;
f[i_,j_,k_]:=Sum[t[a,i]*Sum[s[a,b]*t[b,j]*Sum[s[b,c]*t[c,k],{c,0,1}],{b,0,1}],{a,0,1}];
L := Product[Product[Product[f[i, j, k]^u[i, j, k], {k, 0, 1}], {j, 0, 1}], {i, 0, 1}];
For[n = 1, n <= 3, n++,
   u[0, 0, 0] = Random[Integer, {1, 50}];  u[0, 0, 1] = Random[Integer, {1, 50}];
   u[0, 1, 0] = Random[Integer, {1, 50}];  u[0, 1, 1] = Random[Integer, {1, 50}];
   u[1, 0, 0] = Random[Integer, {1, 50}];  u[1, 0, 1] = Random[Integer, {1, 50}];
   u[1, 1, 0] = Random[Integer, {1, 50}];  u[1, 1, 1] = Random[Integer, {1, 50}];
   Print[u[0, 0, 0], ",", u[0, 0, 1], ",", u[0, 1, 0], ",",u[0, 1, 1], ",",
         u[1, 0, 0], ",", u[1, 0, 1], ",", u[1, 1, 0], ",", u[1, 1, 1]];
   MoviePlot3D[L, {x, 0, 1},{y, 0, 1}, {z, 0, 1}, PlotPoints -> 40,
PlotRange -> All]]
```

We obtained the critical points for the likelihood functions as before. For Example 2 there are three critical points in the unit square $0 \le x, y \le 1$, namely a global maximum at $(0.135, 0.315)$, a local maximum at $(0.783, 0.728)$, and a saddle point at $(0.613, 0.612)$. In Example 3, the unique global maximum occurs on the boundary of the unit square, but there is also a local maximum at $(0.651, 0.709)$ and a saddle point at $(0.278, 0.553)$. In Section 12.4 we will return to these examples.

In Examples 4–6 we look at models with three free parameters. It is no longer possible to obtain the critical values algebraically as before, and we cannot plot a function in three variables. However we can look at level surfaces (3-dimensional contours) of the log-likelihood. For example, the following

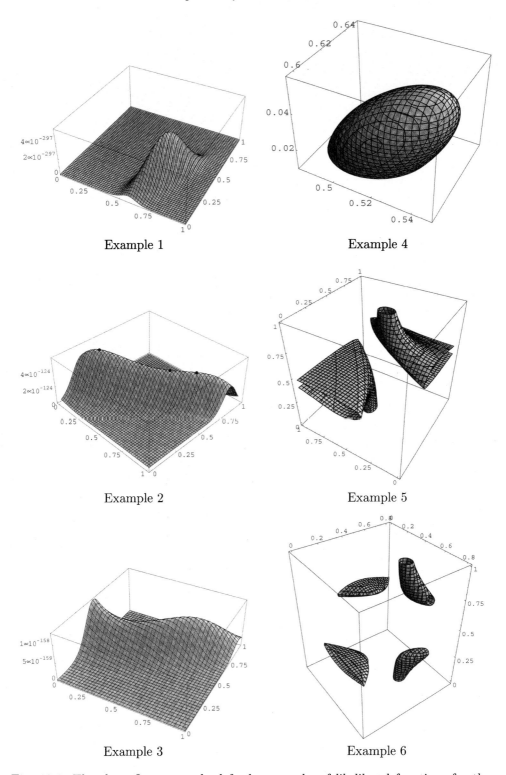

Fig. 12.1. The three figures on the left show graphs of likelihood functions for the 2-parameter models. The figures on the right show the level surfaces of the likelihood function for the 3-parameter models.

`Mathematica` code can be used to plot the level surface of the log-likelihood function in Example 6 at the value -363.5.

```
<< Graphics`ContourPlot3D`
s[0, 0] := x; s[0, 1] := 1 - x; s[1, 0] := 1 - y; s[1, 1] := y;
t[0, 0] := z; t[0, 1] := 1 - z; t[1, 0] := 1 - z; t[1, 1] := z;
f[i_,j_,k_]:=Sum[t[a,i]*Sum[s[a,b]*t[b,j]*Sum[s[b,c]*t[c,k],{c,0,1}],{b,0,1}],{a,0,1}];
logL := Sum[Sum[Sum[u[i, j, k]*Log[f[i, j, k]], {k, 0, 1}], {j, 0, 1}], {i, 0, 1}];
u[0, 0, 0] = 37; u[0, 0, 1] = 20; u[0, 1, 0] = 35; u[0, 1, 1] = 46;
u[1, 0, 0] = 29; u[1, 0, 1] = 13; u[1, 1, 0] = 50; u[1, 1, 1] = 33;
ContourPlot3D[logL, {x, 0.01, 0.99}, {y, 0.01, 0.99}, {z, 0.01, 0.99},
         Contours -> {-363.5}, PlotPoints -> {10, 6}, Axes -> True];
```

Using the level surfaces, we can try to estimate the maximum log likelihood values. A log-likelihood value is larger than the maximum (or smaller than the minimum) if and only if the corresponding level surface is empty. Hence, we can estimate the maxima by picking log likelihood values using binary search. That is, we first find by experiment two values of the log-likelihood function, one above the optimum and one below, so we have a window of possible optimal values. Then we check whether the midpoint of that window is above or below the optimum, hence reducing the window size by half. By repeating this process, we can capture the optima within a small window. A weakness of this method is that software such as `Mathematica` may mistakenly output an empty level surface if the resolution is not high enough.

In Example 4, we restrict w to a constant and let z vary. The level surface at -686 can be seen in Figure 12.1. In Examples 5 and 6 we impose the condition that the emission matrix θ' is symmetric, i.e., $z = w$. The symmetry $L_{\text{obs}}(x, y, z, w) = L_{\text{obs}}(y, x, 1 - w, 1 - z)$ then becomes $L_{\text{obs}}(x, y, z) = L_{\text{obs}}(y, x, 1 - z)$. For Example 5 and 6, the level surfaces move closer to the boundaries $z = w = 0$ and $z = w = 1$ as we gradually increase the likelihood value. Hence there seem to be two global maxima only on the boundaries $z = w = 0$ and $z = w = 1$ in each example. In Example 6, there also seem to be two local maxima on the boundaries $x = 0$ and $y = 0$. Example 3 has the same data vector as Example 6, so it is a "slice" of Example 6.

In the case of full 4-parameter model and the symmetric 3-parameter model, as in Examples 5 and 6, we did not find any examples where there are local maxima inside the (hyper)cube but do not know whether that is true in general. For the full 4-parameter model of length 3 and 4, the coordinate functions $f_{(\tau,\sigma)}(\theta_0, \theta, \theta')$ have global maxima on the boundaries $z = 0$, $z = 1$, $w = 0$, $w = 1$. We do not know if this will hold true for longer HMMs. This is worth pursuing because we would like to know whether the information about the location of global maxima of coordinate functions is useful in determining the location of critical points of the likelihood function L_{obs}.

12.4 The EM algorithm and the gradient of the likelihood

The EM algorithm is usually presented in two steps, the E-step and the M-step. However for the forthcoming discussion it is convenient to view the EM algorithm as a map from the parameter space into itself,

$$M : \Theta \to \Theta, \quad \theta^t \mapsto \theta^{t+1}$$

where

$$\theta^{t+1} = M(\theta^t) = \operatorname{argmax}_\theta E\left[\ell_{obs}(\theta|\tau,\sigma) \,|\, \sigma, \theta^t\right].$$

Here we use θ to denote the triple $(\theta_0, \theta, \theta')$. By applying the mapping repeatedly we get a sequence of parameter estimates $\theta^1, \theta^2, \theta^3, \ldots$ such that $\theta^{t+1} = M^t(\theta^1)$. Local maxima of the likelihood function are fixed points of the mapping M. If $\theta^1, \theta^2, \theta^3, \cdots \to \theta^*$ then, by Taylor expansion,

$$\theta^{t+1} - \theta^* \approx M'(\theta^*)(\theta^t - \theta^*)$$

in the neighborhood of θ^*, where $M'(\theta^*)$ is the first derivative of M evaluated at θ^* [Salakhutdinov *et al.*, 2003]. This implies that the EM algorithm is a linear iteration algorithm with convergence rate matrix $M'(\theta^*)$, and the convergence behavior is controlled by the eigenvalues of $M'(\theta^*)$. The convergence speed is a matter of great practical importance and there is a vast literature available on the subject, but that discussion is outside the scope of this chapter.

More important than the speed of convergence is where the algorithm converges to. The EM algorithm is typically run from a number of different starting points in the hope of finding as many critical points as possible. We can ask questions such as: in what direction (in the parameter space) does the EM algorithm move in each step, how does the choice of starting value affect where it converges to, and along what path does it travel. Since the gradient gives the direction of steepest ascent, it is natural to compare the direction of a step of the EM algorithm to the direction of the gradient. In fact the updated parameter estimates can be written as a function of the parameter estimates from the previous step of the EM algorithm as [Salakhutdinov *et al.*, 2004]:

$$\theta^{t+1} = \theta^t + P(\theta^t)\nabla\ell_{obs}(\theta^t) \tag{12.6}$$

where $\nabla\ell_{obs}(\theta^t)$ is the gradient of the log-likelihood function evaluated at θ^t. The symmetric positive definite matrix P depends on the model, and its form for HMMs was derived in [Salakhutdinov *et al.*, 2004]. Note that each step of the EM algorithm has positive projection onto the gradient of the likelihood function. Also note that a step of a Newton method can be written similarly, except then the gradient is rescaled by the negative inverse Hessian, instead of P. In the EM algorithm the updated parameter estimates are guaranteed to stay within the probability simplex. If other optimization methods are used, it may be necessary to scale the steps taken to ensure that the updated parameters are valid.

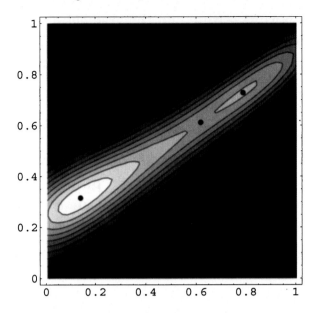

Fig. 12.2. Contour plot of the likelihood function for Example 2.

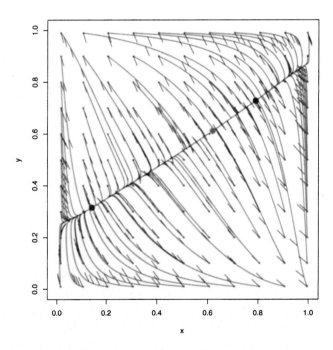

Fig. 12.3. The paths of the EM algorithm for Example 2. The direction of the gradient is indicated with black lines.

For the three examples of 2-parameter models in Section 12.3, the Baum–Welch algorithm was run from 121 different starting points. In the interest of space we will only discuss Example 2 here, but the following discussion also applies to the Examples 1 and 3. In Figure 12.3 we show the paths that the EM algorithm took. Each starting value is indicated by a dot and the two local maxima and the saddle point are indicated with filled circles. None of the paths end at the saddle point, in about one third of the runs Baum–Welch converges to the local maximum at $(0.783, 0.728)$, and for the rest to the (larger) local maximum at $(0.135, 0.315)$. There seems to be a border going from $(0,1)$ to $(1,0)$ through the saddle point, partitioning the parameter space so that if we start the Baum–Welch algorithm from any point in the lower section it converges to $(0.135, 0.315)$, but it converges to $(0.783, 0.728)$ if we start at any point in the top section.

The Baum–Welch was run from each starting point until the change in log-likelihood was $< 10^{-8}$. The number of iterations ranged from 66 to 2785 with an average of 1140. The change in the value of the log-likelihood in each step gets smaller and smaller, in fact for all starting points the ridge was reached in 10 to 75 steps (usually between 20 and 30). The change in log-likelihood in each step along the ridge is very small, usually on the order of 10^{-4}–10^{-5}. It is thus important to set the limit for the change of the log-likelihood low enough, or we would in this case think that we had reached convergence while still somewhere along the ridge.

In Figure 12.3 the direction of the gradient has been indicated with a black line at steps $1, 4, 7$ and 10. For comparison a contour plot of the likelihood is provided. It is clear that although the projection of the EM path onto the gradient is always positive, the steps of the EM are sometimes almost perpendicular to it, especially near the boundary of the parameter space. However, after reaching the ridge all paths are in the direction of the gradient. It is worth noting that since the EM algorithm does not move in the direction of the gradient of the log-likelihood, it will not always converge to the same local maximum as a gradient ascent method with the same starting point.

13

Homology Mapping with Markov Random Fields

Anat Caspi

In this chapter we present a probabilistic approach to the homology mapping problem. This is the problem of identifying regions among genomic sequences that diverged from the same region in a common ancestor. We explore this question as a combinatorial optimization problem, seeking the best assignment of labels to the nodes in a Markov random field. The general problem is formulated using toric models, for which it is unfortunately intractable to find an exact solution. However, for a relevant subclass of models, we find a (non-integer) linear programming formulation that gives us the exact integer solution in polynomial time in the size of the problem. It is encouraging that for a useful subclass of toric models, *maximum a posteriori* inference is tractable.

13.1 Genome mapping

Evolutionary divergence gives rise to different present-day genomes that are related by shared ancestry. Evolutionary events occur at varying rates, but also at different scales of genomic regions. *Local mutation events* (for instance, the point mutations, insertions and deletions discussed in Section 4.5) occur at the level of one or several base-pairs. *Large-scale mutations* can occur at the level of single or multiple genes, chromosomes, or even an entire genome. Some of these mutation mechanisms such as *rearrangement* and *duplication*, were briefly introduced in Section 4.1. As a result, regions in two different genomes could be tied to a single region in the ancestral genome, linked by a series of mutational events. Such genomic regions (be they gene-coding regions, conserved non-coding regions, or other) that are related through divergence from a common region in the ancestral genome are termed *homologs*. *Homology* is a relational term asserting the common ancestry of genomic components. We will refer to two regions, each from a different genome, that diverged from the same region in the ancestor as a *homolog pair*. Our objective is to study the relationships between multiple sequences, and create a *homology mapping*: a mapping which indicates the homolog pairs in the sequences being compared.

A great deal of our understanding of phylogeny and evolution comes from a comparative framework in which homologous regions in related species are contrasted. It is therefore important to accurately identify homolog pairs in order to compare components that are actually linked by common ancestry. More specific relational descriptors exist to delineate the evolutionary events that initiated the divergence between homologs; these are outside the scope of this discussion.

Before whole genome sequences were available, early comparative studies aimed at mapping homology at whichever level of resolution the data would allow. At low resolution, genetic markers were used to map entire genes on long genomic regions like chromosomes. At high resolution, nucleotide sequences of homologous short genomic regions (on the order of one or two genes long) were sequenced for the purpose of nucleotide-by-nucleotide mapping. Nucleotide sequence alignment methods, such as those based on the pair hidden Markov model described in Chapter 7, were designed for this purpose. When aligning short homologous regions, we do not need to consider events such as gene inversions or chromosomal duplications. The earlier alignment models therefore assumed *collinearity* (i.e. that the sequences compared can be mapped sequentially). Assuming collinearity places an ordering constraint on the mapping produced by the methods, and therefore, these models are not suitable for aligning longer genomic regions in which rearrangements, inversions and duplications do occur.

As entire genome sequences become available, we are interested in mapping longer sequence regions, and must offer a model that does not embed the ordering constraint. However, studies suggest that as genomes diverge, functional constraints apply pressure against genome shuffling, inversions, and duplications [Marcotte *et al.*, 1999, Pevzner and Tesler, 2003]. As a result, homologs are more likely to occur in sequential clumps in the genome. This implies that, although we should not constrain our maps to be ordered, we should look at genomic context and expect homologs to extend collinear homolog regions. The model suggested here expresses dependencies among locally interacting portions of the genome. It has an affinity towards locally collinear alignments, but allows mappings that contain rearrangements. The output is a mapping between genomic sequences that is not constrained to preserve the order of the sequences if the most likely mapping contains a rearrangement. Additionally, the mapping may not be one-to-one: a single component in one sequence may be mapped non-uniquely to several components in a second sequence, indicating a genomic duplication event. Due to these differences from the traditional meaning of a *sequence alignment*, we use the term *homology mapping* to describe the desired output.

Let us consider a homolog pair as a single entity, representing a single event in which a particular component in an ancestral sequence split and started mutating in two separate lineages. We call a proposed homolog pair a *match*. A match i poses tentative homology between component a in one sequence, $a \in S1$, and component b in another sequence, $b \in S2$. The *components*, a and b, matched by i need not be atomic subunits of the sequence (such as individual base-pairs). Matches could be proposed between regions of genomic sequences. Using the given genomic sequences, we will propose a collection of homolog pairs (matches). We will then need to devise a method of determining which of these proposed matches fall into which category ("homolog" and "not homolog").

We pose the homology mapping problem one in combinatorial optimization. The *homology mapping* task is: given genomic sequences for comparison and a criterion for globally scoring a homology mapping over those sequences, propose possible homolog pairs (matches), and search for the optimal assignment of the labels to each of them. Specifically, we will construct a graph whose nodes are the proposed collection of matches fitting certain criteria (for instance, all 7-mers in which six positions are exactly the same), and use an optimization method to identify those matches that contribute to the best scoring homology mapping overall for the entire length of the sequences.

Our determination of the homology assignment on the graph (the graph consists of both nodes and edges) will be based on the trade-off between two competing forces: we would like to favor good matches based on the sequences we have observed; and, since related genomes are mostly collinear with occasional breakpoints, we would like neighboring collinear matches to receive similar designations of "homolog" or "not homolog". For example, we would like a weak match that is collinear with, and lodged between, two strong collinear matches to increase the likelihood that all three matches get labeled "homolog". To represent the notion of genomic locale, we will introduce edges among match nodes whose components are located within a specified base-pair distance in the genomic sequence of at least one of the two paired components. Edges represent pairwise relationships or dependencies among the matches. The notion of a *local neighborhood* of a node in the graph, is then defined by the set of matches connected to it by an edge.

We attribute weights to edges to indicate the strength of the dependencies among the matches and their affinities to being assigned together the designation of "homolog" or "not homolog". These weights can quantify associations between matches and a particular labeling based on factors like genomic distance, match density, orientation, or whether considering both pairs "homolog" would necessitate a particular genomic event (like inversion, duplication or shuffling).

Our biological intuition motivates the formulation of a graph. We would now like to perform a classification (discrete labeling) task on a set of objects (the nodes) among which there are undirected dependencies (the edges). Such problems have a rich history in statistics, naturally arising in the formalism of *Markov random fields* (MRFs) [Besag, 1974]. MRFs are stochastic models for discrete labeling problems. In MRFs, a random process generates the underlying distribution of labels. The random processes modeled by MRFs capture properties of local dependencies whereby the probability of a specific label (in our case, the probability of a match being "homolog" or "not homolog") is entirely dependent on a local subset of neighboring labels (in our case, a match is dependent on other matches in its immediate genomic context). The framework allows us to define a joint probability density over all the possible assignments on the graph. We will solve this discrete labeling task by finding the assignment that maximizes the probability of that assignment over the entire length of the sequences.

In summary, we assume a common ancestor between two sequences, and treat evolutionary events as part of a stochastic biological process that results in many homolog pairs distributed according to stochastically defined local dependencies. We will employ optimization methods to search the model for the most likely label assignment for each proposed homolog pair.

13.2 Markov random fields

Our goal here is to introduce the log-linear Markov random field model using algebraic notation (for a more general exposition of MRFs in algebraic statistics, we refer to [Geiger *et al.*, 2005]). The model is specified by an undirected graph $\mathcal{G} = (\mathbf{Y}, E)$. A point in the model is a strictly positive probability distribution which factors according to the graph. The set $\mathbf{Y} = \{Y_1, \ldots, Y_N\}$ contains N nodes representing random variables. A node Y_i has an assignment σ_i from a finite alphabet Σ_i with $l_i = |\Sigma_i|$ values or labels. For example, if our nodes are binary random variables (indicating "not homolog" or "homolog"), then $\Sigma_i = \{0, 1\}$, as there are two possible values for each node factor Y_i, and $|\Sigma_i| = l_i = 2$. The state space is the finite product space of all possible assignments, $\mathbb{Y} = \prod_{Y_i \in \mathbf{Y}} \Sigma_i$. In the case of a uniform alphabet of size l for all the nodes in the graph, this state space is comprised of $m = |\Sigma|^N = l^N$ possible assignments.

We turn our attention to the set E of edges in the graph; it is a subset of $\mathbf{Y} \times \mathbf{Y}$, as it does not necessarily include all possible edges on the graph. Each edge in E, denoted e_{ij}, is an undirected edge between some nodes Y_i and Y_j in $\mathbf{Y}$. The edges define neighborhood associations in the form of direct dependencies among nodes. We define the neighborhood of node Y_i as the set of all nodes Y_j to which Y_i is connected by an edge, $N_i = \{Y_j \mid Y_i, Y_j \in \mathbf{Y}, i \neq j, e_{i,j} \in E\}$. The neighborhood of Y_i is sometimes referred to as the *Markov blanket* of Y_i. While Markov blankets describe the web of associations for a particular node, the entire graph could be factorized into subsets of maximally connected subgraphs in $\mathcal{G}$, known as cliques. Cliques vary in size. The set of cliques of size one, $\mathcal{C}_1(\mathcal{G}) = \{Y_i \mid Y_i \in \mathbf{Y}, N_i = \varnothing\}$ is a subset of the nodes in the graph; the set of cliques of size two is a subset of the adjacent nodes (or *pair-matches*): $\mathcal{C}_2(\mathcal{G}) = \{(Y_i, Y_j) \mid Y_j \in N_i, Y_i \in \mathbf{Y}\}$; the set of cliques of size three enumerates triples of neighboring nodes: $\mathcal{C}_3(\mathcal{G}) = \{(Y_i, Y_j, Y_k) \mid Y_i, Y_j, Y_k \in \mathbf{Y}, e_{ij}, e_{jk}, e_{ik} \in E\}$, etc. The collection of all cliques in the graph is $\mathcal{C}(\mathcal{G}) = \mathcal{C}_1(\mathcal{G}) \bigcup \mathcal{C}_2(\mathcal{G}) \bigcup \mathcal{C}_3(\mathcal{G}) \bigcup \cdots$; it decomposes the graph into *factors*. Decomposing a graph in this way is instrumental in understanding dependencies among the nodes, and in performing efficient inference as will be described below.

The random processes modeled by MRFs preserve the Markov property of the model. Specifically, each edge association is a conditional independence statement which the probability distribution must satisfy. Let σ_i denote a particular realization for the node $Y_i \in \mathbf{Y}$. The Markov property states that the probability of a particular labeling for node Y_i (given the rest of the labeled graph nodes) is only conditioned on the local Markov blanket of node Y_i:

$$\mathbf{P}(\sigma_i | \sigma_{\mathbf{Y} \setminus \{Y_i\}}) = \mathbf{P}(\sigma_i | \sigma_{N_i}). \tag{13.1}$$

That is, to assess a node's conditional probability, we only need the specification of its local neighborhood. Additionally, we see that two nodes, Y_i and Y_j, that do *not* share an edge in E must be conditionally independent because neither Y_i nor Y_j are in $\{N_i, N_j\}$ (the set of nodes defining the probability):

$$\mathbf{P}(\sigma_i, \sigma_j | \sigma_{\mathbf{Y} \setminus \{Y_i, Y_j\}}) = \mathbf{P}(\sigma_i | \sigma_{N_i}) \mathbf{P}(\sigma_j | \sigma_{N_j}), \quad \forall i, j \mid e_{ij} \notin E \cdot$$

By definition, every node Y_j *not* in the Markov blanket of Y_i is conditionally independent of Y_i given the other nodes in the graph:

$$Y_i \perp\!\!\!\perp Y_j \mid \mathbf{Y} \setminus \{Y_i, Y_j\}, \quad \forall Y_j \notin N_i, \quad Y_i, Y_j \in \mathbf{Y} \cdot$$

Note that the set of Markov statements applied to each node ($Y_i \in \mathbf{Y}$) is the full set of conditional independence statements $\mathcal{M}_G$, previously introduced in Equation 1.66, necessary and sufficient to specify the undirected graphical model in its quadratic form. In the context of the homology assignment, this property precisely captures our previously stated biological intuition that matches within genomic locales are interacting.

The Hammersley–Clifford theorem (Theorem 1.30) relates a Markov random field to a set of log-linear *potential functions*, one for each maximal clique in the graph, $\{\phi_{c_i} | c_i \in \mathcal{C}(\mathcal{G})\}$. The *potential function* associates a positive value with each possible assignment on the clique. We denote the potential value of a particular instance of a clique by $\phi_{c_i}(\sigma_{c_i})$, where σ_{c_i} is a realization of the clique c_i.

In this class of models, specifying the factorization and the potential function for each factor is tantamount to specifying the joint probability distribution for the Markov random field. Specifically, the Hammersley–Clifford theorem states that given the MRF and its log-linear potential functions, for any arbitrary realization of the graph σ:

$$\mathbf{P}(\sigma) = \frac{1}{Z} \prod_{c_i \in \mathcal{C}(\mathcal{G})} \phi_{c_i}(\sigma_{c_i}), \tag{13.2}$$

where Z is the *partition function* given by $Z = \sum_{\sigma' \in \Sigma^n} \prod_{c_i \in \mathcal{C}(\mathcal{G})} \phi_{c_i}(\sigma'_{c_i})$. We now see why, in the MRF, the maximally connected subgraphs, $\mathcal{C}(\mathcal{G})$, are referred to as the *factors* of the model. We note that the log-linear potential functions impute strict positivity on the joint probability distribution generated by the cliques. Positivity is required of the MRF joint probability for the Hammersley–Clifford theorem to hold.

In the discrete labeling case, the potential function for each clique can be represented by a contingency matrix, θ^{c_i}, of k dimensions (where k is the size of the clique). The matrix θ^{c_i} has an entry for each possible labeling of the clique, and is indexed by a k-tuple, $(\sigma_1, \sigma_2, \ldots \sigma_k)$, where $\sigma_j \in \Sigma_j$ (Σ_j is the alphabet from which node j takes its value).

Since the joint probability function (13.2) is only dependent on the potential functions and the node assignments in σ, we can equivalently parameterize the model by the full collection of contingency matrices: $\theta = \{\theta^{c_i} | c_i \in \mathcal{C}(\mathcal{G})\}$. We can index each component of the parameter vector θ. For example, $(\theta^{c_i}_{111})$ is the positive value associated with assigning (c_i) the realization 111. In a Markov random field whose nodes take value assignments from a uniform alphabet, Σ, such that $|\Sigma| = l$, the number of parameters in the graph is

$$d = \sum_{c_i \in \mathcal{C}(\mathcal{G})} l^{|c_i|}.$$

The cliques in the factorized graph define d positive model parameters $\theta = \{\theta_1, \ldots, \theta_d\}$. Given θ, the probability of a particular realization of the model σ is defined to be proportional to the product of only those entries in the contingency matrices that correspond to σ,

$$\mathbf{P}_\theta(\sigma) = \frac{1}{Z} \prod_{j=1}^{d} \theta_j^{c_i} \cdot I(\sigma_{c_i} = j), \tag{13.3}$$

where $I(\cdot)$ is the boolean indicator function, and Z is the partition function.

Expression (13.3) fully characterizes a class of *undirected graphical models* with log-linear parameter values. We can define an associated monomial mapping in θ for each assignment σ in the finite state space Σ_i^n:

$$f_\sigma(\theta) = \mathbf{P}_\theta(\sigma) = \frac{1}{Z} \prod_{j=1}^{d} \theta_j^{c_i} \cdot I(\sigma_{c_i} = j).$$

Recall that θ is indexed by a clique factor and its realization, and $I(\cdot)$ is the boolean indicator function. Hence, the degree of each parameter in θ is determined by the realization of the associated factors in the assignment. Since every factor must take on one and only one label, the degree of all m monomials associated with a valid assignment is the same. Pooling all such monomials into a single map $\mathbf{f}$, we have specified a toric model parametrically (for the definition of a *toric model* see Section 1.2):

$$\mathbf{f} : \mathbb{R}^d \to \mathbb{R}^m, \quad \theta \mapsto \frac{1}{\sum_{j=1}^{m} f_j(\theta)} \cdot \left(f_1(\theta), f_2(\theta), \ldots, f_m(\theta) \right). \tag{13.4}$$

We can express each of the monomials, f_i, as a column vector, thereby constructing a $d \times m$ integer design matrix, A. As before, the toric model of A is the image of the orthant $\mathbb{R}^d_{>0}$ under the map:

$$\mathbf{f} : \mathbb{R}^d \to \mathbb{R}^m, \theta \mapsto \frac{1}{\sum_{j=1}^{m} \theta^{a_j}} (\theta^{a_1}, \theta^{a_2}, \ldots, \theta^{a_m})$$

where $\sum_{j=1}^{m} \theta^{a_j}$ is the partition function. By construction, the joint probability distribution, $\mathbf{P}$, is in the image of the mapping $\mathbf{f}$, and we say that $\mathbf{P}$

factors according to the model A. Once again we note the connection to the Hammersley–Clifford theorem since a toric model specified by A coincides with any log-linear MRF as defined above.

13.3 MRFs in homology assignment

Example 13.1 We return to the homology assignment problem and illustrate the application of the MRF model with a simple example. We are given two sequences, $S1$ and $S2$. Our task is to construct a Markov random field in order to find a homology map between the two sequences.

```
Position:      1 2 3 4 5 6 7 8 9 10 11 12 13 14 15 16 17 18 19 20 21 22 23 24 25 26 27 28 29 30

Seq S1:   A A G A C C G C T T G A C T C G G A A A A G G G C T C

Seq S2:   C C G C T A A G A C T C T A T A T A T A G G C T C C C G C C
```

Let us propose a match for every pair of perfectly matching 5-mers in the sequences. We denote a match node as $Y_i = (S1_j, S2_k)$, where i enumerates the nodes, and (j, k) are the indices pointing to the center of the 5-mer in the sequences $S1$ and $S2$, respectively. Let us define an edge between any two proposed matches whose center indices are within 14 base-pairs of one another in at least one of the two sequences. There are four matches for the given sequences. Let $Y_1 = (S1_3, S2_8)$, $Y_2 = (S1_{13}, S2_{10})$, $Y_3 = (S1_7, S2_3)$, $Y_4 = (S1_{26}, S2_{23})$. The full node set is $\mathbf{Y} = \{Y_1, Y_2, Y_3, Y_4\}$, as depicted in Figure 13.1. The graph has four nodes, each taking its values from $\Sigma = \{0, 1\}$, and has a finite state space with $m = 2^4 = 16$ possible outcomes. We introduce four edges among matches whose centers are within 14 base-pairs of one-another: $E = \{e_{\{1,2\}}, e_{\{2,3\}}, e_{\{1,3\}}, e_{\{2,4\}}\}$. Then $\mathcal{G}(\mathbf{Y}, E)$ is a four-node, four-edge graph, as depicted in Figure 13.2.

Note the representation of the local context dependencies among matches in our graph. Match nodes whose assignments are likely to influence each other (those within a specified base-pair distance), are connected by an edge. Each

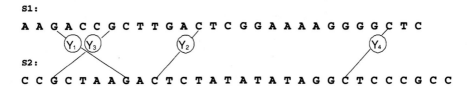

Fig. 13.1. In the given sequences, S_1 and S_2, match nodes are proposed between every two perfectly matching 5-mers.

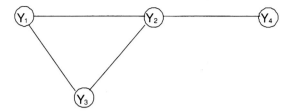

Fig. 13.2. Graph of a Markov random field with two maximal cliques.

node has its own locality, or neighborhood, on the graph:

$$
\begin{aligned}
N_1 &= \{Y_2, Y_3\} \\
N_2 &= \{Y_1, Y_3, Y_4\} \\
N_3 &= \{Y_1, Y_2\} \\
N_4 &= \{Y_2\}
\end{aligned}
$$

This induces a factorization on the graph of two maximally connected cliques, one of size three ($c_1 = \{Y_1, Y_2, Y_3\}$), and another of size two ($c_2 = \{Y_2, Y_4\}$). The complete set of clique factors in the graph is

$$
\mathcal{C}(\mathcal{G}) = \{c_1, c_2\} = \{\{Y_1, Y_2, Y_3\}, \{Y_2, Y_4\}\}.
$$

We parameterize the model with 12 parameters: $d = \sum_{c_i \in \mathcal{C}(\mathcal{G})} l^{|c_i|} = 2^{|c_1|} + 2^{|c_2|} = 2^3 + 2^2 = 12$. The parameter space consists of the contingency matrices for each clique in the graph. Each contingency matrix associates a positive value with each possible assignment on the nodes in the cliques. The parameter matrix contributed by the two-node clique c_2 is denoted θ^{c_2} and is a point in $\mathbb{R}^{2 \times 2}_{>0}$, the space of 2×2 matrices whose four entries are positive. Similarly, the parameters contributed by the three-node clique c_1 is denoted θ^{c_1} and is a point in $\mathbb{R}^{2 \times 2 \times 2}_{>0}$, the space of $2 \times 2 \times 2$ matrices whose eight entries are positive. The model parameter space $\Theta \subset \mathbb{R}^{(2 \times 2 \times 2) + (2 \times 2)}$ consists of all matrices θ whose twelve entries t_i are positive. We associate the clique parameters with the model parameters as follows:

$$
\begin{array}{cccccccc}
t_1 & t_2 & t_3 & t_4 & t_5 & t_6 & t_7 & t_8
\end{array}
$$
$$
\theta^{c_1} = \begin{pmatrix} \theta^{c_1}_{000} & \theta^{c_1}_{001} & \theta^{c_1}_{010} & \theta^{c_1}_{011} & \theta^{c_1}_{100} & \theta^{c_1}_{101} & \theta^{c_1}_{110} & \theta^{c_1}_{111} \end{pmatrix}
$$

$$
\begin{array}{cccc}
t_9 & t_{10} & t_{11} & t_{12}
\end{array}
$$
$$
\theta^{c_2} = \begin{pmatrix} \theta^{c_2}_{00} & \theta^{c_2}_{01} & \theta^{c_2}_{10} & \theta^{c_2}_{11} \end{pmatrix}.
$$

Using this factorization, we can fully characterize the conditional independence relationships among the match nodes. Specifically, we have

$$
\begin{aligned}
P(\sigma_1 | \sigma_2, \sigma_3, \sigma_4) &= P(\sigma_1 | \sigma_2, \sigma_3) \\
P(\sigma_2 | \sigma_1, \sigma_3, \sigma_4) &= P(\sigma_2 | \sigma_1, \sigma_3, \sigma_4) \\
P(\sigma_3 | \sigma_1, \sigma_2, \sigma_4) &= P(\sigma_3 | \sigma_1, \sigma_2) \\
P(\sigma_4 | \sigma_1, \sigma_2, \sigma_3) &= P(\sigma_4 | \sigma_2).
\end{aligned}
$$

Only 2 node pairs are not connected by an edge; the corresponding model is

$$\mathcal{M}_\mathcal{G} = \{\ Y_1 \perp\!\!\!\perp Y_4 \,|\, \{Y_2, Y_3\}, \quad Y_3 \perp\!\!\!\perp Y_4 \,|\, \{Y_1, Y_2\}\ \}.$$

Each such conditional independence statement can be translated into a system of quadratic polynomials in $\mathbb{R}^{\{Y_1, Y_2, Y_3, Y_4\}}$. The representation of independence statements as polynomial equations is an implicit representation of the toric model, where the common zero set of the polynomials represents the model.

For binary alphabets Σ, the following eight quadric forms compose the set $\mathcal{Q}_{\mathcal{M}_\mathcal{G}}$ representing the probability distribution from the example above.

$$p_{0001}p_{1000} - p_{0000}p_{1001}, \ p_{0001}p_{0010} - p_{0000}p_{0011},$$
$$p_{0011}p_{1010} - p_{0010}p_{1011}, \ p_{0101}p_{0110} - p_{0100}p_{0111},$$
$$p_{0101}p_{1100} - p_{0100}p_{1101}, \ p_{1001}p_{1010} - p_{1000}p_{1011},$$
$$p_{0111}p_{1110} - p_{0110}p_{1111}, \ p_{1101}p_{1110} - p_{1100}p_{1111}.$$

This set specifies a model which is the intersection of the zero set of the invariants with the 15-dimensional probability simplex Δ with coordinates $p_{\sigma_1 \sigma_2 \sigma_3 \sigma_4}$. The non-negative points on the variety defined by this set of polynomials represent probability distributions which satisfy the conditional independence statements in $\mathcal{M}_\mathcal{G}$.

The same quadric polynomials can also be determined by the vectors in the kernel of the parametric model matrix $A_\mathcal{G}$. Each quadric form corresponds to a vector in the kernel of $A_\mathcal{G}$. The columns of $A_\mathcal{G}$ are indexed by the m states in the state space, and the rows represent the potential values, indexed by the d model parameters, with a separate potential value for each possible assignment on the clique. Each column in $A_\mathcal{G}$ represents a possible assignment of the graph:

	0000	0001	0010	0011	0100	0101	0110	0111	1000	1001	1010	1011	1100	1101	1110	1111
000·	1	1	0	0	0	0	0	0	0	0	0	0	0	0	0	0
001·	0	0	1	1	0	0	0	0	0	0	0	0	0	0	0	0
010·	0	0	0	0	1	1	0	0	0	0	0	0	0	0	0	0
011·	0	0	0	0	0	0	1	1	0	0	0	0	0	0	0	0
100·	0	0	0	0	0	0	0	0	1	1	0	0	0	0	0	0
101·	0	0	0	0	0	0	0	0	0	0	1	1	0	0	0	0
110·	0	0	0	0	0	0	0	0	0	0	0	0	1	1	0	0
111·	0	0	0	0	0	0	0	0	0	0	0	0	0	0	1	1
·0·0	1	0	1	0	0	0	0	0	1	0	1	0	0	0	0	0
·0·1	0	1	0	1	0	0	0	0	0	1	0	1	0	0	0	0
·1·0	0	0	0	0	1	0	1	0	0	0	0	0	1	0	1	0
·1·1	0	0	0	0	0	1	0	1	0	0	0	0	0	1	0	1

The geometry of the model is embedded in the $(m - 1)$-simplex. Each of the m columns of $A_{l,n}$ represents a distinct point in the d-dimensional space. The convex hull of these points define the polytope $f_{l,n}(\theta)$. Referring back to the toric Markov chain of Chapter 1, we saw that the map $f_{l,n}$ was a k-dimensional object inside the $(m-1)$-dimensional simplex Δ which consisted of all probability distributions in the state space l^n. The dimension k of the polytope is dependent on the cliques in the model.

The equivalence of the three representations of the model: the factorized closed-form joint probability distribution, the implicit representation of the toric model, and the model matrix A_G for log-linear MRF distributions, is guaranteed by the Hammersley–Clifford theorem. In general, however, without the constraint on strictly positive probability density functions, the class of toric models is larger than the log-linear MRF model. □

13.4 Tractable MAP inference in a subclass of MRFs

We return to the homology assignment problem of finding the optimal factor assignment for **Y**. In the case we presented, we have a binary assignment, designating {"homolog", "not homolog"}. We observe biological sequences for which we want a homology assignment, and then estimate the graph topology, conditional dependencies among its factors and potential functions on the graph (the θ vector) as in the example in Section 13.3. We look for the binary assignment to the factors in **Y** that is *maximum A posteriori* (MAP) labeling given the data and the underlying MRF model. We have already described the posterior probability function as a product of the potential functions. We define this to be our optimality criterion and continue to pursue an alternative formulation that is linear in the model parameters.

In Equation 13.3 we expressed $\mathbf{P}_\theta(\sigma)$, the probability of the assignment given the model parameters. The probability of observing σ is the product of only those parameters that correspond to the realization in σ of the maximal cliques on the graph normalized by the partition function. The partition function enumerates ($m = |\Sigma_i|^N$) possible assignments, and despite the model factorizing as the graph, calculating the joint probability is computationally non-trivial. Though mechanisms for estimating the partition function exist, we are more interested in exact tractable MAP solutions.

We can simplify the MAP inference computation by transforming the problem to logarithmic coordinates, where the joint probability density calculation for every assignment becomes a linear sum:

$$-\log(\mathbf{P}_\theta(\sigma)) = -\sum_{c_i \in \mathcal{C}(\mathcal{G})} \log(\theta_{\sigma'_{c_i}}^{c_i}) \cdot I(\sigma'_{c_i} = \sigma_{c_i}) + \log(Z) \cdot$$

This function is called the *energy function* [Boykov *et al.*, 1999] and the MAP inference problem is to minimize it. The log of the partition function, $\log(Z)$, is the same for each possible labeling, and can therefore be ignored for the purpose of optimization. We are left with the first term which was given by the Hammersley–Clifford expansion in terms of the maximal cliques in the graph. This MAP inference problem can be expressed as an integer linear program in which we will have to maximize the negation of this linear term over a variant of the original model polytope. The new polytope is calculated from the m columns of the matrix $\log(A_{\mathcal{G}})$. Computing a vertex on the convex hull of this matrix is the same as tropicalizing the partition function. This reduces the problem to calculating the convex hull of $\log(\mathbf{P}_\theta(\sigma))$ and finding the optimal vertex. Tropicalizing the partition function is intractable beyond small sized problems because the computation is exponential in the number of nodes on the graph (for a more detailed discussion, see Chapter 9). This makes the computation of the joint probability of the graph very difficult. There exists a subclass of log-linear models for which efficient MAP inference exists since the non-integer linear programming relaxation still yields the exact MAP inference solution. This subclass of MRF models is very relevant to the homology assignment problem and other discrete labeling problems.

For many integer and combinatorial optimization problems, some very successful approximation algorithms are based on linear relaxations. In the general case, the solutions to the relaxed programs produce approximations to the desired integer solution [Besag, 1986]. We consider a certain subclass of the log-linear models called *ferromagnetic Ising models*. For the binary labeling case of this model, solving an appropriately constructed linear program gives an exact solution to the MAP inference problem [Greig *et al.*, 1989, Kolmogorov and Zabih, 2003]. Models of this subclass encode situations in which locally related variables (nodes in the same Markov blanket) tend to have the same labeling. This type of value assignment is known as *guilt by association*. In the context of the homology assignment, this is a way of representing an underlying evolutionary process in which homolog matches are related objects whose homology assignment should be consistent with other local matches. This attractive assignment is important in many domains in which different labels have structures of affinities within local neighborhoods. In our formulation, different labels can have different affinities. We will now make precise the restrictions applied to this subclass of MRFs.

(i) *General m-additive interactions.* To avoid having to specify the clique potentials $\phi_{c_i}(\sigma_{c_i})$ for every maximal clique in the graph, we define the maximal clique interaction size to be m, and allow non-zero potential functions only for cliques of size $(1, \ldots, m)$. The maximal cliques on the graph are then defined as some linear combination of the non-zero clique potentials.

(ii) *Attractive potentials.* The contingency matrices defining the clique potentials are restricted to have all the off-diagonal terms identically one, and all the diagonal terms greater than or equal to one. For example, the potential functions for all pair-cliques in $\mathcal{C}_2(\mathcal{G})$ have $\phi_{ij}(k, k) \geq 1$, and $\phi_{ij}(k, l) = 1$, $\forall k \neq l$.

We define our restricted models to be MRFs that satisfy the above two properties, and assume a strictly binary labeling (i.e., $\Sigma = \{0, 1\}$). Those familiar with computer vision literature will recognize that this extends the *Potts model* by admitting non-uniform attractive potentials for different labels, and by allowing non-pairwise interactions between variables, with potentials over cliques of size m [Ferrari *et al.*, 1995]. This class of models is different from the *generalized Potts model* since it has no penalty for assignments that do not have the same label across edges in the graph [Boykov *et al.*, 1999].

The second condition captures some of the previously stated desirable properties for solving the homology assignment problem. When we take the logarithm of the potential terms, the off-diagonal terms vanish, and the diagonal terms are strictly positive. Recall that the diagonal terms are those which assign the same label to all the nodes in the same maximal clique. This explains the term *guilt by association*, since being in the same clique carries a positive potential value associated with being assigned the same value. Additionally, the fact that this property of the model is not restricted to pairwise interactions allows us to encode patterns of transitivity in the homology assignment domain. For example, consider the situation presented above in which we wanted a weak match that is collinear with, and lodged between, two strong collinear matches to increase the likelihood for all three matches to receive the designation of "homolog". The transitivity among the three could be modeled by a ternary clique with a high $\theta_{ijk}(1,1,1)$ for the assignment with all three matches "homolog".

To formulate the integer linear program for the MAP inference task, we will take, for example, the restricted models with the value $m = 2$ (that is, with maximally pairwise interactions). We begin with the set of nodes $\mathbf{Y}$ containing N. We define the variables to be the assignments to the nodes y_i^k as well as to the pairwise edges $y_{(i,j)}^k$. Note that a single node has two variables associated with it and each edge also has two variables associated with it (since we only consider the diagonal terms in the contingency matrix). All variables are restricted to the binary alphabet $\Sigma = \{0, 1\}$, and must satisfy linear normalization and agreement constraints. The optimization criterion is the log of the unnormalized probability of the assignment σ. In the non-integer relaxation, the linear objective function remains the same. The program is:

$$\max \sum_{i=1}^{N} \sum_{k=\{0,1\}} (\log(\theta_k^i) y_i^k) + \sum_{(ij)\in E} \sum_{k=\{0,1\}} (\log(\theta_k^{(i,j)}) y_{(i,j)}^k) \qquad (13.5)$$

$$s.t. \ y_i^k \geq 0, \quad \forall i, k; \qquad \sum_{k=\{0,1\}} y_i^k = 1, \quad \forall i;$$

$$y_{(i,j)}^k \leq y_i^k, \qquad y_{(i,j)}^k \leq y_j^k, \quad \forall (i,j) \in E, k.$$

We merely lifted the integrality constraint and exchanged the consistency constraint $y_{(i,j)}^k = y_i^k \wedge y_j^k$ with two linear conditions $y_{(i,j)}^k \leq y_i^k$ and $y_{(i,j)}^k \leq y_j^k$. This makes sense since $(\log(\theta_{k,k}^{(i,j)}))$ is positive and we are maximizing the objective function. Hence, at the maximum, $y_{(i,j)}^k = \min(y_i^k, y_j^k)$, which is equivalent to $y_{(i,j)}^k = y_i^k \wedge y_j^k$.

The implicitization of the model in the non-integral relaxation has the same quadratic polynomials (i.e., the conditional independence statements persist).

However, we must introduce new polynomials that ensure that a particular value assignment, $(\sigma_{c_i}) = k$, on a clique is only possible when all the component nodes $j \in \{c_i\}$ are assigned the same value, k. The new model is represented by the non-negative set of the added quadratic polynomials (a set of inequalities) rather than the common zero set of all the polynomials as before. These new constraints did not add points to the model outside the convex hull of the original toric model. Hence the polytope remains the same, only the linear formulation incorporates more than just the integer lattice vertices in the set of feasible solutions. Importantly, it has been shown that in the binary labeling case, when a solution exists, the non-integer linear relaxation is guaranteed to produce the integral optimal solution [Kolmogorov and Zabih, 2003]. In practice, this problem conveniently reduces to a problem known as graph min-cut, which can be solved exactly in polynomial time only in the number of nodes on the graph.

This result states that the MAP problem in this restricted class of binary models is tractable, regardless of network topology or clique size. The size of relaxed linear program inference grows only linearly with the number and size of the cliques. This is an encouraging result: for a specific subclass of toric models, MAP inference is tractable.

13.5 The Cystic Fibrosis Transmembrane Regulator

The *greater Cystic Fibrosis Transmembrane Regulator region dataset* (abbreviated CFTR) is a DNA dataset of 12 megabases (Mb) of high-quality sequences from 12 vertebrate genomes [Thomas *et al.*, 2003]. The sequences target a genomic region orthologous to a segment of about 1.8 Mb on human chromosome 7, which encodes 10 genes. One of the genes encodes CFTR, which is the gene mutated in cystic fibrosis.

The original comparative study successfully identified 98% of the exons as well as many conserved non-coding sequences. The gene number and order were found to be mostly conserved across the 12 species, with a strikingly variable amount of interspersed repeats. Additionally, the authors identified three insertions (transposons) that are shared between primates and rodents, which confirms the close relationship of the two lineages and highlights the opportunity such data provides for refining species phylogenies and characterizing the evolutionary process of genomes.

To demonstrate the use of our method, we took the CFTR region from four currently sequenced vertebrate genomes: human, chicken, rat and mouse.

In practice, our method delineates the *sequence matching* aspect (see Chapter 7) from the *homology mapping* aspects of alignment. To produce homolog mappings among multiple sequences, we proceed as follows:

(i) Identify matches (not necessarily identical) between sequences in pairwise comparisons. These matches may be pairs of single base pairs, larger BLAT hits [Kent, 2002], exons, or even complete genes.

(ii) Construct the constrained Ising model based on the matches (as above).

(iii) Find the MAP assignment using linear program solver.

(iv) Output those nodes that were assigned the value of *"homolog"*.

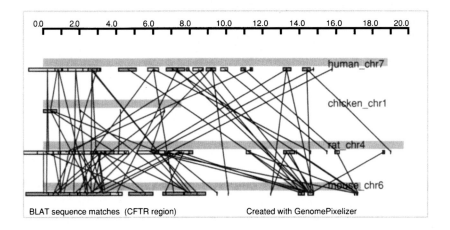

Fig. 13.3. BLAT output

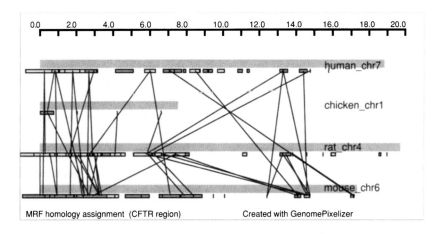

Fig. 13.4. Markov random field MAP output

Figure 13.3 displays the CFTR dataset for the four selected genomes. Horizontal shaded bars represent the CFTR region in a fully sequenced chromosome. The lines between chromosome bars connecting small segments represent BLAT output matches.

Our method takes these matches as nodes, and constructs a Markov network from them. In this particular instance, there are 169 nodes and our network has nearly six thousand edges. On our single processor, 2 GB RAM machine it took 12.18 seconds to solve the corresponding linear program formulation using CPLEX. The results (that is, the nodes that were assigned a value of 1, designated as "homolog") are depicted in Figure 13.4.

We have performed this operation on much larger sets of data, with networks as large as three hundred thousand nodes, and two million edges. Solution time scales polynomially with the number of nodes.

14

Mutagenetic Tree Models

Niko Beerenwinkel

Mathias Drton

Mutagenetic trees are a class of graphical models designed for accumulative evolutionary processes. The state spaces of these models form finite distributive lattices. Using this combinatorial structure, we determine the algebraic invariants of mutagenetic trees. We further discuss the geometry of mixture models. In particular, models resulting from mixing a single tree with an error model are shown to be identifiable.

14.1 Accumulative evolutionary processes

Some evolutionary processes can be described as the accumulation of non-reversible genetic changes. For example, the process of tumor development of several cancer types starts from the set of complete chromosomes and is characterized by the subsequent accumulation of chromosomal gains and losses, or by losses of heterozygosity [Vogelstein *et al.*, 1988, Zang, 2001]. Mutagenetic trees, sometimes also called *oncogenetic trees*, have been applied to model tumor development in patients with different types of cancer, such as renal cancer [Desper *et al.*, 1999, von Heydebreck *et al.*, 2004], melanoma [Radmacher *et al.*, 2001] and ovarian adenocarcinoma [Simon *et al.*, 2000]. For glioblastoma and prostate cancer, tumor progression along the mutagenetic tree has been shown to be an independent cytogenetic marker of patient survival [Rahnenführer *et al.*, 2005].

Amino acid substitutions in proteins may also be modeled as permanent under certain conditions, such as a very strong selective pressure. For example, the evolution of human immunodeficiency virus (HIV) under antiviral drug therapy exhibits this behavior. Mixtures of mutagenetic trees have been used to model the accumulation of drug resistance-conferring mutations of HIV [Beerenwinkel *et al.*, 2004]. This modeling approach has revealed different evolutionary pathways the virus can take to become resistant, which is important for the design of effective therapeutic protocols [Beerenwinkel *et al.*, 2005a].

In general, we consider non-reversible clonal evolutionary processes on a finite set of genetic events. Mutagenetic trees aim at modeling the order and rate of occurrence of these events. A software package for statistical inference with these models is described in [Beerenwinkel *et al.*, 2005b].

14.2 Mutagenetic trees

Consider n binary random variables $X_1, \ldots, X_n$ each indicating the occurrence of an event. We will represent an observation of $X := (X_1, \ldots, X_n)$ as a binary vector $i = (i_1, \ldots, i_n) \in \mathcal{I} := \{0, 1\}^n$, but sometimes use the equivalent representation by the subset $S_i \subset [n] = \{1, \ldots, n\}$ of occurred events, i.e. $S_i = \{v \in [n] \mid i_v = 1\}$. For a subset $A \subset [n]$, we denote by $X_A = (X_v)_{v \in A}$ the corresponding subvector of random variables taking values in $\mathcal{I}_A := \{0, 1\}^A$.

A *mutagenetic tree* T on n events is a connected branching on the set of nodes $V = V(T) = \{0\} \cup [n]$, rooted at node 0. The set of edges in T is denoted by $E(T)$. Since the set of connected rooted branchings on V is in one-to-one correspondence with the set of undirected labeled trees on $n+1$ nodes, Cayley's theorem [Stanley, 1999, Prop. 5.3.2] implies that there are $(n+1)^{n-1}$ different mutagenetic trees on n events. The $4^2 = 16$ mutagenetic trees on $n = 3$ events may be grouped into four classes according to their tree topology (Figure 14.1). For any subset $V' \subset V$, we denote by $T_{V'}$ the induced subgraph of the mutagenetic tree T. In particular, each state $i \in \mathcal{I}$ induces a subgraph $T_i := T_{S_i}$. A subbranching of T is a directed subtree $T_{V'}$ with the same root node 0. Every node $v \in [n]$ has exactly one entering edge $(u, v) \in E(T)$. We call u the *parent* of v, denoted $\mathrm{pa}(v) = u$. For $V' \subset V$, $\mathrm{pa}(V')$ is the vector $(\mathrm{pa}(v))_{v \in V'}$. Finally, the *children* of $u \in V(T)$, denoted by $\mathrm{ch}(u)$, are the nodes $v \in V(T)$ such that $(u, v) \in E(T)$.

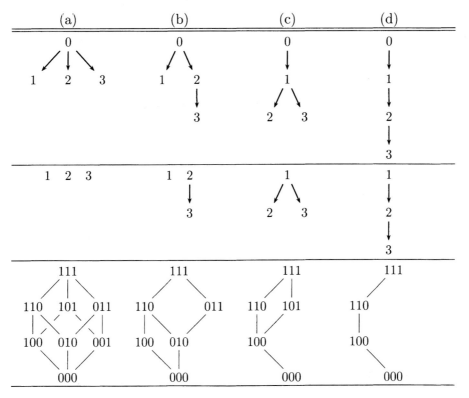

Fig. 14.1. Mutagenetic trees for $n = 3$ events (first row), their induced directed forests (second row) and lattices of compatible states (third row).

A mutagenetic tree T defines a statistical model as follows. With each edge $(\mathrm{pa}(v), v)$, $v \in [n]$, we associate a parameter $\theta_{11}^v \in [0,1]$ and the transition matrix

$$\theta^v = \theta^{\mathrm{pa}(v),v} = \begin{pmatrix} 1 & 0 \\ 1 - \theta_{11}^v & \theta_{11}^v \end{pmatrix}.$$

We will interpret the entry (a,b), $0 \le a, b \le 1$, of this matrix as the conditional probability $\Pr(X_v = b \mid X_{\mathrm{pa}(v)} = a)$. Since $X_v = 1$ indicates the occurrence of the vth genetic event, the first row of the matrix θ^v reflects our assumption that an event in the underlying evolutionary process can only occur if its parent event has already occurred.

Let $\Theta = [0,1]^n$, and let $\Delta = \Delta_{2^n - 1}$ be the $(2^n - 1)$-dimensional probability simplex in $\mathbb{R}^{2^n}$. Let $\theta = (\theta_{11}^1, \dots, \theta_{11}^n)$ and consider the polynomial map

$$\mathbf{f}^{(T)} : \Theta \to \Delta, \quad \theta \mapsto (f_i(\theta))_{i \in \mathcal{I}}, \quad \text{defined by } f_i(\theta) = \prod_{v=1}^n \theta_{i_{\mathrm{pa}(v)}, i_v}^v, \quad (14.1)$$

where we set $i_0 = 1$; compare (1.56).

Definition 14.1 The n-dimensional *mutagenetic tree model* $\mathcal{T} := \mathbf{f}^{(T)}(\Theta) \subset \Delta$ is a submodel of the fully observed tree model defined in Section 1.4, with the root distribution concentrated on $i_0 = 1$. This algebraic statistical model has parameter space Θ and state space $\mathcal{I}$.

In the model $\mathcal{T}$ genetic events are non-reversible and they can occur only if all of their ancestor events have already occurred. This constraint forces some of the polynomials f_i to be zero, which can be seen from the representation

$$f_i(\theta) = \left[\prod_{v: \mathrm{pa}(v) \in S_i \cup \{0\}} (\theta_{11}^v)^{i_v} (1 - \theta_{11}^v)^{1 - i_v} \right] \times \left[\prod_{v: \mathrm{pa}(v) \notin S_i} 0^{i_v} 1^{1 - i_v} \right]. \quad (14.2)$$

Definition 14.2 Let $\Gamma \subset \Delta$ be a statistical model with state space $\mathcal{I}$. A state $i \in \mathcal{I}$ is *compatible* with Γ if there exists $p \in \Gamma$ with $p_i > 0$. Otherwise, i is *incompatible* with Γ. The set of all states compatible with Γ is denoted $\mathcal{C}(\Gamma)$.

Example 14.3 The states 001 and 101 are incompatible with the model $\mathcal{T}$ induced by the tree T in Figure 14.1(b). The non-zero coordinates of $\mathbf{f}^{(T)}$ are

$$\begin{aligned} f_{000}(\theta) &= (1 - \theta_{11}^1)(1 - \theta_{11}^2), & f_{100}(\theta) &= \theta_{11}^1(1 - \theta_{11}^2), \\ f_{010}(\theta) &= (1 - \theta_{11}^1)\theta_{11}^2(1 - \theta_{11}^3), & f_{110}(\theta) &= \theta_{11}^1\theta_{11}^2(1 - \theta_{11}^3), \\ f_{011}(\theta) &= (1 - \theta_{11}^1)\theta_{11}^2\theta_{11}^3, & f_{111}(\theta) &= \theta_{11}^1\theta_{11}^2\theta_{11}^3. \end{aligned}$$

$\square$

Lemma 14.4 *Let T be a mutagenetic tree. For a state $i \in \mathcal{I}$ the following are equivalent:*

(i) $i \in \mathcal{C}(\mathcal{T})$, *i.e. i is compatible with $\mathcal{T}$,*

(ii) *for all $v \in S_i$, $\mathrm{pa}(v) \in S_i \cup \{0\}$,*

(iii) *T_i is a subbranching of T.*

Hence, all states $i \in \mathcal{I}$ are compatible with T if and only if T is the star, i.e. $\mathrm{pa}(T) = 0_{[n]}$. In this case the induced model is the model of complete independence of all events and we refer to it as the star *model, denoted $\mathcal{S}$.*

Proof Immediate from (14.2); see also [Beerenwinkel *et al.*, 2004]. □

The compatible states $\mathcal{C}(T)$ can be constructed from the topology of the mutagenetic tree T.

Algorithm 14.5 (Generation of compatible states)
Input: A mutagenetic tree T.
Output: The set $\mathcal{C}(T)$ of states compatible with T.
Step 1: Sort the nodes of T in any reverse topological order $v_n, v_{n-1}, \ldots, v_1, 0$. (For example, use the reverse breadth-first-search order.)
Step 2: For $v = v_n, v_{n-1}, \ldots, v_1$:
Let $T(v)$ be the subtree of T rooted at v and set

$$
\mathcal{C}_v = \begin{cases} \{0, 1\} & \text{if } v \text{ is a leaf,} \\ \{0_{V(T(v))}\} \cup \left(\{1_v\} \times \prod_{u \in \mathrm{ch}(v)} \mathcal{C}_u\right) & \text{else.} \end{cases}
$$

Step 3: Return the Cartesian product $\mathcal{C}(T) = \prod_{u \in \mathrm{ch}(0)} \mathcal{C}_u$.

The correctness of the algorithm follows from the fact that the states in $\mathcal{I}_{T(v)}$ compatible with $T(v)$ are $(0, \ldots, 0)$ and those states that arise as free combinations of all substates compatible with the subtree models $T(u)$ for each $u \in \mathrm{ch}(v)$ and that have a one in their vth component.

The following Theorem 14.6 connects mutagenetic tree models to directed graphical models (Bayesian networks) and yields, in particular, that maximum likelihood estimates of the parameters θ^v_{11} are rational functions of the data [Lauritzen, 1996, Thm. 4.36]. The theorem is a consequence of the model defining equation (14.1) and Theorem 1.33 and Remark 1.34. The second row of Figure 14.1 illustrates the theorem.

Theorem 14.6 *Let T be a mutagenetic tree and $p \in \Delta$ a probability distribution. Then $p \in T$ if and only if p is in the directed graphical model based on the induced directed forest $T_{[n]}$, and $p_i = 0$ for all incompatible states $i \notin \mathcal{C}(T)$. In particular, if $X = (X_1, \ldots, X_n)$ is distributed according to $\mathbf{f}^{(T)}(\theta)$, $\theta \in \Theta$, i.e. if*

$$
\mathrm{Prob}(X = i) = f_i(\theta), \qquad \forall\, i \in \mathcal{I},
$$

then

$$
\theta^v_{11} = \begin{cases} \mathrm{Prob}(X_v = 1) & \text{if } \mathrm{pa}(v) = 0, \\ \mathrm{Prob}(X_v = 1 \mid X_{\mathrm{pa}(v)} = 1) & \text{otherwise.} \end{cases}
$$

The theorem justifies the term "transition matrix" used earlier for θ^v.

14.3 Algebraic invariants

The power set of $[n]$, which can be written as $\{S_i\}_{i \in \mathcal{I}}$, forms a poset ordered by inclusion. Moreover, $(\{S_i\}_{i \in I}, \cup, \cap)$ is a finite distributive lattice. The corresponding join and meet operations in $\mathcal{I}$ are

$$i \vee j := \big(\max(i_v, j_v)\big)_{v \in [n]} = i_{S_i \cup S_j} \in \mathcal{I} \quad \text{and}$$

$$i \wedge j := \big(\min(i_v, j_v)\big)_{v \in [n]} = i_{S_i \cap S_j} \in \mathcal{I}, \quad i, j \in \mathcal{I},$$

and we will subsequently work with the isomorphic lattice $(\mathcal{I}, \vee, \wedge)$.

Lemma 14.7 *For any mutagenetic tree T, the compatible states $(\mathcal{C}(T), \vee, \wedge)$ form a sublattice of $\mathcal{I}$.*

Proof We need to verify that $i, j \in \mathcal{C}(T)$ implies that both $i \vee j$ and $i \wedge j$ are elements of $\mathcal{C}(T)$. This follows from Lemma 14.4 and the fact that $0 \in V(T_i)$ for all $i \in \mathcal{I}$. □

The third row of Figure 14.1 shows the lattices of compatible states for the respective mutagenetic trees.

Consider now the polynomial ring $R := \mathbb{Q}[p_i, i \in \mathcal{I}]$ whose unknowns are indexed by the states $i \in \mathcal{I}$. The lexicographic order of these binary vectors induces the order

$$p_{0...0000} \succ p_{0...0001} \succ p_{0...0010} \succ p_{0...0011} \succ p_{0...0100} \succ \cdots \succ p_{1...1111}$$

among the ring variables p_i, $i \in \mathcal{I}$. With this order of the indeterminates we use a monomial order "$\succ$" in R that selects the underlined terms in the set

$$\mathcal{G}_T = \{\underline{p_i p_j} - p_{i \vee j} p_{i \wedge j}, \ i, j \in \mathcal{C}(T), \ (i \wedge j) < i < j < (i \vee j)\} \qquad (14.3)$$

as the leading monomials, for example the degree-reverse lexicographic order. This monomial order is fixed for the rest of this chapter and underlies all subsequent results.

Let

$$\mathcal{L}_T := \{p_i \mid i \notin \mathcal{C}(T)\}$$

be the set of unknowns corresponding to the incompatible states. The sets $\mathcal{G}_T$ and $\mathcal{L}_T$ are defined by the lattice structure of $\mathcal{C}(T)$. Their union has a distinguished algebraic property with respect to the fixed term order.

Theorem 14.8 *For any mutagenetic tree T, the set $\mathcal{G}_T \cup \mathcal{L}_T$ is a reduced Gröbner basis for the ideal $\langle \mathcal{G}_T \rangle + \langle \mathcal{L}_T \rangle$ it generates.*

Proof By Buchberger's criterion (Theorem 3.10) we need to show that all S-polynomials $S(f, g)$ of elements $f, g \in \mathcal{G}_T \cup \mathcal{L}_T$ reduce to zero modulo the proposed Gröbner basis. If the leading monomials of f and g are relatively prime, then $S(f, g)$ reduces to zero according to [Cox *et al.*, 1997, Prop. 2.9.4].

The remaining S-polynomials are of the form $S(f_{ij}, f_{il})$ with $j \neq l$ and $f_{ij} := p_i p_j - p_{i \wedge j} p_{i \vee j}$, and can be expressed as

$$
\begin{aligned}
S(f_{ij}, f_{il}) &= p_l p_{i \wedge j} p_{i \vee j} - p_j p_{i \wedge l} p_{i \vee l} \\
&= p_{i \vee j}(f_{l, i \wedge j} + p_{i \wedge j \wedge l} p_{(i \wedge j) \vee l}) - p_{i \vee l}(f_{j, i \wedge l} + p_{i \wedge j \wedge l} p_{(i \wedge l) \vee j}) \\
&= p_{i \vee j} f_{l, i \wedge j} - p_{i \vee l} f_{j, i \wedge l} + p_{i \wedge j \wedge l}(p_{i \vee j} p_{(i \wedge j) \vee l} - p_{i \vee l} p_{(i \wedge l) \vee j}) \\
&= p_{i \vee j} f_{l, i \wedge j} - p_{i \vee l} f_{j, i \wedge l} + p_{i \wedge j \wedge l}(f_{i \vee j, (i \wedge j) \vee l} - f_{i \vee l, (i \wedge l) \vee j}). \quad (14.4)
\end{aligned}
$$

In expression (14.4) not all terms may appear because $f_{ij} = 0$ whenever $(i \wedge j) = i$ or $(i \wedge j) = j$. Since $p_i p_j \succ p_{i \vee j} p_{i \wedge j}$ we find that

$$
\mathrm{LT}(p_{i \vee j} f_{l, i \wedge j}) = p_l p_{i \wedge j} p_{i \vee j}, \qquad \mathrm{LT}(p_{i \wedge j \wedge l} f_{i \vee j, (i \wedge j) \vee l}) \preceq p_l p_{i \wedge j} p_{i \vee j},
$$

$$
\mathrm{LT}(p_{i \vee l} f_{j, i \wedge l}) = p_j p_{i \wedge l} p_{i \vee l}, \qquad \mathrm{LT}(p_{i \wedge j \wedge l} f_{i \vee l, (i \wedge l) \vee j}) \preceq p_j p_{i \wedge l} p_{i \vee l}.
$$

Thus, the multidegree of all four terms is smaller than or equal to the multidegree of $S(f_{ij}, f_{il})$. Hence, $S(f_{ij}, f_{il})$ reduces to zero modulo $\mathcal{G}_T$, and $\mathcal{G}_T \cup \mathcal{L}_T$ is a Gröbner basis [Cox *et al.*, 1997, Thm. 2.9.3].

The Gröbner basis is reduced, because for any $f \in \mathcal{G}_T \cup \mathcal{L}_T$, no monomial of f lies in $\langle \mathrm{LT}((\mathcal{G}_T \cup \mathcal{L}_T) \setminus \{f\}) \rangle$; compare (14.3). $\square$

This result, based on the lattice structure of $\mathcal{C}(T)$, can also be derived from work on algebras with straightening laws on distributive lattices [Hibi, 1987]. However, there is an unexpected relation to mutagenetic tree models, which we will now explore.

The statistical model $\mathcal{T}$ is an algebraic variety in Δ. Let $I_{\mathcal{T}} \subset R$ be the ideal of polynomials vanishing on $\mathcal{T}$ and $I_{\mathcal{T}}^h$ the ideal generated by the homogeneous invariants in $I_{\mathcal{T}}$. Clearly, $\mathcal{L}_T \subset I_{\mathcal{T}}^h$. In addition, Lemma 14.6 implies that conditional independence statements induced by $\mathcal{T}$ yield polynomials in $I_{\mathcal{T}}^h$. The *global Markov property* [Lauritzen, 1996, Sect. 3.2] on the mutagenetic tree T states that $A \perp\!\!\!\perp B \mid C$ if and only if A is separated from B by $C \cup \{0\}$ in T, i.e. if every path from a node $u \in A$ to a node $v \in B$ intersects $C \cup \{0\}$. Note that A is separated from B by $C \cup \{0\}$ in T if and only if A is separated from B by C in the induced forest $T_{[n]}$. According to Remark 1.26, an independence statement $A \perp\!\!\!\perp B \mid C$ induces an ideal of invariants $I_{A \perp\!\!\!\perp B \mid C}$ that is generated by the 2×2 minors

$$
p_{i_A i_B i_C} p_{j_A j_B i_C} - p_{i_A j_B i_C} p_{j_A i_B i_C} = \det \begin{pmatrix} p_{i_A i_B i_C} & p_{i_A j_B i_C} \\ p_{j_A i_B i_C} & p_{j_A j_B i_C} \end{pmatrix}, \quad (14.5)
$$

for all $(i_A, i_B, i_C), (j_A, j_B, i_C) \in \mathcal{I}$. Let $I_{\mathrm{global}(T)}$ be the sum of the independence ideals $I_{A \perp\!\!\!\perp B \mid C}$ over all statements $A \perp\!\!\!\perp B \mid C$ induced by T. In fact, $I_{\mathrm{global}(T)}$ is already generated by the *saturated* independence statements, i.e. with $A \,\dot\cup\, B \,\dot\cup\, C = [n]$ (see [Geiger *et al.*, 2005]). The saturated independence statements translate into quadratic binomials in $I_{\mathcal{T}}^h$. The following proposition connects the statistical ideal $I_{\mathrm{global}(T)}$ to the combinatorial ideal $\langle \mathcal{G}_T \rangle$.

Proposition 14.9 *Let T be a mutagenetic tree. Then*

$$I_{\text{global}(T)} + \langle \mathcal{L}_T \rangle = \langle \mathcal{G}_T \rangle + \langle \mathcal{L}_T \rangle.$$

Proof Consider $i, j \in \mathcal{C}(T)$, $i \neq j$. By Lemma 14.4, the induced subgraphs T_i and T_j are subbranchings of T. If we define $A := S_i \setminus S_j$, $B := S_j \setminus S_i$, and

$$C := [n] \setminus (A \dot\cup B) = (S_i \cap S_j) \dot\cup ([n] \setminus (S_i \cup S_j)),$$

then A and B are separated by $C \cup \{0\}$ in T, hence $A \perp\!\!\!\perp B \mid C$. Setting

$$(i_A, i_B, i_C) = (1_A, 0_B, 1_{S_i \cap S_j}, 0_{[n] \setminus (S_i \cup S_j)}),$$
$$(j_A, j_B, i_C) = (0_A, 1_B, 1_{S_i \cap S_j}, 0_{[n] \setminus (S_i \cup S_j)}),$$

we find that

$$p_i p_j - p_{i \vee j} p_{i \wedge j} = p_{i_A i_B i_C} p_{j_A j_B i_C} - p_{i_A j_B i_C} p_{j_A i_B i_C} \in I_{\text{global}(T)}.$$

This establishes the inclusion $\langle \mathcal{G}_T \rangle \subset I_{\text{global}(T)}$.

To prove the converse it suffices to consider a saturated conditional independence statement $A \perp\!\!\!\perp B \mid C$. Let g be a generator of $I_{A \perp\!\!\!\perp B \mid C}$,

$$g = p_{i_A i_B i_C} p_{j_A j_B i_C} - p_{i_A j_B i_C} p_{j_A i_B i_C}, \quad (i_A, i_B, i_C), (j_A, j_B, i_C) \in \mathcal{I}.$$

First, note that

$$p_{i_A i_B i_C} p_{j_A j_B i_C} \in \langle \mathcal{L}_T \rangle \iff p_{i_A j_B i_C} p_{j_A i_B i_C} \in \langle \mathcal{L}_T \rangle.$$

Indeed, by Lemma 14.4, $k \notin \mathcal{C}(T)$ if and only if there exists $(u, v) \in E(T)$ with $v \in V(T_k)$, but $u \notin V(T_k)$. Since A and B are separated by $C \cup \{0\}$, such an edge cannot connect A and B. Therefore, it must appear in both sets $E(T_{i_A i_B i_C}) \cup E(T_{j_A j_B i_C})$ and $E(T_{i_A j_B i_C}) \cup E(T_{j_A i_B i_C})$.

Assume now that all four states defining g are compatible with T. Then Lemma 14.7 implies that their joins and meets are also compatible. Moreover,

$$i_A i_B i_C \vee j_A j_B i_C = i_A j_B i_C \vee j_A i_B i_C =: i \vee j$$
$$i_A i_B i_C \wedge j_A j_B i_C = i_A j_B i_C \wedge j_A i_B i_C =: i \wedge j$$

and we can write

$$g = (p_{i_A i_B i_C} p_{j_A j_B i_C} - p_{i \vee j} p_{i \wedge j}) + (p_{i \vee j} p_{i \wedge j} - p_{i_A j_B i_C} p_{j_A i_B i_C})$$

as an element of $\langle \mathcal{G}_T \rangle$. $\qquad\square$

In order to characterize the ideal I_T of invariants of T we first draw on previous work on independence ideals in graphical models.

Proposition 14.10

$$V(I_T^h) = V \left(I_{\text{global}(T)} + \langle \mathcal{L}_T \rangle \right).$$

Proof The claim can be derived from Theorem 14.6 and [Garcia *et al.*, 2004, Theorems 6 and 8]. $\qquad\square$

Invariant \ pa(T)	0	0	2	3	3	2	0	2	3	0	0	0	3	0	2	0
	1	3	0	0	1	3	1	0	3	0	3	0	0	1	0	0
	2	1	1	2	0	0	1	2	0	2	0	1	0	0	0	0
p_{000}	·	·	·	·	·	·	·	·	·	·	·	·	·	·	·	·
p_{001}	•	•	•	•	·	·	•	•	·	•	·	•	·	·	·	·
p_{010}	•	•	·	·	•	•	•	·	•	·	•	·	·	•	·	·
p_{011}	•	•	•	·	•	·	·	·	·	·	·	•	·	•	·	·
p_{100}	·	·	•	•	•	•	·	•	•	·	·	·	•	·	•	·
p_{101}	•	·	•	•	·	•	·	•	·	•	·	·	·	·	•	·
p_{110}	·	•	·	•	•	•	·	·	•	·	•	·	·	·	·	·
p_{111}	·	·	·	·	·	·	·	·	·	·	·	·	·	·	·	·
$p_{001}p_{010} - p_{000}p_{011}$	○	○	○	·	○	·	○	·	·	·	·	○	•	•	•	•
$p_{001}p_{100} - p_{000}p_{101}$	○	·	○	○	·	○	·	○	·	○	•	·	·	•	○	•
$p_{001}p_{110} - p_{000}p_{111}$	·	·	·	·	·	·	·	·	·	·	·	·	·	•	•	•
$p_{010}p_{100} - p_{000}p_{110}$	·	○	·	○	○	○	·	·	○	•	○	•	○	·	·	•
$p_{010}p_{101} - p_{000}p_{111}$	·	·	·	·	·	·	·	·	·	·	·	·	•	•	·	•
$p_{011}p_{100} - p_{000}p_{111}$	·	·	·	·	·	·	·	·	·	·	·	·	•	•	·	•
$p_{011}p_{101} - p_{001}p_{111}$	○	○	○	○	·	·	○	○	•	○	•	○	•	·	·	•
$p_{011}p_{110} - p_{010}p_{111}$	○	○	·	·	○	○	○	•	○	•	○	·	·	○	•	•
$p_{101}p_{110} - p_{100}p_{111}$	·	·	○	○	○	○	•	○	○	·	·	•	○	•	○	•
Figure 14.1	(d)							(c)		(b)						(a)

Table 14.1. *Algebraic invariants for the 16 mutagenetic tree models on* $n = 3$ *events. Trees are represented by their parent vector* $\mathrm{pa}(T) = (\mathrm{pa}(v))_{v \in [n]}$. *Polynomials of the degree-reverse lexicographic Gröbner basis of the ideal of invariants are indicated by "•". The polynomials marked by "○" also lie in the ideal generated by the Gröbner basis.*

Theorem 14.11 *The set* $\mathcal{G}_T \cup \mathcal{L}_T$ *is a reduced Gröbner basis for the ideal* I_T^h *of homogeneous invariants of the mutagenetic tree model* T.

Proof By Theorem 14.8 it suffices to show that $\langle \mathcal{G}_T \rangle + \langle \mathcal{L}_T \rangle = I_T^h$. By Proposition 14.9 we have

$$\langle \mathcal{G}_T \rangle + \langle \mathcal{L}_T \rangle = I_{\mathrm{global}(T)} + \langle \mathcal{L}_T \rangle \subset I_T^h.$$

To prove the converse, we use the *strong Nullstellensatz* [Cox et al., 1997, Thm. 4.2.6] (cf. Theorem 3.3), which states that $I(V(I)) = \sqrt{I}$ for any ideal I. Since the initial ideal of $\langle \mathcal{G}_T \rangle + \langle \mathcal{L}_T \rangle$ is square-free, the ideal itself is radical [Sturmfels, 2002, Prop. 5.3]. Therefore,

$$I_T^h \subset \sqrt{I_T^h} = I(V(I_T^h)) = I(V(\langle \mathcal{G}_T \rangle + \langle \mathcal{L}_T \rangle)) = \langle \mathcal{G}_T \rangle + \langle \mathcal{L}_T \rangle,$$

where we also use Propositions 14.9 and 14.10. $\square$

Theorem 14.12 *The ideal of invariants* I_T *of the mutagenetic tree model* T *is generated as*

$$I_T = \langle \mathcal{G}_T \rangle + \langle \mathcal{L}_T \rangle + \left\langle \sum_{i \in \mathcal{I}} p_i - 1 \right\rangle.$$

Proof Consider an algebraic invariant $g \in I_T$ and write $g = \sum_{k=0}^{d} g_k$ as the sum of its homogeneous components g_k of degree k. Using the linear form

$p_+ = \sum_{i \in \mathcal{I}} p_i$, we homogenize g into $g^{(h)} = \sum_{k=0}^{d} g_k p_+^{d-k}$. The difference

$$g^{(h)} - g = \sum_{k=0}^{d-1} g_k \left(p_+^{d-k} - 1 \right) = \sum_{k=0}^{d-1} g_k \left(p_+ - 1 \right) \left(p_+^{d-k-1} + \cdots + p_+ + 1 \right)$$

lies in $\langle \sum_{i \in \mathcal{I}} p_i - 1 \rangle$. By Theorem 14.11 the homogeneous invariant $g^{(h)}$ is in $\langle \mathcal{G}_T \rangle + \langle \mathcal{L}_T \rangle$. Thus, $g = g - g^{(h)} + g^{(h)}$ is in the claimed ideal. $\qquad\square$

Example 14.13 The star model $\mathcal{S}$ (Lemma 14.4, Figure 14.1(a)) is the intersection of the probability simplex with the Segre variety

$$V_{\text{Segre}} = \bigcap_{k=1}^{n} V(p_{i_1 \ldots i_n} p_{j_1 \ldots j_n} - p_{i_1 \ldots i_{k-1} j_k i_{k+1} \ldots i_n} p_{j_1 \ldots j_{k-1} i_k j_{k+1} \ldots j_n}, \ i, j \in \mathcal{I}),$$

i.e. the image of the n-fold Segre embedding $\mathbb{P}^1 \times \cdots \times \mathbb{P}^1 \to \mathbb{P}^{2^n-1}$. To see this note that the minors defining the Segre variety lie in

$$\langle p_i p_j - p_{i \vee j} p_{i \wedge j}, \ i, j \in \mathcal{I} \rangle,$$

because both pairs of states indexing the products of the binomials that define V_{Segre} have the same join and meet. Conversely, let $i, j \in \mathcal{I}$, $i \neq j$. Then there is a sequence of pairs of states

$$(i, j) = (i^{(0)}, j^{(0)}), \ \ldots, \ (i^{(m+1)}, j^{(m+1)}) = (i \vee j, i \wedge j)$$

such that for all $l = 1, \ldots, m$, both $i^{(l+1)}$ and $j^{(l+1)}$ are obtained from $i^{(l)}$ and $j^{(l)}$, respectively, by exchanging at most one index. Hence, the telescoping sum

$$p_i p_j - p_{i \vee j} p_{i \wedge j} = \sum_{l=0}^{m} (p_{i^{(l)}} p_{j^{(l)}} - p_{i^{(l+1)}} p_{j^{(l+1)}})$$

lies in the ideal defining V_{Segre}. $\qquad\square$

Theorem 14.11 together with Algorithm 14.5 provides an efficient method for computing a reduced Gröbner basis for the ideal I_T^h of homogeneous invariants of a mutagenetic tree model. This approach does not require implicitization (cf. Section 3.2) and the computational complexity is linear in the size of the output, i.e. the size of the Gröbner basis.

Proposition 14.14 *Let T be a mutagenetic tree on n events.*

(i) *The cardinality of $\mathcal{L}_T$, i.e. the number of incompatible states is bounded from above by $2^n - n - 1$. The bound is attained if and only if T is a chain.*

(ii) *The cardinality of $\mathcal{G}_T$ is bounded from above by $\binom{2^n+1}{2} - 3^n$. The bound is attained if and only if T is the star $\mathcal{S}$.*

Therefore, the cardinality of the Gröbner basis in Theorem 14.8 is at most of order $O(4^n)$.

Proof The chain model (Figure 14.1(d)) has $n + 1$ compatible states. Any other tree topology has strictly more compatible states (cf. Algorithm 14.5), which proves (i).

The polynomials in $\mathcal{G}_S$ are indexed by the set of pairs $(i, j) \in \mathcal{I}^2$ with $(i \wedge j) < i < j < (i \vee j)$. We write this index set as the difference

$$\{(i, j) \in \mathcal{I}^2 \mid i \leq j\} \setminus \{(i, j) \in \mathcal{I}^2 \mid i \leq j, \; S_i \subset S_j\}. \tag{14.6}$$

The cardinality of the first set is $\binom{2^n + 1}{2}$. For the second set, we group subsets according to their cardinality. A subset of cardinality k has 2^{n-k} supersets. Hence, the second set has cardinality

$$\sum_{k=0}^{n} \binom{n}{n-k} 2^{n-k} = \sum_{k=0}^{n} \binom{n}{k} 2^k 1^{n-k} = (1+2)^n.$$

Since the second set in (14.6) is contained in the first one, the bound in (ii) follows. For the tightness note that if (u, v), $u \neq 0$, is an edge of T, then the polynomial indexed by (i, j) with $S_i = \{u\}$ and $S_j = \{v\}$ is not in $\mathcal{G}_T$, because j is not a compatible state. $\square$

14.4 Mixture models

Let $K \in \mathbb{N}_{>0}$ and $(T_1, \ldots, T_K)$ be a list of K mutagenetic trees. Define the map

$$\mathbf{f}^{(T_1, \ldots, T_K)} : \Delta_{K-1} \times \Theta^K \to \Delta = \Delta_{2^n - 1}$$

$$(\lambda, \theta^{(1)}, \ldots, \theta^{(K)}) \mapsto \sum_{k=1}^{K} \lambda_k \, \mathbf{f}^{(T_k)}(\theta^{(k)}). \tag{14.7}$$

Definition 14.15 The *K-mutagenetic trees mixture model* $(T_1, \ldots, T_K) :=$ $\mathbf{f}^{(T_1, \ldots, T_K)}(\Delta_{K-1} \times \Theta^K) \subset \Delta$ is given by the map $\mathbf{f}^{(T_1, \ldots, T_K)}$. This algebraic statistical model has parameter space $\Delta_{K-1} \times \Theta^K$ and state space $\mathcal{I}$.

A state $i \in \mathcal{I}$ is *compatible* with the K-mutagenetic trees mixture model $(T_1, \ldots, T_K)$ if it is compatible with at least one of the mutagenetic tree models $T_k = \mathbf{f}^{(T_k)}(\Theta)$, i.e. $\mathcal{C}(T_1, \ldots, T_K) = \bigcup_{k=1}^{K} \mathcal{C}(T_k)$.

Example 14.16 Consider the list (S, T), where S is the star in Figure 14.1(a) and T the tree in Figure 14.1(b). All states $i \in \mathcal{I}$ are compatible with the resulting mixture model (S, T). Examples of coordinates of the map $\mathbf{f}^{(S,T)}$ are

$$f_{101}(\lambda, \theta, \bar{\theta}) = \lambda \theta_{11}^1 (1 - \theta_{11}^2) \theta_{11}^3,$$
$$f_{100}(\lambda, \theta, \bar{\theta}) = \lambda \theta_{11}^1 (1 - \theta_{11}^2)(1 - \theta_{11}^3) + (1 - \lambda) \bar{\theta}_{11}^1 (1 - \bar{\theta}_{11}^2). \qquad \square$$

Statistical inference in mutagenetic trees mixture models can be effected via the EM algorithm described in Section 1.3 [Beerenwinkel *et al.*, 2004]. Alternatively the multilinearity of the coordinates of the map $\mathbf{f}^{(T_1, \ldots, T_K)}$ could be exploited (cf. Example 1.7). However, the algebraic study of mutagenetic trees mixture models is difficult. Let us consider an example.

Example 14.17 Let (T_1, T_2) be the list of trees defined by the parent vectors $\text{pa}(T_1) = (2, 0, 0, 3)$ and $\text{pa}(T_2) = (0, 0, 2, 3)$. The resulting mixture model (T_1, T_2) is of dimension 8 and thus non-identifiable. The reduced degree-reverse lexicographic Gröbner basis for the ideal of homogeneous invariants $I^h_{(T_1, T_2)}$ contains the 6 polynomials p_i with $i \notin \mathcal{C}(T_1, T_2) = \mathcal{C}(T_1) \cup \mathcal{C}(T_2)$, which are

$$p_{0001}, \ p_{0101}, \ p_{1001}, \ p_{1010}, \ p_{1011}, \ p_{1101},$$

and a degree-5 polynomial with 15 terms, which can be written as the determinant of the matrix

$$\begin{pmatrix} p_{0000} & p_{0100} & p_{1100} & p_{1000} & 0 \\ p_{0010} & p_{0110} & p_{1110} & 0 & 0 \\ p_{0011} & p_{0111} & p_{1111} & 0 & 0 \\ p_{0110} & 0 & 0 & p_{1110} & p_{0010} \\ p_{0111} & 0 & 0 & p_{1111} & p_{0011} \end{pmatrix}.$$

$\square$

14.4.1 Secant varieties

Consider the list (T, T), in which a single tree is repeated. Every distribution $p \in (T, T)$ is a convex combination of two distributions $p_T, p'_T \in T$, i.e. $p = \lambda p_T + (1 - \lambda)p'_T$, $\lambda \in [0, 1]$. Therefore, (T, T) is a subset of the intersection of the probability simplex Δ with the first secant variety of $\mathbf{f}^{(T)}(\mathbb{C}^{\mathcal{I}})$, i.e. the closure of the set $\{\lambda p_T + (1 - \lambda)p'_T \mid \lambda \in \mathbb{C}, \ p_T \neq p'_T \in \mathbb{C}^{\mathcal{I}}\}$. More generally, the K-mutagenetic trees mixture model $(T, \ldots, T)$ corresponds to the Kth secant variety of T. This is the correspondence between mixture models and secant varieties mentioned in the introduction to Part I; see also Chapter 3.

If T is a chain, then every node in T has at most one child and $|\mathcal{C}(T)| = n + 1$. The chain model T is equal to the n-dimensional variety obtained by intersecting the probability simplex Δ with the $2^n - n - 1$ hyperplanes $p_i = 0$, $i \notin \mathcal{C}(T)$. Since $\mathcal{C}(T, T) = \mathcal{C}(T)$ and $T \subset (T, T)$ it follows that the chain mixture model is trivial in the sense that $(T, T) = T$.

If $T = S$ is the star, then the mixture model (S, S) is also known as a *naive Bayes model* (compare Proposition 14.19). In algebraic terms it is the first secant variety of the Segre variety. It has been shown that $\dim(S, S) = \min(2n + 1, 2^n - 1)$ in this case [Catalisano *et al.*, 2002, Garcia, 2004].

Let us consider an example of a tree that is neither a chain nor the star.

Example 14.18 Let T be the tree over $n = 4$ events with vector of parents $\text{pa}(T) = (2, 0, 0, 3)$. The model (T, T) is non-identifiable and of dimension $\dim(T, T) = 7$. The reduced degree-reverse lexicographic Gröbner basis for the ideal of homogeneous invariants $I^h_{(T, T)}$ contains the 7 polynomials p_i, $i \notin \mathcal{C}(T, T) = \mathcal{C}(T)$, which are

$$p_{0001}, \ p_{0101}, \ p_{1000}, \ p_{1001}, \ p_{1010}, \ p_{1011}, \ p_{1101},$$

and the determinant of the matrix

$$\begin{pmatrix} p_{0000} & p_{0010} & p_{0011} \\ p_{0100} & p_{0110} & p_{0111} \\ p_{1100} & p_{1110} & p_{1111} \end{pmatrix}.$$

□

The following proposition and its proof are analogous to Theorem 14.6.

Proposition 14.19 *Let T be a mutagenetic tree and $K \in \mathbb{N}_{>0}$. Let M be the directed acyclic graph obtained from T by adding the edges $(0, v)$, $v \in [n]$, from the root 0 to every node v. Associating a hidden random variable X_0 with K levels with the root node 0 induces a directed graphical model $\mathcal{M}$ with one hidden variable. Then a probability distribution $p \in \Delta$ is in the K-mutagenetic trees mixture model $(T, \ldots, T)$ if and only if $p \in \mathcal{M}$ and $p_i = 0$ for all $i \notin C(T)$.*

14.4.2 The uniform star as an error model

If the observations contain a state that is incompatible with a tree model T, then the likelihood function of T is constant and equal to zero. Thus, in the presence of false positives and false negatives the maximum likelihood tree will often be the star or have a star-like topology even if other pathways are significantly overrepresented in the data. One way to account for such states is to mix a mutagenetic tree with a uniform star model [Beerenwinkel *et al.*, 2005a]; see [Szabo and Boucher, 2002] for an alternative approach.

Definition 14.20 *Let S be the star over n events. The 1-dimensional uniform star model $S_{\text{uni}} := \mathbf{f}^{(S_{\text{uni}})}([0, 1]) \subset \Delta$ is given by the specialization map $\mathbf{f}^{(S_{\text{uni}})}(\theta_{11}) = \mathbf{f}^{(S)}(\theta_{11}, \ldots, \theta_{11})$.*

In the uniform star model the n events occur independently with the same probability. This is our error model.

Algebraically the uniform star model is the intersection of the probability simplex with the rational normal curve of degree n, i.e. the image of the Veronese map $\mathbb{P}^1 \to \mathbb{P}^n$. To see this, note that the coordinates of $\mathbf{f}^{(S_{\text{uni}})}$ only depend on the number $k = |S_i|$ of occurred events: $f_i(\theta_{11}) = \theta_{11}^k (1 - \theta_{11})^{n-k}$. Hence, we can identify the ideal $I_{S_{\text{uni}}}^h$ generated by the homogeneous model invariants with its image in the ring $\mathbb{Q}[\bar{p}_k]$, $k = 0, \ldots, n$ under the ring homomorphism defined by $p_i \mapsto \bar{p}_{|S_i|}$. In this ring, $I_{S_{\text{uni}}}^h$ is generated by quadrics of the form (cf. Theorem 14.11)

$$\bar{p}_{|S_i|}\bar{p}_{|S_j|} - \bar{p}_{|S_i \vee j|}\bar{p}_{|S_{i \wedge j}|}. \tag{14.8}$$

These quadrics include, after negation, all 2×2 minors of the $2 \times n$ matrix

$$\begin{pmatrix} \bar{p}_0 & \bar{p}_1 & \bar{p}_2 & \cdots & \bar{p}_{n-1} \\ \bar{p}_1 & \bar{p}_2 & \bar{p}_3 & \cdots & \bar{p}_n \end{pmatrix}, \tag{14.9}$$

i.e. the defining polynomials of the rational normal curve of degree n. On the other hand, any quadric in (14.8) can be written as a telescoping sum of the minors in (14.9) as in Example 14.13.

Proposition 14.21 *Let $n \geq 3$. Then, for any n-dimensional mutagenetic tree model $\mathcal{T}$, the mixture model $(\mathcal{S}_{\mathrm{uni}}, \mathcal{T})$ has dimension $n + 2$, and hence is identifiable.*

Proof Clearly, $\dim(\mathcal{S}_{\mathrm{uni}}, \mathcal{T}) \leq n + 2$ because $\dim(\mathcal{T}) = n$ and $\dim(\mathcal{S}_{\mathrm{uni}}) = 1$. Thus, we have to show that the dimension may not drop below $n + 2$.

Consider first a tree $T \neq S$. It is easy to see that $|\mathcal{I} \setminus \mathcal{C}(T)| \geq 2$, because $n \geq 3$. Choosing two states $j, j' \notin \mathcal{C}(T)$ such that $p_j = p_{j'} = 0$ for all $p \in \mathcal{T}$, we obtain that the Jacobian matrix J of the map $\mathbf{f}^{(\mathcal{S}_{\mathrm{uni}}, T)}$ is upper triangular:

$$J = \begin{pmatrix} \lambda \cdot J_T & * \\ 0 & J_{j,j'} \end{pmatrix}, \quad \text{where} \quad J_T = \left(\frac{\partial f_i}{\partial \theta_{11}^v} \right)_{i \in \mathcal{C}(T) \setminus \{j,j'\}, \, v \in [n]}$$

The matrix J_T depends only on $(\theta_{11}^1, \ldots, \theta_{11}^n)$ and up to deletion of two rows of zeros it is the Jacobian matrix of the map $\mathbf{f}^{(T)}$ and, thus, of full rank in the interior of the parameter space of $(\mathcal{S}_{\mathrm{uni}}, \mathcal{T})$. The matrix

$$J_{j,j'} = \begin{pmatrix} \frac{\partial f_j}{\partial \lambda} & \frac{\partial f_j}{\partial \theta_{11}} \\ \frac{\partial f_{j'}}{\partial \lambda} & \frac{\partial f_{j'}}{\partial \theta_{11}} \end{pmatrix}$$

depends only on (λ, θ_{11}) and its determinant equals $1 - \lambda$ times a univariate polynomial $g(\theta_{11})$. Since g has only finitely many roots, it holds almost everywhere that the matrix $J_{j,j'}$ is of full rank 2 and the Jacobian J of full rank $n + 2$. Therefore, $\dim(\mathcal{S}_{\mathrm{uni}}, \mathcal{T}) = n + 2$; compare [Geiger *et al.*, 2001].

If $T = S$ is the star, then we know that the mixture model $(\mathcal{S}_{\mathrm{uni}}, \mathcal{S})$ is obtained by parameter specialization from $(\mathcal{S}, \mathcal{S})$. Since $\dim(\mathcal{S}, \mathcal{S}) = 2n + 1$ the Jacobian of the map $\mathbf{f}^{(\mathcal{S}, \mathcal{S})}$ is of full rank $2n + 1$ almost everywhere. Now it follows from the chain rule that

$$\frac{\partial f_i^{(\mathcal{S}_{\mathrm{uni}}, \mathcal{S})}}{\partial \theta_{11}} = \sum_{v \in [n]} \frac{\partial f_i^{(S,S)}}{\partial \theta_{11}^v} \Big|_{\theta_{11}^v = \theta_{11}},$$

which implies that the Jacobian of the map $\mathbf{f}^{(\mathcal{S}_{\mathrm{uni}}, \mathcal{S})}$ is of full rank $n + 2$. $\qquad \square$

15

Catalog of Small Trees

Marta Casanellas

Luis David Garcia

Seth Sullivant

This chapter is concerned with the description of the *Small Trees website* which can be found at the following web address:

http://bio.math.berkeley.edu/ascb/chapter15/

The goal of the website is to make available in a unified format various algebraic features of different phylogenetic models on small trees. By "small" we mean trees with at most 5 taxa. In the first two sections, we describe a detailed set of notational conventions for describing the parameterizations given a tree, the Fourier transform and the phylogenetic invariants in Fourier coordinates. The third section gives a brief description of each of the types of algebraic information associated with a model and a tree on the Small Trees website. We also include several tables summarizing some of the results that we were able to compute. The fourth section details an example of a page on the website. The final section concerns simulation studies using algebraic invariants for the Kimura 3-parameter model to recover phylogenies. The reader is encouraged to review Section 4.5 for an introduction to the models discussed in this chapter.

15.1 Notation and conventions

The purpose of this section is to establish a notational convention to unify the presentation and automate the process for obtaining all the information associated with any small phylogenetic tree in the website. We assume that each phylogenetic model is presented with a particular tree T together with a figure representing that tree. The figures of trees with up to five leaves will be the ones that can be found on the Small Trees website. The *Newick standard* [Felsenstein, 2004] for representing trees is widely used, in spite of not having a unique representation for a tree. Our conventions fix the labels on a given tree topology and can be used together with the Newick conventions to specify a unique rooted tree.

15.1.1 Labeling trees

We start by fixing all the labels on a tree. If T is a rooted tree, there is a distinguished vertex of T called the root and labeled by the letter r. The tree T should be drawn with the root r at the top of the figure and the edges of the tree below the root. Each edge in the tree is labeled with a lowercase letter $a, b, c, \ldots$ The edges are labeled in alphabetical order starting at the upper left hand corner, proceeding left to right and top to bottom. The leaves are labeled with the numbers $1, 2, 3, \ldots$ starting with the left-most leaf and proceeding left to right. Figure 15.1 shows the "giraffe" tree with four leaves and its labeling.

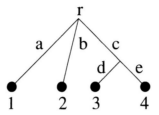

Fig. 15.1. The giraffe tree on four leaves.

If T is an unrooted tree, it should be drawn with the leaves on a circle. The edges of T are labeled with lower-case letters $a, b, c, \ldots$ in alphabetical order starting at the upper left hand corner of the figure and proceeding left to right and top to bottom. The leaves are labeled with the numbers $1, 2, 3, \ldots$ starting at the first leaf "left of 12 o'clock" and proceeding counterclockwise around the perimeter of the tree. Figure 15.2 illustrates this on the "quartet" tree.

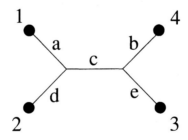

Fig. 15.2. The quartet tree on four leaves.

15.1.2 Joint probability parameterization

For each node in a model, we associate a random variable with two or four states depending on whether we are looking at binary data or DNA data. In the case of binary data these states are $\{0, 1\}$ and for DNA data they are $\{\text{A}, \text{C}, \text{G}, \text{T}\}$ in this order.

The root distribution is a vector of length two or four depending on whether the model is for binary or DNA sequences. The name of this vector is r. Its entries are parameters $r_0, r_1, r_2, \ldots$ and are filled in from left to right and are recycled as the model requires.

Example 15.1 In the general strand symmetric model described in Chapter 16, r always denotes the vector $r = (r_0, r_1, r_1, r_0)$. □

We tacitly assume that the entries in r sum to 1, though we do not eliminate a parameter to take this into account. If the model assumes a uniform root distribution, then r has the form $r = (1/2, 1/2)$ or $r = (1/4, 1/4, 1/4, 1/4)$ according to whether the model is for binary or DNA data.

In each type of model, the letters $a, b, c, \ldots$ that label the edges represent the transition matrices in the model. These are either 2×2 or 4×4 matrices depending on whether the model is for binary data or DNA data. In each case, the matrix is filled from left to right and top to bottom with unknown parameters, recycling a parameter whenever the model requires it. For the transition matrix of the edge labeled with x, these entries are called $x_0, x_1, x_2, \ldots$

Example 15.2 For example, in the Kimura 3-parameter model, the letter a represents the matrix

$$a = \begin{pmatrix} a_0 & a_1 & a_2 & a_3 \\ a_1 & a_0 & a_3 & a_2 \\ a_2 & a_3 & a_0 & a_1 \\ a_3 & a_2 & a_1 & a_0 \end{pmatrix}.$$

The Kimura 2-parameter and Jukes–Cantor models arise from restrictions of the parameters in the Kimura 3-parameter model, and hence the letters denoting the parameters are recycled. For instance, the letter c in the Jukes–Cantor DNA model and the letter d in the Kimura 2-parameter model represent the following matrices

$$c = \begin{pmatrix} c_0 & c_1 & c_1 & c_1 \\ c_1 & c_0 & c_1 & c_1 \\ c_1 & c_1 & c_0 & c_1 \\ c_1 & c_1 & c_1 & c_0 \end{pmatrix}, \qquad d = \begin{pmatrix} d_0 & d_1 & d_2 & d_1 \\ d_1 & d_0 & d_1 & d_2 \\ d_2 & d_1 & d_0 & d_1 \\ d_1 & d_2 & d_1 & d_0 \end{pmatrix}.$$

In the general strand symmetric model the letter e always represents the matrix

$$e = \begin{pmatrix} e_0 & e_1 & e_2 & e_3 \\ e_4 & e_5 & e_6 & e_7 \\ e_7 & e_6 & e_5 & e_4 \\ e_3 & e_2 & e_1 & e_0 \end{pmatrix}.$$

□

We assume that the entries of these matrices satisfy additional linear constraints such that they are transition matrices. For instance, in the Jukes–Cantor DNA model, this constraint is $c_0 + 3c_1 = 1$ and in the general strand symmetric model the two linear relations are $e_0 + e_1 + e_2 + e_3 = 1$ and $e_4 + e_5 + e_6 + e_7 = 1$. We do not, however, use these linear relations to eliminate parameters.

The molecular clock (MC) assumption for a rooted tree T states that for each subtree and each path from the root of that subtree to any leaf i, the products of the transition matrices corresponding to the edges of the path are identical. As the edges are visited along the path, the matrices are multiplied left to right.

Example 15.3 For the giraffe tree in Figure 15.1 the MC assumption translates into the following identities:

$$a = b = cd = ce \text{ and } d = e. \qquad \square$$

These equalities of products of parameter matrices suggest that some parameter matrices should be replaced with products of other parameter matrices and their inverses. This re-parameterization involves rational functions (instead of just polynomials).

Here is a systematic rule for making these replacements. Starting from the bottom of the tree, make replacements for transition matrices. Each vertex in the tree induces equalities among products of transition matrices along all paths emanating downward from this vertex. Among the edges emanating downward from a given vertex, all but one of the transition matrices for these edges will be replaced by a product of other transition matrices and their inverses. When choosing replacements, always replace the transition matrix that belongs to the shorter path to a leaf. If all such paths have the same length, replace the matrices that belong to the left-most edges emanating from a vertex.

Example 15.4 In the 4-leaf giraffe tree from the previous example, we replace the matrix d with e and the matrices a and b with ce. Thus, when writing the parameterization in probability coordinates, only the letters c and e will appear in the parameterizing polynomials. $\qquad \square$

The probabilities of the leaf colorations of a tree with n leaves are denoted by p_W where W is a word of length n in the alphabet $\{0, 1\}$ or $\{\mathtt{A}, \mathtt{C}, \mathtt{G}, \mathtt{T}\}$. Every probability p_W is a polynomial in the parameters of the model. Two of these probabilities p_W and p_U are equivalent if their defining polynomials are identical. This divides the 2^n or 4^n probabilities into equivalence classes. The elements of each class are ordered lexicographically, and the classes are ordered lexicographically by their lexicographically first elements.

Example 15.5 For the Jukes–Cantor DNA model with uniform root distribution on a 3-taxa claw tree there are five equivalence classes:

Class 1: $p_{\mathtt{AAA}}\ p_{\mathtt{CCC}}\ p_{\mathtt{GGG}}\ p_{\mathtt{TTT}}$

Class 2: $p_{\mathtt{AAC}}\ p_{\mathtt{AAG}}\ p_{\mathtt{AAT}}\ p_{\mathtt{CCA}}\ p_{\mathtt{CCG}}\ p_{\mathtt{CCT}}\ p_{\mathtt{GGA}}\ p_{\mathtt{GGC}}\ p_{\mathtt{GGT}}\ p_{\mathtt{TTA}}\ p_{\mathtt{TTC}}\ p_{\mathtt{TTG}}$

Class 3: $p_{\mathtt{ACA}}\ p_{\mathtt{AGA}}\ p_{\mathtt{ATA}}\ p_{\mathtt{CAC}}\ p_{\mathtt{CGC}}\ p_{\mathtt{CTC}}\ p_{\mathtt{GAG}}\ p_{\mathtt{GCG}}\ p_{\mathtt{GTG}}\ p_{\mathtt{TAT}}\ p_{\mathtt{TCT}}\ p_{\mathtt{TGT}}$

Class 4: $p_{\mathtt{ACC}}\ p_{\mathtt{AGG}}\ p_{\mathtt{ATT}}\ p_{\mathtt{CAA}}\ p_{\mathtt{CGG}}\ p_{\mathtt{CTT}}\ p_{\mathtt{GAA}}\ p_{\mathtt{GCC}}\ p_{\mathtt{GTT}}\ p_{\mathtt{TAA}}\ p_{\mathtt{TCC}}\ p_{\mathtt{TGG}}$

Class 5: $p_{\mathtt{ACG}}\ p_{\mathtt{ACT}}\ p_{\mathtt{AGC}}\ p_{\mathtt{AGT}}\ p_{\mathtt{ATC}}\ p_{\mathtt{ATG}}\ p_{\mathtt{CAG}}\ p_{\mathtt{CAT}}\ p_{\mathtt{CGA}}\ p_{\mathtt{CGT}}\ p_{\mathtt{CTA}}\ p_{\mathtt{CTG}}\ p_{\mathtt{GAC}}\ p_{\mathtt{GAT}}\ p_{\mathtt{GCA}}$
$p_{\mathtt{GCT}}\ p_{\mathtt{GTA}}\ p_{\mathtt{GTC}}\ p_{\mathtt{TAC}}\ p_{\mathtt{TAG}}\ p_{\mathtt{TCA}}\ p_{\mathtt{TCG}}\ p_{\mathtt{TGA}}\ p_{\mathtt{TGC}}$ $\qquad \square$

For each class i there will be an indeterminate p_i which denotes the sum of the probabilities in class i. For these N probabilities the expression for the probability p_i as a polynomial or rational function in the parameters appears on the webpage.

Example 15.6 In the 3-taxa claw tree with Jukes–Cantor model and uniform root distribution these indeterminates are:

$$
\begin{aligned}
p_1 &= a_0b_0c_0 + 3a_1b_1c_1 \\
p_2 &= 3a_0b_0c_1 + 3a_1b_1c_0 + 6a_1b_1c_1 \\
p_3 &= 3a_0b_1c_0 + 3a_1b_0c_1 + 6a_1b_1c_1 \\
p_4 &= 3a_1b_0c_0 + 3a_0b_1c_1 + 6a_1b_1c_1 \\
p_5 &= 6a_0b_1c_1 + 6a_1b_0c_1 + 6a_1b_1c_0 + 6a_1b_1c_1
\end{aligned}
$$
$\square$

Note that $p_1 + p_2 + p_3 + p_4 + p_5 = 1$ after substituting $a_0 = 1 - 3a_1$, $b_0 = 1 - 3b_1$ and $c_0 = 1 - 3c_1$.

15.2 Fourier coordinates

Often we will describe these phylogenetic models in an alternate coordinate system called the Fourier coordinates. This change of coordinates happens simultaneously on the parameters and on the probability coordinates themselves. The reader is encouraged to compare the algebraic introduction of the Fourier transform given in this section with the traditional introduction given in Section 17.6 and the references therein.

Each of the 2^n or 4^n Fourier coordinates are denoted by q_W where W is a word in either $\{0, 1\}$ or $\{\mathsf{A}, \mathsf{C}, \mathsf{G}, \mathsf{T}\}$. The Fourier transform from p_U to q_W is given by the following rule:

$$
p_{i_1 \cdots i_n} = \sum_{j_1, \dots, j_n} \chi^{j_1}(i_1) \cdots \chi^{j_n}(i_n) q_{j_1 \cdots j_n}, \tag{15.1}
$$

$$
q_{i_1 \cdots i_n} = \frac{1}{k^n} \sum_{j_1, \dots, j_n} \chi^{i_1}(j_1) \cdots \chi^{i_n}(j_n) p_{i_1 \cdots i_n}. \tag{15.2}
$$

Here χ^i is the character of the group associated to the ith group element. The character tables of the groups we use, namely $\mathbb{Z}_2$ and $\mathbb{Z}_2 \times \mathbb{Z}_2$ are:

	0	1
0	1	1
1	1	−1

and

	A	C	G	T
A	1	1	1	1
C	1	−1	1	−1
G	1	1	−1	−1
T	1	−1	−1	1

In other words, $\chi^i(j)$ is the (i, j)-entry in the appropriate character table. One special feature of this transformation is that the Fourier transform of the joint distribution has a parameterization that can be written in product form; we refer to [Evans and Speed, 1993, Sturmfels and Sullivant, 2005,

Székely *et al.*, 1993] for a detailed treatment of the subject. Equivalently, the Fourier transform simultaneously diagonalizes all transition matrices. Therefore, we replace the transition matrices $a, b, c, \ldots$ with diagonal matrices denoted $A, B, C, \ldots$ The letter A denotes the diagonal matrix

$$A = \begin{pmatrix} A_1 & 0 & 0 & 0 \\ 0 & A_2 & 0 & 0 \\ 0 & 0 & A_3 & 0 \\ 0 & 0 & 0 & A_4 \end{pmatrix}.$$

Furthermore, these parameters satisfy the relations imposed by the corresponding model and the molecular clock assumption. For instance, in the Jukes–Cantor model we have the relations $A_2 = A_3 = A_4$ for the matrix A.

The q_W are polynomials or rational functions in the transformed parameters. They are given parametrically as

$$q_{i_1 \cdots i_n} := \begin{cases} \prod_{e \in E} M_e(k_e) & \text{if } i_n = i_1 + i_2 + \cdots + i_{n-1} \text{ in the group} \\ 0 & \text{otherwise} \end{cases} \tag{15.3}$$

where M_e is the corresponding diagonal matrix associated to edge e, and k_e is the sum (in the corresponding group) of the labels at the leaves that are "beneath" the edge e.

We say that q_W and q_U are equivalent if they represent the same polynomial in terms of these parameters. These Fourier coordinates are grouped into equivalence classes. The elements in the equivalence classes are ordered lexicographically. Most of the Fourier coordinates q_W are zero and these are grouped in class 0. The others are ordered Class 1, Class 2, lexicographically by their lexicographically first element.

Example 15.7 Here we display the classes of Fourier coordinates for the Jukes–Cantor DNA model on the 3-leaf claw tree.

Class 0: q_{AAC} q_{AAT} $\cdots$
Class 1: q_{AAA}
Class 2: q_{ACC} q_{AGG} q_{ATT}
Class 3: q_{CAC} q_{GAG} q_{TAT}
Class 4: q_{CCA} q_{GGA} q_{TTA}
Class 5: q_{CGT} q_{CTG} q_{GCT} q_{GTC} q_{TCG} q_{TGC}　　　　　□

We replace each of the Fourier coordinates in class i by the new Fourier coordinate q_i. We take q_i to be the *average* of the q_W in class i since this operation is better behaved with respect to writing down invariants.

We also record explicitly the linear transformation between the p_i and the q_i by recording a certain rational matrix which describes this transformation. This matrix is called the *specialized Fourier transform*. In general, this matrix will not be square. This is because there may be additional linear relations among the p_i which are encoded in the different q_i classes. Because of this ambiguity, we also explicitly list the inverse map.

It is possible to obtain the matrix that represents the specialized Fourier transform from the matrix that represents the full Fourier transform. If M represents the matrix of the full Fourier transform and N the matrix of the specialized Fourier transform, then the entry N_{ij}, indexed by the ith Fourier class and the jth probability class, is given by the formula

$$N_{ij} = \frac{1}{|C_i||D_j|} \sum_{U \in C_i} \sum_{W \in D_j} M_{UW}, \tag{15.4}$$

where C_i is the ith equivalence class of Fourier coordinates and D_j is the jth equivalence class of probability coordinates. We do not include the 0th equivalence class of Fourier coordinates in the previous formula. The inverse of the specialized Fourier transform is given by the formula

$$N'_{ij} = \frac{1}{k^n} \sum_{U \in C_j} \sum_{W \in D_i} M_{UW}. \tag{15.5}$$

Example 15.8 In the Jukes–Cantor DNA model on the 3-leaf claw tree the specialized Fourier transform matrix is

$$\begin{pmatrix} 1 & 1 & 1 & 1 & 1 \\ 1 & 1 & -\frac{1}{3} & -\frac{1}{3} & -\frac{1}{3} \\ 1 & -\frac{1}{3} & 1 & -\frac{1}{3} & -\frac{1}{3} \\ 1 & -\frac{1}{3} & 1 & 1 & -\frac{1}{3} \\ 1 & -\frac{1}{3} & -\frac{1}{3} & -\frac{1}{3} & \frac{1}{3} \end{pmatrix}.$$

$\square$

15.3 Description of website features

First, we give a brief description of the various algebraic notions described on the Small Trees website.

Dimension (D): The dimension of the model. This corresponds to the least number of parameters needed to specify the model.

Degree (d): The degree of the model. Algebraically, this is defined as the number of points in the intersection of the model and a generic (i.e. "random") subspace of dimension 4^n minus the dimension of the model.

Maximum Likelihood Degree (mld): The maximum likelihood degree of the model. See Section 3.3.

Number of Probability Coordinates (np): Number of equivalences classes of the probability coordinates. See the preceding section.

Number of Fourier Coordinates (nq): Number of equivalence classes of Fourier coordinates without counting Class 0 (see the preceding section). This is also the dimension of the smallest linear space that contains the model.

Specialized Fourier Transform: See the preceding section for a description.

Phylogenetic Invariants: A list of generators of the prime ideal of phylogenetic invariants. These are given in the Fourier coordinates.

Number of minimal generators (μ): The cardinality of the smallest set of generators that define the ideal of phylogenetic invariants.

Singularity Dimension (sD): The dimension of the set of singular points of the model. The singular locus of a model is very important to understand the geometry of the model. In particular, it is necessary for the ML degree computation.

Singularity Degree (sd): The algebraic degree of the set of singular points of the model.

The following tables summarize some of the computational results obtained for the small trees on the website. The '*' symbol means the corresponding entry could not be computed, a '-' symbol means the corresponding entry is not applicable. Table 15.1 describes small trees without molecular clock assumption and Table 15.2 describes 3-taxa trees with the molecular clock condition. The model N71 refers to the Neyman 2-state model [Neyman, 1971], all other models use the same terminology as in Section 4.5. We also list the maximum degree of a minimal generator (m).

	D	d	mld	np	nq	μ	sD	sd
N71	4	8	92	8	8	3	1	24
JC69	3	3	23	5	5	1	1	3
K80	6	12	*	10	10	9	3	22
K81	9	96	*	16	16	33	*	*
N71	5	4	14	8	8	2	2	4
JC69	5	34	*	15	13	33	*	*

Table 15.1. *Trees on three and four taxa without MC.*

	D	d	mld	np	nq	μ	m	sD	sd
N71	2	1	1	3	3	0	-	-	-
JC69	2	3	15	4	4	1	2	1	1
K80	4	6	190	7	7	4	4	2	10
K81	6	12	*	10	10	9	4	3	22
N71	1	1	1	2	2	0	-	-	-
JC69	1	3	7	3	3	1	3	0	2
K80	2	3	15	4	4	1	3	1	1
K81	3	3	40	5	5	1	3	1	3

Table 15.2. *Trees on 3-taxa with MC.*

15.4 Example

Here we describe the Jukes–Cantor model on the quartet tree (see Figure 15.2) in complete detail.

Dimension: $D = 5$ (note that there are only 5 independent parameters, one for each transition matrix.)

Degree: $d = 34$.

Number of Probability Coordinates: $np = 15$ and the classes are:

Class 1: p_{AAAA}

Class 2: $p_{\text{AAAC}}\ p_{\text{AAAT}}\ p_{\text{AAAG}}$

Class 3: $p_{\text{AACA}}\ p_{\text{AAGA}}\ p_{\text{AATA}}$

Class 4: $p_{\text{AACC}}\ p_{\text{AAGG}}\ p_{\text{AATT}}$

Class 5: $p_{\text{AACG}}\ p_{\text{AACT}}\ p_{\text{AAGC}}\ p_{\text{AAGT}}\ p_{\text{AATC}}\ p_{\text{AATG}}$

Class 6: $p_{\text{ACAA}}\ p_{\text{AGAA}}\ p_{\text{ATAA}}$

Class 7: $p_{\text{ACAC}}\ p_{\text{AGAG}}\ p_{\text{ATAT}}$

Class 8: $p_{\text{ACAG}}\ p_{\text{ACAT}}\ p_{\text{AGAC}}\ p_{\text{AGAT}}\ p_{\text{ATAC}}\ p_{\text{ATAG}}$

Class 9: $p_{\text{ACCA}}\ p_{\text{AGGA}}\ p_{\text{ATTA}}$

Class 10: $p_{\text{ACCC}}\ p_{\text{AGGG}}\ p_{\text{ATTT}}$

Class 11: $p_{\text{ACCG}}\ p_{\text{ACCT}}\ p_{\text{AGGC}}\ p_{\text{AGGT}}\ p_{\text{ATTC}}\ p_{\text{ATTG}}$

Class 12: $p_{\text{ACGA}}\ p_{\text{ACTA}}\ p_{\text{AGCA}}\ p_{\text{AGTA}}\ p_{\text{ATCA}}\ p_{\text{ATGA}}$

Class 13: $p_{\text{ACGC}}\ p_{\text{ACTC}}\ p_{\text{AGCG}}\ p_{\text{AGTG}}\ p_{\text{ATCT}}\ p_{\text{ATGT}}$

Class 14: $p_{\text{ACGG}}\ p_{\text{ACTT}}\ p_{\text{AGCC}}\ p_{\text{AGTT}}\ p_{\text{ATCC}}\ p_{\text{ATGG}}$

Class 15: $p_{\text{ACGT}}\ p_{\text{ACTG}}\ p_{\text{AGCT}}\ p_{\text{AGTC}}\ p_{\text{ATCG}}\ p_{\text{ATGC}}$

Here we are using the fact that for any $\mathbb{Z}_2 \times \mathbb{Z}_2$-based model, the four probabilities $p_{(i_1+j)(i_2+j)\cdots(i_n+j)}$ for $j \in \mathbb{Z}_2 \times \mathbb{Z}_2$ are identical. For example, the first class equals $\{p_{\text{AAAA}}, p_{\text{CCCC}}, p_{\text{GGGG}}, p_{\text{TTTT}}\}$.

Joint Probability Parameterization:

$$p_1 = a_0b_0c_0d_0e_0 + 3a_1b_0c_1d_1e_0 + 3a_0b_1c_1d_0e_1 + 3a_1b_1c_0d_1e_1 + 6a_1b_1c_1d_1e_1,$$

$$\begin{aligned}
p_2 = 3(&a_0b_1c_0d_0e_0 + 3a_1b_1c_1d_1e_0 + a_0b_0c_1d_0e_1 + 2a_0b_1c_1d_0e_1 \\
&+ a_1b_0c_0d_1e_1 + 2a_1b_1c_0d_1e_1 + 2a_1b_0c_1d_1e_1 + 4a_1b_1c_1d_1e_1),
\end{aligned}$$

$$\begin{aligned}
p_3 = 3(&a_0b_1c_1d_0e_0 + a_1b_1c_0d_1e_0 + 2a_1b_1c_1d_1e_0 + a_0b_0c_0d_0e_1 \\
&+ 2a_0b_1c_1d_0e_1 + 2a_1b_1c_0d_1e_1 + 3a_1b_0c_1d_1e_1 + 4a_1b_1c_1d_1e_1),
\end{aligned}$$

$$\begin{aligned}
p_4 = 3(&a_0b_0c_1d_0e_0 + a_1b_0c_0d_1e_0 + 2a_1b_0c_1d_1e_0 + a_0b_1c_0d_0e_1 \\
&+ 2a_0b_1c_1d_0e_1 + 2a_1b_1c_0d_1e_1 + 7a_1b_1c_1d_1e_1),
\end{aligned}$$

$$\begin{aligned}
p_5 = 6(&a_0b_1c_1d_0e_0 + a_1b_1c_0d_1e_0 + 2a_1b_1c_1d_1e_0 + a_0b_1c_0d_0e_1 + a_0b_0c_1d_0e_1 \\
&+ a_0b_1c_1d_0e_1 + a_1b_0c_0d_1e_1 + a_1b_1c_0d_1e_1 + 2a_1b_0c_1d_1e_1 + 5a_1b_1c_1d_1e_1),
\end{aligned}$$

$$\begin{aligned}
p_6 = 3(&a_1b_0c_1d_0e_0 + a_0b_0c_0d_1e_0 + 2a_1b_0c_1d_1e_0 + a_1b_1c_0d_0e_1 \\
&+ 2a_1b_1c_1d_0e_1 + 2a_1b_1c_0d_1e_1 + 3a_0b_1c_1d_1e_1 + 4a_1b_1c_1d_1e_1),
\end{aligned}$$

$$\begin{aligned}
p_7 = 3(&a_1b_1c_1d_0e_0 + a_0b_1c_0d_1e_0 + 2a_1b_1c_1d_1e_0 + a_1b_0c_0d_0e_1 + 2a_1b_1c_1d_0e_1 \\
&+ 2a_1b_1c_0d_1e_1 + a_0b_0c_1d_1e_1 + 2a_1b_0c_1d_1e_1 + 2a_0b_1c_1d_1e_1 + 2a_1b_1c_1d_1e_1)
\end{aligned}$$

$$\begin{aligned}
p_8 = 6(&a_1b_1c_1d_0e_0 + a_0b_1c_0d_1e_0 + 2a_1b_1c_1d_1e_0 + a_1b_1c_0d_0e_1 \\
&+ a_1b_0c_1d_0e_1 + a_1b_1c_1d_0e_1 + a_1b_0c_0d_1e_1 + a_1b_1c_0d_1e_1 \\
&+ a_0b_0c_1d_1e_1 + a_1b_0c_1d_1e_1 + 2a_0b_1c_1d_1e_1 + 3a_1b_1c_1d_1e_1),
\end{aligned}$$

$$\begin{aligned}
p_9 = 3(&a_1b_1c_0d_0e_0 + a_0b_1c_1d_1e_0 + 2a_1b_1c_1d_1e_0 + a_1b_0c_1d_0e_1 + 2a_1b_1c_1d_0e_1 \\
&+ a_0b_0c_0d_1e_1 + 2a_1b_1c_0d_1e_1 + 2a_1b_0c_1d_1e_1 + 2a_0b_1c_1d_1e_1 + 2a_1b_1c_1d_1e_1).
\end{aligned}$$

$$\begin{aligned}
p_{10} = 3(&a_1b_0c_0d_0e_0 + a_0b_0c_1d_1e_0 + 2a_1b_0c_1d_1e_0 + 3a_1b_1c_1d_0e_1 \\
&+ a_0b_1c_0d_1e_1 + 2a_1b_1c_0d_1e_1 + 2a_0b_1c_1d_1e_1 + 4a_1b_1c_1d_1e_1),
\end{aligned}$$

$$\begin{aligned}
p_{11} = 6(&a_1b_1c_0d_0e_0 + a_0b_1c_1d_1e_0 + 2a_1b_1c_1d_1e_0 + a_1b_0c_1d_0e_1 \\
&+ 2a_1b_1c_1d_0e_1 + a_1b_0c_0d_1e_1 + a_0b_1c_0d_1e_1 + a_1b_1c_0d_1e_1 \\
&+ a_0b_0c_1d_1e_1 + a_1b_0c_1d_1e_1 + a_0b_1c_1d_1e_1 + 3a_1b_1c_1d_1e_1),
\end{aligned}$$

$$\begin{aligned}
p_{12} = 6(&a_1b_1c_1d_0e_0 + a_1b_1c_0d_1e_0 + a_0b_1c_1d_1e_0 + a_1b_1c_1d_1e_0 \\
&+ a_1b_1c_0d_0e_1 + a_1b_0c_1d_0e_1 + a_1b_1c_1d_0e_1 + a_0b_0c_0d_1e_1 \\
&+ a_1b_1c_0d_1e_1 + 2a_1b_0c_1d_1e_1 + 2a_0b_1c_1d_1e_1 + 3a_1b_1c_1d_1e_1),
\end{aligned}$$

$$\begin{aligned}
p_{13} = 6(&a_1b_1c_1d_0e_0 + a_1b_1c_0d_1e_0 + a_0b_1c_1d_1e_0 + a_1b_1c_1d_1e_0 \\
&+ a_1b_0c_0d_0e_1 + 2a_1b_1c_1d_0e_1 + a_0b_1c_0d_1e_1 + a_1b_1c_0d_1e_1 \\
&+ a_0b_0c_1d_1e_1 + 2a_1b_0c_1d_1e_1 + a_0b_1c_1d_1e_1 + 3a_1b_1c_1d_1e_1),
\end{aligned}$$

$$\begin{aligned}
p_{14} = 6(&a_1b_0c_1d_0e_0 + a_1b_0c_0d_1e_0 + a_0b_0c_1d_1e_0 + a_1b_0c_1d_1e_0 + a_1b_1c_0d_0e_1 \\
&+ 2a_1b_1c_1d_0e_1 + a_0b_1c_0d_1e_1 + a_1b_1c_0d_1e_1 + 2a_0b_1c_1d_1e_1 + 5a_1b_1c_1d_1e_1),
\end{aligned}$$

$$\begin{aligned}
p_{15} = 6(&a_1b_1c_1d_0e_0 + a_1b_1c_0d_1e_0 + a_0b_1c_1d_1e_0 + a_1b_1c_1d_1e_0 + a_1b_1c_0d_0e_1 \\
&+ a_1b_0c_1d_0e_1 + a_1b_1c_1d_0e_1 + a_1b_0c_0d_1e_1 + a_0b_1c_0d_1e_1 + a_0b_0c_1d_1e_1 \\
&+ a_1b_0c_1d_1e_1 + a_0b_1c_1d_1e_1 + 4a_1b_1c_1d_1e_1).
\end{aligned}$$

Number of Fourier Coordinates: $nq = 13$. The classes are

Class 0: $q_{\text{AAAC}}\ q_{\text{AAAG}}\ q_{\text{AAAT}}\ q_{\text{AACA}} \cdots$

Class 1: q_{AAAA}

Class 2: $q_{\text{AACC}}\ q_{\text{AAGG}}\ q_{\text{AATT}}$

Class 3: $q_{\text{ACAC}}\ q_{\text{AGAG}}\ q_{\text{ATAT}}$

Class 4: $q_{\text{ACCA}}\ q_{\text{AGGA}}\ q_{\text{ATTA}}$

Class 5: $q_{\text{ACGT}}\ q_{\text{ACTG}}\ q_{\text{AGCT}}\ q_{\text{AGTC}}\ q_{\text{ATCG}}\ q_{\text{ATGC}}$

Class 6: $q_{\text{CAAC}}\ q_{\text{GAAG}}\ q_{\text{TAAT}}$

Class 7: $q_{\text{CACA}}\ q_{\text{GAGA}}\ q_{\text{TATA}}$

Class 8: $q_{\text{CAGT}}\ q_{\text{CATG}}\ q_{\text{GACT}}\ q_{\text{GATC}}\ q_{\text{TACG}}\ q_{\text{TAGC}}$

Class 9: $q_{\text{CCAA}}\ q_{\text{GGAA}}\ q_{\text{TTAA}}$

Class 10: $q_{\text{CCCC}}\ q_{\text{CCGG}}\ q_{\text{CCTT}}\ q_{\text{GGCC}}\ q_{\text{GGGG}}\ q_{\text{GGTT}}\ q_{\text{TTCC}}\ q_{\text{TTGG}}\ q_{\text{TTTT}}$

Class 11: $q_{\text{CGAT}}\ q_{\text{CTAG}}\ q_{\text{GCAT}}\ q_{\text{GTAC}}\ q_{\text{TCAG}}\ q_{\text{TGAC}}$

Class 12: $q_{\text{CGCG}}\ q_{\text{CGGC}}\ q_{\text{CTCT}}\ q_{\text{CTTC}}\ q_{\text{GCCG}}\ q_{\text{GCGC}}\ q_{\text{GTGT}}\ q_{\text{GTTG}}\ q_{\text{TCCT}}\ q_{\text{TCTC}}\ q_{\text{TGGT}}\ q_{\text{TGTG}}$

Class 13: $q_{\text{CGTA}}\ q_{\text{CTGA}}\ q_{\text{GCTA}}\ q_{\text{GTCA}}\ q_{\text{TCGA}}\ q_{\text{TGCA}}$

Fourier Parameterization:

$$q_1 = A_1 B_1 C_1 D_1 E_1 \qquad q_2 = A_1 B_2 C_1 D_1 E_2 \qquad q_3 = A_1 B_2 C_2 D_2 E_1$$

$$q_4 = A_1 B_1 C_2 D_2 E_2 \qquad q_5 = A_1 B_2 C_2 D_2 E_2 \qquad q_6 = A_2 B_2 C_2 D_1 E_1$$

$$q_7 = A_2 B_1 C_2 D_1 E_2 \qquad q_8 = A_2 B_2 C_2 D_1 E_2 \qquad q_9 = A_2 B_1 C_1 D_2 E_1$$

$$q_{10} = A_2 B_2 C_1 D_2 E_2 \qquad q_{11} = A_2 B_2 C_2 D_2 E_1 \qquad q_{12} = A_2 B_2 C_2 D_2 E_2$$

$$q_{13} = A_2 B_1 C_2 D_2 E_2$$

Specialized Fourier Transform:

```
1    1     1     1     1     1     1     1     1     1     1     1     1     1     1
1  -1/3  -1/3    1   -1/3    1   -1/3  -1/3  -1/3    1   -1/3  -1/3  -1/3    1   -1/3
1  -1/3    1   -1/3  -1/3  -1/3    1   -1/3  -1/3    1   -1/3  -1/3    1   -1/3  -1/3
1    1   -1/3  -1/3  -1/3  -1/3  -1/3  -1/3    1     1     1   -1/3  -1/3  -1/3  -1/3
1  -1/3  -1/3  -1/3    1   -1/3  -1/3    1   -1/3    1   -1/3    1   -1/3  -1/3   1/3
1  -1/3    1   -1/3  -1/3    1   -1/3  -1/3    1   -1/3  -1/3    1   -1/3  -1/3  -1/3
1    1   -1/3  -1/3  -1/3    1     1     1   -1/3  -1/3  -1/3  -1/3  -1/3  -1/3  -1/3
1  -1/3  -1/3  -1/3    1   -1/3  -1/3  -1/3  -1/3   1/3  -1/3   1/3  -1/3   1/3   1/3
1    1     1     1     1   -1/3  -1/3  -1/3  -1/3  -1/3  -1/3  -1/3  -1/3  -1/3  -1/3
1  -1/3  -1/3    1   -1/3  -1/3   1/9   1/9   1/9  -1/3   1/9   1/9   1/9  -1/3   1/9
1  -1/3    1   -1/3  -1/3  -1/3  -1/3   1/3  -1/3  -1/3   1/3  -1/3  -1/3   1/3   1/3
1  -1/3  -1/3   1/3  -1/3   1/3    0    1/3  -1/3    0     0     0    1/3  -1/3
1    1   -1/3  -1/3  -1/3  -1/3  -1/3  -1/3  -1/3  -1/3  -1/3   1/3   1/3   1/3   1/3
```

Inverse of Specialized Fourier Transform:

$$\begin{bmatrix}
\tfrac{1}{64} & \tfrac{3}{64} & \tfrac{3}{64} & \tfrac{3}{64} & \tfrac{3}{32} & \tfrac{3}{64} & \tfrac{3}{64} & \tfrac{3}{32} & \tfrac{3}{64} & \tfrac{9}{64} & \tfrac{3}{32} & \tfrac{3}{16} & \tfrac{3}{32} \\
\tfrac{3}{64} & -\tfrac{3}{64} & -\tfrac{3}{64} & \tfrac{9}{64} & \tfrac{3}{32} & -\tfrac{3}{64} & \tfrac{9}{64} & \tfrac{3}{32} & \tfrac{9}{64} & -\tfrac{9}{64} & -\tfrac{3}{32} & \tfrac{3}{16} & \tfrac{9}{32} \\
\tfrac{3}{64} & -\tfrac{3}{64} & \tfrac{9}{64} & \tfrac{3}{64} & \tfrac{3}{32} & \tfrac{9}{64} & -\tfrac{3}{64} & \tfrac{3}{32} & \tfrac{9}{64} & \tfrac{9}{64} & \tfrac{9}{32} & \tfrac{3}{16} & -\tfrac{3}{32} \\
\tfrac{3}{64} & \tfrac{9}{64} & \tfrac{3}{64} & -\tfrac{3}{64} & \tfrac{3}{32} & \tfrac{3}{64} & \tfrac{3}{64} & \tfrac{3}{32} & \tfrac{9}{64} & \tfrac{27}{64} & -\tfrac{3}{32} & \tfrac{3}{16} & \tfrac{3}{32} \\
\tfrac{3}{32} & -\tfrac{3}{32} & \tfrac{3}{32} & -\tfrac{3}{32} & \tfrac{3}{16} & \tfrac{3}{32} & -\tfrac{3}{32} & \tfrac{3}{16} & \tfrac{3}{32} & -\tfrac{9}{32} & \tfrac{3}{16} & \tfrac{3}{8} & \tfrac{3}{16} \\
\tfrac{3}{64} & \tfrac{9}{64} & \tfrac{3}{64} & \tfrac{3}{64} & \tfrac{3}{32} & \tfrac{9}{64} & \tfrac{9}{64} & \tfrac{9}{32} & -\tfrac{3}{64} & \tfrac{9}{64} & -\tfrac{3}{32} & \tfrac{3}{16} & \tfrac{3}{32} \\
\tfrac{3}{64} & -\tfrac{3}{64} & \tfrac{9}{64} & \tfrac{3}{64} & \tfrac{3}{32} & \tfrac{3}{64} & \tfrac{9}{64} & \tfrac{3}{32} & \tfrac{3}{64} & \tfrac{3}{64} & \tfrac{3}{32} & \tfrac{3}{16} & \tfrac{3}{32} \\
\tfrac{3}{32} & -\tfrac{3}{32} & -\tfrac{3}{32} & -\tfrac{3}{32} & \tfrac{3}{16} & -\tfrac{3}{32} & \tfrac{9}{32} & \tfrac{3}{16} & -\tfrac{3}{32} & \tfrac{3}{32} & \tfrac{3}{16} & 0 & -\tfrac{3}{16} \\
\tfrac{3}{64} & \tfrac{3}{64} & \tfrac{3}{64} & \tfrac{9}{64} & \tfrac{3}{32} & \tfrac{9}{64} & -\tfrac{3}{64} & \tfrac{3}{32} & \tfrac{3}{64} & \tfrac{3}{64} & \tfrac{3}{32} & \tfrac{3}{16} & \tfrac{3}{32} \\
\tfrac{3}{64} & \tfrac{9}{64} & \tfrac{9}{64} & \tfrac{9}{64} & \tfrac{9}{32} & -\tfrac{3}{64} & -\tfrac{3}{64} & \tfrac{3}{32} & \tfrac{3}{64} & -\tfrac{9}{64} & \tfrac{3}{32} & \tfrac{3}{16} & \tfrac{3}{32} \\
\tfrac{3}{32} & -\tfrac{3}{32} & -\tfrac{3}{32} & \tfrac{9}{32} & -\tfrac{3}{16} & -\tfrac{3}{32} & -\tfrac{3}{32} & \tfrac{3}{16} & -\tfrac{3}{32} & \tfrac{3}{32} & \tfrac{3}{16} & 0 & -\tfrac{3}{16} \\
\tfrac{3}{32} & -\tfrac{3}{32} & -\tfrac{3}{32} & -\tfrac{3}{32} & \tfrac{3}{16} & \tfrac{9}{32} & -\tfrac{3}{32} & -\tfrac{3}{16} & -\tfrac{3}{32} & \tfrac{3}{32} & -\tfrac{3}{16} & 0 & \tfrac{3}{16} \\
\tfrac{3}{32} & -\tfrac{3}{32} & \tfrac{9}{32} & -\tfrac{3}{32} & -\tfrac{3}{16} & -\tfrac{3}{32} & -\tfrac{3}{32} & \tfrac{3}{16} & -\tfrac{3}{32} & \tfrac{3}{32} & -\tfrac{3}{16} & 0 & \tfrac{3}{16} \\
\tfrac{3}{32} & \tfrac{9}{32} & -\tfrac{3}{32} & -\tfrac{3}{32} & \tfrac{3}{16} & -\tfrac{3}{32} & -\tfrac{3}{32} & \tfrac{3}{16} & -\tfrac{3}{32} & -\tfrac{9}{32} & \tfrac{3}{16} & \tfrac{3}{8} & \tfrac{3}{16} \\
\tfrac{3}{32} & -\tfrac{3}{32} & -\tfrac{3}{32} & -\tfrac{3}{32} & \tfrac{3}{16} & -\tfrac{3}{32} & -\tfrac{3}{32} & \tfrac{3}{16} & -\tfrac{3}{32} & \tfrac{3}{32} & \tfrac{3}{16} & -\tfrac{3}{8} & \tfrac{3}{16}
\end{bmatrix}$$

Phylogenetic Invariants: We computed the phylogenetic invariants using the results of [Sturmfels and Sullivant, 2005]. The invariants of degree 2 are

$$q_1 q_{10} - q_2 q_9, \qquad q_3 q_7 - q_4 q_6, \qquad q_3 q_8 - q_5 q_6,$$
$$q_4 q_8 - q_5 q_7, \qquad q_3 q_{13} - q_4 q_{11}, \qquad q_3 q_{12} - q_5 q_{11},$$
$$q_4 q_{12} - q_5 q_{13}, \qquad q_6 q_{13} - q_7 q_{11}, \qquad q_6 q_{12} - q_8 q_{11},$$
$$q_7 q_{12} - q_8 q_{13}.$$

The invariants of degree 3 associated to the left interior vertex are

$$q_1 q_{11} q_{11} - q_3 q_6 q_9, \qquad q_1 q_{11} q_{13} - q_3 q_7 q_9, \qquad q_1 q_{11} q_{12} - q_3 q_8 q_9,$$
$$q_1 q_{13} q_{11} - q_4 q_6 q_9, \qquad q_1 q_{13} q_{13} - q_4 q_7 q_9, \qquad q_1 q_{13} q_{12} - q_4 q_8 q_9,$$
$$q_1 q_{12} q_{11} - q_5 q_6 q_9, \qquad q_1 q_{12} q_{13} - q_5 q_7 q_9, \qquad q_1 q_{12} q_{12} - q_5 q_8 q_9,$$
$$q_2 q_{11} q_{11} - q_3 q_6 q_{10}, \qquad q_2 q_{11} q_{13} - q_3 q_7 q_{10}, \qquad q_2 q_{11} q_{12} - q_3 q_8 q_{10},$$
$$q_2 q_{13} q_{11} - q_4 q_6 q_{10}, \qquad q_2 q_{13} q_{13} - q_4 q_7 q_{10}, \qquad q_2 q_{13} q_{12} - q_4 q_8 q_{10},$$
$$q_2 q_{12} q_{11} - q_5 q_6 q_{10}, \qquad q_2 q_{12} q_{13} - q_5 q_7 q_{10}, \qquad q_2 q_{12} q_{12} - q_5 q_8 q_{10}.$$

The invariants of degree 3 associated to the right interior vertex are:

$$q_1 q_5 q_5 - q_3 q_4 q_2, \qquad q_1 q_5 q_8 - q_3 q_7 q_2, \qquad q_1 q_5 q_{12} - q_3 q_{13} q_2,$$
$$q_1 q_8 q_5 - q_6 q_4 q_2, \qquad q_1 q_8 q_8 - q_6 q_7 q_2, \qquad q_1 q_8 q_{12} - q_6 q_{13} q_2,$$
$$q_1 q_{12} q_5 - q_{11} q_4 q_2, \qquad q_1 q_{12} q_8 - q_{11} q_7 q_2, \qquad q_1 q_{12} q_{12} - q_{11} q_{13} q_2,$$
$$q_9 q_5 q_5 - q_3 q_4 q_{10}, \qquad q_9 q_5 q_8 - q_3 q_7 q_{10}, \qquad q_9 q_5 q_{12} - q_3 q_{13} q_{10},$$
$$q_9 q_8 q_5 - q_6 q_4 q_{10}, \qquad q_9 q_8 q_8 - q_6 q_7 q_{10}, \qquad q_9 q_8 q_{12} - q_6 q_{13} q_{10},$$
$$q_9 q_{12} q_5 - q_{11} q_4 q_{10}, \qquad q_9 q_{12} q_8 - q_{11} q_7 q_{10}, \qquad q_9 q_{12} q_{12} - q_{11} q_{13} q_{10}.$$

The maximum likelihood degree for this model is computationally infeasible with the current computer algebra programs. Nevertheless, one could still compute the singular locus of this (toric) model.

15.5 Using the invariants

In this section we report some of the experiments we have made for inferring small trees using phylogenetic invariants. These experiments were made using the invariants for trees with 4 taxa on the Kimura 3-parameter model that can be found on the small trees website which were computed using the techniques in [Sturmfels and Sullivant, 2005]. The results obtained show that phylogenetic invariants can be an efficient method for tree reconstruction.

We implemented an algorithm that performs the following tasks. Given 4 aligned DNA sequences s_1, s_2, s_3, s_4, it first counts the number of occurrences of each pattern for the topology $((s_1, s_2), s_3, s_4)$. Then it changes these absolute frequencies to Fourier coordinates. From this, we have the Fourier transforms in the other two possible topologies for trees with 4 species. We then evaluate all the phylogenetic invariants for the Kimura 3-parameter model in the Fourier coordinates of each tree topology. We call s_f^T the absolute value of this evaluation for the polynomial f and tree topology T. From these values $\{s_f^T\}_f$, we produce a score for each tree topology T, namely $s(T) = \sum_f |s_f^T|$. The algorithm then chooses the topology that has minimum score. There was an attempt to define the score as the Euclidean norm of the values s_f^T, but from our experiments, we deduced that the 1-norm chosen above performs better.

We then tested this algorithm for different sets of sequences. We used the program *evolver* from the package PAML [Yang, 1997] to generate sequences according to the Kimura 2-parameter model with transition/transversion ratio equal to 2 (typical value of mammalian DNA). In what follows we describe the different tests we made and the results we obtained. We generated 4-taxa trees with random branch lengths uniformly distributed between 0 and 1. We performed 600 tests for sequences of lengths between 1000 and 10,000. The percentage of trees correctly reconstructed can be seen in Figure 15.3.

We observed that our method fails to reconstruct the right tree mainly when the length of the interior edge of the tree is small compared to the other branch lengths. More precisely, in the trees that cannot be correctly inferred, the length of the interior edge is about 10% the average length of the other edges.

Our method was also tested by letting the edge lengths be normally distributed with a given mean μ. We chose the values $0.25, 0.05, 0.005$ for the mean μ, following [John et al., 2003]. We also let the standard deviation be 0.1μ. In this case, we tested DNA sequences of lengths ranging from 50 to 10.000. Here, we only display the results for sequences of length up to 1000. because for larger sequences we always inferred the correct tree. For each sequence length in $\{50, 100, 200, 300, 400, 500, 600, 700, 800, 900, 1000\}$, we generated edge lengths normally distributed with mean μ, using the statistical software package R [Ross and Gentleman, 1996] (Section 2.5.1). From this, we

Uniformly distributed edge lengths in (0,1)

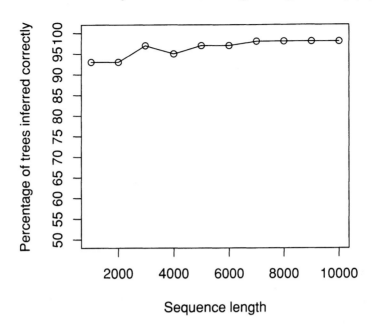

Fig. 15.3. Percentage of trees correctly reconstructed with random branch lengths uniformly distributed between 0 and 1.

constructed 10 sets of four sequences and applied the algorithm to each of them. We repeated this process 10 times; therefore generating 100 sequences for each mean and sequence length. The results are presented in Figure 15.4. We see

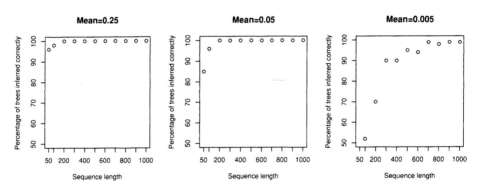

Fig. 15.4. Percentage of trees correctly reconstructed with edge lengths normally distributed with mean equal to 0.25, 0.05, 0.005.

that for $\mu = 0.25$ or $\mu = 0.05$, it is enough to consider sequences of length 200 to obtain a 100% efficiency. A much smaller mean such as $\mu = 0.0005$ was also tested. In this case, an efficiency over 90% was only obtained for sequences of length ≥ 3000.

16

The Strand Symmetric Model

Marta Casanellas

Seth Sullivant

This chapter is devoted to the study of strand symmetric Markov models on trees from the standpoint of algebraic statistics. A strand symmetric Markov model is one whose mutation probabilities reflect the symmetry induced by the double-stranded structure of DNA (see Chapter 4). In particular, a strand symmetric model for DNA must have the following equalities of probabilities in the root distribution:

$$\pi_A = \pi_T \text{ and } \pi_C = \pi_G$$

and the following equalities of probabilities in the transition matrices (θ_{ij}):

$$\theta_{AA} = \theta_{TT}, \theta_{AC} = \theta_{TG}, \theta_{AG} = \theta_{TC}, \theta_{AT} = \theta_{TA},$$

$$\theta_{CA} = \theta_{GT}, \theta_{CC} = \theta_{GG}, \theta_{CG} = \theta_{GC}, \theta_{CT} = \theta_{GA}.$$

Important special cases of strand symmetric Markov models are the group-based phylogenetic models including the Jukes–Cantor model and the Kimura 2 -and 3-parameter models. The *general strand symmetric model* or in this chapter just the *strand symmetric model* (SSM) has only these eight equalities of probabilities in the transition matrices and no further restriction on the transition probabilities. Thus, for each edge in the corresponding phylogenetic model, there are 6 free parameters.

Our motivation for the study of the SSM over the more commonly used group-based models comes from the fact that the SSM captures more biologically meaningful features of real DNA sequences that the group-based models fail to encode. For instance, in any group-based model, the stationary distribution of bases for a single species is always the uniform distribution. On the other hand, computational evidence [Yap and Pachter, 2004] suggests that the stationary distribution of bases for a single species is rarely uniform, but must always be strand symmetric. The SSM has the property that its stationary distributions can be general strand symmetric distributions. We express the SSM as an algebraic statistical model, so we make no mention of rate matrices in this chapter. In this sense, the SSM does not fit into the Felsenstein hierarchy (see Section 4.5), though we still feel it is an important model to study.

For the standard group-based models (i.e. Jukes–Cantor and Kimura), the transition matrices and the entire parameterization can be simultaneously diagonalized by means of the Fourier transform of the group $\mathbb{Z}_2 \times \mathbb{Z}_2$ [Evans and Speed, 1993, Székely *et al.*, 1993]. In addition to the practical uses of the Fourier transform for group-based models (see for example [Semple and Steel, 2003]), this diagonalization makes it possible to compute phylogenetic invariants for these models, by reducing the problem to the claw tree $K_{1,3}$ [Sturmfels and Sullivant, 2005]. Our goal in this chapter is to extend the Fourier transform from group-based models to the strand symmetric model. This is carried out in Section 16.1.

In Section 16.2 we focus on the case of the 3-taxa tree. The computation of phylogenetic invariants for the SSM in the Fourier coordinates is still not complete, though we report on what is known about these invariants. In particular, we describe all invariants of degree three and four. Section 16.4 is concerned with extending known invariants from the 3-taxa tree to an arbitrary tree. In particular, we describe how to extend the given degree three and four invariants from Section 16.2 to an arbitrary binary tree. To do this, we introduce *G-tensors* and explore their properties in Section 16.3.

In Section 16.5, we extend the "gluing" results for phylogenetic invariants in [Allman and Rhodes, 2004a] and [Sturmfels and Sullivant, 2005]. Our exposition and inspiration mainly comes from the work of Allman and Rhodes and we deduce that the problem of determining *defining* phylogenetic invariants for the strand symmetric model reduces to finding phylogenetic invariants for the claw tree $K_{1,3}$. Here *defining* means a set of polynomials which generate the ideal of invariants up to radical; that is, defining invariants have the same zero set as the whole ideal of invariants. This result is achieved by proving some "block diagonal" versions of results which appear in the Allman and Rhodes paper. This line of attack is the heart of Sections 16.3 and 16.5.

16.1 Matrix-valued Fourier transform

In this section we introduce the matrix-valued group-based models and show that the strand symmetric model is one such model. Then we describe the matrix-valued Fourier transform and the resulting simplification in the parameterization of these models, with special emphasis on the strand symmetric model.

Let T be a rooted tree with n-taxa. First, we describe the random variables associated to each vertex v in the tree in the matrix-valued group-based models. Each random variable X_v takes on kl states where k is the cardinality of a finite abelian group G and l is a parameter of the model. The states of the random variable are 2-tuples $\binom{j}{i}$ where $j \in G$ and $i \in \{0, 1, \ldots, l-1\}$.

Associated to the root node R in the tree is the root distribution R_i^j; that is. R_i^j is the probability that the random variable at the root is in state $\binom{j}{i}$. For each edge E of T, the double indexed set of parameters $E_{i_1 i_2}^{j_1 j_2}$ are the entries in the transition matrix associated to this edge. We use the convention that E is both the edge and the transition matrix associated to that edge, to avoid the need for introducing a third index on the matrices. Thus, $E_{i_1 i_2}^{j_1 j_2}$ is the conditional probability of making a transition from state $\binom{j_1}{i_1}$ to state $\binom{j_2}{i_2}$ along the edge E.

Definition 16.1 A phylogenetic model is a *matrix-valued group-based model* if for each edge, the matrix transition probabilities satisfy

$$E_{i_1 i_2}^{j_1 j_2} = E_{i_1 i_2}^{k_1 k_2}$$

when $j_1 - j_2 = k_1 - k_2$ (where the difference is taken in G) and the root distribution probabilities satisfy $R_i^j = R_i^k$ for all $j, k \in G$.

Example 16.2 Consider the strand symmetric model and make the identification of the states $\mathtt{A} = \binom{0}{0}$, $\mathtt{G} = \binom{0}{1}$, $\mathtt{T} = \binom{1}{1}$, and $\mathtt{C} = \binom{1}{0}$. One can check directly from the definitions that the strand symmetric model is a matrix-valued group-based model with $l = 2$ and $G = \mathbb{Z}_2$. $\qquad\square$

To avoid even more cumbersome notation, we will restrict attention to binary trees T and to the strand symmetric model for DNA. While the results of Section 16.2 and 16.4 are exclusive to the case of the SSM, our other results can be easily extended to arbitrary matrix-valued group-based models with the introduction of the more general Fourier transform, though we will not explain these generalizations here.

We assume all edges of T are directed away from the root R. Given an edge E of T let $s(E)$ denote the initial vertex of E and $t(E)$ the trailing vertex. Then the parameterization of the phylogenetic model is given as follows. The probability of observing states $_{i_1 i_2 \dots i_n}^{j_1 j_2 \dots j_n}$ at the leaves is

$$p_{i_1 i_2 \dots i_n}^{j_1 j_2 \dots j_n} = \sum_{\left(\binom{j_v}{i_v}\right) \in H} R_{i_R}^{j_R} \prod_E E_{i_{s(E)} i_{t(E)}}^{j_{s(E)} j_{t(E)}}$$

where the product is taken over all edges E of T and the sum is taken over the set

$$H = \left\{ \left(\binom{j_v}{i_v}\right)_{v \in \mathrm{Int}V(T)} \middle| \, j_v, i_v \in \{0, 1\} \right\}.$$

Here $\mathrm{Int}V(T)$ denotes the interior or non-leaf vertices of T.

Example 16.3 For the 3-leaf claw tree, the parameterization is given by the expression:

$$p_{ijk}^{lmn} = R_0^0 A_{0i}^{0l} B_{0j}^{0m} C_{0k}^{0n} + R_0^1 A_{0i}^{1l} B_{0j}^{1m} C_{0k}^{1n} + R_1^0 A_{1i}^{0l} B_{1j}^{0m} C_{1k}^{0n} + R_1^1 A_{1i}^{1l} B_{1j}^{1m} C_{1k}^{1n}.$$

The study of this particular tree will occupy a large part of the paper. $\qquad\square$

Because of the role of the group in determining the symmetry in the parameterization, the Fourier transform can be applied to make the parameterization simpler. We will not define the Fourier transform in general, only in the specific case of the group $\mathbb{Z}_2$. The Fourier transform applies to all of the probability coordinates, the transition matrices and the root distribution.

Definition 16.4 The *Fourier transform of the probability coordinates is*

$$q_{i_1 i_2 \ldots i_n}^{j_1 j_2 \ldots j_n} = \sum_{k_1, k_2, \ldots, k_n \in \{0,1\}} (-1)^{k_1 j_1 + k_2 j_2 + \cdots + k_n j_n} p_{i_1 i_2 \ldots i_n}^{k_1 k_2 \ldots k_n}.$$

The Fourier transform of the transition matrix E is the matrix e with entries

$$e_{i_1 i_2}^{j_1 j_2} = \frac{1}{2} \sum_{k_1, k_2 \in \{0,1\}} (-1)^{k_1 j_1 + k_2 j_2} E_{i_1 i_2}^{k_1 k_2}.$$

The Fourier transform of the root distribution is

$$r_i^j = \sum_{k \in \{0,1\}} (-1)^{kj} R_i^k.$$

It is easy to check that $e_{i_1 i_2}^{j_1 j_2} = 0$ if $j_1 + j_2 = 1 \in \mathbb{Z}_2$ and similarly that $r_i^j = 0$ if $j = 1$. In particular, writing e as a matrix, we see that the Fourier transform replaces the matrix E with a matrix e that is block diagonal. Generally, when working with our "hands on" the parameters (in particular in Section 16.2) we will write the transition matrices with only one superscript: $e_{i_1 i_2}^j = e_{i_1 i_2}^{jj}$ and the transformed root distribution $r_i = r_i^0$ with no superscript at all, though at other times it will be more convenient to have the extra superscript around, in spite of the redundancy.

Lemma 16.5 *In the Fourier coordinates the parameterization is given by the rule*

$$q_{i_1 i_2 \ldots i_n}^{j_1 j_2 \ldots j_n} = \sum_{(i_v) \in H} r_{i_r}^{j_r} \prod_e e_{i_{s(e)} i_{t(e)}}^{j_{s(e)}}$$

where $j_{s(e)}$ is the sum of j_l such that l is a leaf below $s(e)$ in the tree, $j_r = j_1 + \cdots + j_n$ and H denotes the set

$$H = \{(i_v)_{v \in IntV(T)} \text{ and } i_v \in \{0, 1\}\}.$$

Proof We can rewrite the parameterization in the probability coordinates as

$$p_{i_1 i_2 \ldots i_n}^{k_1 k_2 \ldots k_n} = \sum_{(i_v) \in H} \left(\sum_{(k_v) \in H'} R_{i_R}^{k_R} \prod_E E_{i_{s(E)} i_{t(E)}}^{k_{s(E)} k_{t(E)}} \right)$$

where H is the set defined in the lemma and

$$H' = \{(k_v)_{v \in IntV(T)} \text{ and } k_v \in \mathbb{Z}_2\}.$$

The crucial observation is that for any fixed values of $i_1, \ldots, i_n$ and $(i_v) \in H$, the expression inside the parentheses is a standard group-based model for $\mathbb{Z}_2$. Applying the Fourier transform we have the following expression for $q_{i_1 i_2 \ldots i_n}^{j_1 j_2 \ldots j_n}$:

$$\sum_{k_1,k_2,\ldots,k_n \in \{0,1\}} (-1)^{k_1 j_1 + k_2 j_2 + \cdots + k_n j_n} \sum_{(i_v) \in H} \left(\sum_{(k_v) \in H'} R_{i_R}^{k_R} \prod_E E_{i_{s(E)} i_{t(E)}}^{k_{s(E)} k_{t(E)}} \right)$$

and interchanging summations

$$\sum_{(i_v) \in H} \left(\sum_{k_1,k_2,\ldots,k_n \in \{0,1\}} (-1)^{k_1 j_1 + k_2 j_2 + \cdots + k_n j_n} \sum_{(k_v) \in H'} R_{i_R}^{k_R} \prod_E E_{i_{s(E)} i_{t(E)}}^{k_{s(E)} k_{t(E)}} \right).$$

By our crucial observation above, the expression inside the large parentheses is the Fourier transform of a group-based model and hence by results in [Evans and Speed, 1993] and [Székely *et al.*, 1993] the expression inside the parentheses factors in terms of the Fourier transforms of the transition matrices and root distribution in precisely the way illustrated in the statement of the lemma. $\qquad\square$

Definition 16.6 Given a tree T, the projective variety of the SSM given by T is denoted by $V(T)$. The notation $CV(T)$ denotes the affine cone over $V(T)$.

Proposition 16.7 (Linear Invariants)

$$q_{i_1 i_2 \ldots i_n}^{j_1 j_2 \ldots j_n} = 0 \quad \text{if} \quad j_1 + j_2 + \cdots + j_n = 1 \in \mathbb{Z}_2.$$

Proof The equation $j_1 + j_2 + \cdots + j_n = 1 \in \mathbb{Z}_2$ implies that in the parameterization every summand involves $r_{i_r}^1$ for some i_r. However, all of these parameters are zero. $\qquad\square$

The linear invariants in the previous lemma are equivalent to the fact that

$$p_{i_1 i_2 \ldots i_n}^{j_1 j_2 \ldots j_n} = p_{i_1 i_2 \ldots i_n}^{\bar{j}_1 \bar{j}_2 \ldots \bar{j}_n}$$

where $\bar{j} = 1 - j \in \mathbb{Z}_2$.

Up until now, we have implicitly assumed that all the matrices E involved were actually matrices of transition probabilities and that the root distribution R was an honest probability distribution. If we drop these conditions and look at the parameterization in the Fourier coordinates, we can, in fact, drop r from this representation altogether. That is, the variety parameterized by dropping the transformed root distribution r is the cone over the Zariski closure of the probabilistic parameterization.

Lemma 16.8 *In the Fourier coordinates there is an open subset of $CV(T)$, the cone over the strand symmetric model, that can be parameterized as*

$$q_{i_1 i_2 \ldots i_n}^{j_1 j_2 \ldots j_n} = \sum_{(i_v) \in H} \prod_e e_{i_{s(e)} i_{t(e)}}^{j_{s(e)}}$$

when $j_1 + \cdots + j_n = 0 \in \mathbb{Z}_2$, where $j_{s(e)}$ is the sum of j_l such that l is a leaf below $s(e)$ in the tree and H denotes the set

$$H = \{(i_v)_{v \in IntV(T)} \text{ and } i_v \in \{0,1\}\}.$$

Proof Due to the structure of the reparameterization of the SSM which we will prove in Section 16.3, it suffices to prove the lemma when T is the 3-leaf claw tree $K_{1,3}$. In this case, we are comparing the parameterizations

$$\phi : q_{ijk}^{mno} = r_0 a_{0i}^m b_{0j}^n c_{0k}^o + r_1 a_{1i}^m b_{1j}^n c_{1k}^o$$

and

$$\psi : q_{ijk}^{mno} = d_{0i}^m e_{0j}^n f_{0k}^o + d_{1i}^m e_{1j}^n f_{1k}^o.$$

In the second case, there are no conditions on the parameters. In the first parameterization, the stochastic assumption on the root distribution and transition matrices translates into the following restrictions on the Fourier parameters

$$r_0 = 1, \; a_{l0}^0 + a_{l1}^0 = 1, \; b_{l0}^0 + b_{l1}^0 = 1, \; c_{l0}^0 + c_{l1}^0 = 1$$

for $l = 0, 1$. We must show that for d, e, f belonging to some open subset U we can choose r, a, b, c with the prescribed restrictions which realize the same tensor Q, up to scaling. To do this, define

$$\delta_l = d_{l0}^0 + d_{l1}^0, \; \gamma_l = e_{l0}^0 + e_{l1}^0, \; \lambda_l = f_{l0}^0 + f_{l1}^0$$

for $l = 0, 1$ and take U the subset where these numbers are all non-zero. Set

$$a_{li}^m = \delta_l^{-1} d_{li}^m, \; b_{lj}^n = \gamma_l^{-1} e_{lj}^n, \; c_{lk}^o = \lambda_l^{-1} f_{lk}^o, \; r_0 = 1, \; \text{and} \; r_1 = \frac{\delta_1 \gamma_1 \lambda_1}{\delta_0 \gamma_0 \lambda_0}.$$

Clearly, all the parameters r, a, b, c satisfy the desired prescription. Furthermore, the parameterization with this choice of r, a, b, c differs from the original parameterization by a factor of $(\delta_0 \gamma_0 \lambda_0)^{-1}$. This proves that $\psi(U) \subset \text{Im}(\psi) \subset CV(T)$. On the other hand, we have that $V(T) \subset \text{Im}(\psi)$ because we can always take $d_{li}^m = r_l a_{li}^m$, $e_{lj}^n = b_{lj}^n$, $f_{lk}^o = c_{lk}^o$. Moreover it is clear that $\text{Im}(\psi)$ is a cone and hence $CV(T) \subset \text{Im}(\psi)$. The proof of the lemma is completed by taking the Zariski closure. $\qquad \square$

Example 16.9 In the particular instance of the 3-leaf claw tree the Fourier parameterization of the model is given by the formula

$$q_{ijk}^{mno} = a_{0i}^m b_{0j}^n c_{0k}^o + a_{1i}^m b_{1j}^n c_{1k}^o.$$

$\qquad \square$

16.2 Invariants for the 3-taxa tree

In this section, we will describe the degree 3 and degree 4 phylogenetic invariants for the claw tree $K_{1,3}$ on the strand symmetric model. We originally found these polynomial invariants using the computational algebra package MACAULAY2 [Grayson and Stillman, 2002] though we will give a combinatorial description of these invariants and proofs that they do, in fact, vanish on the strand symmetric model. We do not yet have a good statistical interpretation for the vanishing of these polynomials on the model.

It is an open problem to decide whether or not the 32 cubics and 18 quartics presented here generate the ideal of invariants, or even describe the SSM set theoretically. Computationally, we determined that they generate the ideal up to degree 4. Furthermore, one can show that neither the degree 3 nor the degree 4 invariants alone are sufficient to describe the variety set theoretically.

16.2.1 Degree 3 invariants

Proposition 16.10 *For each $l = 1, 2, 3$ let $m_l, n_l, o_l, i_l, j_l, k_l$ be indices in $\{0, 1\}$ such that $m_l + n_l + o_l = 0$, $m_1 = m_2$, $m_3 = 1 - m_1$, $n_1 = n_3$, $n_2 = 1 - n_1$, $o_2 = o_3$, and $o_1 = 1 - o_2$ in $\mathbb{Z}_2$. Let $f(m_\bullet, n_\bullet, o_\bullet, i_\bullet, j_\bullet, k_\bullet)$ be the polynomial in the Fourier coordinates described as the difference of determinants*

$$
\begin{vmatrix}
q^{m_1 n_1 o_1}_{i_1 j_1 k_1} & q^{m_2 n_1 o_1}_{i_2 j_1 k_1} & 0 \\
q^{m_1 n_2 o_2}_{i_1 j_2 k_2} & q^{m_2 n_2 o_2}_{i_2 j_2 k_2} & q^{m_3 n_3 o_2}_{i_3 j_3 k_2} \\
q^{m_1 n_2 o_3}_{i_1 j_2 k_3} & q^{m_2 n_2 o_3}_{i_2 j_2 k_3} & q^{m_3 n_3 o_3}_{i_3 j_3 k_3}
\end{vmatrix}
-
\begin{vmatrix}
q^{m_1 n_3 o_1}_{i_1 j_3 k_1} & q^{m_2 n_3 o_1}_{i_2 j_3 k_1} & 0 \\
q^{m_1 n_2 o_2}_{i_1 j_2 k_2} & q^{m_2 n_2 o_2}_{i_2 j_2 k_2} & q^{m_3 n_1 o_2}_{i_3 j_1 k_2} \\
q^{m_1 n_2 o_3}_{i_1 j_2 k_3} & q^{m_2 n_2 o_3}_{i_2 j_2 k_3} & q^{m_3 n_1 o_3}_{i_3 j_1 k_3}
\end{vmatrix}.
$$

Then $f(m_\bullet, n_\bullet, o_\bullet, i_\bullet, j_\bullet, k_\bullet)$ is a phylogenetic invariant for $K_{1,3}$ on the SSM.

Remark 16.11 The only non-zero cubic invariants for $K_{1,3}$ which arise from Proposition 16.10 are those satisfying $i_2 = 1 - i_1$, $i_3 = i_2, j_2 = j_1$, $j_3 = 1 - j_1$, $k_2 = 1 - k_1$ and $k_3 = k_1$. We maintain all the indices because they are necessary when we extend invariants to larger trees in Section 16.4. In total, we obtain 32 invariants in this way and we verified in MACAULAY2 [Grayson and Stillman, 2002] that these 32 invariants generate the ideal in degree 3.

Proof In order to prove this result, it is useful to write the parameterization in Fourier coordinates as

$$
q^{mno}_{ijk} =
\begin{vmatrix}
a^m_{0i} b^n_{0j} & -c^o_{1k} \\
a^m_{1i} b^n_{1j} & c^o_{0k}
\end{vmatrix}.
$$

We substitute the Fourier coordinates in $f(m_\bullet, n_\bullet, o_\bullet, i_\bullet, j_\bullet, k_\bullet)$ by their parameterization and we call the first determinant D_1 and the second one D_2 so that

$$
D_1 =
\begin{vmatrix}
\begin{vmatrix} a^{m_1}_{0i_1} b^{n_1}_{0j_1} & -c^{o_1}_{1k_1} \\ a^{m_1}_{1i_1} b^{n_1}_{1j_1} & c^{o_1}_{0k_1} \end{vmatrix} &
\begin{vmatrix} a^{m_2}_{0i_2} b^{n_1}_{0j_1} & -c^{o_1}_{1k_1} \\ a^{m_2}_{1i_2} b^{n_1}_{1j_1} & c^{o_1}_{0k_1} \end{vmatrix} &
0 \\[2em]
\begin{vmatrix} a^{m_1}_{0i_1} b^{n_2}_{0j_2} & -c^{o_2}_{1k_2} \\ a^{m_1}_{1i_1} b^{n_2}_{1j_2} & c^{o_2}_{0k_2} \end{vmatrix} &
\begin{vmatrix} a^{m_2}_{0i_2} b^{n_2}_{0j_2} & -c^{o_2}_{1k_2} \\ a^{m_2}_{1i_2} b^{n_2}_{1j_2} & c^{o_2}_{0k_2} \end{vmatrix} &
\begin{vmatrix} a^{m_3}_{0i_3} b^{n_3}_{0j_3} & -c^{o_2}_{1k_2} \\ a^{m_3}_{1i_3} b^{n_3}_{1j_3} & c^{o_2}_{0k_2} \end{vmatrix} \\[2em]
\begin{vmatrix} a^{m_1}_{0i_1} b^{n_2}_{0j_2} & -c^{o_3}_{1k_3} \\ a^{m_1}_{1i_1} b^{n_2}_{1j_2} & c^{o_3}_{0k_3} \end{vmatrix} &
\begin{vmatrix} a^{m_2}_{0i_2} b^{n_2}_{0j_2} & -c^{o_3}_{1k_3} \\ a^{m_2}_{1i_2} b^{n_2}_{1j_2} & c^{o_3}_{0k_3} \end{vmatrix} &
\begin{vmatrix} a^{m_3}_{0i_3} b^{n_3}_{0j_3} & -c^{o_3}_{1k_3} \\ a^{m_3}_{1i_3} b^{n_3}_{1j_3} & c^{o_3}_{0k_3} \end{vmatrix}
\end{vmatrix}.
$$

Observe that the indices in the first position are the same for each column in both determinants involved in $f(m_\bullet, n_\bullet, o_\bullet, i_\bullet, j_\bullet, k_\bullet)$. Similarly, the indices in the third position are the same for each row in both determinants. Using recursively the formula

$$
\left|\begin{array}{cc} \left|\begin{array}{cc} x_{1,1} & y \\ x_{2,1} & z \\ x_{3,1} \\ x_{4,1} \end{array}\right| & \left|\begin{array}{cc} x_{1,2} & y \\ x_{2,2} & z \\ x_{3,2} \\ x_{4,2} \end{array}\right| & \left|\begin{array}{cc} x_{1,3} & y \\ x_{2,3} & z \\ x_{3,3} \\ x_{4,3} \end{array}\right| \end{array}\right| = \left|\begin{array}{cccc} x_{1,1} & x_{1,2} & x_{1,3} & y \\ x_{2,1} & x_{2,2} & x_{2,3} & z \\ x_{3,1} & x_{3,2} & x_{3,3} & 0 \\ x_{4,1} & x_{4,2} & x_{4,3} & 0 \end{array}\right|
$$

it is easy to see that D_1 can be written as the following 6×6 determinant:

$$
D_1 = \left|\begin{array}{cccccc} a^{m_1}_{0i_1}b^{n_1}_{0j_1} & a^{m_2}_{0i_2}b^{n_1}_{0j_1} & 0 & -c^{o_1}_{1k_1} & 0 & 0 \\ a^{m_1}_{1i_1}b^{n_1}_{1j_1} & a^{m_2}_{1i_2}b^{n_1}_{1j_1} & 0 & c^{o_1}_{0k_1} & 0 & 0 \\ a^{m_1}_{0i_1}b^{n_2}_{0j_2} & a^{m_2}_{0i_2}b^{n_2}_{0j_2} & a^{m_3}_{0i_3}b^{n_3}_{0j_3} & 0 & -c^{o_2}_{1k_2} & 0 \\ a^{m_1}_{1i_1}b^{n_2}_{1j_2} & a^{m_2}_{1i_2}b^{n_2}_{1j_2} & a^{m_3}_{1i_3}b^{n_3}_{1j_3} & 0 & c^{o_2}_{0k_2} & 0 \\ a^{m_1}_{0i_1}b^{n_2}_{0j_2} & a^{m_2}_{0i_2}b^{n_2}_{0j_2} & a^{m_3}_{0i_3}b^{n_3}_{0j_3} & 0 & 0 & -c^{o_3}_{1k_3} \\ a^{m_1}_{1i_1}b^{n_2}_{1j_2} & a^{m_2}_{1i_2}b^{n_2}_{1j_2} & a^{m_3}_{1i_3}b^{n_3}_{1j_3} & 0 & 0 & c^{o_3}_{0k_3} \end{array}\right|.
$$

Now using a Laplacian expansion for the last 3 columns we obtain

$$
D_1 = -c^{o_1}_{0k_1}c^{o_2}_{1k_2}c^{o_3}_{0k_3} \left|\begin{array}{ccc} a^{m_1}_{0i_1}b^{n_1}_{0j_1} & a^{m_2}_{0i_2}b^{n_1}_{0j_1} & 0 \\ a^{m_1}_{1i_1}b^{n_2}_{1j_2} & a^{m_2}_{1i_2}b^{n_2}_{1j_2} & a^{m_3}_{1i_3}b^{n_3}_{1j_3} \\ a^{m_1}_{0i_1}b^{n_2}_{0j_2} & a^{m_2}_{0i_2}b^{n_2}_{0j_2} & a^{m_3}_{0i_3}b^{n_3}_{0j_3} \end{array}\right| -
$$

$$
c^{o_1}_{0k_1}c^{o_2}_{0k_2}c^{o_3}_{1k_3} \left|\begin{array}{ccc} a^{m_1}_{0i_1}b^{n_1}_{0j_1} & a^{m_2}_{0i_2}b^{n_1}_{0j_1} & 0 \\ a^{m_1}_{0i_1}b^{n_2}_{0j_2} & a^{m_2}_{0i_2}b^{n_2}_{0j_2} & a^{m_3}_{0i_3}b^{n_3}_{0j_3} \\ a^{m_1}_{1i_1}b^{n_2}_{1j_2} & a^{m_2}_{1i_2}b^{n_2}_{1j_2} & a^{m_3}_{1i_3}b^{n_3}_{1j_3} \end{array}\right| +
$$

$$
c^{o_1}_{1k_1}c^{o_2}_{1k_2}c^{o_3}_{0k_3} \left|\begin{array}{ccc} a^{m_1}_{1i_1}b^{n_1}_{1j_1} & a^{m_2}_{1i_2}b^{n_1}_{1j_1} & 0 \\ a^{m_1}_{1i_1}b^{n_2}_{1j_2} & a^{m_2}_{1i_2}b^{n_2}_{1j_2} & a^{m_3}_{1i_3}b^{n_3}_{1j_3} \\ a^{m_1}_{0i_1}b^{n_2}_{0j_2} & a^{m_2}_{0i_2}b^{n_2}_{0j_2} & a^{m_3}_{0i_3}b^{n_3}_{0j_3} \end{array}\right| +
$$

$$
c^{o_1}_{1k_1}c^{o_2}_{0k_2}c^{o_3}_{1k_3} \left|\begin{array}{ccc} a^{m_1}_{1i_1}b^{n_1}_{1j_1} & a^{m_2}_{1i_2}b^{n_1}_{1j_1} & 0 \\ a^{m_1}_{0i_1}b^{n_2}_{0j_2} & a^{m_2}_{0i_2}b^{n_2}_{0j_2} & a^{m_3}_{0i_3}b^{n_3}_{0j_3} \\ a^{m_1}_{1i_1}b^{n_2}_{1j_2} & a^{m_2}_{1i_2}b^{n_2}_{1j_2} & a^{m_3}_{1i_3}b^{n_3}_{1j_3} \end{array}\right|.
$$

Doing the same procedure for D_2 we see that its Laplace expansion has exactly the same 4 non-zero terms. □

16.2.2 Degree 4 invariants

We now explain the derivation of some non-trivial degree 4 invariants for the SSM on the $K_{1,3}$ tree. Each of the degree 4 invariants involves 16 of the non-zero Fourier coordinates which come from choosing two possible distinct sets of indices for the group-based indices. Up to symmetry, we may suppose these are from the tensors

$$
q^{mn_1o_1} \text{ and } q^{mn_2o_2}.
$$

Choose the ten indices $i_1, i_2, j_1, j_2, j_3, j_4, k_1, k_2, k_3, k_4$. Define the four matrices $q_i^{mn_1o_1}$ and $q_i^{mn_2o_2}$, $i \in \{i_1, i_2\}$ by

$$q_i^{mn_1o_1} = \begin{pmatrix} q_{ij_1k_1}^{mn_1o_1} & q_{ij_1k_2}^{mn_1o_1} \\ q_{ij_2k_1}^{mn_1o_1} & q_{ij_2k_2}^{mn_1o_1} \end{pmatrix} \text{ and } q_i^{mn_2o_2} = \begin{pmatrix} q_{ij_3k_3}^{mn_2o_2} & q_{ij_3k_4}^{mn_2o_2} \\ q_{ij_4k_3}^{mn_2o_2} & q_{ij_4k_4}^{mn_2o_2} \end{pmatrix}.$$

For any of these matrices, adding an extra subindex j means taking the jth row of the matrix, e.g. q_{1j}^{011} is the vector $(q_{1j0}^{011} \quad q_{1j1}^{011})$.

Theorem 16.12 *The 2×2 minors of the following 2×3 matrix are all degree 4 invariants of the SSM on the 3 leaf claw tree:*

$$\begin{pmatrix} |q_{i_1}^{mn_1o_1}| & \begin{vmatrix} q_{i_1j_1}^{mn_1o_1} \\ q_{i_2j_2}^{mn_1o_1} \end{vmatrix} + \begin{vmatrix} q_{i_1j_2}^{mn_1o_1} \\ q_{i_2j_1}^{mn_1o_1} \end{vmatrix} & |q_{i_2}^{mn_1o_1}| \\ |q_{i_1}^{mn_2o_2}| & \begin{vmatrix} q_{i_1j_3}^{mn_2o_2} \\ q_{i_2j_4}^{mn_2o_2} \end{vmatrix} + \begin{vmatrix} q_{i_1j_4}^{mn_2o_2} \\ q_{i_2j_3}^{mn_2o_2} \end{vmatrix} & |q_{i_2}^{mn_2o_2}| \end{pmatrix}.$$

These degree 4 invariants are not in the radical of the ideal generated by the degree 3 invariants above. Up to symmetry of the SSM on $K_{1,3}$, the 18 degree 4 invariants which arise in this way are the only minimal generators of the degree 4 piece of the ideal.

Proof The third sentence was proved computationally using **MACAULAY2** [Grayson and Stillman, 2002]. The second sentence follows by noting that all of the degree 3 invariants above use 3 different superscripts whereas the degree 4 invariants use only 2 different superscripts and the polynomials are multi-homogeneous in these indices. Thus, an assignment of arbitrary values to the tensors q^{000} and q^{011} and setting $q^{101} = q^{110} = 0$ creates a set of Fourier values which necessarily satisfy all degree 3 invariants but do not satisfy the degree 4 polynomials described in the statement of Theorem 16.12.

Now we will prove that these polynomials are, in fact, invariants of the SSM on $K_{1,3}$. The parameterization of $q_i^{mn_1o_1}$ and $q_i^{mn_2o_2}$ can be rewritten as

$$q_i^{mn_1o_1} = a_{0i}^m M_0^0 + a_{1i}^m M_1^0 \quad \text{and} \quad q_i^{mn_2o_2} = a_{0i}^m M_0^1 + a_{1i}^m M_1^1$$

where each of the four matrices M_0^0, M_1^0, M_0^1, and M_1^1 are arbitrary 2×2 matrices of rank 1. This follows by noting that the $q^{mn_1o_1}$ uses b^{n_1} and c^{o_1} in its description, $q^{mn_2o_2}$ uses b^{n_2} and c^{o_2} in its description, and simply rewriting these descriptions into matrix notation.

In particular, $q_{i_1}^{mn_1no_1}$ and $q_{i_2}^{mn_1o_1}$ lie in the plane spanned by the rank 1 matrices M_0^0 and M_1^0 and $q_{i_1}^{mn_2o_2}$ and $q_{i_2}^{mn_2o_2}$ lie in the plane spanned by the rank 1 matrices M_0^1 and M_1^0. Furthermore, the coefficients used to write these linear combinations are the same for the pair $\{q_{i_1}^{mn_1o_1}, q_{i_1}^{mn_2o_2}\}$, and for the pair $\{q_{i_2}^{mn_1o_1}, q_{i_2}^{mn_2o_2}\}$.

If we are given a general point on the variety of the SSM, none of the matrices $q_{i_1}^{mn_1o_1}$, $q_{i_2}^{mn_1o_1}$, $q_{i_1}^{mn_2o_2}$ or $q_{i_2}^{mn_2o_2}$ will have rank 1. This implies that, generically, each set of matrices

$$\mathcal{M}_1 = \{\lambda q_{i_1}^{mn_1o_1} + \gamma q_{i_2}^{mn_1o_1} | \lambda, \gamma \in \mathbb{C}\}$$

$$\mathcal{M}_2 = \{\lambda q_{i_1}^{mn_2o_2} + \gamma q_{i_2}^{mn_2o_2} | \lambda, \gamma \in \mathbb{C}\}$$

contains precisely 2 lines of rank 1 matrices. This is because the variety of 2×2 rank 1 matrices has degree 2. The set of values of λ and γ which produce these lines of rank 1 matrices are the same because of the way that $q_{i_1}^{mn_1o_1}$, $q_{i_2}^{mn_1o_1}$, $q_{i_1}^{mn_2o_2}$ or $q_{i_2}^{mn_2o_2}$ were written in terms of M_0^0, M_1^0, M_0^1 and M_1^1. In the first case, this set of λ and γ is the solution set of the quadratic equation

$$\left| \lambda q_{i_1}^{mn_1o_1} + \gamma q_{i_2}^{mn_1o_1} \right| = 0$$

and in the second case this set is the solution to the quadratic equation

$$\left| \lambda q_{i_1}^{mn_2o_2} + \gamma q_{i_2}^{mn_2o_2} \right| = 0.$$

These two quadrics having the same zero set is equivalent to the vanishing of the three 2×2 minors in the statement of the theorem. Since the minors in the statement of the theorem vanish for a general point on the parameterization they must vanish on the entire variety and hence are invariants of the SSM. $\qquad \square$

16.3 G-tensors

In this section we introduce the notion of a G-tensor which should be regarded as a multi-dimensional analog of a block diagonal matrix. We describe G-tensor multiplication and a certain variety defined for G-tensors which will be useful for extending invariants from the 3-leaf claw tree to arbitrary trees. This variety $^GV(4^{r_1}, 4^{r_2}, 4^{r_3})$ generalizes in a natural way the SSM on the claw tree $K_{1,3}$.

Notation. Let G be a group. For an n-tuple $\mathbf{j} = (j_1, \ldots, j_n) \in G^n$ we denote by $\sigma(\mathbf{j})$ the sum $j_1 + \cdots + j_n \in G$.

Definition 16.13 Let $q_{\mathbf{i_1} \cdots \mathbf{i_n}}^{\mathbf{j_1} \cdots \mathbf{j_n}}$ define a $4^{r_1} \times \cdots \times 4^{r_n}$ tensor Q where the upper indices are r_i-tuples in the group $G = \mathbb{Z}_2$. We say that Q is G-tensor if whenever $\sigma(\mathbf{j_1}) + \cdots + \sigma(\mathbf{j_n}) \neq 0$ in G, we have $q_{\mathbf{i_1} \cdots \mathbf{i_n}}^{\mathbf{j_1} \cdots \mathbf{j_n}} = 0$. If $n = 2$, then Q is called a G-matrix.

Lemma 16.14 *If Q is a $4 \times \cdots \times 4$ tensor arising from the SSM in the Fourier coordinates, then Q is a G-tensor. All the Fourier parameter matrices $e_{i_1 i_2}^{j_1 j_2}$ are G-matrices.*

Proof This is immediate from Proposition 16.7 and the comments following Definition 16.4. $\qquad \square$

Convention. Henceforth, we order any set of indices $\{(\begin{smallmatrix} j_1 \cdots j_t \\ i_1 \cdots i_t \end{smallmatrix})\}_{j_1,\ldots,j_t,i_1,\ldots,i_t}$ so that we put first those indices whose upper sum $\sigma(\mathbf{j}) = j_1 + \cdots + j_t$ is equal to zero.

Henceforth, we will use only Fourier coordinates and we will refer to the corresponding tensor of Fourier coordinates as Q.

An operation on tensors that we will use frequently is the tensor multiplication $*$ which is defined as follows. If R and Q are n-dimensional and m-dimensional tensors so that R (resp. Q) has κ states at the last index (resp. first index), the $(m + n - 2)$-dimensional tensor $R * Q$ is defined as

$$(R * Q)_{i_1,\ldots,i_{n+m-2}} = \sum_{j=1}^{\kappa} R_{i_1,\ldots,i_{n-1},j} \cdot Q_{j,i_n,\ldots,i_{n+m-2}}.$$

If R and Q are matrices this is the usual matrix multiplication. Note that if R and Q are G-tensors then $R * Q$ is also a G-tensor. We can also perform the $*$ operation on two varieties: if V and W are varieties of tensors then $V * W = \overline{\{R * Q | R \in V, Q \in W\}}$. If T' is a tree with taxa $v_1, \ldots, v_n$ and T'' is a tree with taxa $w_1, \ldots, w_m$ we call $T' * T''$ the tree obtained by identifying the vertices v_n and w_1, deleting this new vertex, and replacing the two corresponding edges by a single edge. This is a useful tool for constructing a reparameterization of the variety associated to an n-leaf tree T_n in terms of the parameterization for two smaller trees.

Proposition 16.15 *Let T_n be an n-leaf tree. Let $T_n = T_{n-1} * T_3$ be a decomposition of T_n into an $(n-1)$-leaf tree and a 3-leaf tree at a cherry. Then*

$$CV(T_n) = CV(T_{n-1}) * CV(T_3).$$

Proof Consider the parameterization for T_{n-1} written in the usual way as

$$q_{i_1 i_2 \ldots i_{n-2} k}^{j_1 j_2 \ldots j_{n-2} l} = \sum_{(i_v) \in H} \prod_e e_{i_{s(e)} i_{t(e)}}^{j_{s(e)}}.$$

and the parameterization for the 3-leaf tree T_3

$$r_{k i_{n-1} i_n}^{l j_{n-1} j_n} = \sum_{i_u \in \{0,1\}} \prod_f f_{i_{s(f)} i_{t(f)}}^{j_{s(f)}}$$

where u is the interior vertex of T_3. Writing the first tensor as Q and the second as R, we have an entry of $P = Q * R$ given by the formula

$$p_{i_1 i_2 \ldots i_n}^{j_1 j_2 \ldots j_n} = \sum_{k \in \{0,1\}} q_{i_1 i_2 \ldots i_{n-2} k}^{j_1 j_2 \ldots j_{n-2} l} r_{k i_{n-1} i_n}^{l j_{n-1} j_n}$$

where l satisfies $j_{n-1} + j_n + l = 0 \in \mathbb{Z}_2$. Let $\mathbf{e}$ and $\mathbf{f}$ denote the distinguished edges of T_{n-1} and T_3 respectively which are joined to make the tree T_n. Expanding the expression and regrouping, the right hand side becomes

$$\sum_{k \in \{0,1\}} \left(\sum_{(i_v) \in H} \prod_e e_{i_{s(e)} i_{t(e)}}^{j_{s(e)}} \right) \left(\sum_{i_u \in \{0,1\}} \prod_f f_{i_{s(f)} i_{t(f)}}^{j_{s(f)}} \right)$$

$$= \sum_{(i_v)\in H} \sum_{i_u\in\{0,1\}} \sum_{k\in\{0,1\}} \prod_e e^{j_{s(e)}}_{i_{s(e)}i_{t(e)}} \prod_f f^{j_{s(f)}}_{i_{s(f)}i_{t(f)}}.$$

$$= \sum_{(i_v)\in H} \sum_{i_u\in\{0,1\}} \prod_{e\neq\mathbf{e}} e^{j_{s(e)}}_{i_{s(e)}i_{t(e)}} \prod_{f\neq\mathbf{f}} f^{j_{s(f)}}_{i_{s(f)}i_{t(f)}} \left(\sum_{k\in\{0,1\}} e^l_{i_{s(\mathbf{e})}i_k} f^l_{i_k i_{t(\mathbf{f})}} \right).$$

The parenthesized expression is the product of the G-matrices $\mathbf{e}$ and $\mathbf{f}$. Replacing this expression with a new single G-matrix of parameters along the conjoined edge $\mathbf{ef}$ proves that $CV(T_{n-1}) * CV(T_3) \subseteq CV(T_n)$. Now expanding the reparameterization given in Lemma 16.8 as a sum on the vertex u we obtain the other inclusion. $\qquad\square$

Now we define a variety $^GV(4^{r_1}, 4^{r_2}, 4^{r_3})$ which plays a large role when we extend invariants.

Definition 16.16 For $l = 1, 2, 3$ let $\binom{j_l}{i_l}$ be a string of indices of length r_l. Let $_lM$ be an arbitrary G-matrix of size 4^{r_l} where the rows are indexed by $\{\binom{0}{0}, \binom{0}{1}, \binom{1}{0}, \binom{1}{1}\}$ and the columns are indexed by the 4^{r_l} indices $\binom{j_l}{i_l}$. Define the parameterization $Q = \psi_{r_1,r_2,r_3}(_1M, _2M, _3M)$ by

$$Q^{\mathbf{j_1 j_2 j_3}}_{\mathbf{i_1 i_2 i_3}} = \sum_{i\in\{0,1\}} {_1M}^{\sigma(\mathbf{j_1})\mathbf{j_1}}_{i\mathbf{i_1}} {_2M}^{\sigma(\mathbf{j_2})\mathbf{j_2}}_{i\mathbf{i_2}} {_3M}^{\sigma(\mathbf{j_3})\mathbf{j_3}}_{i\mathbf{i_3}}$$

if $\sigma(\mathbf{j_1}) + \sigma(\mathbf{j_2}) + \sigma(\mathbf{j_3}) = 0$ and $Q^{\mathbf{j_1 j_2 j_3}}_{\mathbf{i_1 i_2 i_3}} = 0$ if $\sigma(\mathbf{j_1}) + \sigma(\mathbf{j_2}) + \sigma(\mathbf{j_3}) = 1$. The projective variety that is the Zariski closure of the image of ψ_{r_1,r_2,r_3} is denoted $^GV(4^{r_1}, 4^{r_2}, 4^{r_3})$. The affine cone over this variety is $C^GV(4^{r_1}, 4^{r_2}, 4^{r_3})$.

Remark 16.17 By the definition of $^GV(4^{r_1}, 4^{r_2}, 4^{r_3})$, we have that any $Q \in$ $^GV(4^{r_1}, 4^{r_2}, 4^{r_3})$ is a G-tensor. Furthermore $^GV(4, 4, 4)$ is equal to the variety defined by the SSM on the 3-leaf claw tree $K_{1,3}$.

Besides the fact that $^GV(4^{r_1}, 4^{r_2}, 4^{r_3})$ is equal to the SSM when $r_1 = r_2 = r_3 = 1$, the importance of this variety for the strand symmetric model comes from the fact that $^GV(4^{r_1}, 4^{r_2}, 4^{r_3})$ contains the SSM for any binary tree as illustrated by the following proposition.

Proposition 16.18 Let T be a binary tree and v an interior vertex. Suppose that removing v from T partitions the leaves of T into the three sets $\{1, \ldots, r_1\}$, $\{r_1 + 1, \cdots, r_1 + r_2\}$, and $\{r_1 + r_2 + 1 \ldots, r_1 + r_2 + r_3\}$. Then the SSM on T is a subvariety of $^GV(4^{r_1}, 4^{r_2}, 4^{r_3})$.

In the proposition, the indices in the Fourier coordinates for the SSM are grouped in the natural way according to the tripartition of the leaves.

Proof In the parametric representation

$$q^{j_1 j_2 \ldots j_n}_{i_1 i_2 \ldots i_n} = \sum_{(i_v)\in H} \prod_e e^{j_{s(e)}}_{i_{s(e)} i_{t(e)}}$$

perform the sum associated to the vertex v first. This realizes the G-tensor Q as the sum over the product of entries of three G-tensors. □

Our goal for the remainder of this section is to prove a result analogous to Theorem 7 in [Allman and Rhodes, 2004a]. This theorem will provide a method for explicitly determining the ideal of invariants for $^G V(4^{r_1}, 4^{r_2}, 4^{r_3})$ from the ideal of invariants for $^G V(4, 4, 4)$. Denote by $^G M(2l, 2m)$ the set of $2l \times 2m$ G-matrices. A fundamental observation is that if $r_3' \geq r_3$ then

$$C^G V(4^{r_1}, 4^{r_2}, 4^{r_3'}) = C^G V(4^{r_1}, 4^{r_2}, 4^{r_3}) * {}^G M(4^{r_3}, 4^{r_3'}).$$

Thus, we need to understand the $*$ operation when V and W are "well-behaved" varieties.

Lemma 16.19 *Let* $V \subset {}^G M(2l, 4)$ *be a variety with* $V * {}^G M(4, 4) = V$. *Let* I *be the vanishing ideal of* V. *Let* K *be the ideal of* 3×3 G-*minors of the* $2l \times 2m$ G-*matrix of indeterminates* Q. *Let* Z *be* $2m \times 4$ G-*matrix of indeterminates and*

$$L \; = \; \langle \operatorname{coeff}_Z(f(Q * Z)) | f \in \operatorname{gens}(I) \rangle.$$

Then $K + L$ *is the vanishing ideal of* $W = V * {}^G M(4, 2m)$.

By a G-*minor* we mean a minor which involves only the non-zero entries in the G-matrix Q.

Proof A useful fact is that

$$L \; = \; \langle f(Q * A) | f \in I, A \in {}^G M(2m, 4) \rangle.$$

Let J be the vanishing ideal of W. By the definition of W, all the polynomials in K must vanish on it. Moreover, if $f(Q * A)$ is a polynomial in L, then it vanishes at all the points of the form $P * B$, for any $P \in V$ and $B \in {}^G M(4, 2m)$. Indeed, as $P * B * A \in V$ and $f \in I$ we have $f(P * B * A) = 0$. As all the points of W are of this form, we obtain the inclusion $K + L \subseteq J$. Our goal is to show that $J \subseteq K + L$.

Since $V * {}^G M(4, 4) = V$, we must also have $W * {}^G M(2m, 2m) = W$. This implies that there is an action of $Gl(\mathbb{C}, m) \times Gl(\mathbb{C}, m)$ on W and hence, any graded piece of J, the vanishing ideal of W, is a representation of $Gl(\mathbb{C}, m) \times Gl(\mathbb{C}, m)$. Let J_d be the dth graded piece of J. Since $Gl(\mathbb{C}, m) \times Gl(\mathbb{C}, m)$ is reductive, we just need to show each irreducible subspace M of J_d belongs to $K + L$. By construction, $K + L$ is also invariant under the action of $Gl(\mathbb{C}, m) \times Gl(\mathbb{C}, m)$ and, hence, it suffices to show that there exists a polynomial $f \in M$ such that $f \in K + L$.

Let $f \in M$ be an arbitrary polynomial in the irreducible representation M. Let P be a $2l \times 4$ G-matrix of indeterminates. Suppose that for all $B \in {}^G M(4, 2m)$, $f(P * B) \equiv 0$. This implies that f vanishes when evaluated at any G-matrix Q which has rank 2 in both components. Hence, $f \in K$.

If $f \notin K$ there exists a $B \in {}^G M(4, 2m)$ such that $f_B(P) := f(P * B) \not\equiv 0$.

Renaming the P indeterminates we can take D to be a matrix in $^G(2m, 4)$ formed by 1s and 0s such that $f_B(Q * D) \not\equiv 0$. Since $f \in J$, we must have $f_B(P) \in I$. Therefore $f_B(Q * D) \in L$. Let $B' = D * B \in {}^GM(2m, 2m)$. Although $B' \notin Gl(\mathbb{C}, m) \times Gl(\mathbb{C}, m)$, the representation M must be closed and hence $f(Q * B') = f_B(Q * D) \in M$ which completes the proof. $\qquad\square$

Proposition 16.20 *Generators for the vanishing ideal of $^GV(4^{r_1}, 4^{r_2}, 4^{r_3})$ are explicitly determined by generators for the vanishing ideal of $^GV(4, 4, 4)$.*

Proof Starting with $^GV(4, 4, 4)$, apply the preceding lemma three times. Now we will explain how to compute these polynomials explicitly. For $l = 1, 2, 3$ let Z_l be a $4^{r_l} \times 4$ G-matrix of indeterminates. This G-matrix Z_l acts on the $4^{r_1} \times 4^{r_2} \times 4^{r_3}$ tensor Q by G-tensor multiplication in the lth coordinate. For each $f \in \text{gens}(I)$, where I is the vanishing ideal of $^GV(4, 4, 4)$, we construct the polynomials $\text{coeff}_Z f(Q * Z_1 * Z_2 * Z_3)$. That is, we construct the $4 \times 4 \times 4$ G-tensor $Q * Z_1 * Z_2 * Z_3$, plug this into f and expand, and extract, for each Z monomial, the coefficient, which is a polynomial in the entries of Q. Letting f range over all the generators of I determines an ideal L.

We can also flatten the 3-way G-tensor Q to a G-matrix in three different ways. For instance, we can flatten it to a $4^{r_1} \times 4^{r_2 + r_3}$ G-matrix grouping the last two coordinates together. Taking the ideal generated by the 3×3 G-minors in these three flattenings yields an ideal K. The ideal $K + L$ generates the vanishing ideal of $^GV(4^{r_1}, 4^{r_2}, 4^{r_3})$. $\qquad\square$

16.4 Extending invariants

In this section we will show how to derive invariants for arbitrary trees from the invariants introduced in Section 16.2. We also introduce the degree 3 determinantal flattening invariants which arise from flattening the n-way G-tensor associated to a tree T under the SSM along an edge of the tree. The idea behind all of our results is to use the embedding of the SSM into the variety $^GV(4^{r_1}, 4^{r_2}, 4^{r_3})$.

Let T be a tree with n-taxa on the SSM and let v be any interior vertex. Removing v creates a tripartition of the leaves into three sets of cardinalities r_1, r_2 and r_3, which we may suppose, without loss of generality, are the sets $\{1, \ldots, r_1\}$, $\{r_1 + 1, \ldots, r_1 + r_2\}$, and $\{r_1 + r_2 + 1, \ldots, r_1 + r_2 + r_3\}$.

Proposition 16.21 *Let $f(m_\bullet, n_\bullet, o_\bullet, i_\bullet, j_\bullet, k_\bullet)$ be one of the degree 3 invariants for the 3-taxa tree $K_{1,3}$ introduced in Proposition 16.10. For each $l = 1, 2, 3$ we choose sets of indices $\mathbf{m_l}, \mathbf{i_l} \in \{0, 1\}^{r_1}$, $\mathbf{n_l}, \mathbf{j_l} \in \{0, 1\}^{r_2}$, and $\mathbf{o_l}, \mathbf{k_l} \in \{0, 1\}^{r_3}$ such that $\sigma(\mathbf{m_l}) = m_l$ $\sigma(\mathbf{n_l}) = n_l$ and $\sigma(\mathbf{o_l}) = o_l$. Then*

$$
f(\mathbf{m_\bullet}, \mathbf{n_\bullet}, \mathbf{o_\bullet}, \mathbf{i_\bullet}, \mathbf{j_\bullet}, \mathbf{k_\bullet})
$$
$$
= \begin{vmatrix} q_{i_1 j_1 k_1}^{m_1 n_1 o_1} & q_{i_2 j_1 k_1}^{m_2 n_1 o_1} & 0 \\ q_{i_1 j_2 k_2}^{m_1 n_2 o_2} & q_{i_2 j_2 k_2}^{m_2 n_2 o_2} & q_{i_3 j_3 k_2}^{m_3 n_3 o_2} \\ q_{i_1 j_2 k_3}^{m_1 n_2 o_3} & q_{i_2 j_2 k_3}^{m_2 n_2 o_3} & q_{i_3 j_3 k_3}^{m_3 n_3 o_3} \end{vmatrix} - \begin{vmatrix} q_{i_1 j_3 k_1}^{m_1 n_3 o_1} & q_{i_2 j_3 k_1}^{m_2 n_3 o_1} & 0 \\ q_{i_1 j_2 k_2}^{m_1 n_2 o_2} & q_{i_2 j_2 k_2}^{m_2 n_2 o_2} & q_{i_3 j_1 k_2}^{m_3 n_1 o_2} \\ q_{i_1 j_2 k_3}^{m_1 n_2 o_3} & q_{i_2 j_2 k_3}^{m_2 n_2 o_3} & q_{i_3 j_1 k_3}^{m_3 n_1 o_3} \end{vmatrix}
$$

is a phylogenetic invariant for T.

Proof The polynomial $f(\mathbf{m_\bullet}, \mathbf{n_\bullet}, \mathbf{o_\bullet}, \mathbf{i_\bullet}, \mathbf{j_\bullet}, \mathbf{k_\bullet})$ must vanish on the variety $^{G}V(4^{r_1}, 4^{r_2}, 4^{r_3})$. This is because choosing $\mathbf{m_1}, \mathbf{m_2}, \ldots$ in the manner specified corresponds to choosing a $3 \times 3 \times 3$ subtensor of Q which belongs to a $4 \times 4 \times 4$ G-subtensor of Q (after flattening to a 3-way tensor). Since $^{G}V(4, 4, 4)$ arises as a projection of $^{G}V(4^{r_1}, 4^{r_2}, 4^{r_3})$ onto this G-subtensor, $f(\mathbf{m_\bullet}, \mathbf{n_\bullet}, \mathbf{o_\bullet}, \mathbf{i_\bullet}, \mathbf{j_\bullet}, \mathbf{k_\bullet})$ belongs to the corresponding elimination ideal. Since the variety of the SSM for T is contained in the variety $^{G}V(4^{r_1}, 4^{r_2}, 4^{r_3})$, $f(\mathbf{m_\bullet}, \mathbf{n_\bullet}, \mathbf{o_\bullet}, \mathbf{i_\bullet}, \mathbf{j_\bullet}, \mathbf{k_\bullet})$ is an invariant for the SSM on T. $\qquad\square$

Similarly, we can extend the construction of degree four invariants to arbitrary trees T by replacing the indices in their definition with vectors of indices. We omit the proof which follows the same lines as the preceding proposition.

Proposition 16.22 *Let* $\mathbf{m}, \mathbf{i_l} \in \{0, 1\}^{r_1}$, $\mathbf{n_l}, \mathbf{j_l} \in \{0, 1\}^{r_2}$, *and* $\mathbf{o_l}, \mathbf{k_l} \in \{0, 1\}^{r_3}$. *Then the three 2×2 minors of*

$$
\begin{pmatrix}
\left| q_{\mathbf{i_1}}^{\mathbf{mn_1o_1}} \right| & \begin{vmatrix} q_{\mathbf{i_1j_1}}^{\mathbf{mn_1o_1}} \\ q_{\mathbf{i_2j_2}}^{\mathbf{mn_1o_1}} \end{vmatrix} + \begin{vmatrix} q_{\mathbf{i_1j_2}}^{\mathbf{mn_1o_1}} \\ q_{\mathbf{i_2j_1}}^{\mathbf{mn_1o_1}} \end{vmatrix} & \left| q_{\mathbf{i_2}}^{\mathbf{mn_1o_1}} \right| \\[4ex]
\left| q_{\mathbf{i_1}}^{\mathbf{mn_2o_2}} \right| & \begin{vmatrix} q_{\mathbf{i_1j_3}}^{\mathbf{mn_2o_2}} \\ q_{\mathbf{i_2j_4}}^{\mathbf{mn_2o_2}} \end{vmatrix} + \begin{vmatrix} q_{\mathbf{i_1j_4}}^{\mathbf{mn_2o_2}} \\ q_{\mathbf{i_2j_3}}^{\mathbf{mn_2o_2}} \end{vmatrix} & \left| q_{\mathbf{i_2}}^{\mathbf{mn_2o_2}} \right|
\end{pmatrix}
$$

are are all degree 4 invariants of the SSM on the tree T.

We now describe the determinantal edge invariants which arise by flattening the G-tensor Q to a matrix along each edge of the tree. As we shall see, their existence is already implied by Proposition 16.20. We make the special point of describing them here because they will be useful in the next section.

Let e be an edge in the tree T. Removing this edge partitions the leaves of T into two sets of size r_1 and r_2. The G-tensor Q flattens to a $4^{r_1} \times 4^{r_2}$ G-matrix R. Denote by $\mathcal{F}_e$ the set of 3×3 G-minors of R.

Proposition 16.23 *The 3×3 G-minors $\mathcal{F}_e$ are invariants of the SSM on T.*

Proof The edge e is incident to some interval vertex v of T. These 3×3 G-minors are in the ideal of $^{G}V(4^{r_1}, 4^{r'_2}, 4^{r'_3})$ associated to flattening the tensor Q to a 3-way G tensor at this vertex. Then by Proposition 16.18, $\mathcal{F}_e$ are invariants of the SSM on T. $\qquad\square$

16.5 Reduction to $K_{1,3}$

In this section, we explain how the problem of computing defining invariants for the SSM on a tree T reduces to the problem of computing them on the claw tree $K_{1,3}$. Our statements and proofs are intimately related to the results of Allman and Rhodes [Allman and Rhodes, 2004a].

Given an internal vertex v of T, denote by $^{G}V_{v}$ the variety $^{G}V(4^{r_1}, 4^{r_2}, 4^{r_3})$ associated with flattening the G-tensor Q to a 3-way tensor according to the tripartition induced by v.

Theorem 16.24 *Let T be a binary tree. For each $v \in \mathrm{Int}V(T)$, let $\mathcal{F}_v$ be a set of invariants which define the variety $^{G}V_v$ set-theoretically. Then*

$$CV(T) = \bigcap_{v \in \mathrm{Int}V(T)} {}^{G}V_v$$

and hence

$$\mathcal{F}_{flat}(T) = \bigcup_{v \in \mathrm{Int}V(T)} \mathcal{F}_v$$

are a defining set of invariants for the SSM on T.

The theorem reduces the computation of defining invariants to $K_{1,3}$ since a defining set of invariants for $^{G}V(4^{r_1}, 4^{r_2}, 4^{r_3})$ can be determined from a set of defining invariants for $^{G}V(4, 4, 4) = V(K_{1,3})$. Given the reparameterization result of Section 16.3, it will suffice to prove the following lemma about the $*$ operation on G-matrix varieties.

Lemma 16.25 *Let $V \subseteq {}^{G}M(2l, 4)$ and $W \subseteq {}^{G}M(4, 2m)$ be two varieties such that $V = V * {}^{G}M(4, 4)$ and $W = {}^{G}M(4, 4) * W$. Then*

$$V * W = \left(V * {}^{G}M(4, 2m) \right) \cap \left({}^{G}M(2l, 4) * W \right).$$

Proof Denote by U the variety on the right hand side of the equality. Since both of the component varieties of U contain $V * W$, we must have $V * W \subseteq U$. Our goal is to show the reverse inclusion. Let $Q \in U$. This matrix can be visualized as a block diagonal matrix:

$$Q = \begin{pmatrix} Q_0 & 0 \\ 0 & Q_1 \end{pmatrix}.$$

Since $Q \in U$, it must be the case that the rank of Q_0 and Q_1 are both less than or equal to 2. Thus, we can factorize Q as $Q = R * S$ where $R \in {}^{G}M(2l, 4)$ and $S \in {}^{G}M(4, 2m)$. Without loss of generality, we may suppose that the factorization $Q = R * S$ is non-degenerate in the sense that the rank of each of the matrices R and S has only rank(Q) non-zero rows. Our goal is to show that $R \in V$ and $S \in W$ as this will imply the theorem.

By our assumption that the factorization $Q = R * S$ is non-degenerate, there exists a G-matrix $A \in {}^{G}M(2m, 4)$ such that $Q * A = R * S * A = R$ (We call A the *pseudo-inverse* of S). Augmenting the matrix A with extra 0-columns, we get a G-matrix $A' \in {}^{G}M(2m, 2m)$. Then $Q * A' \in V * {}^{G}M(4, 2m)$ since Q is in $V * {}^{G}M(4, 2m)$ and $V * {}^{G}M(4, 2m)$ is closed under multiplication by G-matrices on the right. On the other hand, the natural projection of $Q * A'$ to $^{G}M(2l, 4)$ is $Q * A = R$. Since the projection $V * {}^{G}M(4, 2m) \rightarrow {}^{G}M(2l, 4)$ is the variety V because $V = V * {}^{G}M(4, 4)$, we have $R \in V$. A similar argument yields $S \in W$ and completes the proof. $\square$

Now we are in a position to give the proof the Theorem 16.24.

Proof We proceed by induction on n the number of leaves of T. If $n = 3$ there is nothing to show since this is the 3-leaf claw tree $K_{1,3}$. Let T by a binary n-taxa tree. The tree T has a cherry T_3, and thus we can represent the tree $T = T_{n-1} * T_3$ and the resulting variety as $V(T) = V(T_{n-1}) * V(T_3)$ by the reparameterization. Now we apply the induction hypothesis to T_{n-1} and T_3. The varieties $V(T_{n-1})$ and $V(T_3)$ have the desired representation as intersections of $^G V_v$. By the preceding lemma, it suffices to show that this representation extends to the variety $V(T_{n-1}) * {^G M}(4, 16)$ and $^G M(4^{n-1}, 4) * V(T_3)$. This is almost immediate, since

$$^G V(4^{r_1}, 4^{r_2}, 4^{r_3}) * {^G M}(4, 4^s) = {^G V}(4^{r_1}, 4^{r_2}, 4^{r_3+s-1})$$

where $^G M(4, 4^s)$ acts on a *single index* of $^G V(4^{r_1}, 4^{r_2}, 4^{r_3})$ (recall that $^G V(4^{r_1}, 4^{r_2}, 4^{r_3})$ can be considered as either a 3-way tensor or an n-way $4 \times \cdots \times 4$ tensor). This equation of varieties applies to each of the component varieties in the intersection representation of $V(T_{n-1})$ and $V(T_3)$ and completes the proof. $\qquad\square$

17

Extending Tree Models to Splits Networks

David Bryant

In this chapter we take statistical models designed for trees and adapt them for *splits networks*, a more general class of mathematical structures. The models we propose provide natural swing-bridges between trees, filling in gaps in the probability simplex. There are many reasons why we might want to do this. Firstly, the splits networks provide a graphical representation of phylogenetic uncertainty. Data that is close to tree-like produces a network that is close to a tree, while noisy or badly modeled data produce complex networks. Secondly, models that incorporate several trees open up possibilities for new tests to assess the relative support for different trees, in both likelihood and Bayesian frameworks. Thirdly, by searching through network space rather than tree space we may well be able to avoid some of the combinatorial headaches that make searching for trees so difficult.

17.1 Trees, splits and splits networks

Splits are the foundation of phylogenetic combinatorics, and they will be the building blocks of our general statistical model. Recall (from Chapter 2) that a *split* $S = \{A, B\}$ of a finite set X is an unordered partition of X into two non-empty blocks. A *phylogenetic tree* for X is a pair $\mathcal{T} = (T, \phi)$ such that T is a tree with no vertices of degree two and ϕ is a bijection from X to the leaves of T.

Removing an edge e from a phylogenetic tree divides the tree into two connected components, thereby inducing a split of the set of leaves that we say is the *split associated to e*. We use splits($\mathcal{T}$) to denote the sets associated to edges of T. The phylogenetic tree $\mathcal{T}$ can be reconstructed from the collection splits($\mathcal{T}$). The *Splits Equivalence Theorem* (Theorem 2.35) tells us that a collection $\mathcal{S}$ of splits is contained in splits($\mathcal{T}$) for some phylogenetic tree $\mathcal{T}$ if and only if the collection is *pairwise compatible*, that is, for all pairs of splits $\{A, B\}, \{A', B'\}$ at least one of the intersections

$$A \cap A', \ A \cap B', \ B \cap A', \ B \cap B'$$

is empty.

If we think of phylogenetic trees as collections of compatible splits then it becomes easy to generalize trees: we simply consider collections of splits that may not be pairwise compatible. This is the approach taken by Split Decomposition [Bandelt and Dress, 1992], Median Networks [Bandelt *et al.*, 1995], SpectroNet [Huber *et al.*, 2002], Neighbor-Net [Bryant and Moulton, 2004], Consensus Networks [Holland *et al.*, 2004] and Z-networks [Huson *et al.*, 2004].

The usefulness of these methods is due to a particularly elegant graphical representation for general collections of splits: the splits network. This has been the basis of `SplitsTree4` [Huson and Bryant, 2005] which implements many of the tree generalizations mentioned above (see Figure 2.8).

To define splits networks, we first need to discuss splits graphs. These graphs have multiple characterizations. We will work with three of these here.

For a graph G let d_G denote the (unweighted) shortest path metric. A map ψ from a graph H to a graph G is an *isometric embedding* if $d_H(u, v) = d_G(\psi(u), \psi(v))$ for all $u, v \in V(H)$. A graph G is a *partial cube* if there exists an isometric embedding from G to a hypercube. [Wetzel, 1995] called these graphs *splits graphs*. This terminology has persisted in the phylogenetics community, despite the potential for confusion with the graph-theoretic term 'splits graph' (a special class of perfect graphs). Refer to [Imrich and Klavžar, 2000] for a long list of characterizations for partial cubes.

[Wetzel, 1995] (see also [Dress and Huson, 2004]) characterized splits graphs in terms of isometric colorings. Let κ be an edge coloring of the graph. For each pair $u, v \in V(G)$ let $C_\kappa(u, v)$ denote the set of colors that appear on *every* shortest path from u to v. We say that κ is an *isometric coloring* if $d_G(u, v) = |C_\kappa(u, v)|$ for all pairs $u, v \in V(G)$. In other words, κ is isometric if the edges along any shortest path all have different colors, while any two shortest paths between the same pair of vertices have the same set of edge colors. A connected graph is a splits graph if and only if it has an isometric coloring [Wetzel, 1995].

A third characterization of splits graphs is due to [Winkler, 1984]. We define a relation Θ on pairs of edges $e_1 = \{u_1, v_1\}$ and $e_2 = \{u_2, v_2\}$ in a graph G by

$$e_1 \Theta e_2 \Leftrightarrow d_G(u_1, u_2) + d_G(v_1, v_2) \neq d_G(u_1, v_2) + d_G(v_1, u_2). \qquad (17.1)$$

This relation is an equivalence relation if and only if G is a splits graph.

Two edges e_1 and e_2 in a splits graph have the same color in an isometric coloring if and only if the isometric embedding of the splits graph into the hypercube maps e_1 and e_2 to parallel edges, if and only if $e_1 \Theta e_2$. Thus, a splits graph has, essentially, a unique isometric coloring and a unique isometric embedding into the hypercube. The partition of edges into color classes is completely determined by the graph.

Suppose now that we have a splits graph G and a map $\phi \colon X \to V(G)$. Using the isometric embedding, one can quickly prove that removing all edges in a particular color class partitions the graph into exactly two connected (and convex) components. This in turn induces a split of X, via the map ϕ. A *splits network* is a pair $\mathcal{N} = (G, \phi)$ such that

    (i) G is a splits graph,

    (ii) each color class induces a distinct split of X.

The set of splits induced by the different color classes is denoted splits($\mathcal{N}$).

    It is time for two examples. The splits network on the left of Figure 17.1 corresponds to a collection of compatible splits – it is a tree. In this network, every edge is in a distinct color class. If we add the split $\{\{2,6\},\{1,3,4,5\}\}$ we obtain the splits network on the right. There are four color classes in this graph that contain more than a single edge. These are the three horizontal pairs of parallel edges and the four edges marked in bold that induce the extra split.

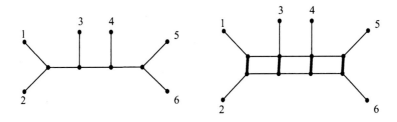

Fig. 17.1. Two splits networks. On the left, a splits network for compatible splits (i.e. a tree). On the right, the same network with the split $\{\{2,6\},\{1,3,4,5\}\}$ included.

    It is important to realize that the splits network for a collection of splits may not be unique. Figure 17.2 reproduces an example in [Wetzel, 1995]. Both graphs are splits networks for the set

$$\mathcal{S} = \Big\{\{\{1,2,3\},\{4,5,6,7\}\},\{\{2,3,4\},\{1,5,6,7\}\},$$

$$\{\{1,2,7\},\{3,4,5,6\}\},\{\{1,2,6,7\},\{3,4,5\}\}\Big\}.$$

Each is minimal, in the sense that no subgraph of either graph is also a splits network for $\mathcal{S}$. In both graphs, the edges in the color class inducing the split $\{\{1,2,3\},\{4,5,6,7\}\}$ are in bold. In this example the two minimal graphs are isomorphic, but this is generally not the case.

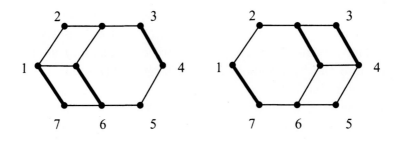

Fig. 17.2. Two different, and minimal, splits networks for the same set of splits.

17.2 Distance-based models for trees and splits graphs

In molecular phylogenetics, the length of an edge in a tree is typically measured in terms of the average (or expected) number of mutations that occurred, per site, along that edge. The *evolutionary distance* between two sequences equals the sum of the lengths of the edges along the unique path that connects them in the unknown "true" phylogeny. There is a host of methods for estimating the evolutionary distance starting from the sequences alone. These form the basis of distance-based approaches to phylogenetics.

The oldest statistical methods for phylogenetics use models of how evolutionary distances estimated from pairwise comparisons of sequences differ from the true evolutionary distances (or *phyletic distances*) in the true, but unknown, phylogenetic tree [Cavalli-Sforza and Edwards, 1967, Farris, 1972, Bulmer, 1991]. It is assumed that the pairwise estimates are distributed, at least approximately, according to a multivariate normal density centered on the true distances. The variance–covariance matrix for the density, here denoted by $\mathbf{V}$, can be estimated from the data [Bulmer, 1991, Susko, 2003], though early papers used a diagonal matrix, or the identity, for $\mathbf{V}$.

Once we have a variance–covariance matrix, and the observed distances, we can begin maximum likelihood estimation of the true distances δ_T, from which we can construct the maximum likelihood tree. Note that the term maximum likelihood here refers only to our approximate distance-based model, not to the maximum likelihood estimation introduced by [Felsenstein, 1981]. Let n be the number of leaves. The maximum likelihood estimator is the tree metric $\widehat{\delta_T}$ that maximizes the likelihood function

$$L(\widehat{\delta_T}) = \Phi_{\binom{n}{2}}(d - \delta_T | \mathbf{V}),$$

where Φ_m is the probability density function for the m-dimensional multivariate normal:

$$\Phi_m(x|\mathbf{V}) = \frac{1}{(2\pi)^{\frac{m}{2}}\sqrt{\det(\mathbf{V})}} e^{-\frac{1}{2}x^T \mathbf{V}^{-1} x}.$$

Equivalently, we can minimize the least squares residue

$$\sum_{w<x}\sum_{y<z} \left(\widehat{\delta_T}(w,x) - d(w,x)\right) \mathbf{V}^{-1}_{(wx)(yz)} \left(\widehat{\delta_T}(y,z) - d(y,z)\right).$$

In either formulation, the optimization is carried out over all tree metrics in $\mathcal{T}_X$, the space of trees (Chapter 2).

We can describe tree metrics in terms of linear combinations of split metrics. The split metric for a split $\{A, B\}$ is the pseudo-metric on X given by

$$\delta_{\{A,B\}}(x,y) = \begin{cases} 0 & \text{if } \{x,y\} \subseteq A \text{ or } \{x,y\} \subseteq B; \\ 1 & \text{otherwise.} \end{cases}$$

Since every split is associated with an edge, it is natural to denote by $w_{\{A,B\}}$

the length of the edge associated to a split $\{A, B\} \in \text{splits}(T)$. Then

$$\delta_T = \sum_{\{A,B\}\in\text{splits}(T)} w_{\{A,B\}}\delta_{\{A,B\}}. \tag{17.2}$$

This formulation can be used to estimate edge lengths on a fixed topology.

Equation (17.2) generalizes immediately to splits networks. Suppose that the lengths of the edges in a splits network $\mathcal{N}$ are given by the split weights $w_{\{A,B\}}$. Hence, all edges in the same color class have the same length. The distance between two labeled vertices x, y is the length of the shortest path between them, which in turn equals the sum of the weights of the splits separating x and y. We can therefore define a *network metric* $\mathcal{N}$ by

$$\delta_{\mathcal{N}} = \sum_{\{A,B\}\in\text{splits}(\mathcal{N})} w_{\{A,B\}}\delta_{\{A,B\}}.$$

The statistical model for distances from splits networks then works exactly as it did for phylogenetic trees. We assume that the observed distances d are distributed according to a multivariate normal centered on the network metric $\delta_{\mathcal{N}}$. The covariance matrix can be estimated using the non-parametric method of [Susko, 2003]. The likelihood of a network metric $\widehat{\delta_{\mathcal{N}}}$ is, as before, given by $L(\widehat{\delta_{\mathcal{N}}}) = \Phi_{\binom{n}{2}}(d - d_{\mathcal{N}}|\mathbf{V})$.

We immediately encounter the problem of identifiability. Phylogenetic trees, together with their edge lengths, are determined uniquely from their tree metrics. The same does not apply for network distances. The split metrics $\delta_{\{A,B\}}$ associated to splits of a network will not, in general, be linearly independent.

In practice, identifiability has not been too much of a problem. Split decomposition produces *weakly compatible* collections of splits. These have linearly independent split metrics and are uniquely determined from their network metrics [Bandelt and Dress, 1992]. Neighbor-Net produces networks based on *circular collections of splits* which, as a subclass of weakly compatible splits, are also uniquely determined from their network metrics.

However the most important shortcoming of distance-based methods, for either trees or networks, is that they lack the statistical efficiency of likelihood methods based on full stochastic models (see, e.g. [Felsenstein, 2003]). When we estimate distances from pairwise sequence comparisons we are effectively ignoring the joint probabilities of larger sets of sequences. What we gain in speed, we lose in accuracy.

17.3 A graphical model on a splits network

The Markov model for trees outlined in Chapter 2 and Chapter 4 is just a special case in a general class of *graphical models*. Given the vast literature on graphical models, it seems that the logical generalization of the hidden tree model would be a graphical model defined on the splits network. This was the approach taken by [Strimmer and Moulton, 2000, Strimmer *et al.*, 2001].

Let $\mathcal{N}$ be a splits network. The first step is to choose a root and direct all edges away from the root (Figure 17.3). We now can apply a directed graphical model. The probability that a node is assigned a particular state depends on the states assigned to its parents: Strimmer and Moulton suggest several ways that this may be done.

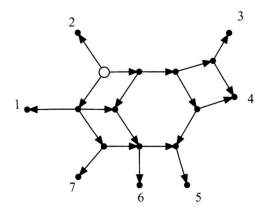

Fig. 17.3. Edge directions induced by placing the root at the white vertex.

There are significant problems with this general approach. Firstly, the probability of observing the data changes for different positions of the root, even when the mutation process is a time-reversible model. It was claimed that this permitted estimation of the root, but there is no indication that the differences in distributions corresponded to any evolutionary phenomenon.

Secondly, different splits networks for the *same* set of splits give different pattern probabilities, even though the networks represent exactly the same information.

Thirdly, the internal nodes in splits networks do not represent hypothetical ancestors. They are products of an embedding in a hypercube.

[Strimmer *et al.*, 2001] eventually concluded that splits networks may not provide a suitable underlying graph for a stochastic network. It is true that graphical model technology can not be applied 'straight-off-the-shelf' to splits networks. We need to be more sensitive to their particular properties. In the following section we will develop a model for splits networks that avoids the problems encountered in this graphical model approach. The downside, however, is that we must first restrict ourselves to a special class of mutation models: group-based models.

17.4 Group-based mutation models

A mutation model on state space $\{1, 2, \ldots, r\}$ is said to be a *group-based model* if there exists an Abelian group G with elements $g_1, \ldots, g_r$ and a function $\psi \colon G \to \mathbb{R}$ such that the instantaneous rate matrix Q satisfies

$$Q_{ij} = \psi(g_j - g_i)$$

for all i, j. The group operation on G is denoted using addition and we will use 0 for the identity element.

Let f be a homomorphism from G to the multiplicative group if complex numbers with modulus one. Thus $f(g + g') = f(g)f(g')$ for all $g, g' \in G$. The set of these homomorphisms forms a group $\widehat{G}$ that is isomorphic to G. This is called the *character group* of G. We label the elements of $\widehat{G}$ so that the map $g \mapsto \widehat{g}$ taking $g \in G$ to $\widehat{g} \in \widehat{G}$ is an isomorphism. If $g = 0$ then $\widehat{g}$ is the function taking every element of G to 1. As usual, the conjugate of a complex number x is written $\overline{x}$.

Lemma 17.1 *Suppose that* $g, h, h' \in G$, $a \in \mathbb{Z}$. *Then we have the following identities:*

$$
\begin{aligned}
\widehat{g}(-h) &= \overline{\widehat{g}(h)}; \\
\widehat{g}(h + h') &= \widehat{g}(h)\widehat{g}(h'); \\
\widehat{(h + h')}(g) &= \widehat{h}(g)\widehat{h}'(g); \\
\widehat{ag}(h) &= \widehat{g}(ah);
\end{aligned}
$$

as well as the orthogonality property

$$
\sum_{h \in G} \widehat{g}(h) = \begin{cases} |G| & \text{if } g = h; \\ 0 & \text{otherwise.} \end{cases}
$$

Proof See, for example, [Körner, 1989]. □

Some indexing conventions will make our life easier. Since the elements of G are in one-to-one correspondence with $\{1, 2, \ldots, r\}$ we will index Q and $P(t)$ by group elements. So $Q_{g_i g_j}$ is equivalent to Q_{ij}.

We start with some basic observations about group-based models.

Lemma 17.2   (i) *The eigenvalues of* Q *are given by*

$$
\lambda_g = \sum_{h \in G} \overline{\widehat{g}(h)} \psi(h).
$$

(ii) *The transition probabilities are given by*

$$
P_{gg'}(t) = \frac{1}{r} \sum_{h \in G} \widehat{h}(g' - g) e^{\lambda_h t}.
$$

(iii) *The uniform distribution is a stationary distribution.*

(iv) *If the process is ergodic and time-reversible then* $\psi(g) = \psi(-g)$ *for all* $g \in G$.

Proof Define the $r \times r$ matrix K by $K_{ij} = \widehat{g}_i(g_j)$. Then

$$
\begin{aligned}
(KQ)_{gg'} &= \sum_{h \in G} \widehat{g}(h)\psi(g' - h) \\
&= \sum_{h \in G} \widehat{g}(g' - h)\psi(h) \qquad \text{[replacing h by $g' - h$]} \\
&= \widehat{g}(g') \sum_{h \in G} \overline{\widehat{g}(h)}\psi(h) \\
&= K_{gg'}\lambda_g.
\end{aligned}
$$

Thus the rows of K are left-eigenvectors for Q. This proves (i). Let Λ be the diagonal matrix with $\Lambda_{gg} = \lambda_g$. Then $Q = K^{-1}\Lambda K$. By the orthogonality property in Lemma 17.1 we have $K^{-1} = \frac{1}{|G|}K^*$. Thus

$$
\begin{aligned}
P_{gg'}(t) &= (e^{Qt})_{gg'} \\
&= \frac{1}{|G|}(K^* e^{\Lambda t} K)_{gg'} \\
&= \frac{1}{r}\sum_{h \in G} \overline{\widehat{h}(g)}e^{\lambda_h t}\widehat{h}(g') \\
&= \frac{1}{r}\sum_{h \in G} \widehat{h}(g' - g)e^{\lambda_h t}
\end{aligned}
$$

proving (ii). For (iii), observe that the first row of K gives a left-eigenvector that is all ones. Finally, if the process is ergodic then the uniform distribution is the unique stationary distribution. This, together with the assumption that the process is time-reversible, implies that both Q and $P(t)$ are symmetric and that $\psi(g) = \psi(-g)$ for all g. □

As an example, consider the case when $r = 4$. There are two (up to isomorphism) Abelian groups on four elements: $\mathbb{Z}_4$ and $\mathbb{Z}_2 \times \mathbb{Z}_2$. If $G = \mathbb{Z}_4$ then the condition that $\psi(g) = \psi(-g)$ implies that Q must have the form

$$
Q = \begin{pmatrix}
-2a - b & a & b & a \\
a & -2a - b & a & b \\
b & a & -2a - b & a \\
a & b & a & -2a - b
\end{pmatrix}
$$

which is Kimura's 2-parameter (K80) model (Chapter 4). If $G = \mathbb{Z}_2 \times \mathbb{Z}_2$ then we always have $g = -g$ so there are 3-parameters available for Q:

$$
Q = \begin{pmatrix}
-a - b - c & a & b & c \\
a & -a - b - c & c & b \\
b & c & -a - b - c & a \\
c & b & a & -a - b - c
\end{pmatrix}.
$$

In this case we obtain Kimura's three parameter model (K81) [Kimura, 1981].

17.5 Group-based models for trees and splits

Suppose that we have an ergodic, time-reversible, group-based mutation model with state set $\Sigma = \{1, 2, \ldots, r\}$ and Abelian group G, $|G| = r$, where $Q_{ij} = \psi(g_j - g_i)$ for all i, j. Let $P(t) = e^{Qt}$ denote the corresponding transition probabilities (see Section 4.5). Let $\mathcal{T} = (T, \phi)$ be a phylogenetic tree with n leaves. We use $t_e = t_{kl}$ to denote the length of an edge $e = kl \in E(T)$ which joins the vertices k and l. In terms of the tree models of Section 4.5, $\theta^{kl} = P(t_{kl})$ for all $kl \in E(T)$.

We define

$$\rho_t(g) = \frac{1}{r} \sum_{h \in G} \widehat{h}(g) e^{\lambda_h t}$$

so that by Lemma 17.2, $P_{gg'}(t) = \rho_t(g' - g)$ for all $g, g' \in G$ and $t \geq 0$.

Lemma 17.3 Let σ be a map from $V(T)$ to Σ. For each edge $e = kl$ define $x_e = g_{\sigma_l} - g_{\sigma_k}$. Then

$$p_\sigma = \frac{1}{r} \prod_{e \in E(T)} \rho_{t_e}(x_e).$$

Proof By Lemma 17.2 the mutation model has a uniform stationary distribution. We can therefore apply (1.56), giving

$$
\begin{aligned}
p_\sigma &= \frac{1}{|\Sigma|} \prod_{kl \in E(T)} \theta^{kl}_{\sigma_k \sigma_l} \\
&= \frac{1}{r} \prod_{kl \in E(T)} \rho_{t_{kl}}(g_{\sigma_l} - g_{\sigma_k}) \\
&= \frac{1}{r} \prod_{e \in E(T)} \rho_{t_e}(x_e).
\end{aligned}
$$
$\qquad\square$

Let χ be a map from the leaves of T to Σ. We say that $\sigma : V(T) \to \Sigma$ extends χ if $\sigma_i = \chi_i$ for all leaves i. Under the *hidden tree model* the probability of observing χ is given by

$$p_\chi = \sum_{\sigma : \sigma \text{ extends } \chi} p_\sigma.$$

Suppose that $E(T) = \{e_1, e_2, \ldots, e_q\}$, let $\{A_k, B_k\}$ be the split associated to edge k and let $\mathbf{A}$ be the $(n - 1) \times q$ matrix defined by

$$\mathbf{A}_{ik} = \begin{cases} 1 & i \text{ and } n \text{ are on opposite sides of } \{A_k, B_k\} \\ 0 & \text{otherwise.} \end{cases} \tag{17.3}$$

The next observation is crucial, since it allows us to re-express the likelihood in a form that extends immediately to arbitrary collections of splits.

Theorem 17.4 Define the vector $y = y[\chi] \in G^{n-1}$ by $y_i = g_{\chi_n} - g_{\chi_i}$. Let σ be a map from $V(T)$ to Σ. For each edge $e = kl$ define $x_e = g_{\sigma_l} - g_{\sigma_k}$. Then

σ extends χ if and only if $\mathbf{A}x = y$. Furthermore, the probability of observing χ is given by

$$p_\chi = \sum_{x \in G^q : \mathbf{A}x = y} \prod_{e \in E(T)} \rho_{t_e}(x_e). \tag{17.4}$$

Proof We prove that $\mathbf{A}x = y$ if and only if σ extends χ. The second claim then follows from Lemma 17.3.

For each leaf i, let E_i be the edges on the path from leaf n to leaf i. We will assume that T is rooted at leaf n, so all edges in E_i are directed away from n. Then

$$\begin{aligned} (\mathbf{A}x)_i &= \sum_{kl \in E_i} x_{kl} \\ &= \sum_{kl \in E_i} (g_{\sigma_l} - g_{\sigma_k}) \\ &= g_{\sigma_n} - g_{\sigma_i}. \end{aligned}$$

Thus $\mathbf{A}x = y$ if and only if $g_{\sigma_i} = g_{\chi_i}$ for all leaves i, if and only if $\sigma_i = \chi_i$ for all leaves i, if and only if σ extends χ. $\qquad\square$

The importance of Theorem 17.4 so far as we are concerned is that p_χ is not expressed in terms of the tree structure: it is defined in terms of splits. We can therefore generalize the definition of pattern probabilities to any collection of splits.

Let $\mathcal{N}$ be a weighted splits network with splits $\{A_1, B_1\}, \ldots, \{A_q, B_q\}$ and let t_k be the length assigned to split $\{A_k, B_k\}$. Let A be the matrix defined by (17.3). The *probability of a phylogenetic character* χ *given* $\mathcal{N}$ is then defined by

$$p_\chi = \sum_{x \in G^q : \mathbf{A}x = y} \prod_{k=1}^{q} \rho_{t_k}(x_k). \tag{17.5}$$

The uncanny similarity between (17.4) and (17.5) now gives

Theorem 17.5 *Let $\mathcal{N}$ be a weighted splits network. If the splits of $\mathcal{N}$ are compatible then the character probabilities correspond to exactly those given by the tree-based model.*

We can rephrase this model in terms of graphical models on the splits network. We say that a map $\sigma : V(\mathcal{N}) \to \Sigma$ is *concordant* if $\sigma_l - \sigma_k = \sigma_j - \sigma_i$ for all pairs of edges $ij, kl \in E(\mathcal{N})$ in the same color class. The probability of a map σ is just the product of $P_{\sigma_k \sigma_l}(t_{kl})$ over all edges $kl \in E(T)$, where t_{kl} is the length of the edge. We then have that p_χ equals the probability that a map σ extends χ, conditional on σ being concordant.

17.6 A Fourier calculus for splits networks

[Székely *et al.*, 1993] describe a *Fourier calculus on evolutionary trees* that generalizes the Hadamard transform [Hendy and Penny, 1989, Steel *et al.*, 1992]. Using their approach, we can take the observed character frequencies, apply a transformation, and obtain a vector of values from which we can read off the support for different splits. They show that if the observed character frequencies correspond exactly to the character probabilities determined by some phylogenetic tree then the split supports will correspond exactly to the splits and branch lengths in the phylogenetic tree. Conversely, the inverse transformation gives a single formula for the character probabilities in any tree.

This theory generalizes seamlessly from trees to splits networks—in fact so seamlessly that the proofs of [Székely *et al.*, 1993] require almost no modifications to establish the general case. Their approach was to prove that their transform worked when applied to character probabilities from a tree. The correctness of the inverse formula then followed by applying a Fourier transformation. In this section, we will prove the same results but working in the opposite direction. We show that, starting with weights on the splits, a single invertible formula gives the character probabilities. Our motivation is that, at some point in the future, we will need to generalize these results beyond Abelian group-based models, and the elegant Fourier inversion formula may not exist in this context.

For $x, y \in G^m$ we define

$$\widehat{y}(x) = \prod_{i=1}^{m} \widehat{y}_i(x_i).$$

The set $\{\widehat{y} : y \in G^m\}$ forms a group under multiplication that is isomorphic to G^m.

Lemma 17.6 *Suppose that $z \in G^q$ and $y \in G^{n-1}$. Let $\mathbf{A}$ be an $(n-1) \times q$ integer matrix. Either*

$$\sum_{x \in G^q : \mathbf{A}x = y} \widehat{z}(x) = 0,$$

or there is $u \in G^{n-1}$ such that $z = \mathbf{A}^T u$ and so

$$\sum_{x \in G^q : \mathbf{A}x = y} \widehat{z}(x) = r^{q-(n-1)} \widehat{z}(u).$$

Proof Suppose that $\sum_{x \in G^q : \mathbf{A}x = y} \widehat{z}(x) \neq 0$. For any v such that $\mathbf{A}v = 0$ we have

$$\sum_{x \in G^q : \mathbf{A}x = y} \widehat{z}(x) = \sum_{x \in G^q : \mathbf{A}x = y} \widehat{z}(x + v) = \widehat{z}(v) \sum_{x \in G^q : \mathbf{A}x = y} \widehat{z}(x)$$

so $\widehat{z}(v) = 1$.

For every $x, y \in G^{n-1}$ we have

$$\mathbf{A}x = \mathbf{A}y \Leftrightarrow \mathbf{A}(x - y) = 0 \Leftrightarrow \widehat{z}(x - y) = 1 \Leftrightarrow \widehat{z}(x) = \widehat{z}(y).$$

Define $H = \{\mathbf{A}x : x \in G^q\}$, so that H forms a normal subgroup of G^{n-1}. Define the map $f : H \to \mathbb{C}$ by setting $f(\mathbf{A}x) = \widehat{z}(x)$ for all $x \in G^q$. This is a homomorphism from H to the unit circle, since $f(\mathbf{A}x + \mathbf{A}y) = f(\mathbf{A}(x + y)) = \widehat{z}(x + y) = \widehat{z}(x)\widehat{z}(y) = f(\mathbf{A}x)f(\mathbf{A}y)$. By Lemma 104.3 of [Körner, 1989] we can extend f to the rest of G. Thus there is u such that $f = \widehat{u}$ and, for all $x \in G^q$, $\widehat{z}(x) = \widehat{u}(\mathbf{A}x)$. The result now follows by expanding $\widehat{u}(\mathbf{A}x)$. $\qquad\square$

We now arrive at our main theorem. It provides the formula linking weights on splits to pattern probabilities, for trees *and* splits networks.

Theorem 17.7 *Let $\mathcal{S}$ be a collection of splits, $|\mathcal{S}| = q$, and $\mathbf{A}$ be the $(n-1) \times q$ matrix defined by*

$$\mathbf{A}_{ik} = \begin{cases} 1 & i \text{ and } n \text{ are on opposite sides of } \{A_k, B_k\} \\ 0 & otherwise. \end{cases} \tag{17.6}$$

Let b be the real-valued vector with entries indexed by G^{n-1} so that for all $z \in G^{n-1}$,

$$b_z = \begin{cases} \psi(h)t_k & \text{if there is } h \in G \text{ and } k \text{ s.t. } z_i = h\mathbf{A}_{ik} \text{ for all } i \\ -\sum_{v \in G^{n-1} - \{0\}} b_v & if z = 0 \\ 0 & otherwise. \end{cases}$$

Let $\mathbf{H}$ be the matrix with rows and columns indexed by G^{n-1} and $\mathbf{H}_{gg'} = \widehat{g}(g)$.

Given any map χ from the leaves to the set of states $\{1, 2, \dots, r\}$, define $y \in G^{n-1}$ by $y_i = g_{\chi_n} - g_{\chi_i}$ for all leaves $i = 1, \dots, n-1$. Then the probability of χ is given by

$$p_\chi = \left[\mathbf{H}^{-1} \exp[\mathbf{H}b]\right]_y. \tag{17.7}$$

Proof From (17.5) we have

$$p_\chi = \sum_{\substack{x:\mathbf{A}x=y \\ x \in G^q}} \prod_{k=1}^{q} \rho_{t_k}(x_k)$$

$$= \sum_{\substack{x:\mathbf{A}x=y \\ x \in G^q}} \prod_{k=1}^{q} \frac{1}{r} \sum_{h \in G} \widehat{h}(x_k) e^{\lambda_h t_k}$$

$$= \frac{1}{r^q} \sum_{\substack{x:\mathbf{A}x=y \\ x \in G^q}} \sum_{z \in G^q} \prod_{k=1}^{q} \widehat{z_k}(x_k) e^{\lambda_{z_k} t_k}$$

$$= \frac{1}{r^q} \sum_{z \in G^q} \left(\sum_{\substack{x:\mathbf{A}x=y \\ x \in G^q}} \widehat{z}(x) \right) \exp\left[\sum_{k=1}^{q} \lambda_{z_k} t_k \right].$$

So far we have just applied the definitions, reversed a summation and product, and regrouped. From Lemma 17.6 we have that $\sum_{x \in G^q : \mathbf{A}x=y} \widehat{z}(x)$ equals zero unless $z = -\mathbf{A}^T u$ for some $u \in G^{n-1}$. We therefore ignore all z for which this does not hold. Substituting and using $\widehat{-u} = \overline{\widehat{u}}$ we obtain

$$p_\chi \;=\; \frac{1}{r^{n-1}} \sum_{u \in G^{n-1}} \overline{\widehat{u}}(y) e^{\beta_u} \;=\; \Big[\mathbf{H}^{-1} \exp[\beta]\Big]_y$$

where

$$\beta_u \;=\; \sum_{k=1}^{q} \lambda_{(\mathbf{A}^T \overline{u})_k} t_k \;=\; \sum_{k=1}^{q} \sum_{h \in G} \widehat{\mathbf{A}^T u_k}(h) \psi(h) t_k$$

$$=\; \sum_{k=1}^{q} \sum_{h \in G} \widehat{u}(\eta_{kh}) \psi(h) t_k \;=\; \sum_{v \in G^{n-1}} \widehat{u}(v) b_v$$

$$=\; \mathbf{H}b. \qquad\qquad\qquad\qquad \square$$

We have proven, more or less, Theorem 6 of [Székely *et al.*, 1993] without any reference to trees. In the special case that $r = 2$, (17.7) becomes the classical *Hadamard transform* of [Hendy and Penny, 1989, Steel *et al.*, 1992]. This is comforting: [Felsenstein, 2003] describes the Hadamard type approach as "one of the nicest applications of mathematics to phylogenies so far."

Note that the formula $\mathbf{H}^{-1} \exp[\mathbf{H}b]$ is invertible. This means that every splits network gives a different character distribution. We cannot recover splits networks from their distance metrics d_N but we can recover them from their character probabilities. A maximum likelihood estimator based on (17.7) will be statistically consistent.

One key problem remains. The constraint that we only use group-based mutation models is too much of a restriction. For nucleotide data, and especially for protein data, a uniform stationary distribution is unrealistic. It is reasonable to believe that some reasonable generalization of these results exists for more general mutation models: after all there is no such restriction on distance-based methods. What exact form these generalizations will take is, at the moment, anybody's guess.

18

Small Trees and Generalized Neighbor-Joining

Mark Contois

Dan Levy

Direct reconstruction of phylogenetic trees by maximum likelihood methods is computationally prohibitive for trees with many taxa; however, by computing all trees for subsets of taxa of size m, we can infer the entire tree. In particular, if $m = 2$, the traditional distance-based methods such as neighbor-joining [Saitou and Nei, 1987] and UPGMA [Sneath and Sokal, 1973] are applicable. Under distance-based methods, 2-leaf subtrees are completely determined by the total length between each pair of leaves. We extend this idea to m leaves by developing the notion of m-dissimilarity [Pachter and Speyer, 2004]. By building trees on subsets of size m of the taxa and finding the total length, we can obtain an m-dissimilarity map. We will explain the *generalized neighbor-joining* (GNJ) algorithm [Levy *et al.*, 2005] for obtaining a phylogenetic tree with edge lengths from an m-dissimilarity map.

This algorithm is consistent: given an m-dissimilarity map D^T that comes from a tree T, GNJ returns the correct tree. However, in the case of data that is "noisy", e.g., when the observed dissimilarity map does not lie in the space of trees, the accuracy of GNJ depends on the reliability of the subtree lengths. Numerical methods may run into trouble when models are of high degree (Section 1.3); exact methods for computing subtrees, therefore, could only serve to improve the accuracy of GNJ. One family of such methods consists of algorithms for finding critical points of the ML equations as discussed in Chapter 15 and in [Hoşten *et al.*, 2005]. We explore the results of this method for the Jukes–Cantor DNA model on three taxa and conjecture that, for any 3-subtree, there is at most one critical point yielding edge lengths that are positive and real.

18.1 From alignments to dissimilarity

Any method for phylogenetic tree reconstruction begins with a multiple sequence alignment. Distance-based methods then proceed by comparing pairs of taxa from this alignment to find the distances between them. Let X be the set of taxa and $\binom{X}{k}$ the set of all subsets $R \subset X$ such that $|R| = k$. Then $\binom{X}{2}$ is the set of all pairs of taxa and $D \colon \binom{X}{2} \to \mathbb{R}_{>0}$ assigns a distance to each pair of taxa. We call D a *dissimilarity map on* X. Where there is no confusion, we will write $D(\{a, b\})$ as $D(a, b)$.

If a and b are taxa in X, we may compare their aligned sequences to find the Jukes–Cantor corrected distance $D_{\mathrm{JC}}(a, b)$. If the alignment has a length of L and a and b differ in k places, then

$$D_{\mathrm{JC}}(a, b) = -\frac{3}{4} \log \left(1 - \frac{4}{3}p \right), \tag{18.1}$$

where $p = \frac{k}{L}$. It has been shown in Example 4.25 that $D_{\mathrm{JC}}(a, b)$ is the maximum likelihood branch length estimate of the alignment of a and b, with respect to the JC model on the simple two-leaf tree. The concept of distance-based methods is that the branch length estimate on the two-leaf tree is a good estimate of the total path length from a to b in the maximum likelihood tree T on X. Stated in the terms of Section 2.4, distance based methods are effective when D_{JC} is close to δ_T, the tree metric on X induced by T. (Recall that $\delta_T(a, b) = \sum_{e \in P_{ab}} l_T(e)$, where P_{ab} is the path from a to b in T and l_T is the length function associated to the edges of T.)

We can extend the notion of dissimilarity maps to subsets of X larger than two. We define an *m-dissimilarity map on* X as a function $D : \binom{X}{m} \to \mathbb{R}_{>0}$. That is, D assigns a positive real value to every subset of X of size m. In particular, a dissimilarity map is a 2-dissimilarity map. Again, where there is no confusion, we will write $D(\{x_1, ..., x_m\})$ as $D(x_1, ..., x_m)$. For a subset $R \subseteq X$, we define $[R]$ as the spanning subtree of R in T: $[R]$ is the smallest subtree of T containing R. For two leaves $a, b \in X$, the path from a to b, P_{ab}, is equivalent to the spanning subtree $[\{a, b\}]$.

Just as we defined tree metrics induced by a tree T, we can define the m-dissimilarity map induced by T, $D_m^T : \binom{X}{m} \to \mathbb{R}_{>0}$, by

$$D_m^T(R) = \sum_{e \in [R]} l_T(e)$$

for $R \subset X, |R| = m$. So $D_m^T(R)$ is the sum of all the edge lengths in the spanning subtree of R. We call $D_m^T(R)$ the *m-subtree length* of $[R]$. Note that this is also a measure of the phylogenetic diversity (PD) of the subtree R [Faith, 1992]. Since $P_{ab} = [\{a, b\}]$, $D_2^T(a, b) = \delta_T(a, b)$, and the 2-dissimilarity map induced by T is a tree metric.

Just as we used the JC corrected distance to approximate δ_T, we can employ analytical or numerical methods to find maximum likelihood estimates for the total branch lengths of subtrees with m leaves. To find an approximate m-dissimilarity map D, for all $R \in \binom{X}{m}$, we find the MLE tree for R and sum the branch lengths:

$$D(R) = \sum_{e \in T(R)} l_{T(R)}(e),$$

where $T(R)$ is the MLE tree for R.

18.2 From dissimilarity to trees

The neighbor-joining algorithm and its variants are methods that take an approximate dissimilarity map and construct a "nearby" tree. Similarly, we would like to define an algorithm that takes as its input an m-dissimilarity map and constructs a "nearby" tree, and we would like this method to be consistent: given an m-dissimilarity map D_m^T induced by a tree T, our algorithm should return T.

The crux of our method is the construction of a 2-dissimilarity map S_D from the m-dissimilarity map D. In the case that D is induced by a tree T, then S_D will be the tree metric induced by a tree T' which is isomorphic to a contraction of certain "deep" edges of T. If $|X| > 2m - 1$, then T and T' are topologically equivalent. Further, there exists an invertible linear map from edge lengths of T to the edge lengths of T'.

The deletion of an edge e in T divides T into two subtrees $T_1(e)$ and $T_2(e)$. If $L(T_1(e))$ and $L(T_2(e))$ denote the leaves in each subtree, then we define the *depth of e, $d(e)$* by

$$d(e) = \min(|L(T_1(e))|, |L(T_2(e))|).$$

Observe that the pendant edges are exactly those edges of depth 1, and, if T has n leaves, then $d(e) \leq \frac{n}{2}$ for any $e \in E(T)$. We define $E_{>k}(T) = \{e \in E : d(e) > k\}$. For example, $E_{>1}(T)$ is all the interior edges of T.

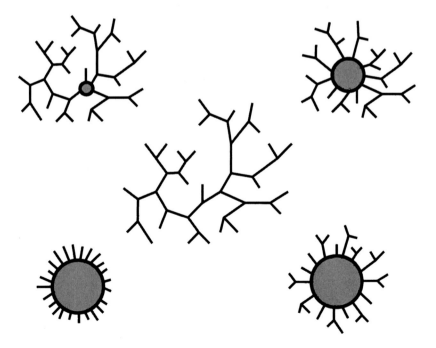

Fig. 18.1. In the center, we have a tree T surrounded by $T/E_{>k}$ for $k = 11, 4, 2$, and 1 starting at the top left and proceeding clockwise. The grey circle represents the vertex to which the edges of $E_{>k}$ have collapsed. Notice that T has 24 leaves and so $E_{>12} = \emptyset$ and $T/E_{>12} = T$.

For any edge e in T, we may contract e by deleting the edge and identifying its vertices. We write the resulting tree as T/e. For a set of edges E', the contraction of each edge in that set is denoted by T/E'. For example, $T/E_{>1}(T)$ contracts all interior edges and has the star topology (Figure 18.1). We may now state our claim explicitly.

Theorem 18.1 *Let D be an m-dissimilarity map on a set X of size n. We define*

$$S_D(i,j) = \sum_{R \in \binom{X \setminus \{i,j\}}{m-2}} D(i,j,R). \qquad (18.2)$$

If $D = D_m^T$ then $S_D = D^{T'}$ where T' is isomorphic to $T/E_{>n-m}$. Further, there exists an invertible linear transformation from the interior edge lengths of T' to $T/E_{>n-m}$. If $E_{>n-m}$ is empty, then there is also an invertible linear transformation between the pendant edges; otherwise, the pendant edges may not be determined uniquely.

For a fixed phylogenetic tree T and an integer m, we let $S(i,j) = S_{D_m^T}(i,j)$. Observe that any linear combination of the m-subtree lengths is a linear combination of the edge lengths $l_T(e)$ in the tree. This is because $D_m^T(R) = \sum_{e \in [R]} l_T(e)$; i.e. every m-subtree length is the sum of the edge lengths in the spanning subtree.

For a linear function on the m-subtree lengths $F : \mathbb{R}^{\binom{n}{m}} \to \mathbb{R}$, let $v_F(e)$ denote the coefficient of $l_T(e)$ in F. For instance, $v_{S(i,j)}(e)$ denotes the coefficient of $l_T(e)$ in $S(i,j)$. Note that $v_{F+G}(e) = v_F(e) + v_G(e)$. We will also use the notation $L_i(e)$ to denote the set of leaves in the component of $T \setminus e$ that contains leaf i.

Lemma 18.2 *Given a pair of leaves a, b and any edge e we have*

$$v_{S(a,b)}(e) = \begin{cases} \binom{n-2}{m-2} & e \in P_{ab}; \\[2mm] \binom{n-2}{m-2} - \binom{|L_a(e)|-2}{m-2} & e \notin P_{ab}. \end{cases}$$

Proof If e is on the path from a to b, then it will be included in all the subtrees $[a, b, Y]$. If e is not on the path from a to b, then the only way it will be excluded is if all the other leaves fall on the a side of e (which is the same as the b side); that is, if $Y \subset L_a(e) \setminus \{a, b\}$. There are $\binom{|L_a(e)|-2}{m-2}$ such sets. $\square$

Lemma 18.3 *Given a quartet $(a_1, a_2; a_3, a_4)$ in T with interior vertices b_1 and*

b_2 *(Figure 18.2),*

$$v_{S(a_1,a_2)+S(a_3,a_4)}(e) = \begin{cases} 2\binom{n-2}{m-2} - \binom{n-|L_{a_i}(e)|-2}{m-2} & e \in P_{a_ib_{\lceil i/2 \rceil}}; \\[2mm] 2\binom{n-2}{m-2} - \binom{|L_{a_1}(e)|-2}{m-2} - \binom{|L_{a_3}(e)|-2}{m-2} & e \in P_{b_1b_2}; \\[2mm] 2\binom{n-2}{m-2} - 2\binom{|L_{a_1}(e)|-2}{m-2} & e \notin [a_1,a_2,a_3,a_4 \end{cases}$$

$$v_{S(a_1,a_3)+S(a_2,a_4)}(e) = \begin{cases} 2\binom{n-2}{m-2} - \binom{n-|L_{a_i}(e)|-2}{m-2} & e \in P_{a_ib_{\lceil i/2 \rceil}}; \\[2mm] 2\binom{n-2}{m-2} & e \in P_{b_1b_2}; \\[2mm] 2\binom{n-2}{m-2} - 2\binom{|L_{a_1}(e)|-2}{m-2} & e \notin [a_1,a_2,a_3,a_4]; \end{cases}$$

and

$$v_{S(a_1,a_4)+S(a_2,a_3)}(e) = v_{S(a_1,a_3)+S(a_2,a_4)}(e).$$

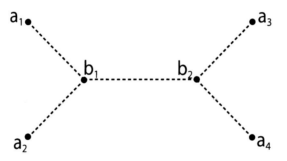

Fig. 18.2. A quartet $(a_1, a_2; a_3, a_4)$.

Proof We use the fact that $v_{S(a_1,a_2)+S(a_3,a_4)}(e) = v_{S(a_1,a_2)}(e) + v_{S(a_3,a_4)}(e)$ and apply Lemma 18.2. We also note that $L_{a_1}(e) = L_{a_i}(e)$ holds for all $e \notin [\{a_1,a_2,a_3,a_4\}]$ and all i. $\qquad\square$

Corollary 18.4 *For a quartet $(a_1, a_2; a_3, a_4)$, we define*

$$S(a_1, a_2; a_3, a_4) = S(a_1, a_2) + S(a_3, a_4) - S(a_1, a_3) - S(a_2, a_4).$$

Then,

$$v_{S(a_1,a_2;a_3,a_4)}(e) = \begin{cases} -\binom{|L_{a_1}(e)|-2}{m-2} - \binom{n-|L_{a_1}(e)|-2}{m-2} & e \in P_{b_1b_2}; \\[2mm] 0 & otherwise. \end{cases}$$

Corollary 18.4 implies that S satisfies the four-point condition (as discussed in Section 2.4 and Theorem 2.36) although it may be that $v_{S(a_1a_2;a_3a_4)}(e) = 0$ which means that the edge e has been contracted in T'. In particular, this

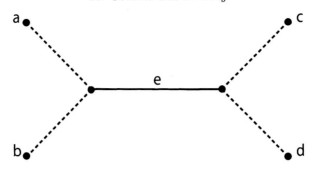

Fig. 18.3. The quartet $(a, b; c, d)$ has only the one edge e on its splitting path.

happens when $|L_{a_1}(e)| < m$ and $n - |L_{a_1}(e)| < m$ which is equivalent to $d(e) > n - m$. So for any quartet $(a_1, a_2; a_3, a_4)$, if the splitting path contains at least one edge e such that $d(e) \leq n - m$, then T' has the same quartet. However, if every edge in the splitting path has $d(e) > n - m$, then T' does not contain that quartet. Consequently, T' is isomorphic to $T/E_{>n-m}$.

It remains to show that there is an invertible linear map between the edge lengths in T' and $T/E_{>n-m}$.

Lemma 18.5 *If e is an internal edge of $T/E_{>n-m}$ with e' the corresponding edge in T' then*

$$l_{T'}(e') = \frac{1}{2} \left(\binom{|L_a(e)| - 2}{m - 2} + \binom{|L_c(e)| - 2}{m - 2} \right) l_T(e)$$

where a is a leaf in one component of $T - e$ and c a leaf in the other.

Proof Since e is an internal edge, we may choose a, b, c and d such that e is the only edge on the splitting path of $(a, b; c, d)$ (Figure 18.3). Then

$$
\begin{aligned}
l_{T'}(e') &= \frac{1}{2} S(a, b; c, d) \\
&= \frac{1}{2} \left(\binom{|L_a(e)| - 2}{m - 2} + \binom{|L_c(e)| - 2}{m - 2} \right) l_T(e) \\
&= \frac{1}{2} \left(\binom{d(e) - 2}{m - 2} + \binom{n - d(e) - 2}{m - 2} \right) l_T(e).
\end{aligned}
$$

$\square$

Corollary 18.6

$$l_T(e) = \frac{2 l_{T'}(e')}{\left(\binom{d(e)-2}{m-2} + \binom{n-d(e)-2}{m-2} \right)},$$

which is well defined if $d(e) \leq n - m$.

Lemma 18.7 *Denote the edges adjacent to the leaves by $e_1, \ldots, e_n$ (with the*

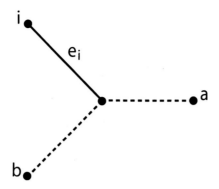

Fig. 18.4. The pendant edge e_i is incident on two other edges. We may choose leaves a and b such that $P_{ia} \cap P_{ib} = e_i$.

corresponding edges in T' denoted by $e'_1, \ldots, e'_n$) and the set of internal (non-pendant) edges by $\mathrm{int}(T) = E_{>1}$. *Let*

$$C_i = \sum_{e \in \mathrm{int}(T)} \left(\binom{n-2}{m-2} - \binom{|L_i(e)| - 2}{m-2} \right) l_T(e)$$

and let $\mathbf{A}$ *be the matrix* $2\binom{n-3}{m-2}\mathbf{I} + \binom{n-3}{m-3}\mathbf{J}$, *where* $\mathbf{I}$ *is the identity matrix and* $\mathbf{J}$ *is the matrix with every entry equal to one. Then*

$$\begin{pmatrix} l_{T'}(e'_1) \\ \vdots \\ l_{T'}(e'_n) \end{pmatrix} = \frac{1}{2}\mathbf{A} \begin{pmatrix} l_T(e_1) \\ \vdots \\ l_T(e_n) \end{pmatrix} + \frac{1}{2} \begin{pmatrix} C_1 \\ \vdots \\ C_n \end{pmatrix}.$$

Proof The interior vertex of a pendant edge e_i is incident to two other edges. Choose a leaf a such that P_{ia} intersects one of the edges, and b such that P_{ib} intersects the other (Figure 18.4). Then

$$l_{T'}(e') = \frac{1}{2} \left(S(i,a) + S(i,b) - S(a,b) \right)$$

which after some algebra gives the above lemma. □

Corollary 18.8

$$\begin{pmatrix} l_T(e_1) \\ \vdots \\ l_T(e_n) \end{pmatrix} = \mathbf{A}^{-1} \begin{pmatrix} 2l_{T'}(e'_1) - C_1 \\ \vdots \\ 2l_{T'}(e'_n) - C_n \end{pmatrix}$$

where $\mathbf{A}^{-1} = \frac{1}{2\binom{n-3}{m-2}} \left(\mathbf{I} - \frac{m-2}{(m-1)(n-2)}\mathbf{J} \right)$.

In order to recover $l_T(e)$ for every edge, we start by calculating the interior edge lengths, after which we can calculate the values C_i. The matrix $\mathbf{A}$ is always invertible if $m \leq n - 1$; however, calculating C_i requires that $\mathrm{int}(T) = \mathrm{int}(T')$. If $n < 2m-1$, then while we can determine all the interior edge lengths of $T/E_{>n-m}$ from T'. But, if $E_{>n-m}$ is non-empty, then some interior edges of T have been contracted in T'. If we delete the edges of $E_{>n-m}$ from T to form the forest $T \setminus E_{>n-m}$, then every connected component of $T \setminus E_{>n-m}$ has strictly fewer than m leaves. As a result, every m-subtree length will include at least one undetermined edge, and so there is no way to uniquely determine the lengths of the pendant edges. This completes the proof of Theorem 18.1.

18.3 The need for exact solutions

These observations about S_D allow us to employ traditional distance-based methods, such as neighbor-joining (algorithm 2.41), to construct trees from m-dissimilarity maps. For an alignment on a set of taxa X with $|X| = n$, we can proceed as follows:

(i) For each $R \in \binom{X}{m}$ use a (ML) method to find $D(R)$.
(ii) For each pair (i, j), compute $S_D(i, j)$.
(iii) Apply neighbor-joining to S_D to find a tree T'.
(iv) Apply linear transformation from Corollary 18.8 to the edge lengths of T' to find T.

We refer to this class of algorithms as *generalized neighbor-joining* (GNJ).

In Section 2.4, we encountered the neighbor-joining method and the cherry-picking matrix

$$Q_D(i, j) \quad = \quad (n - 2) \cdot D(i, j) - \sum_{k \neq i} D(i, k) - \sum_{k \neq j} D(j, k). \qquad (18.3)$$

If D is a tree metric for a tree T, then the minimum entry $Q_D(x, y)$ identifies $\{x, y\}$ as a cherry in T. We also encountered the quartet split length

$$w_D(i, j; k, l) = \frac{1}{4}[2D(i, j) + 2D(k, l) - D(i, k) - D(i, l) - D(j, k) - D(j, l)],$$

which was instrumental in constructing an analog to the cherry-picking matrix. Here, we will see that it also plays a crucial rule in describing the effect of agglomeration on the cherry-picking matrix.

Having chosen a pair $\{x, y\}$ that minimizes Q_D, the neighbor-joining algorithm proceeds to agglomerate x and y to a single vertex z. We form a new set $X' = X \setminus \{x, y\} \cup \{z\}$ and define D' as a dissimilarity map on X' such that, for $k, l \in X' \setminus z$,

$$D'(z, k) \quad = \quad \frac{1}{2}[D(x, k) + D(y, k) - D(x, y)],$$
$$D'(k, l) \quad = \quad D(k, l).$$

A brief computation allows us to explicitly relate the agglomeration update to the cherry-picking matrix. Specifically:

$$Q_{D'}(z, k) \quad = \quad \frac{1}{2}[Q_D(x, k) + Q_D(y, k)] + 2D(x, y),$$
$$Q_{D'}(k, l) \quad = \quad Q_D(k, l) + 2w_D(x, y; k, l) + 2D(x, y), \qquad (18.4)$$

where $Q_{D'}$ is the $(n-1) \times (n-1)$ cherry-picking matrix after agglomeration.

Given a 2-dissimilarity map D, we may construct a 3-dissimilarity map induced by D as follows:

$$D_3(i,j,k) = \frac{1}{2}[D(i,j) + D(i,k) + D(j,k)]. \qquad (18.5)$$

Theorem 18.9 *Given* $D : \binom{X}{2} \to \mathbb{R}_{>0}$, *define* D_3 *as in equation (18.5). If we let* $S = S_{D_3}$, *then* Q_S *is related to* Q_D *by an affine transformation. If* S' *and* D' *are the 2-dissimilarity maps on* X' *after agglomeration of* S *and* D *respectively, then* $Q_{S'}$ *and* $Q_{D'}$ *are also related by an affine transformation.*

Proof We first note that

$$\begin{aligned}
S(i,j) &= \sum_{x \in X \backslash \{i,j\}} D_3(i,j,x) \\
&= \frac{1}{2} \sum_{x \in X \backslash \{i,j\}} D(i,j) + D(i,x) + D(j,x) \\
&= \frac{1}{2}[(n-4)D(i,j) + \sum_{x \in X \backslash \{i\}} D(i,x) + \sum_{x \in X \backslash \{j\}} D(j,x)]. \quad (18.6)
\end{aligned}$$

We substitute equation (18.6) into the cherry-picking equation (18.3) and find that

$$Q_S = \frac{n-4}{2} Q_D - 2T,$$

where $T = \sum D(x,y)$ over all $\{x,y\} \in \binom{X}{2}$.

To compute the effect on the agglomeration update, we agglomerate x and y to a vertex z as above. We first note that

$$w_S(i,j;k,l) = \frac{n-4}{2} w_D(i,j;k,l).$$

Applying this to the aggolmeration update function (18.4), we obtain

$$Q_{S'} = \frac{n-4}{2} Q_{D'} - 2T + \sum_{i \in X \backslash x} D(i,x) + \sum_{y \in X \backslash x} D(i,y). \qquad \square$$

Corollary 18.10 *Given a 2-dissimilarity map* D *and induced 3-dissimilarity map* D_3, *neighbor-joining applied to* D *is equivalent to GNJ applied to* D_3.

Corollary 18.10 provides a means for comparing GNJ with neighbor-joining. In particular, given an MLE method for generating trees on alignments of three taxa, we may compare its fidelity to the true total length with the fidelity of the 3-dissimilarity map induced by neighbor-joining's 2-dissimilarity map.

To this end, we compared the HKS method applied to Jukes–Cantor claw trees ([Hoşten *et al.*, 2005], described in the following section) with the 3-dissimilarity map D_3 constructed from Jukes–Cantor corrected 2-distances (as in equation (18.1)). We used seq-gen [Rambaut and Grassly, 1997] to generate pseudorandom alignments of triplets evolved from the Jukes–Cantor DNA

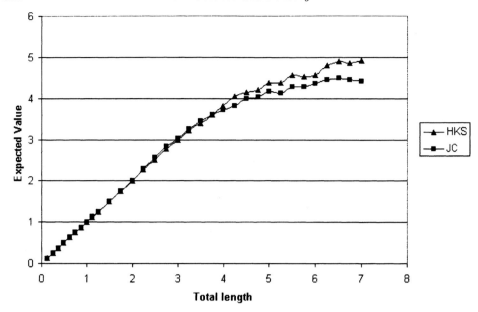

Fig. 18.5. The expected value of the total length for claw trees with edge ratios (1:1:1), (1:1:4), (1:2:4), and (1:4:4) as a function of the true total length. HKS indicates the HKS claw tree method and JC indicates D_3 computed from the Jukes–Cantor corrected 2-distance.

model ($\alpha = \frac{1}{4}$) on four different tree shapes over a range of total tree lengths. The total lengths varied from .125 to 7 and for each total length and each tree shape, we generated 100 alignments of length 500. Figure 18.5 shows that, as the total length increases, HKS generated values are, on average, closer to the true value than those generated from the Jukes–Cantor corrected 2-distance.

We may infer from this and Corollary 18.10 that generalized neighbor-joining with the HKS claw tree will be more accurate than neighbor-joining with Jukes–Cantor corrected distances, particularly in trees with large pairwise distances. Experiments coupling GNJ with fastDNAml, a numerical approximation method, show an improvement over traditional neighbor-joining [Levy *et al.*, 2005]. If Conjecture 18.12 (which proposes an upper bound of 1 on the number of ML solutions corresponding to biologically meaningful trees) is true, then methods such as fastDNAml, which locate such meaningful local maxima, will be assured to have found the only global maximum for a given 3-leaf tree, and therefore be demonstrably compatible with global maxima methods such as HKS.

18.4 Jukes–Cantor triples

A method is described in [Hoşten *et al.*, 2005] (here called the *HKS method*) by which local maxima of likelihood functions may be identified using computer algebra systems such as Singular and Macaulay 2; theapplication of this method to finding an ML parameter set for the Jukes–Cantor DNA model on the three-taxon "claw" phylogenetic tree with uniform root distribution is also demonstrated. (This tree and its model invariants are described in Chapter 15 as well as in [Pachter and Sturmfels, 2005].) The parameter μ_i represents the probability of a character's remaining the same along branch i, and $\pi_i = \frac{1}{3}(1 - \mu_i)$ the probability of its changing, according to the following transition matrix (whose rows and columns are indexed by $\{A, C, G, T\}$):

$$\begin{pmatrix} \mu_i & \pi_i & \pi_i & \pi_i \\ \pi_i & \mu_i & \pi_i & \pi_i \\ \pi_i & \pi_i & \mu_i & \pi_i \\ \pi_i & \pi_i & \pi_i & \mu_i \end{pmatrix}.$$

We define the members of $K = \{[123], [dis], [12], [13], [23]\}$ to be equivalence classes of observed characters at homologous loci for our three taxa: $[123]$ indicates that all three characters are identical, and $[dis]$ that all three are distinct; $[12]$ corresponds to identical characters at taxa 1 and 2 and a different character at taxon 3; and so on. As described in Chapter 1, an observation here is just a vector $u = (u_{123}, u_{dis}, u_{12}, u_{13}, u_{23})$ generated by tallying character similarities over some number of loci.

The probability of making an observation in class k at a given locus is denoted by p_k. For the claw tree, we may write down each of these as a trilinear form in $\mu_1, \mu_2, \mu_3, \pi_1, \pi_2, \pi_3$:

$$\begin{aligned} p_{123} &= \mu_1\mu_2\mu_3 + 3\pi_1\pi_2\pi_3 \\ p_{dis} &= 6\mu_1\pi_2\pi_3 + 6\pi_1\mu_2\pi_3 + 6\pi_1\pi_2\mu_3 + 6\pi_1\pi_2\pi_3 \\ p_{12} &= 3\mu_1\mu_2\pi_3 + 3\pi_1\pi_2\mu_3 + 6\pi_1\pi_2\pi_3 \\ p_{13} &= 3\mu_1\pi_2\mu_3 + 3\pi_1\mu_2\pi_3 + 6\pi_1\pi_2\pi_3 \\ p_{23} &= 3\pi_1\mu_2\mu_3 + 3\mu_1\pi_2\pi_3 + 6\pi_1\pi_2\pi_3. \end{aligned}$$

Our wish, as the reader may have anticipated, is to find a parameter $\widehat{\mu} = (\widehat{\mu}_1, \widehat{\mu}_2, \widehat{\mu}_3)$ that maximizes the likelihood function $L(\mu) = \prod_{k \in K} p_k^{u_k}$, or equivalently the log-likelihood function $\ell(\mu) = \sum_{k \in K} u_k \log p_k$, given an observation u.

Algebraic statistical methods, including HKS, tend to operate in vector spaces of arbitrary dimension over algebraically closed fields such as $\mathbb{C}$. For many problems susceptible to ML methods, there is an obvious motivation to constrain the solution set to a region (or regions) of the parameter space corresponding to inputs that are in some sense "reasonable" or "meaningful"— this has already been seen in Example 1.6, where only parameters lying in the preimage of the probability simplex are admissible.

Specifically, for phylogenetic tree models it is natural to define the following restricted parameter space:

Definition 18.11 Given an algebraic statistical model of a phylogenetic tree T, a *biologically meaningful parameter* for the model is a parameter $\mu \in \mathbb{R}^d$ for which all branch lengths in the tree corresponding to μ under the model are non-negative and real.

In the case of the claw tree, branch lengths may be uniquely determined from transition probabilities by the Jukes–Cantor corrected distance map as given in equation (18.1) (the transition probability p there is our μ_i). Now consider the inverse map:

$$\mu_i = \frac{1 + 3e^{-4b_i/3}}{4}, \tag{18.7}$$

where b_i is the length of branch i. Note that this map uniquely determines (μ_1, μ_2, μ_3) given (b_1, b_2, b_3). Furthermore, $\mu_i \in \mathbb{R}$ if and only if $\mathrm{Im}(b_i) \in \{3\pi n/4 \mid n \in \mathbb{Z}\}$; when b_i is real, branch lengths in $[0, \infty)$ go to μ_is in $(\frac{1}{4}, 1]$.

On this tree, therefore, a biologically meaningful parameterization is one in which $\frac{1}{4} < \mu_1, \mu_2, \mu_3 \leq 1$.

We applied the HKS algorithm, using `Maple` and `Singular` code due to Garcia and Hoşten, to find local maxima of the likelihood function for this model. (The portion of this code that has been made publically available may be found at `http://bio.math.berkeley.edu/ascb/chapter18/`). Given some 20,000 observation vectors—half with independent randomly generated components and half derived from `seq-gen` alignments—the computation, which took approximately 50 CPU hours on a 400MHz Pentium II, yielded at most one biologically meaningful solution for each observation. Preliminary exploration suggested the following conjecture as reasonable:

Conjecture 18.12 *For any observation (u_k), the Jukes–Cantor DNA claw tree model admits at most one local maximum (p_k) that gives a biologically meaningful parameterization $\widehat{\mu}$.*

This conjecture, if true, would have a number of interesting consequences: even if approximate numerical methods are used to target local maxima of the likelihood function, we could be assured that any maximum occurring in the biologically meaningful region $(\frac{1}{4}, 1]^3$ was unique. Given continuing advances in hardware and software for computational biology, we look forward to applying the GNJ and HKS method to other models of small phylogenetic trees described in Chapter 15, and to identifying bounds on the number of biologically meaningful ML solutions for these models.

19

Tree Construction using Singular Value Decomposition

Nicholas Eriksson

We present a new, statistically consistent algorithm for phylogenetic tree construction that uses the algebraic theory of statistical models (as developed in Chapters 1 and 3). Our basic tool is *Singular Value Decomposition* (SVD) from numerical linear algebra.

Starting with an alignment of n DNA sequences, we show that SVD allows us to quickly decide whether a split of the taxa occurs in their phylogenetic tree, assuming only that evolution follows a tree Markov model. Using this fact, we have developed an algorithm to construct a phylogenetic tree by computing only $O(n^2)$ SVDs.

We have implemented this algorithm using the SVDLIBC library (available at http://tedlab.mit.edu/~dr/SVDLIBC/) and have done extensive testing with simulated and real data. The algorithm is fast in practice on trees with 20–30 taxa.

We begin by describing the general Markov model and then show how to flatten the joint probability distribution along a partition of the leaves in the tree. We give rank conditions for the resulting matrix; most notably, we give a set of new rank conditions that are satisfied by non-splits in the tree. Armed with these rank conditions, we present the tree-building algorithm, using SVD to calculate how close a matrix is to a certain rank. Finally, we give experimental results on the behavior of the algorithm with both simulated and real-life (ENCODE) data.

19.1 The general Markov model

We assume that evolution follows a tree Markov model, as introduced in Section 1.4, with evolution acting independently at different sites of the genome. We do not assume that the transition matrices for the model are stochastic. Furthermore, we do not assume the existence of a global rate matrix (as in Section 4.5).

This model is called the general Markov model. It is a more general model than any in the Felsenstein hierarchy (Figure 4.7). The main results in this chapter therefore hold no matter what model in the Felsenstein hierarchy one works with.

Under the general dogma that statistical models are algebraic varieties, the polynomials (called "phylogenetic invariants") defining the varieties are of great interest. Phylogenetic invariants have been studied extensively since [Lake, 1987, Cavender and Felsenstein, 1987]. Linear invariants for the Jukes–Cantor model have been used to infer phylogenies on four and five taxa; see [Sankoff and Blanchette, 2000]. Sturmfels and Sullivant finished the classification of the invariants for group-based models [Sturmfels and Sullivant, 2005]; see Chapter 15 for an application of these invariants for constructing trees on four taxa. Invariants for the general Markov model have been studied in [Allman and Rhodes, 2003, Allman and Rhodes, 2004a].

The main problem with invariants is that there are exponentially many polynomials in exponentially many variables to test on exponentially many trees. Because of this, they are currently considered impractical by many and have only been applied to small problems. However, we solve the problem of this combinatorial explosion by only concentrating on invariants which are given by rank conditions on certain matrices, called "flattenings".

19.2 Flattenings and rank conditions

Recall from Chapters 2 and 17 that a *split* $\{A, B\}$ in a tree is a partition of the leaves obtained by removing an edge of the tree. We will say that $\{A, B\}$ is a *partition* of the set of leaves if it is not necessarily a split but merely a disjoint partition of the set of leaves into two sets.

Throughout, all trees will be assumed to be binary with n leaves. We let m denote the number of states in the alphabet Σ. Usually $m = 4$ and $\Sigma = \{A, C, G, T\}$ or $m = 2$ and $\Sigma = \{0, 1\}$. We will write the joint probabilities of an observation on the leaves as $p_{i_1 \ldots i_n}$. That is, $p_{i_1 \ldots i_n}$ is the probability that leaf j is observed to be in state i_j for all $j \in \{1, \ldots, n\}$. We write P for the entire probability distribution.

Although the descriptions of tree-based models in this book all deal with rooted trees, we will mostly consider unrooted tree models, which are equivalent to them for the general Markov model; see [Allman and Rhodes, 2004a] for details on this technical point. Our tree-building algorithm constructs an unrooted tree, additional methods would be required to find the root.

Definition 19.1 A *flattening* along a partition $\{A, B\}$ is the $m^{|A|}$ by $m^{|B|}$ matrix where the rows are indexed by the possible states for the leaves in A and the columns are indexed by the possible states for the leaves in B. The entries of this matrix are given by the joint probabilities of observing the given pattern at the leaves. We write $\mathrm{Flat}_{A,B}(P)$ for this matrix.

Example 19.2 (Flattening a partition on 4 taxa) Let T be a tree with 4 leaves and let $m = 4$, $\Sigma = \{A, C, G, T\}$. The partition $\{1, 3\}, \{2, 4\}$ flattens to the 16×16 matrix $\mathrm{Flat}_{\{1,3\},\{2,4\}}(P)$ where the rows are indexed by bases of taxa 1 and 3 and the columns by bases of taxa 2 and 4:

$$
\text{Flat}_{\{1,3\},\{2,4\}}(P) = \begin{array}{c} \\ \text{AA} \\ \text{AC} \\ \text{AG} \\ \text{AT} \\ \text{CA} \\ \vdots \end{array}
\begin{array}{ccccccc}
\text{AA} & \text{AC} & \text{AG} & \text{AT} & \text{CA} & \text{CC} & \cdots \\
\begin{pmatrix} p_{AAAA} & p_{AAAC} & p_{AAAG} & p_{AAAT} & p_{ACAA} & p_{ACAC} & \cdots \\
p_{AACA} & p_{AACC} & p_{AACG} & p_{AACT} & p_{ACCA} & p_{ACCC} & \cdots \\
p_{AAGA} & p_{AAGC} & p_{AAGG} & p_{AAGT} & p_{ACGA} & p_{ACGC} & \cdots \\
p_{AATA} & p_{AATC} & p_{AATG} & p_{AATT} & p_{ACTA} & p_{ACTC} & \cdots \\
p_{CAAA} & p_{CAAC} & p_{CAAG} & p_{CAAT} & p_{CCAA} & p_{CCAC} & \cdots \\
\vdots & \vdots & \vdots & \vdots & \vdots & \vdots & \vdots \end{pmatrix}
\end{array}.
$$

$\square$

Next we define a measure of how close a general partition of the leaves is to being a split. If A is a subset of the leaves of T, we let T_A be the subtree induced by the leaves in A (in Chapter 18 this subtree is denoted by $[A]$). That is, T_A is the minimal set of edges needed to connect the leaves in A.

Definition 19.3 Suppose that $\{A, B\}$ is a partition of $[n]$. The *distance* between the partition $\{A, B\}$ and the nearest split, written $e(A, B)$, is the number of edges that occur in $T_A \cap T_B$.

Notice that $e(A, B) = 0$ exactly when $\{A, B\}$ is a split.

Consider $T_A \cap T_B$ as a subtree of T_A. Color the nodes in $T_A \cap T_B$ red, the nodes in $T_A \setminus (T_A \cap T_B)$ blue. Say that a node is *monochromatic* if it and all of its neighbors are of the same color. We let $\text{mono}(A)$ be the number of monochromatic red nodes. That is:

Definition 19.4 Define $\text{mono}(A)$ as the number of nodes in $T_A \cap T_B$ that do not have a node in $T_A \setminus (T_A \cap T_B)$ as a neighbor.

See Figure 19.1 for an example of $e(A, B)$ and $\text{mono}(A)$.

Our main theorem ties together how close a partition is to being a split with the rank of the flattening associated to that partition.

Theorem 19.5 *Let $\{A, B\}$ be a partition of $[n]$, let T be a binary, unrooted tree with leaves labeled by $[n]$, and assume that the joint probability distribution P comes from a Markov model on T with an alphabet with m letters. Then the generic rank of the flattening $\text{Flat}_{A,B}(P)$ is given by*

$$
\min\left(m^{e(A,B)+1-\text{mono}(A)}, m^{e(A,B)+1-\text{mono}(B)}, m^{|A|}, m^{|B|} \right). \tag{19.1}
$$

Proof We claim that $\text{Flat}_{A,B}(P)$ can be thought of as the joint distribution for a simple graphical model. Pick all the nodes that are shared by the induced subtrees for A and B: call this set R. If R is empty, then $\{A, B\}$ is a split; in that case let R be one of the vertices of the edge separating A and B. Notice that $|R| = e(A, B) + 1$. Think of these vertices as a single hidden random variable which we will also call R with $m^{|R|} = m^{e(A,B)+1}$ states. Group the states of the nodes in A together into one $m^{|A|}$-state observed random variable;

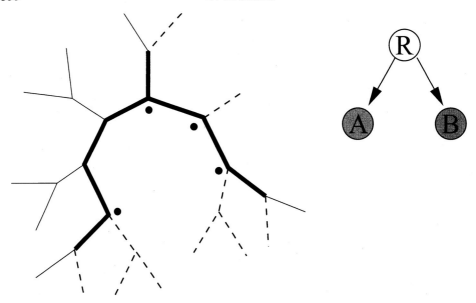

Fig. 19.1. If A is given by the 8 dashed leaves and B by the 7 solid leaves, then $e(A, B) = 8$ (shown in bold) and $\mathrm{Flat}_{A,B}(P)$ is the joint distribution for a 3-state graphical model where the root R has m^9 states and the descendents A and B have m^8 and m^7 states, respectively. Here $\mathrm{mono}(B) = 4$ (indicated by the dots), so the $m^9 \times m^8$ matrix M_A has rank $m^{9-4} = m^5$, which is the rank of $\mathrm{Flat}_{A,B}(P)$.

similarly the nodes in B are grouped into a $m^{|B|}$-state random variable. Then create the graphical model with one hidden $m^{|R|}$-state random variable and two descendent observed variables with $m^{|A|}$ and $m^{|B|}$ states. Notice that $\mathrm{Flat}_{A,B}(P)$ is the joint distribution for this model. See Figure 19.1 for an example.

Furthermore, the distribution for this simplified model factors as

$$\mathrm{Flat}_{A,B}(P) = M_A^{\mathrm{T}}\, \mathrm{diag}(\pi(R)) M_B \qquad (19.2)$$

where $\pi(R)$ is the distribution of R and M_A and M_B are the $m^{|R|} \times m^{|A|}$ and $m^{|R|} \times m^{|B|}$ transition matrices. That is, the (i,j)th entry of M_A is the probability of transitioning from state i at the root R to state j at A.

To say the tree distribution factors as (19.2) just means that

$$\mathrm{Prob}(A = i, B = j) =$$
$$\sum_k \mathrm{Prob}(R = k)\mathrm{Prob}(A = i \mid R = k)\mathrm{Prob}(B = j \mid R = k).$$

Notice that all of the terms in this expression can be written as polynomials in the edge parameters (after choosing a rooting). Therefore the rank of $\mathrm{Flat}_{A,B}(P)$ is at most $m^{\min(|R|,|A|,|B|)}$.

However, the matrices in this factorization do not necessarily have full rank. For example, if one of the nodes in R has only neighbors that are also in R, then the $m^{|R|} \times m^{|A|}$ transition matrices from R to A have many rows that are the same, since the transition from a state of R to a state of A does not depend on the value of this one node. More generally, if a node of R has no neighbors in $T_A \setminus (T_A \cap T_B)$, then the entries of the transition matrix M_A do not depend on the value of this node. But the entries do depend on the values of all other nodes of R (that is, those with neighbors in $T_A \setminus (T_A \cap T_B)$). So R really behaves like a model with $m^{|R|-\text{mono}(A)}$ states on the transition to A and $m^{|R|-\text{mono}(B)}$ states for the transition to B. There are enough parameters so that after canceling out these equal rows, all other rows are linearly independent. Therefore, the rank of M_A is $\min\left(m^{|R|-\text{mono}(A)}, m^{|A|}\right)$ (and similarly for M_B), so the theorem follows. □

This theorem gives rise to a well-known corollary upon noticing that if $\{A, B\}$ is a split, then $e(A, B) = 0$ (see [Allman and Rhodes, 2004a], for example).

Corollary 19.6 *If $\{A, B\}$ is a split in the tree, the generic rank of* $\text{Flat}_{A,B}(P)$ *is m.*

A partial converse of Corollary 19.6 will be used later.

Corollary 19.7 *If $\{A, B\}$ is not a split in the tree, and we have $|A|, |B| \geq 2$ then the generic rank of* $\text{Flat}_{A,B}(P)$ *is at least m^2.*

Proof Since we have $|A|, |B| \geq 2$, we must show that the two other exponents in (19.1) are at least 2. That is, we have to show that $e(A, B) + 1 - \text{mono}(A) \geq 2$ (the case for B is symmetric). This term counts the number of nodes in $T_A \cap T_B$ that are directly connected to a part of T_A outside of $T_A \cap T_B$. Since $\{A, B\}$ is not a split, we know that $|T_A \cap T_B| = e(A, B) + 1 \geq 2$. Consider $T_A \cap T_B$ as a subtree with at least 2 nodes of T_A. The only way for all but one of these nodes to be isolated from the rest of the tree is to have the two consist of a leaf and its parent. However, this is impossible since $\{A, B\}$ is a disjoint partition of the set of leaves, so $T_A \cap T_B$ contains no leaves. □

Example 19.8 In Example 19.2, the 16×16 matrix $\text{Flat}_{\{1,3\},\{2,4\}}(P)$ has rank 4 if the split $\{\{1,3\}, \{2,4\}\}$ occurs in the tree, otherwise, it has rank 16. □

In fact, if $m = 2$, it has recently been shown [Allman and Rhodes, 2004b] that the rank conditions in Corollary 19.6 generate the ideal of invariants for the general Markov model. However, they do not suffice if $m = 4$, since in that case a polynomial of degree 9 lies in the ideal of invariants (see [Strassen, 1983, Garcia *et al.*, 2004]) but this polynomial is not generated by the degree 5 rank conditions (see [Landsberg and Manivel, 2004]).

19.3 Singular Value Decomposition

Singular Value Decomposition provides a method to compute the distance between a matrix and the nearest rank k matrix. In this section, we briefly introduce the basic properties of SVD for real matrices. See [Demmel, 1997] for a thorough treatment.

Definition 19.9 A *singular value decomposition* of a $m \times n$ matrix A (with $m \geq n$) is a factorization $A = U\Sigma V^{\mathrm{T}}$ where U is $m \times n$ and satisfies $U^{\mathrm{T}}U = I$, V is $n \times n$ and satisfies $V^{\mathrm{T}}V = I$ and $\Sigma = \mathrm{diag}(\sigma_1, \sigma_2, \ldots, \sigma_n)$, where $\sigma_1 \geq \sigma_2 \geq \cdots \geq \sigma_n \geq 0$ are called the *singular values* of A.

Definition 19.10 Let a_{ij} be the (i,j)th entry of A. The *Frobenius norm*, written $\|A\|_{\mathrm{F}}$, is the root-sum-of-squares norm on $\mathbb{R}^{m \cdot n}$. That is,

$$\|A\|_{\mathrm{F}} = \sqrt{\sum a_{ij}^2}.$$

The L_2 *norm* (or operator norm), written $\|A\|_2$, is given by

$$\|A\|_2 = \max_{\substack{x \in \mathbb{R}^n \\ x \neq 0}} \left\{ \frac{\|Ax\|}{\|x\|} \right\},$$

where $\|x\|$ is the usual root-sum-of-squares vector norm.

The following is Theorem 3.3 of [Demmel, 1997]:

Theorem 19.11 *The distance from A to the nearest rank k matrix is*

$$\min_{\mathrm{Rank}(B)=k} \|A - B\|_{\mathrm{F}} = \sqrt{\sum_{i=k+1}^{m} \sigma_i^2}$$

in the Frobenius norm and

$$\min_{\mathrm{Rank}(B)=k} \|A - B\|_2 = \sigma_{k+1}$$

in the L_2 norm.

One way of computing the singular values is to compute the eigenvalues of $A^{\mathrm{T}}A$; the singular values are the square roots of these eigenvalues. Therefore, general techniques for solving the real symmetric eigenvalue problem can be used to compute the SVD. These various methods, both iterative and direct, are implemented by many software packages for either sparse or general matrices. We will discuss the computational issues with SVD after we describe how to use it to construct phylogenetic trees.

19.4 Tree-construction algorithm

Now that we know how to tell how close a matrix is to being of a certain rank, we can test whether a given split comes from the underlying tree or not by using the SVD to tell how close a flattening matrix is to being rank m. However, since there are exponentially many possible splits, we must carefully search through this space. Following a suggestion by S. Snir, we do this by building the tree bottom up, at each step joining cherries together, in a method reminiscent of neighbor-joining (Algorithm 2.41).

It is an interesting open question whether the additional information in Theorem 19.5 about non-splits that are almost splits can be harnessed to produce an improved algorithm.

Algorithm 19.12 (Tree construction with SVD)

Input: A multiple alignment of genomic data from n species, from the alphabet Σ with m states.

Ouput: An unrooted binary tree with n leaves labeled by the species.

Initialization: Compute empirical probabilities $p_{i_1 \ldots i_n}$. That is, count occurrences of each possible column of the alignment, ignoring columns with characters not in Σ. Store the results in a sparse format.

Loop: For k from n down to 4, perform the following steps.

For each of the $\binom{k}{2}$ pairs of species compute the SVD for the split $\{\{\text{pair}\}, \{\text{other } k - 2 \text{ species}\}\}$. Pick the pair whose flattening is closest to rank m according to the Frobenius norm and join this pair together in the tree. That is, consider this pair as a single element when picking pairs at the next step.

Proposition 19.13 *Algorithm 19.12 needs the computation of at most $(n - 1)^2 - 3$ SVDs.*

Proof At the first step, we compute an SVD $\binom{n}{2}$ times. At each subsequent step, we only need to compute those splits involving the pair that we just joined together. Thus we compute $(n - 2) + (n - 3) + \cdots + 3 = \binom{n-1}{2} - 3$ total SVDs after the first step for $\binom{n}{2} + \binom{n-1}{2} = (n - 1)^2 - 3$ SVD computations in total. In fact, not all of these are even necessary; some steps will involve computing both partitions $\{A, B\}$ and $\{B, A\}$, in which case one can be ignored. $\square$

The flattenings are very large (size $m^{|A|} \times m^{|B|}$), yet they are typically very sparse. If an alignment is of length L, at most L entries of the flattening, typically many fewer, are non-zero. Generally, computing all singular values of an $a \times b$ matrix takes $O(a^2 b + ab^2)$ time. However, Lanczos iterative methods (cf. Chapter 7 of [Demmel, 1997]) allow singular values to be computed quickly individually, starting with the largest. Furthermore, sparse matrix techniques allow us to take advantage of this structure without having to deal with matrices of exponential size.

Since we will be comparing the SVD from different sized splits, we need to compute distances in the Frobenius norm, which does not change as the dimensions of the matrices change (as long as the number of entries is constant). This means that we should compute all singular values; however that is difficult computationally. But in practice, the singular values typically decrease very quickly, so it suffices to compute only the largest singular values to estimate the Frobenius norm.

By exploiting the sparsity and only computing singular values until they become sufficiently small, we find that we are able to very quickly compute the SVD for flattenings coming from trees with at most 31 leaves with binary data ($m = 2$) and up to 15 leaves with DNA data ($m = 4$). This limitation is due to limits on the size of array indices in **SVDLIBC** and can probably be exceeded. Furthermore, there are good approximation algorithms for SVD that could make very large problems practical [Frieze *et al.*, 1998].

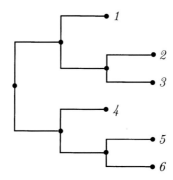

Fig. 19.2. The 6-taxa tree constructed in Example 19.15.

Theorem 19.14 *Algorithm 19.12 is statistically consistent. That is, as the probability distribution converges to a distribution that comes from the general Markov model on a binary tree T, the probability that Algorithm 19.12 outputs T goes to 1.*

Proof We must show that the algorithm picks a correct split at each step; that is, as the empirical distribution approaches the true distribution, the probability of choosing a bad split goes to zero. By Corollary 19.6, we see that a true split will lead to a flattening that approaches a rank m matrix, while Corollary 19.7 shows that other partitions will approach a matrix of rank at least m^2 (except for partitions where one set contains only one element; however, these are never considered in the algorithm). Therefore, as the empirical distribution approaches the true one, the distance of a split from rank m will go to zero while the distance from rank m of a non-split will not.   □

Example 19.15 We begin with an alignment of DNA data of length 1000 for 6 species, labeled $1, \ldots, 6$, simulated from the tree in Figure 19.2 with all branch lengths equal to 0.1. For the first step, we look at all pairs of the 6 species. The score column is the distance in the Frobenius norm from the flattening to the nearest rank 4 matrix:

```
Partition          Score
2 3 | 1 4 5 6      5.8374
5 6 | 1 2 3 4      6.5292
1 2 | 3 4 5 6      20.4385
1 3 | 2 4 5 6      20.5153
4 6 | 1 2 3 5      23.1477
4 5 | 1 2 3 6      23.3001
1 4 | 2 3 5 6      44.9313
3 4 | 1 2 5 6      52.1283
2 4 | 1 3 5 6      52.6763
1 6 | 2 3 4 5      52.9438
1 5 | 2 3 4 6      53.1727
```

```
3 6 | 1 2 4 5    59.5006
3 5 | 1 2 4 6    59.7909
2 6 | 1 3 4 5    59.9546
2 5 | 1 3 4 6    60.3253
picked split 1 4 5 6 | 2 3
tree is 1 4 5 6 (2,3)
```

After the first step, we see that the split $\{\{2,3\},\{1,4,5,6\}\}$ is the best, so we join nodes 2 and 3 together in the tree and continue. Notice that the scores of the partitions roughly correspond to how close they are to being splits:

```
Partition         Score
1 2 3 | 4 5 6     5.8534
5 6 | 1 2 3 4     6.5292
4 6 | 1 2 3 5     23.1477
4 5 | 1 2 3 6     23.3001
1 4 | 2 3 5 6     44.9313
2 3 4 | 1 5 6     45.1427
1 6 | 2 3 4 5     52.9438
2 3 6 | 1 4 5     53.0300
1 5 | 2 3 4 6     53.1727
2 3 5 | 1 4 6     53.3838
picked split 1 2 3 | 4 5 6
tree is 4 5 6 (1,(2,3))
```

After the second step, we join node 1 to the $\{2,3\}$ cherry and continue:

```
Partition         Score
5 6 | 1 2 3 4     6.5292
4 6 | 1 2 3 5     23.1477
4 5 | 1 2 3 6     23.3001
picked split 1 2 3 4 | 5 6
tree is 4 (1,(2,3)) (5,6)


Final tree is (4,(1,(2,3)),(5,6))
```

We have found the last cherry, leaving us with 3 remaining groups which we join together to form an unrooted tree. □

19.5 Performance analysis

19.5.1 Building trees with simulated data

The idea of simulation is that we first pick a tree and simulate a model on that tree to obtain aligned sequence data. Then we build a tree using Algorithm 19.12 and other methods from that data and compare the answers to the original tree.

We used the program seq-gen [Rambaut and Grassly, 1997] to simulate data of various lengths for the tree in Figure 19.3 with the two sets of branch

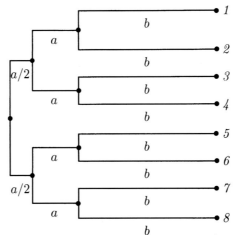

Fig. 19.3. The 8-taxa tree used for simulation with $(a, b) = (0.01, 0.07)$ and $(0.02, 0.19)$.

lengths given in Figure 19.3. This tree was chosen as a particularly difficult tree [Strimmer and von Haeseler, 1996, Ota and Li, 2000].

We simulated DNA data under the general reversible model (the most general model supported by seq-gen). Random numbers uniformly distributed between 1 and 2 were chosen on each run for the six rate matrix parameters (see Figure 4.7). The root frequencies were all set to $1/4$.

Next, the data was collapsed to binary data (that is, A and G were identified, similarly C and T). We used binary data instead of DNA data because of numerical instability with SVD using the much larger matrices from the DNA data. It should be noted that Algorithm 19.12 performed better on binary data than on DNA data. This may be due to the instability, but it may also be because the rank conditions define the entire ideal for binary data.

We ran all tests using our Algorithm 19.12 as well as two algorithms from the PHYLIP package (see Section 2.5): neighbor-joining (i.e., Algorithm 2.41), and a maximum likelihood algorithm (dnaml). We used Jukes–Cantor distance estimation for neighbor-joining and the default settings for dnaml. All three algorithms took approximately the same amount of time, except for dnaml, which slowed down considerably for long sequences.

Figures 19.4 and 19.5 show the results of the simulations. Each algorithm was run 1000 times for each tree and sequence length. While SVD performed slightly worse than the others, it showed very comparable behavior. It should be noted that SVD constructs trees according to a much more general model than the other two methods, so it should be expected to have a higher variance.

19.5.2 Building trees with real data

For data, we use the October 2004 freeze of the ENCODE alignments. For detailed information on these, see Section 4.3, Chapters 21 and 22.

As in Chapter 21, we restricted our attention to 8 species: human, chimp,

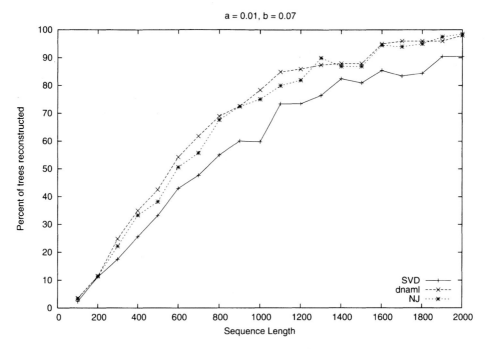

Fig. 19.4. Percentage of trees reconstructed correctly (for the 8-taxa tree with branch lengths $(a, b) = (0.01, 0.07)$) using our SVD algorithm and two `PHYLIP` packages.

galago, mouse, rat, cow, dog, and chicken. We processed each of the 44 EN-CODE regions to obtain 3 data sets. First, for each region, all of the ungapped columns were chosen. Second, within each region, all ungapped columns that corresponded to RefSeq annotated human genes were chosen. Third, we restricted even further to only the human exons within the genes. Bins without all 8 species and bins with less than 100 ungapped positions in the desired class were removed from consideration. This left us with 33 regions for the entire alignment, and 28 for both the gene and exon regions, of lengths between 302 and over 100000 base pairs. See Chapter 21 for a more thorough discussion of these data sets.

As is discussed in Section 21.4, tree construction methods that use genomic data usually misplace the rodents on the tree. The reasons for this are not entirely known, but it could be because tree construction methods generally assume the existence of a global rate matrix (cf. Section 4.5) for all the species. However, rat and mouse have mutated faster than the other species. Our method does not assume anything about the rate matrix and thus is promising for situations where additional assumptions beyond the Markov process of evolution at independent sites are not feasible.

In fact, Table 19.1 shows that our algorithm performs better than `dnaml` on the ENCODE data sets. Note that the measure used is the *symmetric distance* on trees, which counts the number of splits present in one tree that aren't present in the other.

While neither algorithm constructed the correct tree a majority of the time,

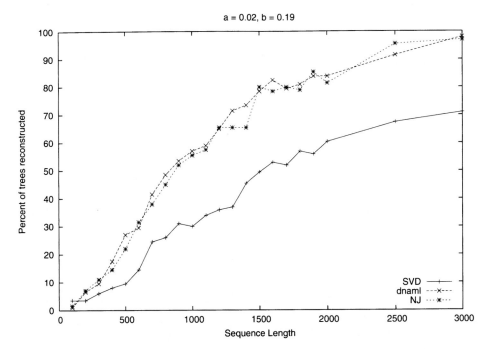

Fig. 19.5. Percentage of trees reconstructed correctly (for the 8-taxa tree with branch lengths $(a, b) = (0.02, 0.019)$) using our SVD algorithm and two PHYLIP packages.

	SVD		dnaml	
	Ave. distance	% correct	Ave. distance	% correct
All	2.06	5.8	3.29	2.9
Gene	1.93	10.3	3.21	0.0
Exon	2.43	21.4	3.0	3.5

Table 19.1. *Comparison of the SVD algorithm and **dnaml** on data from the ENCODE project. Distance between trees is given by the symmetric distance, % correct gives the percentage of the regions which had the correct tree reconstructed.*

the SVD algorithm came much closer on average and constructed the correct tree much more often than **dnaml**, which almost never did (see Figure 21.4 for the correct tree and a common mistake).

20

Applications of Interval Methods to Phylogenetics

Raazesh Sainudiin

Ruriko Yoshida

When statistical inference is conducted in a maximum likelihood (ML) framework as discussed in Chapter 1, we are interested in the global maximum of the likelihood function over the parameter space. In practice we settle for a local optimization algorithm to numerically approximate the global solution since explicit analytical solutions for the maximum likelihood estimates (MLEs) are typically difficult to obtain. See Chapter 3 or 18 for algebraic approaches to solving such ML problems. In this chapter we will take a rigorous numerical approach to the ML problem for phylogenetic trees via interval methods. We accomplish this by first constructing an interval extension of the recursive formulation for the likelihood function of an evolutionary model on an unrooted tree. We then we use an adaptation of a widely applied global optimization algorithm using interval analysis for the phylogenetic context to rigorously enclose ML values as well as MLEs for branch lengths. The method is applied to enclose the most likely 2- and 3-taxa trees under the *Jukes–Cantor model* of DNA evolution. The method is general and can provide rigorous estimates when coupled with standard phylogenetic algorithms. Solutions obtained with such methods are equivalent to computer-aided proofs, unlike solutions obtained with conventional numerical methods.

Statistical inference procedures that obtain MLEs through conventional numerical methods may suffer from several major sources of errors. To fully appreciate the sources of errors we need some understanding of a number screen. Computers can only support a finite set of numbers, usually represented as fixed-length binary floating point quantities of the form, $x = \pm m \cdot 2^e = \pm 0.m \cdot 2^e$, where $m = (m_1 m_2 \ldots m_p)$ is the signed mantissa ($m_1 = 1$, $m_i \in \{0, 1\}, \forall i, 1 < i \leq p$) with base 2; p is the precision; and e is the exponent ($\underline{e} \leq e \leq \overline{e}$) [IEEE Task P754, 1985]. Thus, the smallest and largest machine-representable numbers in absolute value are $\underline{x} = 0.10 \ldots 0 \cdot 2^{\underline{e}}$ and $\overline{x} = 0.11 \ldots 1 \cdot 2^{\overline{e}}$, respectively. Therefore, the binary floating-point system of most machines $\mathcal{R} = \mathcal{R}(2, p, \underline{e}, \overline{e})$ is said to form a screen of the real numbers in the interval $[-\overline{x}, +\overline{x}]$ with 0 uniquely represented by $0.00 \ldots 0 \cdot 2^{\underline{e}}$. When numerical inference procedures rely on inexact computer arithmetic with a number screen they may suffer from at least five types of errors: *roundoff error*, the difference between computed and exact result [Cuyt *et al.*, 2001,

Loh and Walster, 2002]; *truncation error*, from having to truncate an infinite sequence of operations; *conversion error*, inability to machine-represent decimals with infinite binary expansion; and *ill-posed statistical experiment*, presence of unknown non-identifiable subspaces.

The verified global optimization method [Hansen, 1980] sketched below rigorously encloses the global maximum of the likelihood function through interval analysis [Moore, 1967]. Such interval methods evaluate the likelihood function over a continuum of points including those that are not machine-representable and account for all sources of errors described earlier. In this chapter we will see that interval methods, in contrast to heuristic local search methods, can enclose the global optimum with guaranteed accuracy by exhaustive search within any compact set of the parameter space. We begin with a brief introduction to analysis in the space of all compact real intervals, our basic platform for rigorous numerics.

20.1 Brief introduction to interval analysis

Lowercase letters denote *real numbers*, e.g., $x \in \mathbb{R}$. Uppercase letters represent compact *real intervals*, e.g., $X = [\underline{x}, \overline{x}] = [\inf(X), \sup(X)]$. Any compact interval X belongs to the set of all compact real intervals $\mathbb{IR} := \{[a,b] : a \leq b,\ a, b \in \mathbb{R}\}$. The *diameter* and the *midpoint* of X are $d(X) := \overline{x} - \underline{x}$ and $m(X) := (\underline{x} + \overline{x})/2$, respectively. The *smallest* and *largest absolute value* of an interval X are the real numbers given by $\langle X \rangle := \min\{|x| : x \in X\} = \min\{|\underline{x}|, |\overline{x}|\}$, if $0 \notin X$, and 0 otherwise, and $|X| := \max\{|x| : x \in X\} = \max\{|\underline{x}|, |\overline{x}|\}$, respectively. The *absolute value* of an interval X is $|X|_{[\ ]} := \{|x| : x \in X\} = [\langle X \rangle, |X|]$. The *relative diameter* of an interval X, denoted by d_{rel}, is the diameter $d(X)$ itself if $0 \in X$ and $d(X)/\langle X \rangle$ otherwise. An interval X with zero diameter is called a *thin interval* with $\underline{x} = \overline{x} = x$ and thus $\mathbb{R} \subset \mathbb{IR}$. The *hull* of two intervals is $X \underline{\cup} Y := [\min\{\underline{x}, \underline{y}\}, \min\{\overline{x}, \overline{y}\}]$. By the notation $X \Subset Y$, we mean that X is *strictly contained* in Y, i.e., $\underline{x} > \underline{y}$ and $\overline{x} < \overline{y}$. No notational distinction is made between a real number $x \in \mathbb{R}$ and a real vector $x = (x_i, \ldots, x_n)^{\mathrm{T}} \in \mathbb{R}^n$. Similarly, a real interval X and a *real interval vector* or *box* $X = (X_1, \ldots, X_n)^{\mathrm{T}} \in \mathbb{IR}^n$, i.e., $X_i = [\underline{x_i}, \overline{x_i}] = [\inf(X_i), \sup(X_i)] \in \mathbb{IR}$, where $i = 1, \ldots, n$ share the same notation. The diameter, relative diameter, midpoint, and hull operations for boxes are defined componentwise to yield vectors. The maximum over the components is taken to obtain the maximal diameter and the maximal relative diameter, $d_\infty(X) = \max_i d(X_i)$ and $d_{\mathrm{rel},\infty}(X) = \max_i d_{\mathrm{rel}}(X_i)$, respectively, for a box X. Also $\mathbb{IR}$ under the metric $\mathfrak{h}$, given by $\mathfrak{h}(X, Y) := \max\{|\underline{x} - \underline{y}|, |\overline{x} - \overline{y}|\}$, is a complete metric space. Convergence of a sequence of intervals $\{X^{(i)}\}$ to an interval X under the metric $\mathfrak{h}$ is equivalent to the sequence $\mathfrak{h}(X^{(i)}, X)$ approaching 0 as i approaches ∞, which in turn is equivalent to both $\underline{x}^{(i)} \to \underline{x}$ and $\overline{x}^{(i)} \to \overline{x}$. Continuity and differentiability of a function $F : \mathbb{IR}^n \to \mathbb{IR}^k$ are defined in the usual way. Let $\circ$ denote a binary operation. An interval arithmetic (IA) operation $X \circ Y := \{x \circ y : x \in X, y \in Y\}$ thus yields the set containing the result of the operation performed on every real pair $(x, y) \in (X, Y)$. Although there are uncountably many real operations to consider during an interval operation, the properties of continuity, monotonicity, and compactness imply that:

$$X + Y = [\underline{x} + \underline{y}, \overline{x} + \overline{y}], \; X \cdot Y = [\min\{\underline{x}\underline{y}, \underline{x}\overline{y}, \overline{x}\underline{y}, \overline{x}\overline{y}\}, \max\{\underline{x}\underline{y}, \underline{x}\overline{y}, \overline{x}\underline{y}, \overline{x}\overline{y}\}],$$

$$X - Y = [\underline{x} - \overline{y}, \overline{x} - \underline{y}], \; X/Y = X \cdot [1/\overline{y}, 1/\underline{y}], \; 0 \notin Y.$$

This definition of IA leads to the *property of inclusion isotony* which stipulates that $X \circ Y$ contains $V \circ W$ provided $V \subseteq X$ and $W \subseteq Y$. Note that continuous functions of compact sets are necessarily inclusion isotonic. The identity elements of $+$ and $\cdot$ are the thin intervals 0 and 1, respectively. Multiplicative and additive inverses do not exist except when X is also thin. IA is commutative and associative but not distributive. However, $X \cdot (Y + Z) \subseteq (X \cdot Y) + (X \cdot Z)$. For any real function $f(x) : \mathbb{R}^n \to \mathbb{R}$ and some box $X \in \mathbb{IR}^n$, let the image of f over X be denoted by $f(X) := \{f(x) : x \in X\}$. Inclusion isotony also holds for interval evaluations that are compositions of arithmetic expressions and the elementary functions. When real constants, variables, and operations in f are replaced by their interval counterparts, we obtain $F(X) : \mathbb{IR}^n \to \mathbb{R}$, the natural interval extension of f. Guaranteed enclosures of the image of $f(X)$ are obtained by $F(X)$ due to the *inclusion property*, which states that if $x \in X$, then $f(x) \in F(X)$. The natural interval extension $F(X)$ often overestimates the image $f(X)$, but can be shown under mild conditions to linearly approach the image as the maximal diameter of the box X goes to zero, i.e., $\mathfrak{h}(F(X), f(X)) \leq \alpha \cdot d_\infty(X)$ for some $\alpha \geq 0$. This implies that a partition of X into smaller boxes $\{X^{(1)}, \ldots, X^{(m)}\}$ gives better enclosures of $f(X)$ through the union $\bigcup_{i=1}^m F(X^{(i)})$. This is illustrated by the gray rectangles of a given shade that enclose the image of the nonlinear function shown in Figure 20.1. The darker the shade of the image enclosure the finer the corresponding partition on the domain $[-10, 6]$.

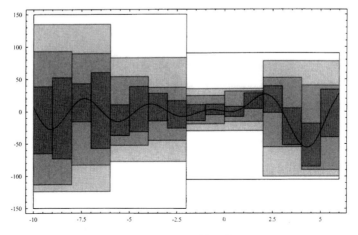

Fig. 20.1. Image enclosure of $-\sum_{k=1}^5 k\, x \sin\left(\frac{k(x-3)}{3}\right)$ linearly tightens with the mesh.

Some interval extensions of f are better at enclosing the true image than others. Figure 20.2 exhibits three functions. These functions are equivalent as real maps but their natural interval extensions yield successively tighter range enclosures: $F^{(1)} \in F^{(2)} \in F^{(3)}$. Note that $F^{(3)} \subset F^{(2)}$ since $X^2 \subset X \cdot X$ in IA. If X appears only once in the expression and all parameters are thin intervals, it was shown by [Moore, 1979] that the natural interval extension does indeed yield a tight enclosure, i.e., $F(X) = f(X)$. In general, we can tighten an enclosure by minimizing the occurrence of X in the expression.

There is another way to improve the tightness of the image enclosure. Let $\nabla f(x)$ and $\nabla^2 f(x)$ denote the gradient and Hessian of f, respectively. Now let $\nabla F(x)$ and $\nabla^2 F(x)$ represent their corresponding interval extensions. A better enclosure of $f(X)$ over all $x \in X$ with a fixed center $c = m(X) \in X$ is possible for a differentiable f with the following centered form:

$$f(x) = f(c) + \nabla f(b) \cdot (x - c) \in f(c) + \nabla f(X) \cdot (x - c) \subseteq F_c(X),$$

for some $b \in X$. Here $F_c(X) := f(c) + \nabla F(X) \cdot (X - c)$ is the interval extension of the centered form of f with center $c = m(X)$ and decays quadratically to $f(X)$ as the maximal diameter of X approaches 0. Next we introduce automatic differentiation (AD) to obtain gradients, Hessians, and their enclosures for a twice-differentiable function.

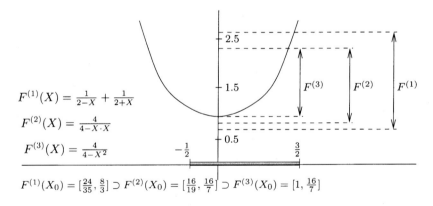

$$F^{(1)}(X) = \frac{1}{2-X} + \frac{1}{2+X}$$

$$F^{(2)}(X) = \frac{4}{4-X \cdot X}$$

$$F^{(3)}(X) = \frac{4}{4-X^2}$$

$$F^{(1)}(X_0) = [\tfrac{24}{35}, \tfrac{8}{3}] \supset F^{(2)}(X_0) = [\tfrac{16}{19}, \tfrac{16}{7}] \supset F^{(3)}(X_0) = [1, \tfrac{16}{7}]$$

Fig. 20.2. Extension-specific dependence of image enclosures.

When it becomes too cumbersome or impossible to explicitly compute $\nabla f(x)$ and $\nabla^2 f(x)$ of a function $f : \mathbb{R}^n \to \mathbb{R}$, we may employ a Hessian differentiation arithmetic, also known as second-order AD [Rall, 1981]. This approach defines an arithmetic on a set of ordered triples. Consider a twice-continuously differentiable function $f : \mathbb{R}^n \to \mathbb{R}$ with the gradient vector $\nabla f(x) := (\partial f(x)/\partial x_1, \ldots, \partial f(x)/\partial x_n)^{\mathrm{T}} \in \mathbb{R}^n$, and Hessian matrix $\nabla^2 f(x) := ((\partial^2 f(x)/\partial x_i \partial x_j))_{i,j=\{1,\ldots,n\}} \in \mathbb{R}^{n \times n}$. For every f, consider its corresponding ordered triple $(f(x), \nabla f(x), \nabla^2 f(x))$. The ordered triples corresponding to a constant function, $c(x) = c : \mathbb{R}^n \to \mathbb{R}$, and a component identifying function (or variable), $I_j(x) = x_j : \mathbb{R}^n \to \mathbb{R}$, are $(c, 0, 0)$ and $(x_j, e^{(j)}, 0)$, respectively, where $e^{(j)}$ is the jth unit vector and the 0s are additive identities in their appropriate spaces. To perform an elementary operation $\circ \in \{+, -, \cdot, /\}$ on a pair of such triples to obtain another, as in $(h(x), \nabla h(x), \nabla^2 h(x)) := (f(x), \nabla f(x), \nabla^2 f(x)) \circ (g(x), \nabla g(x), \nabla^2 g(x))$, or to compose the triples of two elementary functions, we use the chain rule of Newtonian calculus. The AD process may be extended from real functions to interval-valued functions. By replacing the real xs above by interval Xs and performing all operations in the real IA with the interval extension F of f, we can rigorously enclose the components of the triple $(F(X), \nabla F(X), \nabla^2 F(X))$ through an interval-extended Hessian differentiation arithmetic so that, for every $x \in X \in \mathbb{IR}^n$, we have $f(x) \in F(X) \in \mathbb{IR}$, $\nabla f(x) \in \nabla F(X) \in \mathbb{IR}^n$, and $\nabla^2 f(x) \in \nabla^2 F(X) \in \mathbb{IR}^{n \times n}$. We can now apply interval AD to find the roots of nonlinear functions.

The interval version of Newton's method computes an enclosure of the zero x^* of a continuously differentiable function $f(x)$ in the interval X through the following dynamical system in $\mathbb{IR}$:

$$X^{(j+1)} = \left(m(X^{(j)}) - \frac{f(m(X^{(j)}))}{F'(X^{(j)})} \right) \cap X^{(j)}, \qquad j = 0, 1, 2, \ldots$$

In this system $X^{(0)} = X$, $F'(X^{(j)})$ is the enclosure of $f'(x)$ over $X^{(j)}$, and $m(X^{(j)})$ is the mid-point of $X^{(j)}$. The interval Newton method will never diverge provided that $0 \notin F'(X^{(0)})$, or equivalently that a unique zero of f lies in $X^{(0)}$. The interval Newton method was derived by [Moore, 1967]. If there is only one root x^* of a continuously differentiable f in a compact $X^{(0)}$, then the sequence of compact sets $X^{(0)} \supseteq X^{(1)} \supseteq X^{(2)} \ldots$ can be shown to converge quadratically to x^* [Alefeld and Herzberger, 1983]. We can derive the above dynamical system in $\mathbb{IR}$ via the mean value theorem. Let $f(x)$ be continuously differentiable and $f'(x) \neq 0$ for all $x \in X$ such that x^* is the only zero of f in X. Then, by the mean value theorem, there exists $c \in (x, x^*)$ such that $f(x) - f(x^*) = f'(c)(x - x^*)$ for every x. Since $f'(c) \neq 0$ by assumption, and since $f(x^*) = 0$, it follows that

$$x^* = x - \frac{f(x)}{f'(c)} \in x - \frac{f(x)}{F'(X)} =: N(X), \quad \forall x \in X.$$

We call $N(X)$ the Newton operator; it contains x^*. Since our root of interest lies in X, we have $x^* \in N(X) \cap X$. Note that the above dynamical system in $\mathbb{IR}$ is obtained by replacing x with $m(X)$ and X with $X^{(j)}$ in the previous expression. The usual Newton method lends itself to an intuitive geometric interpretation: in the jth iteration, think of shining a beam of light onto the domain from the point $(x^{(j)}, f(x^{(j)}))$ along the tangent to $f(x)$ at $x^{(j)}$. The intersection of this beam (the white line in Figure 20.3) with the domain provides $x^{(j+1)}$, which is where the next iteration is resumed. In the interval Newton method, then, we shine a set of beams from the point $(x^{(j)}, f(x^{(j)}))$ along the directions of all the tangents to $f(x)$ on the entire interval X. The intersection of these beams (the gray floodlight region of Figure 20.3) with the domain is $N(X^{(j)})$. The iteration is resumed with the new interval $X^{(j+1)} = $

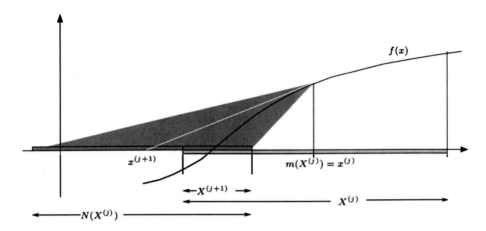

Fig. 20.3. Geometric interpretation of the interval Newton method.

$N(X^{(j)}) \cap X^{(j)}$. Next we extend the interval Newton method in order to allow $F'(X)$ to contain 0.

By including the points $+\infty$ and $-\infty$ to $\mathbb{R}$, it becomes possible to define extended interval arithmetic (EIA) on $\mathbb{IR}^* := \mathbb{IR} \cup \{(-\infty, \overline{x}] : \overline{x} \in \mathbb{R}\} \cup \{[\underline{x}, +\infty) : \underline{x} \in \mathbb{R}\} \cup (-\infty, +\infty)$, the set of intervals with end points in the complete lattice $\mathbb{R}^* := \mathbb{R} \cup \{+\infty\} \cup \{-\infty\}$, with respect to the ordering relation $\leq$. Let $[\,]$ denote the empty interval. Division by intervals containing 0 becomes possible with the following rules:

$$
X/Y := \begin{cases}
(-\infty, +\infty) & \text{if } 0 \in X, \text{or } Y = [0,0] \\
[\,] & \text{if } 0 \notin X, \text{and } Y = [0,0] \\
[\overline{x}/\underline{y}, +\infty) & \text{if } \overline{x} \leq 0, \text{ and } \overline{y} = 0 \\
[\underline{x}/\overline{y}, +\infty) & \text{if } 0 \leq \underline{x}, \text{ and } 0 = \underline{y} < \overline{y} \\
(-\infty, \overline{x}/\overline{y}] & \text{if } \overline{x} \leq 0, \text{ and } 0 = \underline{y} < \overline{y} \\
(-\infty, \underline{x}/\underline{y}] & \text{if } 0 \leq \underline{x}, \text{ and } \underline{y} < \overline{y} = 0 \\
(-\infty, \overline{x}/\overline{y}] \cup [\overline{x}/\underline{y}, +\infty) & \text{if } \overline{x} \leq 0, \text{ and } [0,0] \in Y \\
(-\infty, \underline{x}/\underline{y}] \cup [\underline{x}/\overline{y}, +\infty) & \text{if } 0 \leq \underline{x}, \text{ and } [0,0] \in Y.
\end{cases}
$$

When X is a thin interval with $x = \underline{x} = \overline{x}$ and Y has $+\infty$ or $-\infty$ as one of its bounds, then extended interval subtraction is also necessary for the extended interval Newton algorithm, and is defined as follows:

$$
[\underline{x}, \overline{x}] - Y := \begin{cases}
(-\infty, +\infty) & \text{if } Y = (-\infty, +\infty) \\
(-\infty, x - \underline{y}] & \text{if } Y = (\underline{y}, +\infty) \\
[x - \overline{y}, +\infty) & \text{if } Y = (-\infty, \overline{y}].
\end{cases}
$$

The extended interval Newton method uses the EIA described above and is a variant of the method based on [Hansen and Sengupta, 1981] with Ratz's modifications [Ratz, 1992] as implemented in [Hammer $et\ al.$, 1995]. It can be used to enclose the roots of a continuously differentiable function $f : \mathbb{R}^n \to \mathbb{R}^n$

in a given box $X \in \mathbb{IR}^n$. Let $J_f(x) := ((\partial f_i(x)/\partial x_j))_{i,j=\{1,\dots,n\}} \in \mathbb{R}^{n \times n}$ denote the Jacobian matrix of f at x. Let $J_F(X) \supset J_f(X)$ denote the Jacobian of the interval extension of f. The Jacobian can be computed via AD by computing the gradient of each component f_i of f. By the mean value theorem, $f(m(X)) - f(x^*) = J_f(w) \cdot (m(X) - x^*)$, for some $x^* \in X, w = (w_1, w_2, \dots, w_n)$, where $w_i \in X$ for all $i \in \{1, 2, \dots, n\}$. Setting $f(x^*) = 0$ yields the relation $x^* \in \mathcal{N}(X) \cap X$, where $\mathcal{N}(X) := m(X) - (J_F(X))^{-1} \cdot F(m(X))$, for all $x \in X$ such that $J_F(x)$ is invertible. An iteration scheme $X^{(j+1)} := \mathcal{N}(X^{(j)}) \cap X^{(j)}$ for $j = 0, 1, \dots$, and $X^{(0)} := X$ will enclose the zeros of f contained in X. We may relax the requirement that every matrix in $J_F(X)$ be invertible by using the inverse of the midpoint of $J_F(X)$, i.e., $(m(J_F(X)))^{-1} =: p \in \mathbb{R}^{n \times n}$, as a matrix preconditioner. The extended interval Gauss–Seidel iteration, which is also applicable to singular systems [Neumaier, 1990], is used to solve the preconditioned interval linear equation

$$p \cdot F(m(X)) = p \cdot J_F(X) \cdot (m(X) - x^*)$$
$$a = G \cdot (c - x^*),$$

where $a \in A := p \cdot F(m(X)), G := p \cdot J_F(X)$, and $c := m(X)$. Thus the solution set $\mathbf{S} := \{x \in X : g \cdot (c - x) = a \text{ for all } g \in G\}$ of the interval linear equation $a = G \cdot (c - x)$ has the componentwise solution set $\mathbf{S}_i = \{x_i \in X_i : \sum_{j=1}^n (g_{i,j} \cdot (c_j - x_j)) = a_i, \forall g \in G\}, \forall i \in \{1, \dots, n\}$. Now set $Y = X$, and solve the ith equation for the ith variable iteratively for each i as follows:

$$y_i = c_i - \frac{1}{g_{i,i}}\left(a_i + \sum_{j=1,j\neq i}^n (g_{i,j} \cdot (y_j - c_j))\right)$$
$$\in \left(c_i - \frac{1}{G_{i,i}}\left(A_i + \sum_{j=1,j\neq i}^n (G_{i,j} \cdot (Y_j - c_j))\right)\right) \cap Y_i.$$

Then $\mathcal{N}_{GS}(X)$, the set resulting from one extended interval Newton Gauss–Seidel step such that $\mathbf{S} \subseteq \mathcal{N}_{GS}(X) \subseteq X$, contains interval vector(s) Y obtained by this iteration. Thus the roots of f are enclosed by the discrete dynamical system $X^{(j)} = \mathcal{N}_{GS}(X^{(j)})$ in $\mathbb{IR}^n$. Every 0 of f that lies in X also lies in $\mathcal{N}_{GS}(X)$. If $\mathcal{N}_{GS}(X) = []$, the empty interval, then f has no solution in X. If $\mathcal{N}_{GS}(X) \in X$, then f has a unique solution in X [Hansen, 1992]. When $G_{ii} \supset 0$, the method is applicable with EIA that allows for division by 0. In such cases, we may obtain up to two disjoint compact intervals for Y_i subsequent to EIA and intersection with the previous compact interval X_i. In such cases, the iteration is applied to each resulting sub-interval.

All the interval arithmetic demonstrated up to this point involved real intervals. However, $\mathcal{R}$, the set of floating-point numbers available on a computing machine, is finite. A *machine interval* is a real interval with bounds in $\mathcal{R}$, the set of floating-point numbers described in the introduction. We can perform IA on $\mathbb{IR} = \{X \in \mathbb{IR} : \underline{x}, \overline{x} \in \mathcal{R}\}$, the set of all machine intervals, in a computer. In spite of the finiteness of $\mathbb{IR}$, the strength of IA lies in a machine interval X being able to enclose a segment of the entire continuum of reals between its machine-representable boundaries. Operations with real intervals can be tightly enclosed by the *rounding directed* operations, provided by the

IEEE arithmetic standard, with the smallest machine intervals containing them [Hammer *et al.*, 1995, Kulisch *et al.*, 2001].

20.2 Enclosing the likelihood of a compact set of trees

Let $\mathcal{D}$ denote a homologous set of distinct DNA sequences of length v from n taxa. We are interested in the branch lengths of the most likely tree under a particular topology. Let b denote the number of branches and s denote the number of nodes of a tree with topology τ. Thus, for a given unrooted topology τ with n leaves and b branches, the unknown parameter $\theta = (\theta_1, \ldots, \theta_b)$ is the real vector of branch lengths in the positive orthant ($\theta_q \in \mathbb{R}_+$). An explicit model of DNA evolution is needed to construct the likelihood function which gives the probability of observing data $\mathcal{D}$ as a function of the parameter θ. The simplest such continuous time Markov chain model (JC69) on the state space Σ is due to Jukes and Cantor [Jukes and Cantor, 1969]. We may compute $\ell^{(k)}(\theta)$, the log-likelihood at site $k \in \{1, \ldots, v\}$, through the following post-order traversal [Felsenstein, 1981]:

(i) Associate with each node $q \in \{1, \ldots, s\}$ with m descendents, a partial likelihood vector, $\mathbf{l}_q := (\mathbf{l}_q^A, \mathbf{l}_q^C, \mathbf{l}_q^G, \mathbf{l}_q^T) \in \mathbb{R}^4$, and let the length of the branch leading to its ancestor be θ_q.

(ii) For a leaf node q with nucleotide i, set $\mathbf{l}_q^i = 1$ and $\mathbf{l}_q^j = 0$ for all $j \neq i$. For any internal node q, set $\mathbf{l}_q := (1, 1, 1, 1)$.

(iii) For an internal node q with descendents $s_1, s_2, \ldots, s_m$,

$$\mathbf{l}_q^i = \sum_{j_1, \ldots, j_m \in \Sigma} \{ \mathbf{l}_{s_1}^{j_1} \cdot P_{i,j_1}(\theta_{s_1}) \cdot \mathbf{l}_{s_2}^{j_2} \cdot P_{i,j_2}(\theta_{s_2}) \ldots \mathbf{l}_{s_m}^{j_m} \cdot P_{i,j_m}(\theta_{s_m}) \}.$$

(iv) Compute $\mathbf{l}_q$ for each sub-terminal node q, followed by those of their ancestors recursively to finally compute $\mathbf{l}_r$ for the root node r to obtain the log-likelihood for site k: $\ell^{(k)}(\theta) = \mathbf{l}_r = \log \sum_{i \in \Sigma} (\pi_i \cdot \mathbf{l}_r^i)$.

Assuming independence across sites we obtain $\ell(\theta) = \sum_{k=1}^{v} \ell^{(k)}(\theta)$, the natural logarithm of the likelihood function for the data $\mathcal{D}$, by multiplying the site-specific likelihoods. The problem of finding the global maximum of this likelihood function is equivalent to finding the global minimum of $l(\theta) := -\ell(\theta)$. Replacing every constant c by its corresponding constant triple $(C, 0, 0)$, every variable θ_j by its triple $(\Theta_j, e^{(j)}, 0)$, and every real operation or elementary function by its counterpart in interval-extended Hessian differentiation arithmetic in the above post-order traversal yields a rigorous enclosure of the negative log-likelihood triple $(\mathcal{L}(\Theta), \nabla\mathcal{L}(\Theta), \nabla^2\mathcal{L}(\Theta))$ of the negative log-likelihood function $l(\theta)$ over Θ.

20.3 Global optimization

20.3.1 Branch-and-bound

The most basic strategy in global optimization through enclosure methods is to employ rigorous branch-and-bound techniques. Such techniques recursively partition (branch) the original compact space of interest into compact sub-spaces and discard (bound) those subspaces that are guaranteed to not contain the global optimizer(s). For the real scalar-valued multi-dimensional objective function $l(\theta)$, the interval branch-and-bound technique can be applied to its natural interval extension $\mathcal{L}(\Theta)$ to obtain an interval enclosure $\mathcal{L}^*$ of the global minimum value l^* as well as the set of minimizer(s) to a specified accuracy ϵ. Note that this set of minimizer(s) of $\mathcal{L}(\theta)$ is the set of maximizer(s) of the likelihood function for the observed data $\mathcal{D}$. The strength of such methods arises from the algorithmic ability to discard large sub-boxes from the original search region:

$$\Theta^{(0)} = (\Theta_1^{(0)}, \ldots, \Theta_b^{(0)}) := ([\underline{\theta}_1^{(0)}, \overline{\theta}_1^{(0)}], \ldots, [\underline{\theta}_b^{(0)}, \overline{\theta}_b^{(0)}]) \subset \mathbb{R}^b,$$

that are not candidates for global minimizer(s). Four tests that help discard sub-regions are described below. Let $\mathcal{L}$ denote a list of ordered pairs of the form $(\Theta^{(i)}, \mathcal{L}_{\Theta^{(i)}})$, where $\Theta^{(i)} \subseteq \Theta^{(0)}$, and $\mathcal{L}_{\Theta^{(i)}} := \min(\mathcal{L}(\Theta^{(i)}))$ is a lower bound for the image of the negative log-likelihood function l over $\Theta^{(i)}$. Let $\tilde{l}$ be an upper bound for l^* and $\nabla\mathcal{L}(\Theta^{(i)})_k$ denote the kth interval of the gradient box $\nabla\mathcal{L}(\Theta^{(i)})$. If no information is available for $\tilde{l}$, then $\tilde{l} = \infty$.

20.3.1.1 Midpoint cutoff test

The basic idea of the *midpoint cutoff test* is to discard sub-boxes of the search space $\Theta^{(0)}$ with the lower bound for their image enclosures above $\tilde{l}$, the current best estimate of an upper bound for l^*. Figure 20.4 shows a multi-modal l as a function of a scalar θ over $\Theta^{(0)} = \bigcup_{i=1}^{16} \Theta^{(i)}$. For this illustrative example, $\tilde{l}$ is set as the upper bound of the image enclosure of l over the smallest machine interval containing the midpoint of $\Theta^{(15)}$, the interval with the smallest lower bound of its image enclosure. The shaded rectangles show the image enclosures over intervals that lie strictly above $\tilde{l}$. In this example the *midpoint cutoff test* would discard all other intervals except $\Theta^{(1)}$, $\Theta^{(2)}$, and $\Theta^{(4)}$. Given a list $\mathcal{L}$ and candidate upper bound $\tilde{l}$, the midpoint cutoff test works as follows:

(i) Given a list $\mathcal{L}$ and $\tilde{l}$.
(ii) Choose an element j of $\mathcal{L}$, such that $j = \operatorname{argmin} \mathcal{L}_{\Theta^{(i)}}$, since $\Theta^{(j)}$ is likely to contain a minimizer.
(iii) Find its midpoint $c = m(\Theta^{(j)})$ and let C be the smallest machine interval containing c.
(iv) Compute a possibly improved $\tilde{l} = \min\{\tilde{l}, \overline{\mathcal{L}}_C\}$, where $\overline{\mathcal{L}}_C := \max(\mathcal{L}(C))$.
(v) Discard any ith element of $\mathcal{L}$ for which $\mathcal{L}_{\Theta^{(i)}} > \tilde{l} \geq l^*$.

20.3.1.2 Monotonicity test

For a continuously differentiable function $l(\theta)$, the *monotonicity test* determines whether $l(\theta)$ is strictly monotone over an entire sub-box $\Theta^{(i)} \subset \Theta^{(0)}$. If l is strictly monotone over $\Theta^{(i)}$, then a global minimizer cannot lie in the interior of $\Theta^{(i)}$. Therefore, $\Theta^{(i)}$ can only contain a global minimizer as a boundary point if this point also lies in the boundary of $\Theta^{(0)}$. Figure 20.5 illustrates

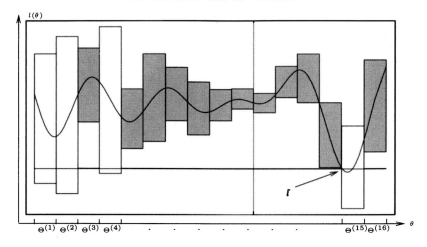

Fig. 20.4. Midpoint cutoff test.

the *monotonicity test* for the one-dimensional case. In this example the search space of interest, $\Theta^{(0)} = [\underline{\theta}^{(0)}, \overline{\theta}^{(0)}] = \cup_{i=1}^{8} \Theta^{(i)}$, can be reduced considerably. In the interior of $\Theta^{(0)}$, we may delete $\Theta^{(2)}$, $\Theta^{(5)}$, and $\Theta^{(7)}$, since $l(\theta)$ is monotone over them as indicated by the enclosure of the derivative $l'(\theta)$ being bounded away from 0. Since $l(\theta)$ is monotonically decreasing over $\Theta^{(1)}$ we may also delete it, since we are only interested in minimization. The sub-box $\Theta^{(8)}$ may be pruned to its right boundary point $\theta^{(8)} = \overline{\theta}^{(8)} = \overline{\theta}^{(0)}$ due to the strictly decreasing nature of $l(\theta)$ over it. Thus the *monotonicity test* has pruned $\Theta^{(0)}$ to the smaller candidate set $\{\overline{\theta}^{(0)}, \Theta^{(3)}, \Theta^{(4)}, \Theta^{(6)}\}$ for a global minimizer.

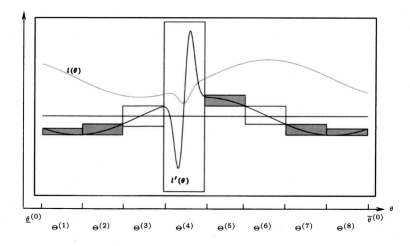

Fig. 20.5. Monotonicity test.

- Given $\Theta^{(0)}$, $\Theta^{(i)}$, and $\nabla \mathcal{L}(\Theta^{(i)})$.
- Iterate for $k = 1, \ldots, b$
  - If $0 \in \nabla \mathcal{L}(\Theta^{(i)})_k$, then leave $\Theta_k^{(i)}$ unchanged, as it may contain a stationary point of l.

– Otherwise, $0 \notin \nabla \mathcal{L}(\Theta^{(i)})_k$. This implies that $\Theta^{(i)}$ can be pruned, since $l^* \notin \Theta^{(i)}$ except possibly at the boundary points, as follows:

(i) if $\min (\nabla \mathcal{L}(\Theta^{(i)})_k) > 0$ and $\underline{\theta}\,_k^{(0)} = \underline{\theta}\,_k^{(i)}$, then $\Theta_k^{(i)} = [\underline{\theta}\,_k^{(i)}, \underline{\theta}\,_k^{(i)}]$;

(ii) else if $\max(\nabla \mathcal{L}(\Theta^{(i)})_k) < 0$ and $\overline{\theta}\,_k^{(0)} = \overline{\theta}\,_k^{(i)}$, then $\Theta_k^{(i)} = [\overline{\theta}\,_k^{(i)}, \overline{\theta}\,_k^{(i)}]$;

(iii) else, delete the ith element of $\mathfrak{L}$ and stop the iteration.

20.3.1.3 Concavity test

Given $\Theta^{(i)} \in \Theta^{(0)}$, and the diagonal elements $(\nabla^2 \mathcal{L}(\Theta^{(i)}))_{kk}$ of $\nabla^2 \mathcal{L}(\Theta^{(i)})$, note that if $\min ((\nabla^2 \mathcal{L}(\Theta^{(i)}))_{kk}) < 0$ for some k, then $\nabla^2 \mathcal{L}(\Theta^{(i)})$ cannot be positive semidefinite, and therefore $l(\theta)$ cannot be convex over $\Theta^{(i)}$ and thus cannot contain a minimum in its interior. In the one-dimensional example shown in Figure 20.5, an application of the *concavity test* to the candidate set $\{\underline{\theta}^{(0)}, \Theta^{(4)}, \Theta^{(6)}\}$ for a global minimizer returned by the *monotonicity test* would result in the deletion of $\Theta^{(6)}$ due to the concavity of $l(\theta)$ over it.

- Given $\Theta^{(i)} \in \Theta^{(0)}$ and $\nabla^2 \mathcal{L}(\Theta^{(i)})$
- If $\min ((\nabla^2 \mathcal{L}(\Theta^{(i)}))_{kk}) < 0$ for any $k \in \{1, \ldots, b\}$, then delete the ith element of $\mathfrak{L}$.

20.3.1.4 Interval Newton test

Given $\Theta^{(i)} \in \Theta^{(0)}$, and $\nabla \mathcal{L}(\Theta^{(i)})$, we attempt to solve the system, $\nabla \mathcal{L}(\theta) = 0$ in terms of $\theta \in \Theta^{(i)}$.

- Apply one extended interval Newton Gauss–Seidel step to the linear interval equation $a = G \cdot (c - \theta)$, where $a := p \cdot \mathcal{L}(m(\Theta^{(i)}))$, $G := p \cdot \nabla^2 \mathcal{L}(\Theta^{(i)})$, $c := m(\Theta^{(i)})$, and $p := (m(\nabla^2 F(X)))^{-1}$, in order to obtain $\mathcal{N}'_{\mathrm{GS}}(\Theta^{(i)})$.
- One of the following can happen,

(i) If $\mathcal{N}'_{\mathrm{GS}}(\Theta^{(i)})$ is empty, then discard $\Theta^{(i)}$.

(ii) If $\mathcal{N}'_{\mathrm{GS}}(\Theta^{(i)}) \in \Theta^{(i)}$, then replace $\Theta^{(i)}$ by the contraction $\mathcal{N}'_{\mathrm{GS}}(\Theta^{(i)}) \cap \Theta^{(i)}$.

(iii) If $0 \in G_{jj}$, and the extended interval division splits $\Theta_j^{(i)}$ into a non-empty union of $\Theta_j^{(i),1}$ and $\Theta_j^{(i),2}$, then the iteration is continued on $\Theta_j^{(i),1}$, while $\Theta_j^{(i),2}$, if non-empty, is stored in $\mathfrak{L}$ for future processing. Thus, one extended interval Newton Gauss–Seidel step can add at most $b + 1$ sub-boxes to $\mathfrak{L}$.

20.3.2 Verification

Given a collection of sub-boxes $\{\Theta^{(1)}, \ldots, \Theta^{(n)}\}$, each of width $\leq \epsilon$, that could not be discarded by the tests in Section 20.3.1, one can attempt to verify the existence and uniqueness of a local minimizer within each sub-box $\theta^{(i)}$ by checking whether the conditions of the following two theorems are satisfied. For proof of these two theorems see [Hansen, 1992] and [Ratz, 1992].

(i) If $\mathcal{N}'_{\mathrm{GS}}(\Theta^{(i)}) \in \Theta^{(i)}$, then there exists a unique stationary point of $\mathcal{L}$, i.e., a unique zero of $\nabla \mathcal{L}$ exists in $\Theta^{(i)}$.

(ii) If $(I + \frac{1}{\kappa} \cdot (\nabla^2 \mathcal{L}(\Theta^{(i)}))) \cdot Z \Subset Z$, where $(\nabla^2 \mathcal{L}(\Theta^{(i)}))_{d,\infty} \leq \kappa \in \mathbb{R}$ for some $Z \in \mathbb{IR}^n$, then the spectral radius $\rho(s) < 1$ for all $s \in (I - \frac{1}{\kappa} \cdot (\nabla^2 \mathcal{L}(\Theta^{(i)})))$ and all symmetric matrices in $\nabla^2 \mathcal{L}(\Theta^{(i)})$ are positive definite.

If the conditions of the above two theorems are satisfied by some $\Theta^{(i)}$, then a unique stationary point exists in $\Theta^{(i)}$ and this stationary point is a local minimizer. Therefore, if exactly one candidate sub-box for minimizer(s) remains after pruning the search box $\Theta^{(0)}$ with the tests in Section 20.3.1, and if this sub-box satisfies the above two conditions for the existence of a unique local minimizer within it, then we have rigorously enclosed the global minimizer in the search interval. On the other hand, if there are two or more sub-boxes in our candidate list for minimizer(s) that satisfy the above two conditions, then we may conclude that each sub-box contains a candidate for a global minimizer which may not necessarily be unique (as in the case of disconnected sub-boxes each of which contains a candidate). Observe that failure to verify the uniqueness of a local minimizer in a sub-box can occur if it contains more than one point, or even a continuum of points, that are stationary.

20.3.3 Algorithm

- *Initialization:*

  Step 1 Let the search region be a single box $\Theta^{(0)}$ or a collection of not necessarily connected, but pairwise disjoint boxes, $\Theta^{(i)}$, $i \in \{1, \ldots, r\}$.

  Step 2 Initialize the list $\mathfrak{L}$ which may just contain one element $(\Theta^{(0)}, \mathcal{L}_{\Theta^{(0)}})$ or several elements

  $$\{(\Theta^{(1)}, \mathcal{L}_{\Theta^{(1)}}), (\Theta^{(2)}, \mathcal{L}_{\Theta^{(2)}}), \ldots, (\Theta^{(r)}, \mathcal{L}_{\Theta^{(r)}})\}.$$

  Step 3 Let ϵ be a specified tolerance.

  Step 4 Let $\max_{\mathfrak{L}}$ be the maximal length allowed for list $\mathfrak{L}$.

  Step 5 Set the non-informative lower bound for l^*, i.e., $\tilde{l} = \infty$

- *Iteration:*

  Step1 Perform the following operations:

  Step 1.1 Improve $\tilde{l} = \min\{\tilde{l}, \max(\mathcal{L}(m(\Theta^{(j)})))\}$,
  $j = \operatorname{argmin}\{\mathcal{L}_{\Theta^{(i)}}\}$.

  Step 1.2 Perform the *midpoint cutoff test* on $\mathfrak{L}$.

  Step 1.3 Set $\mathcal{L}^* = [\mathcal{L}_{\Theta^{(j)}}, \tilde{l}]$.

  Step 2 Bisect $\Theta^{(j)}$ along its longest side k, i.e., $d(\Theta^{(j)}_k) = d_\infty(\Theta^{(j)})$, to obtain sub-boxes $\Theta^{(j_q)}$, $q \in \{1, 2\}$.

  Step 3 For each sub-box $\Theta^{(j_q)}$, evaluate $(\mathcal{L}(\Theta^{(j_q)}), \nabla\mathcal{L}(\Theta^{(j_q)}), \nabla^2\mathcal{L}(\Theta^{(j_q)}))$, and do the following:

  Step 3.1 Perform the *monotonicity test* to possibly discard $\Theta^{(j_q)}$.

Step 3.2 *Centered form cutoff test:*

Improve the image enclosure of $\mathcal{L}(\Theta^{(j_q)})$ by replacing it with its centered form $\mathcal{L}_c(\Theta^{(j_q)}) :=$

$$\{\mathcal{L}(m(\Theta^{(j_q)})) + \nabla\mathcal{L}(\Theta^{(j_q)}) \cdot (\Theta^{(j_q)} - m(\Theta^{(j_q)}))\} \cap \mathcal{L}(\Theta^{(j_q)}),$$

and then discarding $\Theta^{(j_q)}$, if $\tilde{l} < \underline{\mathcal{L}}_{\Theta^{(j_q)}}$.

Step 3.3 Perform the *concavity test* to possibly discard $\Theta^{(j_q)}$.

Step 3.4 Apply an *extended interval Newton Gauss–Seidel step* to $\Theta^{(j_q)}$, in order to either entirely discard it or shrink it into v sub-sub-boxes, where v is at most $2s - 2$.

Step 3.5 For each one of these sub-sub-boxes $\Theta^{(j_q,u)}$, $u \in \{1, \dots, v\}$

Step 3.5.1 Perform the *monotonicity test* to possibly discard $\Theta^{(j_q,u)}$.

Step 3.5.2 Try to discard $\Theta^{(j_q,u)}$ by applying the *centered form cutoff test* in **Step 3.2** to it.

Step 3.5.3 Append $(\Theta^{(j_q,u)}, \underline{\mathcal{L}}_{\Theta^{(j_q,u)}})$ to $\mathfrak{L}$ if $\Theta^{(j_q,u)}$ could not be discarded by **Step 3.5.1** and **Step 3.5.2**.

- *Termination:*

Step 1 Terminate iteration if $d_{rel,\infty}(\Theta^{(j)}) < \epsilon$, or $d_{rel,\infty}(\mathcal{L}^*) < \epsilon$, or $\mathfrak{L}$ is empty, or $\mathrm{Length}(\mathfrak{L}) > \max_{\mathfrak{L}}$.

Step 2 Verify uniqueness of minimizer(s) in the final list $\mathfrak{L}$ by applying algorithm given in Section 20.3.2 to each of its elements.

20.4 Applications to phylogenetics

By way of example, we apply our enclosure method to identifying the global maximum of the log-likelihood function for the JC69 model of DNA evolution on the 3-taxa unrooted tree. The homologous sequences used were taken from the mitochondrial DNA of the chimpanzee (*Pan troglodytes*), gorilla (*Gorilla gorilla*), and orangutan (*Pongo pygmaeus*) [Brown *et al.*, 1982]. There is only one unrooted multifurcating topology for three species with all three branches emanating from the root like a star. The data set for this problem is summarized in [Sainudiin, 2004] by 29 data patterns. The sufficient statistic for this data is $(7, 100, 42, 46, 700)$. Details on obtaining this sufficient statistic can be found in Chapter 18. The parameter space is three-dimensional, corresponding to the three branch lengths of the 3-leaved star tree τ_1. The algorithm is given a large search box $\Theta^{(0)}$. The results are summarized in Table 20.1. The notation x_a^b means the interval $[xa, xb]$ (e.g., $5.9816221384_0^2 \times 10^{-2} = [5.98162213840 \times 10^{-2}, 5.98162213842 \times 10^{-2}]$). Figure 20.6 shows the parameter space being rigorously pruned as the algorithm progresses. When there are four taxa, the phylogeny estimation problem is more challenging as there are four distinct topologies to consider in addition to the branch lengths. A similar method was used to solve the most likely phylogeny of four primates with data from their mitochondria [Sainudiin, 2004].

Table 20.1. *Machine interval MLEs of a log-likelihood function for a phylogenetic tree on 3-taxa.*
Chimpanzee (1), Gorilla (2), and Orangutan (3).

$\Theta^{(0)}$ and Tree	$\Theta^* \supset \theta^*$	$-\mathcal{L}(\Theta^*) \supset -l(\theta^*)$
$[1.0 \times 10^{-11}, 10.0]^{\otimes 3}$	$5.9816221384_0^2 \times 10^{-2}$	
$\tau_1 = (1,2,3)$	$5.4167416794_0^2 \times 10^{-2}$	
	$1.3299089685_8^9 \times 10^{-1}$	$-2.1503180658656_6^5 \times 10^3$

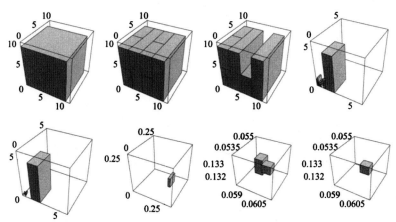

Fig. 20.6. Progress of the algorithm as it prunes $[0.001, 10.0]^{\otimes 3}$.

The running time of the global optimization algorithm depends on where the MLEs lie in the parameter space. For trees with smaller branch lengths, the running time is faster, while larger trees have a much longer running time. The Table 20.2 shows the mean and 95% confidence intervals of the number of calls to the likelihood function $\mathcal{L}$ and the CPU time in seconds for each of four trees with different weights. The results summarized in Table 20.2 are from 100 data sets, each of sequence length 1000, simulated under the JC69 model upon each one of the four trees shown in the first column.

The enclosure of an MLE by means of interval methods is equivalent to a proof of maximality. The method is robust in the presence of multiple local maxima or non-identifiable manifolds with the same ML value. For example, when a time-reversible Markov chain, such as JC69, is superimposed on a rooted tree, only the sum of the branch lengths emanating from the root is identifiable. Identifiability is a prerequisite for statistical consistency of estimators. To demonstrate the ability of interval methods to enclose the non-identifiable ridge along $\theta_1 + \theta_2$ in the simplest case of a two-leaved tree, we formulated a non-identifiable negative log-likelihood function $l(\theta_1, \theta_2)$ with its global minimizers along $\theta_1 + \theta_2 = \frac{3}{4} \log(45/17) = 0.730087$ under a fictitious dataset for which 280 out of 600 sites are polymorphic. Figure 20.7 shows the contours of $l(\theta_1, \theta_2)$ in gray scale and the solutions of the interval method (gray and black rectangles) and those of 10 quasi-Newton searches with random initializations

Table 20.2. *Computational efficiency for four different 3-taxa trees.*

True Tree	Calls to $\mathcal{L}(\Theta^*)$	CPU time
$(1:0.01, 2:0.07, 3:0.07)$	$1272\ [1032, 1663]$	$0.55\ [0.45, 0.72]$
$(1:0.02, 2:0.19, 3:0.19)$	$3948\ [2667, 6886]$	$1.75\ [1.17, 3.05]$
$(1:0.03, 2:0.42, 3:0.42)$	$20789\ [12749, 35220]$	$9.68\ [5.94, 16.34]$
$(1:0.06, 2:0.84, 3:0.84)$	$245464\ [111901, 376450]$	$144.62\ [64.07, 232.94]$

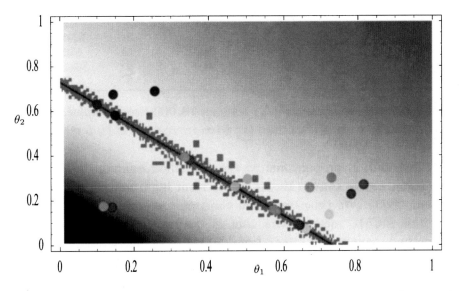

Fig. 20.7. The non-identifiable subspace of minimizers $\theta_1 + \theta_2 = \frac{3}{4}\log(45/17)$ of $l(\theta_1, \theta_2)$ under the JC69 model evolving on a rooted two-leaved tree is enclosed by a union of up to 30,000 rectangles. The larger gray, and smaller black rectangles have tolerances of $\epsilon = 1.0 \times 10^{-4}$ and $\epsilon = 1.0 \times 10^{-6}$, respectively. The 10 pairs of colored ovals are the initial and final points of 10 local quasi-Newton searches with random initializations.

(10 pairs of colored ovals). Observe that the basin of attraction for each point on $\theta_1 + \theta_2 = 0.730087$ under a quasi-Newton local search algorithm is the line running orthogonal to it.

Interval methods can be slow on currently available processors that are optimized for floating-point arithmetic, especially when applied naively. Efficiency can be gained by pre-enclosing the likelihood function over a fine mesh and accessing it via hash tables. Algebraic techniques can be also be used to reduce the data into sufficient statistics. Interval methods are particularly suited for solving a high dimensional problem by amalgamating the solutions of several lower-dimensional problems. For instance, we can apply the rigorously enclosed MLEs to the generalized neighbor-joining (GNJ) method discussed in Chapter 18. We call this the *numerically rigorous generalized neighbor-joining method (NRGNJ)*. Using `fastDNAml` which implements a gradient flow algorithm with floating-point arithmetic, [Levy *et al.*, 2005] computed dissimilarity maps that are needed for the GNJ method. The NRGNJ method uses, instead, the rigorously enclosed MLEs. We applied this method to find the NJ tree for 21 *S-locus receptor kinase* (SRK) sequences [Sainudiin *et al.*, 2005] involved in the self/non-self discriminating self-incompatibility system of the mustard family [Nasrallah, 2002]. We sampled 10,000 trees from a Markov chain with

Δ	NRGNJ	fastDNAml	DNAml(A)	DNAml(B)	TrExML
0	0	0	2	3608	0
2	0	0	1	471	0
4	171	6	3619	5614	0
6	5687	5	463	294	5
8	4134	3987	5636	13	71
10	8	5720	269	0	3634
12	0	272	10	0	652
14	0	10	0	0	5631
16	0	0	0	0	7

Table 20.3. *Symmetric difference (Δ) between 10,000 trees sampled from the likelihood function via MCMC and the trees reconstructed by 5 methods.*

stationary distribution proportional to the likelihood function by means of a Markov chain Monte Carlo (MCMC) algorithm implemented in PHYBAYES [Aris-Brosou, 2003]. We then compared the tree topology of each tree generated by this MCMC method with that of the reconstructed trees via the NRGNJ method, fastDNAml, DNAml from PHYLIP package [Felsenstein, 2004], and TrExML [Wolf *et al.*, 2000] under their respective default settings with the JC69 model. We used treedist [Felsenstein, 2004] to compare two tree topologies. If the symmetric difference Δ between two topologies is 0, then the two topologies are identical. Larger Δs are reflective of a larger distance between the two compared topologies. Table 20.3 summarizes the distance between a reconstructed tree and the MCMC samples from the normalized likelihood function. For example, the first two elements in the third row of Table 20.3 mean that 171 out of the 10,000 MCMC sampled trees are at a symmetric difference of 4 ($\Delta = 4$) from the tree reconstructed via the NRGNJ method. DNAml was used in two ways: DNAml(A) is a basic search with no global rearrangements, whereas DNAml(B) applies a broader search with global rearrangements and 100 jumbled inputs. The fruits of the broader search are reflected by the accumulation of MCMC sampled trees over small Δ values from the DNAml(B) tree. Although the NRGNJ tree is identical to the Saito and Nei NJ tree (with pairwise distance) [Saitou and Nei, 1987] as well as to the fastDNAml-based NJ tree with 3 leaves for this dataset, we now have the guarantee from the NRGNJ method that the MLEs for each triplet was enclosed.

21

Analysis of Point Mutations in Vertebrate Genomes

Jameel Al-Aidroos

Sagi Snir

Using homologous sequences from eight vertebrates, we present a concrete example of the estimation of mutation rates in the models of evolution introduced in Chapter 4. We detail the process of data selection from a multiple alignment of the ENCODE regions, and compare rate estimates for each of the models in the Felsenstein hierarchy of Figure 4.7. We also address a standing problem in vertebrate evolution, namely the resolution of the phylogeny of the Eutherian orders, and discuss several challenges of molecular sequence analysis in inferring the phylogeny of this subclass. In particular, we consider the question of the position of the rodents relative to the primates, carnivores and artiodactyls; we affectionately dub this question the *rodent problem*.

21.1 Estimating mutation rates

Given an alignment of sequence homologs from various taxa, and an evolutionary model from Section 4.5, we are naturally led to ask the question, "what tree (with what branch lengths) and what values of the parameters in the rate matrix for that model are suggested by the alignment?" One answer to this question, the so-called maximum-likelihood solution, is, "the tree and rate parameters which maximize the probability that the given alignment would be generated by the given model." (See also Sections 1.3 and 3.3.)

There are a number of available software packages which attempt to find, to varying degrees, this maximum-likelihood solution. For example, for a few of the most restrictive models in the Felsenstein hierarchy, the package PHYLIP [Felsenstein, 2004] will very efficiently search the tree space for the maximum-likelihood tree and rate parameters. The commercially available software PAUP* [Swofford, 1998] effectively searches the tree space, and implements a wider range of models than PHYLIP. We chose to use another package, PAML [Yang, 1997]. Although it does not reliably search the tree space, PAML is flexible enough, given a particular tree topology, to find the maximum-likelihood branch lengths and rate parameters for *any* of the models described in the Felsenstein hierarchy.

An evolutionary model for a set of taxa consists of a tree-topology T, with an assignment of the taxa to the leaves of T, an unknown branch length, t_e, for each edge $e \in E(T)$, and rate matrix Q. The entries of Q depend on a certain number of rate parameters $\alpha, \beta, \ldots$, and often on the base frequencies $\pi_A, \pi_C, \pi_G, \pi_T$, as described, for example, in Figure 4.7. These entries are a measure of the instantaneous rates of mutation among nucleotides. For a particular edge $e \in E(T)$, the probability of transition from nucleotide i to j along e is the entry $[P_e]_{ij}$ of the transition matrix $P_e = e^{Qt_e}$. The probability of generating a particular alignment of sequences from a specific tree and fixed parameters is given by the likelihood function

$$
\prod_{\substack{\text{all alignment columns C} \\ \text{specifying } \{\texttt{A},\texttt{C},\texttt{G},\texttt{T}\}\text{-labels} \\ \text{for tree leaves}}} \left(\sum_{\substack{\text{all labelings} \\ \text{by } \{\texttt{A},\texttt{C},\texttt{G},\texttt{T}\} \\ \text{of internal} \\ \text{vertices}}} \left(\prod_{\substack{e = (k,l), \\ e \in E(T)}} \substack{[P_e]_{ij} \\ \text{for label } i \text{ at } k \\ \text{and label } j \text{ at } l} \right) \right).
$$

Thus, for a given tree topology, the maximum-likelihood solution consists of the values of the branch lengths, the base frequencies and the other rate parameters that maximize the likelihood function.

Table 21.1 contains sample inputs for PAML, and its output of the maximum-likelihood solution for the HKY85 model. Notice that PAML returns one rate parameter (in the third last line of output in Table 21.1), κ, whereas the HKY85 model in Figure 4.7 has two, α and β. The product Qt_e in the likelihood function forces the branch lengths and rate matrix to be determined only up to scaling. Thus, for example, in the HKY85 model of Figure 4.7 we could multiply all the branch lengths by β, divide the rate matrix by β, and take $\kappa = \alpha/\beta$ to overcome the missing parameter. Notice also that whereas the nucleotides are ordered alphabetically in Chapter 4, PAML orders them T-C-A-G when it reports base frequencies (in the last line of the output in Table 21.1), and in columns and rows of rate matrices. The values for $\pi_A, \pi_C, \pi_G, \pi_T$ and κ in the output in Table 21.1 and the form of the HKY85 model from Figure 4.7 allow us to write the rate matrix

$$
Q = \begin{bmatrix} \cdot & \pi_C & \kappa\pi_G & \pi_T \\ \pi_A & \cdot & \pi_G & \kappa\pi_T \\ \kappa\pi_A & \pi_C & \cdot & \pi_T \\ \pi_A & \kappa\pi_C & \pi_G & \cdot \end{bmatrix} = \begin{bmatrix} \cdot & 0.28929 & 1.63618 & 0.22577 \\ 0.20244 & \cdot & 0.28250 & 1.30761 \\ 1.75746 & 0.28929 & \cdot & 0.22577 \\ 0.20244 & 1.67551 & 0.28250 & \cdot \end{bmatrix}.
$$

PAML represents trees in the *Newick format*. This is a recursive definition of rooted trees in which vertices (i.e., species, extant or extinct) grouped in parentheses share a common ancestor. If two vertices are grouped in the same set of parentheses, they are sister taxa. For example the tree $((1, 2), 3)$ is a rooted triplet over the species $\{1, 2, 3\}$ such that 1 and 2 are sister taxa and their most recent common ancestor (represented by the vertex $(1, 2)$) is a sister of 3. This format is extended to weighted trees by attaching a number to the right of a vertex. This number is the length of the edge entering that vertex. Note that in the example in Table 21.1, the input tree is unweighted, while the

INPUT: the sequence file "seq.txt"

```
8 78306
chimp     GGGGAAGGGGAACCGGGGCCGGGGCCGGAACCGGAAGGGGGTTTT...
chicken   GGGGGGGGGGGGGAAGGGGCCGGGGCCGGAACCGGGGAAGGGGTTTT...
human     GGGGAAGGGGAACCGGGGCCGGGGCCGGAACCGGAAGGGGGTTTT...
galago    GGGGAAGGGGGGGTTGGGGCCGGGGCCGGAACCGGAAGGGGGTTTT...
cow       GGGGAAGGGGAAAAGGGGCCGGGGCCGGAATTGGAAGGGGGTTTT...
dog       GGGGAAGGGGAACCGGGGCCGGGGCCGGAACCGGAAGGGGGTTTT...
rat       GGGGGGGGGGGAAAAGGGGTTGGGGAAGGAACCGGAAGGGGGTTTT...
mouse     GGGGGGGGGGGAAAAGGGGAAAAGGGGGGGAACCGGAAGGGGGTTTT...
```

INPUT: the tree structure file "tree.txt"

```
((((human,chimp),galago),(mouse,rat)),(cow,dog),chicken);
```

INPUT: the PAML control file "baseml.ctl"

```
     model    = 4         * 0:JC69, 1:K80, 2:F81, ...
     nhomo    = 1         * 0 or 1:  homogeneous, 2:...
  treefile    = tree.txt  * tree structure file name
   seqfile    = seq.txt   * sequence data file name
  cleandata   = 0         * remove ambiguous data (1:y,0:n)
   outfile    = out.txt   * main result file
     noisy    = 3         * 0,1,2,3:  how much junk on screen
   verbose    = 1         * 1:  detailed output, 0:  concise output
     getSE    = 0         * 0:  omit; 1:  get S.E.s of estimates
   runmode    = 0         * 0:  user tree; 1 or 2:  find tree...
 Small_Diff   = 1e-6      * step value for derivative estimates
    method    = 0         * 0:  simult.; 1:  one branch at a time
     clock    = 0         * 0:no clock, 1:clock; 2:local clock...
  fix_kappa   = 0         * 0:  estimate kappa; 1:  fix kappa
     kappa    = 2.5       * initial or fixed kappa
```

OUTPUT: excerpt from the PAML output file "out.txt"

```
lnL(ntime:  14 np:  18):-396419.669383 +0.000000
 :
 :
((((((human:  0.004484, chimp:  0.005159):  0.068015, galago:
0.102113):  0.014305, (mouse:  0.068227, rat:  0.062353):  0.182379):
0.011786, (cow:  0.114727, dog:  0.095417):  0.018334):  0.014105,
chicken:0.555154);

Detailed output identifying parameters kappa under HKY85:  5.79179
base frequency parameters
0.22577 0.28929 0.20244 0.28250
```

Table 21.1. PAML *input and output for mutation rate estimation in the HKY85 model.*

tree returned by PAML contains the edge lengths. The probability of obtaining the input alignment with the calculated parameters and lengths is given in the output from PAML as a log-likelihood in the first line of the excerpt.

The next two sections of this chapter discuss the ENCODE project, from which the sequences for this study are taken, information about a refinement of the alignment which isolates only synonymous substitution sites, and the results of the implementation of each of the models of the Felsenstein hierarchy in PAML. Table 21.2 describes how to implement each of the models of the Felsenstein hierarchy in PAML by making small adjustments in the baseml.ctl file to the options model and nhomo (which controls whether the base frequencies are

Model	model =	nhomo =
JC69	0	0
K80	1	0
K81	10 [2 (AC CA GT TG) (AG GA CT TC)]	0
CS05	9 [1 (AC AT CG GT)]	1
F81	2	1
HKY85	4	1
F84	3	1
TN93	6	1
SYM	10 [5 (AC CA) (AG GA) (AT TA) (CG GC) (CT TC)]	0
REV	7	1

Table 21.2. *Implementing the models of the Felsenstein hierarchy in* PAML.

uniform or parameters). These examples demonstrate only some of the versatility of that software in dealing with more sophisticated models. Section 21.4 introduces the main biological problem addressed, namely the position of the rodents in the phylogeny of the mammals considered in this chapter.

21.2 The ENCODE data

Our analysis of mutation rates is based on alignments of the human genome from regions identified by the ENCODE Pilot Project which was described briefly in Section 4.3. The Berkeley ENCODE Group, working in collaboration with the genome analysis team at UCSC has used the MERCATOR mapping program [Dewey, 2005], LiftOver [Kent *et al.*, 2002], and Shuffle-LAGAN [Brudno *et al.*, 2003b] to map these regions to homologous regions from other vertebrates. These regions were aligned using MAVID [Bray and Pachter, 2004]. In this study we have used the alignments of the Stanford re-ordering of the October 2004 freeze of the homologous regions, which are available at http://bio.math.berkeley.edu/encode/.

ENCODE's pilot project [Consortium, 2004] identifies 44 regions in the human genome for extensive study. The first 14 of these are manually selected regions, ENm001 to ENm014, chosen to include well-understood functional regions of the genome. For example, ENm001 contains the gene *CFTR*, associated with cystic fibrosis, which has been studied extensively since its discovery in 1989. The remaining 30 regions, the so-called "random" regions, were chosen to represent varying degrees of non-exonic conservation with respect to orthologous sequence from the mouse genome, and varying degrees of gene density. Table 21.3 describes the manual and random ENCODE regions.

A primary goal of the Berkeley ENCODE Group is to generate mappings from the human genome in the ENCODE regions to homologous regions in assemblies of sequence from other vertebrates, and to align the homologs. Such alignments have been generated for each ENCODE region, although the set of taxa in which homologs have been identified varies from region to region. In this study, we restricted our attention to eight vertebrates: human (*Homo sapiens*), galago monkey (*Otolemur garnettii*), chimp (*Pan troglodytes*), rat

Manual ENCODE Regions

Bin	Importance or Expected function
[ENm001]	*CFTR* – Cystic Fibrosis
ENm002	Interleukin Cluster – immune system regulation
ENm003	*Apo* Cluster – liver and intestinal function
ENm004	region from Chromosome 22
ENm005	region from Chromosome 21
ENm006	region from Chromosome X
[ENm007]	region from Chromosome 19
ENm008	α-globin – implicated in α-thalassemia (anemia)
ENm009	β-globin – implicated in Sickle Cell Anemia
ENm010	*HOXA* Cluster embryonic development: body axis and limb patterning
[ENm011]	*IGF2/H19* – insulin growth factor: growth and early development
ENm012	*FOXP2* - language and speech development
ENm013	region from Chromosome 7 – selected to balance stratification
ENm014	region from Chromosome 7 – selected to balance stratification

Random ENCODE Regions

density of conserved non-exonic bases	density of genic bases low (0-50%-ile)	medium (60-80%-ile)	high (80-100%-ile)
low (0-50%-ile)	ENr111, [ENr112] ENr113, ENr114	ENr121, ENr122 ENr123	ENr131, ENr132 ENr133
medium (50-80%-ile)	ENr211, ENr212 [ENr213]	ENr221, ENr222 ENr223	ENr231. ENr232 ENr233
high (80-100%-ile)	ENr311, [ENr312] ENr313	ENr321. ENr322 ENr323, ENr324	ENr331, ENr332 ENr333,[ENr334]

Table 21.3. *Summary of the manual and random ENCODE regions. Regions marked [..] were omitted from our analysis; at the time of our analysis the homologous sequence for one of the vertebrates had yet to be identified or aligned in these regions.*

(*Rattus norvegicus*), mouse (*Mus musculus*), dog (*Canis familiaris*), cow (*Bos taurus*) and chicken (*Gallus gallus*). We considered only the 37 ENCODE regions which had alignments of homologs for all eight taxa.

21.3 Synonymous substitutions

Starting with the MAVID alignment of the homologs of the human ENCODE regions, we refined the alignment by isolating only those columns corresponding to *synonymous substitution* sites in exons. In this section we define synonymous substitution, and describe the process used to identify such sites.

Recall from Table 4.1 in Chapter 4 that every amino acid is coded by a sequence of three nucleotides, called a codon. As there are four different nucleotides, this scheme allows for $4^3 = 64$ different codons. However, since there are only twenty amino acids, the above implies that some of thme are encoded by more than one codon, giving some redundancy to the amino acid coding scheme. Nucleotide mutations in the gene (which necessarily change the codons in the gene) are divided into three types, depending on which amino acid is encoded by the new codon:

(i) synonymous mutations: mutations that alter a particular codon, but **do not** alter the encoded amino acid;

a.a.	GlnMetGlnGlnLeuGlnGlnGlnGlnHisLeuLeu...LeuGln...GlnGlyLeuIle
human	CAGATGCAACAACTCCAGCAGCAGCAGCATCTGCTCAGCCTTCAGCGTCAGGGACTCATC
galago	CAGATGCAACAACTCCAGCAGCAGCAGCATCTGCTCAGCCTTCAGCGTCAGGGACTCATC
mouse	CAAATGCAGCAGCTACAGCAGCAACAACATCTGCTCAGCCTTCAGCGCCAGGGCCTCATC
rat	CAGATGCAGCAACTACAGCAGCAGCAGCATCTGCTCAGCCTTCAGTGTCAGGGCCTCATC
cow	CAGATGCAACAACTCCAGCAGCAGCAGCATCTGCTCAGCCTTCAGCGTCAGGGACTCATC
chicken	CAGATGCAACAACTTCAGCAGCAGCAACATCTGCTGAACCTTCAGCGTCAGGGACTCATT
chimp	CAGATGCAACAACTCCAGCAGCAGCAGCATCTGCTCAGCCTTCAGCGTCAGGGACTCATC
dog	CAGATGCAACAACTCCAGCAGCAGCAGCATCTGCTCAGCCTTCAGCGTCAGGGACTCATC

```
..*.....*..*..*..*..*..*..*..*..*..*.....*...*.....*..*..*..*
123123123123123123123123123123123123123123123123123123123123
```

Fig. 21.1. Excerpt of alignment from *FOXP2* (RefSeq annot. NM_148900) in ENm012. The top row indicates the amino acid translation of the codon (where there is agreement among all the taxa). The stars denote columns corresponding to the third positions in codons that all translate to the same amino acid. The bottom row indicates the frame dictated by the human gene annotation, where 1 denotes the first position in a codon.

(ii) missense mutations: mutations that alter the codon so as to produce a different amino acid;

(iii) non-sense mutations: mutations that change a codon that encodes an amino acid into one of the STOP codons (TAA, TAG, or TGA).

Because synonymous mutations do not alter the encoded amino acid, they do not alter the synthesized protein. Such mutations produce no functional changes, and are thus considered to be free from selective pressure. By removing the selective pressure, we restrict our attention to those sites whose mutation is more likely to behave according to a random Markov process. Furthermore, by isolating these synonymous substitution sites, we impose a level of homogeneity on the data. For example, although PAML implements models that allow for the rate matrix to vary among sites, we believe that by selecting the neutral mutation sites, we substantially reduce the need for this introduction of extra parameters into the models.

The procedure for refinement of the data consisted of mapping annotations of human genes to the alignment, and identifying triples of columns containing synonymous codons. For each of the 37 ENCODE regions, we consulted the *refGene* table in the UCSC annotation database (http://hgdownload. cse.ucsc.edu/downloads.html) to find annotations of genes within the region. The *refGene* table lists the start and end positions in the chromosome of the transcription region, the coding region, and the exons for each gene. It contains annotations only for the so-called *refSeq* genes, namely those that have been manually verified and catalogued in the Reference Sequence collection of NCBI (http://www.ncbi.nlm.nih.gov/RefSeq/). For each gene in the region, we mapped the annotation of its coding region to the MAVID alignment, and extracted the relevant columns. Below each codon triple in the human sequence, we identified the amino acids encoded by the triple in each of the homologs. We discarded all triples which contained gaps in any of the homologs, and extracted the third column from each triple in which each

Model	Biological Tree	ML Tree	difference
JC69	−420360	−419662	698
K80	−398246	−397791	455
K81	−398242	−397785	457
CS05	−396420	−395987	433
F81	−419291	−418618	673
HKY85	−396420	−395987	433
F84	−396479	−396045	434
TN93	−396308	−395880	428
SYM	−397702	−397237	465
REV	−395649	−395195	454

Table 21.4. *Log-likelihoods for the Felsenstein hierarchy models on the biologically correct tree and on the maximum-likelihood tree.*

homolog's codon encoded the amino acid in the human sequence. We note here that because of gaps in the alignment and because we use only the frame of the human annotation to identify coding triples, it is possible that some sites chosen correspond to a second or third position in a codon of one of the other vertebrates. However, the high degree of conservation in the aligned exons and the stringent requirement of agreement in the preceding two columns meant that extracted columns were very unlikely to suffer from this fault; a visual inspection of the alignments revealed that extracted columns very reliably consisted of third positions. An excerpt of the alignment from the speech and language development gene, *FOXP2*, is given in Figure 21.1.

With the refined sequence data in hand, we used PAML to find the maximum-likelihood solutions for the mutation rate parameters, base frequencies and branch lengths. However, as was indicated in the introduction, PAML does not reliably search the tree space, so we restricted ourselves to only the two trees in Figure 21.4, namely the *biological tree* (that is, the accepted tree representing the evolutionary history of those species), and the maximum likelihood tree (the tree attaining the highest likelihood value under all models studied). The rate matrices for each model are displayed in Figures 21.2 and 21.3. The associated log-likelihoods are found in Table 21.4.

21.4 The rodent problem

Rodents have the special characteristic that although their molecular information closely resembles that of the primates, they exhibit very different morphological features. This discrepancy has attracted a lot of attention and has formed the basis of much research.

In this section we point out the phenomenon that current tree reconstruction methods misplace the two rodents, mouse and rat, on the tree, with respect to other mammals. The phylogenetic tree describing the evolution of the taxa in this study is now widely accepted among evolutionary biologists. It is supported by both fossil records and molecular data. See, for example, [Madsen *et al.*, 2001, Murphy *et al.*, 2001, Phillips and Penny, 2003,

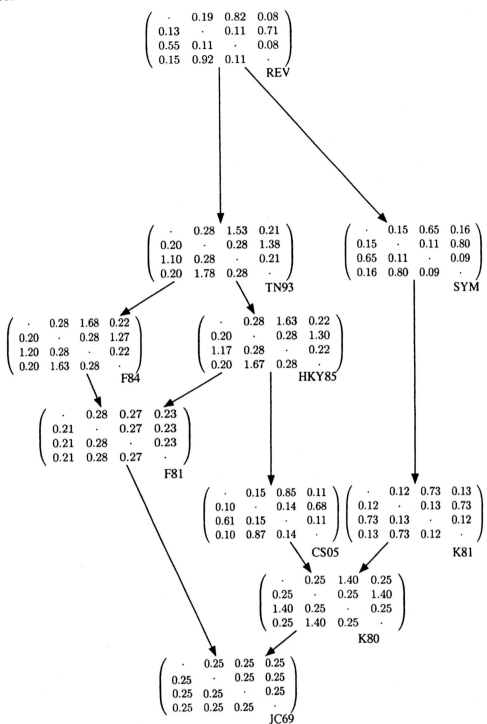

Fig. 21.2. The Felsenstein hierarchy for the ML Tree.

Lin *et al.*, 2002, Schmitz and Zischler, 2003]. For this reason we call it the "biologically correct" in this chapter (Figure 21.4, top), with the caveat that there is no guarantee of certainty it is correct, despite the evidence cited. In the biologically correct tree we have the primate clade, composed of the sibling

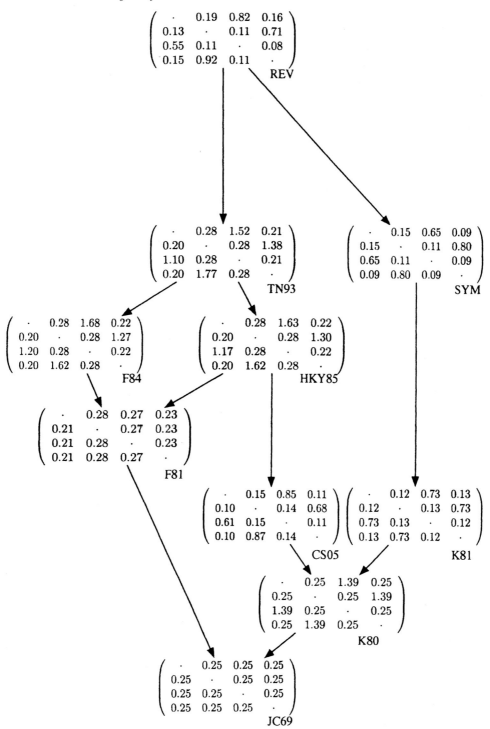

Fig. 21.3. The Felsenstein hierarchy for the Biologically Correct Tree.

human and chimpanzee and then the galago as an outgroup to these two. The rodent (mouse,rat) clade is a sister group to the primates and an artiodactyls-carnivores clade is an outgroup to the former species. By using the chicken as an outgroup to all these, we get a rooting of the tree. However, the currently

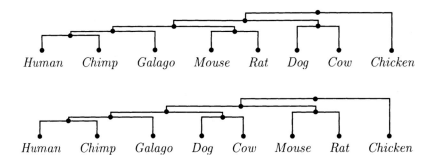

Fig. 21.4. The trees evaluated in the study. (Top) The accepted tree representing the evolution of our eight vertebrates. (Bottom) The tree obtaining the maximum likelihood in all substitution models; note that the (mouse, rat) clade is an outgroup to the other mammals.

available substitution-based phylogenetic reconstruction methods, regardless of the evolutionary model, misplace the rodents and place them as an outgroup to the primates and artiodactyls–carnivores (Figure 21.4, bottom). A partial explanation of this phenomenon is given by the fact that rodents have shorter generation time. This causes the rate of synonymous substitutions in the rodents to be 2.0 times faster than in human and the rate of non-synonymous substitutions to be 1.3 times faster [Wu and Li, 1985].

This question of rodent divergence time (and the relative location of this event in the phylogeny) has a long history and is still an active research topic with the increasing availability of complete mammalian genomes (e.g., [Human Genome Sequencing Consortium, 2001, Waterston *et al.*, 2002, Hillier *et al.*, 2004, Gibbs *et al.*, 2004]). Recent work [Adkins *et al.*, 2001, Thomas *et al.*, 2003] has also addressed the concern that few methods can find the correct tree from genome alignments. [Adkins *et al.*, 2001] addressed the question of monophyly of the rodent order and noticed that some of the genes investigated returned a false tree. The work of [Thomas *et al.*, 2003] revealed that analysis of transposable element insertions in a highly-conserved region (containing, in particular, the cystic–fibrosis gene *CFTR*) supports the biological tree. They used a subset of the models used in this chapter. Specifically, they used the GTR model (with constant rates along lineages), and HKY85 with variable base frequencies across ancestral vertices (c.f. the "nhomo" parameter) which allows for a different rate matrix on every lineage. An important contribution they made to the problem was the identification of the two specific rival topologies, and the observation that the accepted topology obtains a lesser likelihood score. However, none of these studies raised the question of why all existing evolutionary models fail to reconstruct the biological tree, and specifically, why they misplace the order of rodents.

A common feature of many of the most popular models used to reconstruct the evolutionary tree is the assumption of a constant rate matrix along all branches. This leaves time as the only free parameter between the different branches of the tree. As a consequence, distortion in the tree topology is reflected by placement of the rodent speciation event higher in the tree, i.e., closer to the root (Figure 21.4, bottom). This behavior is shared among all models. However, it is not entirely clear to us why the accelerated rate of evolution should change the topology of the tree, rather than extending the rodent branch within the topology of the biological tree. One possible explanation is the phenomenon of *long branch attraction* (LBA) [Felsenstein, 1978, Hendy and Penny, 1989]. This is where two fast evolving taxa are grouped together although they belong to different clades. It is caused mainly when a distant species is used as an outgroup for tree rooting, or when the number of taxa is relatively small. We assert that the rodent problem cannot be explained simply by the argument of LBA for three main reasons: (a) the outgroup used in our study, the chicken, is not that distant a species; (b) the number of sites in the input data (over 78,000) is relatively big considering the number of species [Holland *et al.*, 2003]; and (c) [Thomas *et al.*, 2003, Supplement] and [Huelsenbeck *et al.*, 2000] examined an even larger set of species, and yet arrived at a tree supporting the ML tree.

In nature, a mutation rate might increase or decrease during the course of evolution. The models described in Section 4.5 allow rate heterogeneity among sites; that is, different sites might exhibit different rates. However, at a single site, the rate along all lineages is constant. More advanced models allow in-site rate heterogeneity which accounts for rate changes throughout evolution. Some of these [Galtier and Gouy, 1998, Yang and Roberts, 1995] build on previous models such as HKY85 or TN93 by using the same rate matrices (and therefore also the assumption of evolutionary model), but enhance them by allowing the parameters of the matrix to change at every vertex. Other models (e.g., [Huelsenbeck *et al.*, 2000]) allow the rate to behave as a Poisson process, enabling non-homogeneity even along a single tree branch. All these models strive to imitate the natural mutation process. However, more descriptive models require more parameters, and additional parameters can turn maximum-likelihood estimation into a computationally hard task, even on a fixed tree topology.

In addition to the models outlined above, PAML implements several more advanced models to be detailed below. These models enable the further flexibility of different rates among the branches. In order to check whether these advanced features resolve the discrepancy between the biological tree and the ML tree, we tested them on our data.

For the first benchmark, we relaxed the homogeneity constraint assumed in the former set of tests. We note here that even the model described by [Huelsenbeck *et al.*, 2000], which allows greater flexibility, inferred an incorrect tree with respect to the rodent divergence event. For certain models, PAML allows κ to vary along the tree branches. Naturally, this option applies only to the models involving κ; F84, HKY85 and T92. This allows the rate matrix to vary between branches in order to obtain an optimal solution. This flexibility is also found in the models described by [Yang and Roberts, 1995], who used

Model	Biological Tree	ML Tree	difference
F84	−397851	−397388	463
HKY85	−396144	−395704	440
TN92	−396108	−395668	440

Table 21.5. *Log-likelihoods for models allowing κ to vary from branch to branch.*

Model	Biological Tree	ML Tree	difference
HKY85	−396604	−396339	265
T92	−396711	−396443	268
TN93	−396493	−396232	261
REV	−395834	−395559	275
UNREST	−395646	−395365	281

Table 21.6. *Log-likelihoods for models using local clock option.*

F84 or HKY85, and [Galtier and Gouy, 1998] who used T92. Since we were interested to see if the distance between the two trees is decreased, we only measured the likelihood obtained for the two trees. The results are displayed in Table 21.5.

In the next trial, we tried to partition the clades of the trees into different rate groups. This approach was motivated by [Mindell and Honeycutt, 1990], who showed that opossums, artiodactyls and primates possess very similar mutation rates, while rodents evolve at a significantly higher rate. This calls for a model that discriminates between the different branches of the tree according to their clade. The *local clock* option of PAML described in [Yoder and Yang, 2000] allows for such a partition. In this model, the same rate matrix is assumed along all lineages in the tree. However, when the transition matrix for branch e is computed, the rate matrix is multiplied by the branch length, t_e, and another scalar, the rate, r_e, along that branch. This change in the calculation provides the model with the property that the inferred tree satisfies the molecular clock property, while allowing rate heterogeneity. Indeed in [Douzery *et al.*, 2003] this model was employed with a very similar grouping (as the set of taxa was different) for the study of the discrepancy between fossil calibration and estimation based on molecular data. In their study, however, only the biological tree was considered. We arrange the leaves into 3 groups:

(i) the chicken clade (i.e., the branch from the root to the chicken);
(ii) the rodent clade (comprises all edges joining the ancestral rodent);
(iii) the remaining taxa.

The results obtained under this model are depicted in Table 21.6. It can be seen that, under this model, the difference in the log-likelihood is diminished; nevertheless, the ML tree and biologically correct tree remain different.

22

Ultra-Conserved Elements in Vertebrate and Fly Genomes

Mathias Drton

Nicholas Eriksson

Garmay Leung

Ultra-conserved elements in an alignment of multiple genomes are consecutive nucleotides that are in perfect agreement across all the genomes. For aligned vertebrate and fly genomes, we give descriptive statistics of ultra-conserved elements, explain their biological relevance, and show that the existence of ultra-conserved elements is highly improbable in neutrally evolving regions.

22.1 The data

Our analyses of ultra-conserved elements are based on multiple sequence alignments produced by MAVID [Bray and Pachter, 2004]. Prior to the alignment of multiple genomes, homology mappings (from Mercator [Dewey, 2005]) group into bins genomic regions that are anchored together by neighboring homologous exons. A multiple sequence alignment is then produced for each of these alignment bins. MAVID is a global multiple alignment program, and therefore homologous regions with more than one homologous hit to another genome may not be found aligned together. Table 22.1 shows an example of Mercator's output for a single region along with the beginning of the resulting MAVID multiple sequence alignment.

Species	Chrom.	Start	End		Alignment
Dog	chrX	752057	864487	+	A----AACCAAA---------
Chicken	chr1	122119382	122708162	−	TGCTGAGCTAAAGATCAGGCT
Zebra fish	chr9	19018916	19198136	+	------ATGCAACATGCTTCT
Puffer fish	chr2	7428614	7525502	+	---TAGATGGCAGACGATGCT
Fugu fish	asm1287	21187	82482	+	---TCAAGGG-----------

Table 22.1. *Mercator output for a single bin, giving the position and orientation on the chromosome. Notice that the Fugu fish genome has not been fully assembled into chromosomes (cf. Section 4.2).*

The vertebrate dataset consists of 10,279 bins over 9 genomes (Table 22.2). A total of 4,368 bins (42.5%) contain alignments across all 9 species. The evolutionary relationships among these species (which first diverged about 450 million years ago) are shown in Figure 22.1. For a discussion of the problem of placing the rodents in the phylogenetic tree, see Section 21.4 and Figure 21.4.

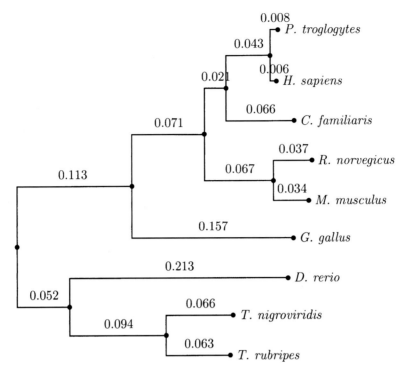

Fig. 22.1. Phylogenetic tree for whole genome alignment of 9 vertebrates.

With the exception of the probability calculations in phylogenetic tree models, our subsequent findings on ultra-conserved elements do not depend on the form of this tree.

Species	Genome Size	Genome Release Date
Zebra fish (*Danio rerio*)	1.47 Gbp	11/27/2003
Fugu fish (*Takifugu rubripes*)	0.26 Gbp	04/02/2003
Puffer fish (*Tetraodon nigroviridis*)	0.39 Gbp	02/01/2004
Dog (*Canis familiaris*)	2.38 Gbp	07/14/2004
Human (*Homo sapiens*)	2.98 Gbp	07/01/2003
Chimp (*Pan troglogytes*)	4.21 Gbp	11/13/2003
Mouse (*Mus musculus*)	2.85 Gbp	05/01/2004
Rat (*Rattus norvegicus*)	2.79 Gbp	06/19/2003
Chicken (*Gallus gallus*)	1.12 Gbp	02/24/2004

Table 22.2. *Genomes in the nine-vertebrate alignment with size given in billion base pairs.*

The fruit fly dataset consists of 8 *Drosophila* genomes (Table 22.3). Of the 3,731 alignment bins, 2,985 (80.0%) contain all 8 species, which reflects the smaller degree of evolutionary divergence. A phylogenetic tree for these 8 species, which diverged at least 45 million years ago, is illustrated in Figure 22.2.

The pilot phase of the ENCODE project (cf. Sections 4.3 and 21.2) provides an additional dataset of vertebrate sequences homologous to 44 regions of the human genome. There are 14 manually selected regions of particular biological

Species	Genome Size	Genome Release Date
D. melanogaster	118 Mbp	04/21/2004
D. simulans	119 Mbp	08/29/2004
D. yakuba	177 Mbp	04/07/2004
D. erecta	114 Mbp	10/28/2004
D. ananassae	136 Mbp	12/06/2004
D. pseudoobscura	125 Mbp	08/28/2003
D. virilis	152 Mbp	10/29/2004
D. mojavensis	177 Mbp	12/06/2004

Table 22.3. *Genomes in the eight-Drosophila alignment with size given in million base pairs.*

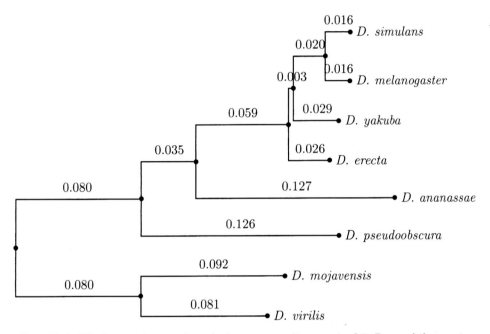

Fig. 22.2. Phylogenetic tree for whole genome alignment of 8 *Drosophila* species.

interest and 30 randomly selected regions with varying degrees of non-exonic conservation and gene density. Each manually selected region consists of 0.5–1.9 Mbp, while each randomly selected region is 0.5 Mbp in length. This gives a total of about 30 Mbp, approximately 1% of the human genome.

Varying with the region under consideration, a subset of the following 11 species is aligned along with the human genome in the preliminary October 2004 freeze: chimp, baboon (*Papiocynocephalus anubis*), marmoset (*Callithrix jacchus*), galago (*Otolemur garnettii*), mouse, rat, dog, armadillo (*Dasypus novemcintus*), platypus (*Ornithorhynchus anatinus*), and chicken. This collection of species lacks the three fish of the nine-vertebrate alignment. Armadillo and platypus sequences are only available for the first manually picked ENCODE region, and sequences for every region are only available for human, mouse, rat, dog and chicken. The number of species available for each region varies between 6 and 11 for manually selected regions, and between

8 and 10 for randomly selected regions. For each region, `Shuffle-LAGAN` [Brudno *et al.*, 2003b] was applied between the human sequence and each of the other available sequences to account for rearrangements. Based on these re-shuffled sequences, a multiple sequence alignment for each region was produced with `MAVID`.

The three sets of multiple alignments are available for download at `http://bio.math.berkeley.edu/ascb/chapter22/`.

22.2 Ultra-conserved elements

A position in a multiple alignment is *ultra-conserved* if for all species the same nucleotide appears in the position. An *ultra-conserved element* of length ℓ is a sequence of consecutive ultra-conserved positions $(n, n+1, \ldots, n+\ell-1)$ such that positions $n-1$ and $n+\ell$ are not ultra-conserved.

Example 22.1 Consider a subset of length 24 of a three-genome alignment:

```
G--ACCCAATAGCACCTGTTGCGG
CGCTCTCCA---CACCTGTTCCGG
CATTCT---------CTGTTTTGG
    *           ***** **
```

where ultra-conserved positions are marked by a star `*`. This alignment contains three ultra-conserved elements, one of length 1 in position 5, one of length 5 covering positions 16–20, and one of length 2 in positions 23–24.   □

22.2.1 Nine-vertebrate alignment

We scanned the entire nine-vertebrate alignment described in Section 22.1 and extracted 1,513,176 ultra-conserved elements, whose lengths are illustrated in Figure 22.3. The median and the mean length of an ultra-conserved element is equal to 2 and 1.918, respectively.

We will focus on the 237 ultra-conserved elements of length at least 20, covering 6,569 bp in sum. These 237 elements are clustered together; they are only found in 113 of the 4,368 bins containing all 9 species. The length distribution is heavily skewed toward shorter sequences as seen in Figure 22.3, with 75.5% of these regions shorter than 30 bp and only 10 regions longer than 50 bp.

The longest ultra-conserved element in the alignment is 125 bp long:

```
CTCAGCTTGT CTGATCATTT ATCCATAATT AGAAAATTAA TATTTTAGAT GGCGCTATGA
TGAACCCATT ATGGTGATGG GCCCCGATAT CAATTATAAC TTCAATTTCA ATTTCACTTA
CAGCC.
```

The next-longest ultra-conserved elements are two elements of length 85, followed by one element for each one of the lengths 81, 66, 62, 60, 59, 58, and 56. In particular, there is exactly one ultra-conserved element of length 42, which is the *"meaning of life"* element discussed in [Pachter and Sturmfels, 2005].

A number of the ultra-conserved elements are separated only by a few (less

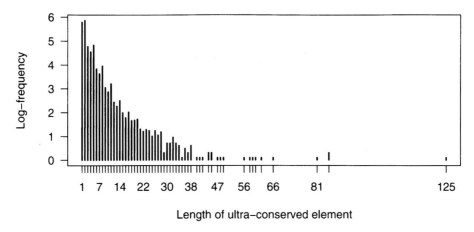

Fig. 22.3. Frequencies of vertebrate ultra-conserved elements ($\log_{10}$-scale).

than 10), ungapped, intervening positions. In 18 cases, there is a single intervening position. Typically, these positions are nearly ultra-conserved, and display differences only between the fish and the other species. Collapsing the ultra-conserved elements separated by fewer than 10 bases reduces the number of ultra-conserved elements to 209, increases the base coverage to 6,636 bp, and brings the total number of regions greater than 50 bp in length to 26.

In the human genome, the GC-ratio (proportion of G and C among all nucleotides) is 41.0%. The ultra-conserved elements are slightly more AT-rich; for the 237 elements of length 20 or longer, the GC-ratio is 35.8%. However, GC-content and local sequence characteristics were not enough to identify ultra-conserved regions using data from only one genome.

22.2.2 ENCODE alignment

The 44 ENCODE regions contain 139,043 ultra-conserved elements, 524 of which are longer than 20 bp. These long elements cover 17,823 bp. By base coverage, 73.5% of the long elements are found in the manually chosen regions. The longest one is in region ENm012, of length 169 and consists of the DNA sequence:

```
AAGTGCTTTG TGAGTTTGTC ACCAATGATA ATTTAGATAG AGGCTCATTA CTGAACATCA
CAACACTTTA AAAACCTTTC GCCTTCATAC AGGAGAATAA AGGACTATTT TAATGGCAAG
GTTCTTTTGT GTTCCACTGA AAAATTCAAT CAAGACAAAA CCTCATTGA.
```

This sequence does not contain a subsequence of length 20 or longer that is ultra-conserved in the nine-vertebrate alignment, but the 169 bp are also ultra-conserved in the nine-vertebrate alignment if the three fish are excluded from consideration. The only overlap between the nine-vertebrate and ENCODE ultra-conserved elements occurs in the regions ENm012 and ENm005, where there are 3 elements that are extensions of ultra-conserved elements in the nine-vertebrate alignment.

Table 22.4 shows the number of species aligned in the 44 ENCODE alignments and the respective five longest ultra-conserved elements that are of length 20 or larger. Omitted randomly selected regions do not contain any ultra-conserved elements of length at least 20.

Manually selected			Randomly selected		
Region	Spec.	Ultra-lengths	Region	Spec.	Ultra-lengths
ENm001	11	$28, 27, 23, 20_2$	ENr122	9	22
ENm002	8	$39, 28, 27, 26_4$	ENr213	9	$30, 27, 26, 24, 23_2$
ENm003	9	$38, 28_2, 26, 25_2$	ENr221	10	$36_2, 32_2, 29$
ENm004	8	$35, 26_2, 25, 20$	ENr222	10	$29, 22$
ENm005	10	$114, 62, 38, 34, 32$	ENr231	8	$26, 23, 20$
ENm006	8	$-$	ENr232	8	$26, 25, 20$
ENm007	6	$-$	ENr233	9	$25, 24, 20$
ENm008	9	$23, 22$	ENr311	10	$42, 31, 25, 21$
ENm009	10	$-$	ENr312	9	$60, 31, 22, 20_4$
ENm010	8	$86, 68, 63, 61, 60_2$	ENr313	9	27
ENm011	7	$-$	ENr321	10	$68, 44, 38, 37, 35$
ENm012	9	$169, 159, 125_2, 123$	ENr322	9	$126, 80, 79, 61, 55$
ENm013	10	$30, 26, 23, 22$	ENr323	8	$53, 50, 45, 42, 29$
ENm014	10	$41_2, 39, 26_2$	ENr331	9	26
			ENr332	10	26
			ENr334	8	$79, 50, 44, 37, 32$

Table 22.4. *Number of species and lengths of ultra-conserved elements in ENCODE alignments. Subindices indicate multiple occurrences.*

22.2.3 Eight-Drosophila alignment

There are 5,591,547 ultra-conserved elements in the *Drosophila* dataset with 1,705 elements at least 50 bp long and the longest of length 209 bp. We focused on the 255 *Drosophila* ultra-conserved elements of length at least 75 bp, covering 23,567 bp total. These regions are also found clustered together, occurring over 163 bins out of the 2,985 bins with all 8 species aligned together. The shortest distance between consecutive ultra-conserved elements is 130 bp, and therefore regions were not collapsed for this dataset. The mean and median length of ultra-conserved elements are 2.605 and 2, respectively. The length distribution of all ultra-conserved elements is shown in Figure 22.4. This set of ultra-conserved elements is also somewhat more AT-rich, with a GC-ratio of 38.8% (for those elements of length at least 75 bp) compared with a GC-ratio of 42.4% across the entire *D. melanogaster* genome.

22.3 Biology of ultra-conserved elements

22.3.1 Nine-vertebrate alignment

Using the UCSC genome browser annotations of known genes for the July 2003 (hg16) release of the human genome, we investigated which ultra-conserved elements overlap known functional regions. Intragenic regions cover 62.6% of the

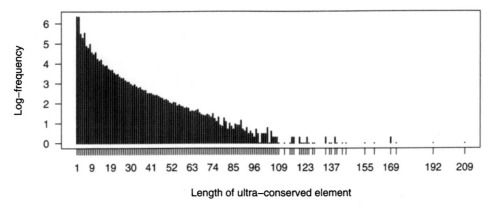

Fig. 22.4. Frequencies of *Drosophila* ultra-conserved elements ($\log_{10}$-scale).

bases of the 209 collapsed ultra-conserved elements described in Section 22.2.1. However, intragenic coverage increases to 67.6% for short elements (less than 30 bp) and drops to 56.3% for longer elements (at least 30 bp), as shown in Figures 22.5(a) and 22.5(b). While shorter ultra-conserved elements tend to correspond to exons, longer ones are generally associated with introns and unannotated regions. Nine ultra-conserved elements cover a total of 306 bp in the intronic regions of *POLA*, the alpha catalytic subunit of DNA polymerase. Six other genes are associated with more than 100 bp of ultra-conserved elements. Four of these genes are transcription factors involved in development (*SOX6, FOXP2, DACH1, TCF7L2*). In fact, elements near *DACH* that were highly conserved between human and mouse and also present in fish species have been shown to be *DACH* enhancers; see [Nobrega *et al.*, 2003].

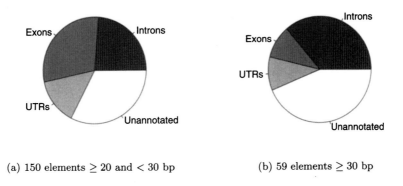

(a) 150 elements ≥ 20 and < 30 bp   (b) 59 elements ≥ 30 bp

Fig. 22.5. Functional base coverage of collapsed vertebrate ultra-conserved elements based on annotations of known human genes.

Among the 237 uncollapsed ultra-conserved elements of length at least 20, 151 are in intragenic regions of 96 genes. The remaining 86 elements did not overlap any annotated gene. However, by grouping together elements that have the same upstream and downstream flanking genes, there are only 27 super-regions to consider, with 51 unique flanking genes. There are 6 super-regions with at least 99 bp overlapping with ultra-conserved elements. At least one of the flanking genes for each of these 6 super-regions is a transcription factor located 1–314 kb away (*IRX3, IRX5, IRX6, HOXD13, DMRT1, DMRT3, FOXD3, TFEC*). The overall average distance to the closest flanking gene on either side is 138 kb and ranges from 312 bp to 1.2 Mbp.

It is a natural question whether the genes near or overlapping ultra-conserved elements tend to code for similar proteins. We divided the set of 96 genes with ultra-conserved overlap into 3 groups based on where in the gene the overlap occurred: exon, intron or untranslated region (UTR). If ultra-conserved elements overlap more than one type of genic region, then the gene is assigned to each of the appropriate groups. The 51 genes flanking ultra-conserved elements in unannotated regions form a fourth group of genes.

The Gene Ontology (GO) Consortium (`http://www.geneontology.org`) provides annotations for genes with respect to the molecular function of their gene products, the associated biological processes, and their cellular localization [Ashburner *et al.*, 2000]. For example, the human gene *SOX6* is annotated for biological process as being involved in cardioblast differentiation and DNA-dependent regulation of transcription. Mathematically, each of the three ontologies can be considered as a partially ordered set (*poset*) in which the categories are ordered from most to least specific. For example, cardioblast differentiation is more specific than cardiac cell differentiation, which in turn is more specific than both cell differentiation and embryonic heart tube development. If a gene possesses a certain annotation, it must also possess all more general annotations; therefore GO consists of a map from the set of genes to order ideals in the three posets. We propose that this mathematical structure is important for analyzing the GO project.

In this study, we only considered molecular function and biological process annotations. These annotations are available for 46 of the 54 genes with exonic overlap, for all of the 28 with intronic overlap, for 14 of the 20 with UTR overlap, and for 30 of the 51 genes flanking unannotated elements. Considering one GO annotation and one of the 4 gene groups at a time, we counted how many of the genes in the group are associated with the considered annotation. Using counts of how often this annotation occurs among all proteins found in Release 4.1 of the Uniprot database (`http://www.uniprot.org`), we computed a p-value from Fisher's exact test for independence of association with the annotation and affiliation with the considered gene group. Annotations associated with at least 3 genes in a group and with an unadjusted p-value smaller than $3.0 \cdot 10^{-2}$ are reported in Table 22.5. DNA-dependent regulation of transcription and transcription factor activity are found to be enriched in non-exonic ultra-conserved elements, corresponding to previously reported findings [Bejerano *et al.*, 2004, Boffelli *et al.*, 2004, Sandelin *et al.*, 2004, Woolfe *et al.*, 2005]. Conserved exonic elements tend to be involved in protein modification.

We scanned the human genome for repeated instances of these ultra-

GO Annotation	p-value
Exons (14)	
protein serine/threonine kinase activity	$4.545 \cdot 10^{-3}$
transferase activity	$1.494 \cdot 10^{-2}$
neurogenesis	$1.654 \cdot 10^{-2}$
protein amino acid phosphorylation	$2.210 \cdot 10^{-2}$
Introns (10)	
regulation of transcription, DNA-dependent	$8.755 \cdot 10^{-4}$
transcription factor activity	$2.110 \cdot 10^{-3}$
protein tyrosine kinase activity	$4.785 \cdot 10^{-3}$
protein amino acid phosphorylation	$1.584 \cdot 10^{-2}$
protein serine/threonine kinase activity	$2.806 \cdot 10^{-2}$
UTRs (3)	
regulation of transcription, DNA-dependent	$1.403 \cdot 10^{-4}$
transcription factor activity	$3.971 \cdot 10^{-3}$
Flanking within 1.2 Mbp (4)	
transcription factor activity	$3.255 \cdot 10^{-11}$
regulation of transcription, DNA-dependent	$2.021 \cdot 10^{-8}$
development	$5.566 \cdot 10^{-3}$

Table 22.5. *GO annotations of genes associated with vertebrate ultra-conserved elements. The number of GO annotations tested for each group are in parentheses. For each group, only GO annotations associated with at least 3 genes in the group were considered.*

conserved elements and found that 14 of the original 237 elements have at least one other instance within the human genome. Generally, the repeats are not ultra-conserved except for some of the seven repeats that are found both between *IRX6* and *IRX5* and between *IRX5* and *IRX3* on chromosome 16. These genes belong to a cluster of Iroquois homeobox genes involved in embryonic pattern formation [Peters *et al.*, 2000]. These repeated elements include two 32 bp sequences that are perfect reverse complements of each other and two (of lengths 23 bp and 28 bp) that are truncated reverse complements of each other. Overall, there are 5 distinct sequences within 226 bp regions on either side of *IRX5* that are perfect reverse complements of each other. The reverse complements are found in the same relative order (Figure 22.6). Furthermore, exact copies of the two outermost sequences are found both between *IRX4* and *IRX2* and between *IRX2* and *IRX1* on chromosome 5. Both of these regions are exactly 226 bp long. The repetition of these short regions and the conservation of their relative ordering and size suggests a highly specific coordinated regulatory signal with respect to these Iroquois homeobox genes and strengthens similar findings reported by [Sandelin *et al.*, 2004].

The longest ultra-conserved element that is repeated in the human genome is of length 35 and is found 18 additional times. None of these 18 instances are ultra-conserved, but this sequence is also found multiple times in other vertebrate genomes: 13 times in chimp, 10 times in mouse, 5 times in both rat and dog, 4 times in tetraodon, 3 times in zebra fish, and twice in both fugu and chicken. Of the 19 instances found in the human genome, two are found in

```
54102348 TGTAATTACAATCTTACAGAAACCGGGCCGATCTGTATATAAATCTCACCATCCAATTAC
54102408 AAGATGTAATAATTTTGCACTCAAGCTGGTAATGAGGTCTAATACTCGTGCATGCGATAA
54102468 TCCCCTCTGGATGCTGGCTTGATCAGATGTTGGCTTTGTAATTAGACGGGCAGAAAATCA
54102528 TTATTTCATGTTCAAATAGAAAATGAGGTTGGTGGGAAGTTAATTT

55002049 AAATTAACTTCCCACCAACCTAATTTTTTCCTGAACATGAAATAATGATTTTCTGCCCGT
55002109 CTAATTACAAAGCCAACATCTGATCAAGCCAGCATCCAGAGGGGATTATCGCATGCACGA
55002169 GTATTAGACCTCATTACCAGCTTGAGTGCAAAATTATTACATCTTGTAATTGGATGGTGA
55002229 GATTTATATACAGATCGGCCCGGTTTCTGTAAGATTGTAATTACA
```

Fig. 22.6. Sequences found on either side of *IRX5*. Positions underlined with a thick line are ultra-conserved with respect to the nine-vertebrate alignment. Sequences underlined with a thin line are not ultra-conserved but their reverse complement is. Indices are with respect to human chromosome 16.

well-studied actin genes, *ACTC* and *ACTG*, and the remainder are found in predicted retroposed pseudogenes with actin parent genes. These predictions are based on the retroGene track of the UCSC genome browser. Retroposed pseudogenes are the result of the reverse transcription and integration of the mRNA of the original functional gene. Actins are known to be highly conserved proteins, and β- and γ-actins have been shown to have a number of non-functional pseudogenes [Ng *et al.*, 1985, Pollard, 2001]. The precise conservation of this 35 bp sequence across a number of human actin pseudogenes may suggest that these integration events may be relatively recent changes in the human genome.

22.3.2 ENCODE alignment

Based on the annotations of known human genes provided by the UCSC Genome Browser, 69.2% of the bases of the ultra-conserved elements of length at least 20 in the ENCODE alignment overlap intragenic regions. Shorter sequences (less than 50 bp) have far more overlap with exons and UTRs than longer sequences (at least 50 bp), as illustrated in Figures 22.7(a) and 22.7(b). These longer sequences are heavily biased towards intronic overlap, accounting for 67.7% of these sequences by base coverage.

Values for the gene density and non-exonic conservation level (human–mouse) are available for the randomly selected ENCODE regions (see Section 21.2). For these regions, the base coverage by ultra-conserved elements is not correlated with gene density (Pearson correlation = -0.0589) and is moderately correlated with non-exonic conservation (Pearson correlation = 0.4350).

While we do not repeat the gene ontology analysis from the previous section, we note that the regions with the greatest number of ultra-conserved elements by base coverage are regions with well-known genes involved in DNA-dependent transcriptional regulation (Table 22.6). The elements in these 5 regions account for 80.3% of the bases of the ultra-conserved elements found in this dataset.

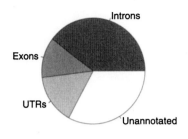

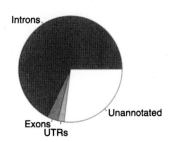

(a) 445 elements ≥ 20 and < 50 bp   (b) 79 elements ≥ 50 bp

Fig. 22.7. Functional base coverage of ultra-conserved elements found in ENCODE regions based on annotations of known human genes.

The 35 longest ultra-conserved elements, of length at least 69 bp, are also all found in these 5 regions.

	Ultra Coverage (bp)	Transcription Factor Genes	# Aligned Species
ENm012	9,086	*FOXP2*	9
ENr322	2,072	*BC11B*	9
ENm010	1,895	*HOXA1-7,9-11,13*; *EVX1*	8
ENm005	718	*GCFC*; *SON*	10
ENr334	549	*FOXP4*; *TFEB*	8

Table 22.6. *ENCODE regions with the greatest number of ultra-conserved elements by base coverage and their associated transcription factor genes.*

22.3.3 Eight-Drosophila alignment

We analyzed the 255 ultra-conserved elements of length at least 75 bp using the Release 4.0 annotations of *D. melanogaster*. These elements overlap 95 unique genes. Although the intragenic overlap for shorter elements (less than 100 bp) is only 42.9%, this proportion increases to 68.2% for the elements that are at least 100 bp in length (Figures 22.8(a) and 22.8(b)). Unlike the vertebrate dataset, longer regions are associated with exons, while shorter regions tend to correspond to unannotated elements.

The three genes with the greatest amount of overlap with ultra-conserved elements are *para* (765 bp), *nAcRα-34E* (426 bp) and *nAcRα-30D* (409 bp). All three of these genes are involved in cation channel activity, and the ultra-conserved elements correspond mostly with their exons. As with the nine-vertebrate dataset, the full set of 95 *D. melanogaster* genes is assessed for GO annotation enrichment, using all Release 4.0 *D. melanogaster* genes as the background set (Table 22.7). GO annotations exist for 78 of these 95 genes, which we did not differentiate further according to where in the gene overlap with an ultra-conserved element occurred. Genes involved in synaptic trans-

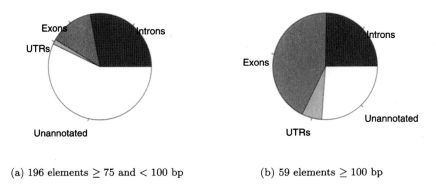

(a) 196 elements $\geq$ 75 and $<$ 100 bp   (b) 59 elements $\geq$ 100 bp

Fig. 22.8. Functional base coverage of ultra-conserved elements found in the *Drosophila* alignment based on annotations of known *D. melanogaster* genes.

mission are strongly over-represented in genes that have an ultra-conserved element overlap with their exons, introns and UTRs. These genes include those involved with ion channel activity, signal transduction and receptor activity, playing roles in intracellular signaling cascades, muscle contraction, development, and behavior. RNA binding proteins are also found to be over-represented. Another group of over-represented genes are those involved in RNA polymerase II transcription factor activity. These genes are strongly associated with development and morphogenesis.

The 130 ultra-conserved elements found in unannotated regions are grouped together into 109 regions by common flanking genes. These regions are flanked by 208 unique genes, 134 of which have available GO annotations. The distance from these ultra-conserved elements to their respective nearest gene ranges from 0.2–104 kb and is 16 kb on average. A number of transcription factors involved with development and morphogenesis are found within this set of genes. Five of the 10 flanking genes with ultra-conserved sequences both upstream and downstream are transcription factors (*SoxN*, *salr*, *toe*, *H15*, *sob*). In total, 44 unique transcription factors are found across the intragenic and flanking gene hits.

Ten of the original 255 ultra-conserved elements are repeated elsewhere in the *D. melanogaster* genome. However, all of these repeats correspond to annotated tRNA or snRNA, but not to homologous exons or regulatory regions. There are 10 ultra-conserved elements that overlap with tRNA (757 bp in sum), two that overlap with snRNA (191 bp in sum), and one that overlaps with ncRNA (81 bp). None of the ultra-conserved elements correspond to annotated rRNA, regulatory regions, transposable elements, or pseudogenes.

22.3.4 Discussion

We studied ultra-conserved elements in three very different datasets: an alignment of nine distant vertebrates, an alignment of the ENCODE regions in

GO Annotation	p-value
Exons, Introns, and UTRs (41)	
synaptic transmission	$3.290 \cdot 10^{-9}$
specification of organ identity	$1.044 \cdot 10^{-6}$
ventral cord development	$3.674 \cdot 10^{-6}$
RNA polymerase II transcription factor activity	$4.720 \cdot 10^{-6}$
muscle contraction	$8.714 \cdot 10^{-6}$
voltage-gated calcium channel activity	$3.548 \cdot 10^{-5}$
RNA binding	$7.650 \cdot 10^{-5}$
synaptic vesicle exocytosis	$3.503 \cdot 10^{-4}$
leg morphogenesis	$3.503 \cdot 10^{-4}$
calcium ion transport	$6.401 \cdot 10^{-4}$
Flanking within 104 kb (58)	
regulation of transcription	$8.844 \cdot 10^{-7}$
neurogenesis	$5.339 \cdot 10^{-6}$
ectoderm formation	$8.285 \cdot 10^{-6}$
endoderm formation	$2.125 \cdot 10^{-5}$
salivary gland morphogenesis	$5.870 \cdot 10^{-5}$
Notch signaling pathway	$1.591 \cdot 10^{-4}$
leg joint morphogenesis	$1.788 \cdot 10^{-4}$
RNA polymerase II transcription factor activity	$2.381 \cdot 10^{-4}$
salivary gland development	$4.403 \cdot 10^{-4}$
signal transducer activity	$5.308 \cdot 10^{-4}$
foregut morphogenesis	$8.004 \cdot 10^{-4}$

Table 22.7. *GO annotations of genes associated with Drosophila ultra-conserved elements. The number of GO annotations tested for each group are in parentheses. For each group, each tested GO annotation is associated with at least 3 genes in the group.*

mammals, and an alignment of eight fruit flies. As Figures 22.5, 22.7, and 22.8 show, ultra-conserved elements overlap with genes very differently in the three datasets. In particular, in the *Drosophila* dataset, exonic conservation is much more substantial. This conservation at the DNA level is very surprising, as the functional constraint on coding regions is expected to be at the amino acid level. Therefore, the degeneracy of the genetic code should allow synonymous mutations (see Section 21.3) to occur without any selective constraint.

The GO analysis showed that non-coding regions near or in genes associated with transcriptional regulation tended to contain ultra-conserved elements in all datasets. In *Drosophila*, ultra-conserved elements overlapped primarily with genes associated with synaptic transmission. While the exonic conservation in *Drosophila* is due in part to a much shorter period of evolution, the exact conservation of exons whose gene products are involved in synaptic transmission may be fly-specific.

Non-coding regions that are perfectly conserved across all 9 species may be precise regulatory signals for highly specific DNA-binding proteins. In particular, repeated ultra-conserved elements such as those found near the Iroquois homeobox genes on chromosome 16 are excellent candidates for such regulatory elements. Of course, it is interesting to note that the degree of conservation in our ultra-conserved elements exceeds what is observed for other known functional elements, such as splice sites. We discuss the statistical significance of ultra-conservation in Section 22.4.

Many of our results mirror those of previous studies. However, these studies have considered long stretches of perfectly conserved regions across shorter evolutionary distances [Bejerano *et al.*, 2004], or aligned regions above some relatively high threshold level of conservation [Boffelli *et al.*, 2004, Sandelin *et al.*, 2004, Woolfe *et al.*, 2005]. We have focused on ultra-conserved elements across larger evolutionary distances. As a result, we have not captured all regions containing high levels of conservation, but have identified only those regions that appear to be under the most stringent evolutionary constraints.

22.4 Statistical significance of ultra-conservation

Which ultra-conserved elements are of a length that is statistically significant? In order to address this question, we choose a model and compute the probability of observing an ultra-conserved element of a given length for the nine-vertebrate and *Drosophila*-alignments. First we consider phylogenetic tree models. These models allow for dependence of the occurrence of nucleotides in the genomes of different species at any given position in the aligned genomes, but make the assumption that evolutionary changes to DNA at one position in the alignment occur independently from changes at all other, and in particular, neighboring positions. Later we also consider a Markov chain, which does not model evolutionary changes explicitly but incorporates a simple pattern of dependence among different genome positions.

Before being able to compute a probability in a phylogenetic tree model, we must build a tree and estimate the parameters of the associated model. The tree for the nine-vertebrate alignment is shown in Figure 22.1. The topology of this tree is well-known, so we assume it fixed and use PAML [Yang, 1997] to estimate model parameters by maximum likelihood. As input to PAML, we choose the entire alignments with all columns containing a gap removed. The resulting alignment was 6,300,344 positions long for the vertebrates and 26,216,615 positions long for the *Drosophila*. Other authors (see Chapter 21 or [Pachter and Sturmfels, 2005]) have chosen to focus only on synonymous substitutions in coding regions, since they are likely not selected for or against and thus give good estimates for neutral substitution rates. However, our independence model does not depend on the functional structure of the genome; that is, it sees the columns as i.i.d. samples. Thus, we believe that it is more appropriate to use all the data available to estimate parameters.

There are many phylogenetic tree models (Section 4.5) and we concentrate here on the Jukes–Cantor and HKY85 models. With the parameter estimates from PAML, we can compute the probability p_{cons} of observing an ultra-conserved position in the alignment. Recall that the probability $p_{i_1 \ldots i_s}$ of seeing the nucleotide vector $(i_1, \ldots, i_s) \in \{\texttt{A}, \texttt{C}, \texttt{G}, \texttt{T}\}^s$ in a column of the alignment of s species is given by a polynomial in the entries of the transition matrices $P_e(t)$, which are obtained as $P_e(t) = \exp(Qt_e)$ where t_e is the length of the edge e in the phylogenetic tree and Q is a rate matrix that depends on the model selected.

Under the Jukes–Cantor model for the nine-vertebrate alignment, the maximum likelihood (ML) branch lengths are shown in Figure 22.1 and give the probabilities

$$p_{\mathtt{AAAAAAAAA}} = \cdots = p_{\mathtt{TTTTTTTTT}} = 0.01139...$$

Thus, the probability of a conserved column under this model is $p_{\mathrm{cons}} = 0.0456$. If we require that the nucleotides are identical not only across present-day species but also across ancestors, then the probability drops slightly to 0.0434.

Under the HKY85 model for the nine-vertebrate alignment, the ML branch lengths are very similar to those in Figure 22.1 and the additional parameter is estimated as $\kappa = 2.4066$ (in the notation of Figure 4.7, $\kappa = \alpha/\beta$). The root distribution is estimated to be almost uniform. These parameters give the probabilities

$$p_{\mathtt{AAAAAAAAA}} = \cdots = p_{\mathtt{TTTTTTTTT}} = 0.00367,$$

which are much smaller than their counterpart in the Jukes–Cantor model. The HKY85 probability of a conserved column is $p_{\mathrm{cons}} = 0.014706$. If we assume that nucleotides must also be identical in ancestors, this probability drops to 0.01234.

The binary indicators of ultra-conservation are independent and identically distributed according to a Bernoulli distribution with success probability p_{cons}. The probability of seeing an ultra-conserved element of length at least ℓ starting at a given position in the alignment therefore equals p_{cons}^{ℓ}. Moreover, the probability of seeing an ultra-conserved element of length at least ℓ anywhere in a genome of length N can be bounded above by $Np_{\mathrm{cons}}^{\ell}$. Recall that the length of the human genome is roughly 2.8 Gbp and the length of *D. melanogaster* is approximately 120 Mbp. Table 22.8 contains the evaluated probability bound for different values of ℓ.

	Nine-vertebrate (human)			*Drosophila (D. melanogaster)*	
	Jukes–Cantor	HKY85		Jukes–Cantor	HKY85
p_{cons}	0.0456	0.0147	p_{cons}	0.1071	0.05969
10	0.0001	$1.3 \cdot 10^{-9}$	15	$7.8 \cdot 10^{-6}$	$1.2 \cdot 10^{-9}$
20	$4.1 \cdot 10^{-18}$	$6.2 \cdot 10^{-28}$	75	$4.6 \cdot 10^{-64}$	$4.3 \cdot 10^{-83}$
125	$6.0 \cdot 10^{-159}$	$2.4 \cdot 10^{-220}$	209	$4.3 \cdot 10^{-194}$	$4.1 \cdot 10^{-247}$

Table 22.8. *Probabilities of seeing ultra-conserved elements of certain lengths in an independence model with success probability p_{cons} derived from two phylogenetic tree models.*

However, 46% of the ungapped columns in the nine-vertebrate alignment are actually ultra-conserved. This fraction is far greater than the 5% we would expect with the JC model and the 1% under the HKY85 model. This suggests that the model of independent alignment positions is overly simplistic. If we collapse the alignment to a sequence of binary indicators of ultra-conserved positions, then a very simple non-independence model for this binary sequence is a Markov chain model (cf. Section 1.4 and Chapter 10).

In a Markov chain model, the length of ultra-conserved elements is geometrically distributed. That is, the probability that an ultra-conserved element is of length ℓ equals $\theta^{\ell-1}(1 - \theta)$, where θ is the probability of transitioning from one ultra-conserved position to another. The expected value of the length of an ultra-conserved element is equal to $1/(1 - \theta)$. The probability that an ultra-conserved element is of length ℓ or longer equals

$$\sum_{k=\ell}^{\infty} \theta^{k-1}(1 - \theta) = \theta^{\ell-1}.$$

Therefore, the probability that at least one of U ultra-conserved elements found in a multiple alignment is of length at least ℓ is equal to

$$1 - (1 - \theta^{\ell-1})^U \approx U \cdot \theta^{\ell-1} \quad \text{for large } \ell.$$

Restricting ourselves to the nine-vertebrate alignment (computations for the *Drosophila* alignment are qualitatively similar), we used the mean length of the ultra-conserved elements described in Section 22.3.1 to estimate the transition probability θ to 0.4785. Then the probability that at least one of the 1,513,176 ultra-conserved elements of the nine-vertebrate alignment is of length 25 or longer equals about 3%. The probability of seeing one of the U ultra-conserved elements being 30 or more bp long is just below 1/1000. However, the dependence structure in a Markov chain model cannot explain the longest ultra-conserved elements in the alignment. For example, the probability of one of the U elements being 125 or more bp long is astronomically small (0.3×10^{-33}). This suggests that the Markov chain model does not capture the dependence structure in the binary sequence of ultra-conservation indicators. At a visual level, this is clear from Figure 22.3. Were the Markov chain model true then, due to the resulting geometric distribution for the length of an ultra-conserved element, the log-scale frequencies should fall on a straight line, which is not the case in Figure 22.3. Modeling the process of ultra-conservation statistically requires more sophisticated models. The phylogenetic hidden Markov models that appear in [McAuliffe et al., 2004, Siepel and Haussler, 2004] provide a point of departure.

Despite the shortcomings of the calculations, it is clear that it is highly unlikely that the ultra-conserved elements studied in this chapter occur by chance. The degree of conservation strongly suggests extreme natural selection in these regions.

References

[Abril *et al.*, 2005] JF Abril, R Castelo, and R Guigó. Comparison of splice sites in mammals and chicken. *Genome Research*, 15:111–119, 2005. Cited on p. 144

[Adkins *et al.*, 2001] RM Adkins, EL Gelke, D Rowe, and RL Honeycutt. Molecular phylogeny and divergence time estimates for major rodent groups: evidence from multiple genes. *Molecular Biology and Evolution*, 18(5):777–791, 2001. Cited on p. 384

[Agresti, 1990] A Agresti. *Categorical Data Analysis*. Wiley Series in Probability and Mathematical Statistics: Applied Probability and Statistics. John Wiley & Sons Inc., New York, 1990. A Wiley-Interscience Publication. Cited on p. 14

[Aji and McEliece, 2000] SM Aji and RJ McEliece. The generalized distributive law. *IEEE Transactions on Information Theory*, 46(2):325–343, 2000. Cited on p. 41

[Alefeld and Herzberger, 1983] G Alefeld and J Herzberger. *An introduction to interval computations*. Academic press, New York, 1983. Cited on p. 363

[Alexandersson *et al.*, 2003] M Alexandersson, S Cawley, and L Pachter. SLAM: cross-species gene finding and alignment with a generalized pair hidden Markov model. *Genome Research*, 13(3):496–502, 2003. Cited on p. 145

[Allman and Rhodes, 2003] ES Allman and JA Rhodes. Phylogenetic invariants for the general Markov model of sequence mutation. *Mathematical Biosciences*, 186(2):113–144, 2003. Cited on p. 348

[Allman and Rhodes, 2004a] ES Allman and JA Rhodes. Phylogenetic ideals and varieties for the general Markov model. Preprint, 2004. Cited on p. 153, 306, 317, 319, 348, 351

[Allman and Rhodes, 2004b] ES Allman and JA Rhodes. Quartets and parameter recovery for the general Markov model of sequence mutation. *AMRX Applied Mathematics Research Express*, 2004(4):107–131, 2004. Cited on p. 102, 351

[Altschul *et al.*, 1990] SF Altschul, W Gish, W Miller, EW Myers, and DJ Lipman. Basic local alignment search tool. *Journal of Molecular Biology*, 215:403–410, 1990. Cited on p. 80

[Andrews, 1963] G Andrews. A lower bound on the volume of strictly convex bodies with many boundary points. *Transactions of the American Mathematical Society*, 106:270–279, 1963. Cited on p. 174

[Apostol, 1976] TM Apostol. *Introduction to Analytic Number Theory*. Springer-Verlag, New York, 1976. Undergraduate Texts in Mathematics. Cited on p. 221

[Ardila, 2005] F Ardila. A tropical morphism related to the hyperplane arrangement of the complete bipartite graph. *Discrete and Computational Geometry*, 2005. To appear. Cited on p. 116

[Aris-Brosou, 2003] S Aris-Brosou. How Bayes tests of molecular phylogenies compare with frequentist approaches. *Bioinformatics*, 19:618–624, 2003. Cited on p. 374

[Ashburner *et al.*, 2000] M Ashburner, CA Ball, JA Blake, D Botstein, H Butler, JM Cherry, K Dolinski, SS Dwight, JT Eppig, MA Harris, et al. Gene Ontology: tool for the unification of biology. The Gene Ontology Consortium. *Nature*

Genetics, 25:25–29, 2000. Cited on p. 394

[Balding *et al.*, 2003] DJ Balding, M Bishop, and C Cannings, editors. *Handbook of Statistical Genetics (2 Volume Set)*. John Wiley & Sons, second edition, 2003. Cited on p. 126

[Bandelt and Dress, 1992] HJ Bandelt and AWM Dress. A canonical decomposition theory for metrics on a finite set. *Advances in Mathematics*, 92:47–105, 1992. Cited on p. 323, 326

[Bandelt *et al.*, 1995] HJ Bandelt, P Forster, BC Sykes, and MB Richards. Mitochondrial portraits of human population using median networks. *Genetics*, 141:743–753, 1995. Cited on p. 323

[Basu *et al.*, 2003] S Basu, R Pollack, and MF Roy. *Algorithms in Real Algebraic Geometry*, volume 10 of *Algorithms and Computation in Mathematics*. Springer-Verlag, Berlin, 2003. Cited on p. 95

[Bayer and Mumford, 1993] D Bayer and D Mumford. What can be computed in algebraic geometry? In *Computational Algebraic Geometry and Commutative Algebra, Sympos. Math., XXXIV*, pages 1–48. Cambridge University Press, Cortona, 1991, 1993. Cited on p. 239

[Beerenwinkel *et al.*, 2004] N Beerenwinkel, J Rahnenführer, M Däumer, D Hoffmann, R Kaiser, J Selbig, and T Lengauer. Learning multiple evolutionary pathways from cross-sectional data. In *Proceedings of the 8th Annual International Conference on Research in Computational Biology (RECOMB '04), 27–31 March 2004, San Diego, CA*, pages 36–44, 2004. To appear in *Journal of Computational Biology*. Cited on p. 278, 281, 287

[Beerenwinkel *et al.*, 2005a] N Beerenwinkel, M Däumer, T Sing, J Rahnenführer, T Lengauer, J Selbig, D Hoffmann, and R Kaiser. Estimating HIV evolutionary pathways and the genetic barrier to drug resistance. *Journal of Infectious Diseases*, 2005. To appear. Cited on p. 278, 289

[Beerenwinkel *et al.*, 2005b] N Beerenwinkel, J Rahnenführer, R Kaiser, D Hoffmann, J Selbig, and T Lengauer. Mtreemix: a software package for learning and using mixture models of mutagenetic trees. *Bioinformatics*, 2005. To appear. Cited on p. 278

[Bejerano *et al.*, 2004] G Bejerano, M Pheasant, I Makunin, S Stephen, WJ Kent, JS Mattick, and D Haussler. Ultraconserved elements in the human genome. *Science*, 304:1321–1325, 2004. Cited on p. 394, 400

[Besag, 1974] J. Besag. Spatial interaction and the statistical analysis of lattice systems. *Journal of the Royal Statistical Society*, B,36:192–236, 1974. Cited on p. 266

[Besag, 1986] J Besag. On the statistical analysis of dirty pictures. *Journal of the Royal Statistical Society*, B 48 No. 3:259–302, 1986. Cited on p. 274

[Bickel and Doksum, 2000] PJ Bickel and KA Doksum. *Mathematical statistics: Basic Ideas and Selected Topics, Vol I (2nd Edition)*. Prentice Hall, 2000. Cited on p. 6, 8

[Billera *et al.*, 2001] LJ Billera, SP Holmes, and K Vogtmann. Geometry of the space of phylogenetic trees. *Advances in Applied Mathematics*, 27(4):733–767, 2001. Cited on p. 68

[Blanchette *et al.*, 2004] M Blanchette, WJ Kent, C Riemer, L Elnitski, AFA Smit, KM Roskin, R Baertsch, K Rosenbloom, H Clawson, ED Green, et al. Aligning multiple genomic sequences with the threaded blockset aligner. *Genome Research*, 14:708–715, 2004. Cited on p. 137

[Boffelli *et al.*, 2003] D Boffelli, J McAuliffe, D Ovcharenko, KD Lewis, I Ovcharenko, L Pachter, and EM Rubin. Phylogenetic shadowing of primate sequences to find functional regions of the human genome. *Science*, 299(5611):1391–4, 2003. Cited on p. 130, 157

[Boffelli *et al.*, 2004] D Boffelli, MA Nobrega, and EM Rubin. Comparative genomics at the vertebrate extremes. *Nature Reviews Genetics*, 5:456–465, 2004. Cited on p. 394, 400

[Bosma *et al.*, 1997] W Bosma, J Cannon, and C Playoust. The MAGMA algebra system I: the user language. *Journal of Symbolic Computation*, 24(3-4):235–265,

1997. Cited on p. 77

[Bourque et al., 2004] G Bourque, PA Pevzner, and G Tesler. Reconstructing the genomic architecture of ancestral mammals: lessons from human, mouse, and rat genomes. *Genome Research*, 14(4):507–16, 2004. Cited on p. 131

[Boykov et al., 1999] Y Boykov, O Veksler, and R Zabih. Fast approximate energy minimization via graph cuts. *Intl. Conf. on Computer Vision*, 1999. Cited on p. 273, 275

[Bray and Pachter, 2004] N Bray and L Pachter. MAVID: constrained ancestral alignment of multiple sequences. *Genome Research*, 14(4):693–9, 2004. Cited on p. 82, 137, 378, 387

[Bray et al., 2003] N Bray, I Dubchak, and L Pachter. AVID: A global alignment program. *Genome Research*, 13(1):97–102, 2003. Cited on p. 137

[Brown et al., 1982] WM Brown, EM Prager, A Wang, and AC Wilson. Mitochondrial DNA sequences of primates, tempo and mode of evolution. *Journal of Molecular Evolution*, 18:225–239, 1982. Cited on p. 371

[Brown, 2002] TA Brown. *Genomes 2*. John Wiley & Son, Inc., 2002. Cited on p. 2, 130

[Brudno et al., 2003a] M Brudno, C Do, G Cooper, MF Kim, E Davydov, ED Green, A Sidow, and S Batzoglou. LAGAN and Multi-LAGAN: efficient tools for large-scale multiple alignment of genomic DNA. *Genome Research*, 13:721–731, 2003. Cited on p. 137

[Brudno et al., 2003b] M Brudno, S Malde, A Poliakov, C Do, O Couronne, I Dubchak, and S Batzoglou. Glocal alignment: finding rearrangements during alignment. *Special issue on the Proceedings of the ISMB 2003, Bioinformatics*, 19:54i–64i, 2003. Cited on p. 378, 390

[Bryant and Moulton, 2004] D Bryant and V Moulton. NeighborNet: An agglomerative algorithm for the construction of planar phylogenetic networks. *Molecular Biology And Evolution*, 21:255–265, 2004. Cited on p. 323

[Buchberger, 1965] B Buchberger. *An algorithm for finding a basis for the residue class ring of a zero-dimensional polynomial ideal (in German)*. PhD thesis, Univ. Innsbruck, Dept. of Math., Innsbruck, Austria, 1965. Cited on p. 93

[Bulmer, 1991] D Bulmer. Use of the method of generalized least squares in reconstructing phylogenies from sequence data. *Molecular Biology and Evolution*, 8(6):868–883, 1991. Cited on p. 325

[Burge and Karlin, 1997] C Burge and S Karlin. Prediction of complete gene structures in human genomic DNA. *Journal of Molecular Biology*, 268(1):78–94, 1997. Cited on p. 145, 236

[Campbell et al., 1999] A Campbell, J Mrazek, and S Karlin. Genome signature comparisons among prokaryote, plasmid and mitochondrial DNA. *Proceedings of the National Academy of Sciences USA*, 96(16):9184–9189, 1999. Cited on p. 128, 129

[Catalisano et al., 2002] MV Catalisano, AV Geramita, and A Gimigliano. Ranks of tensors, secant varieties of Segre varieties and fat points. *Linear Algebra Appl.*, 355:263–285, 2002. Cited on p. 288

[Catanese et al., 2005] F Catanese, S Hoşten, A Khetan, and B Sturmfels. The maximum likelihood degree. *American Journal of Mathematics*, 2005. To appear. Cited on p. 10, 106

[Cavalli-Sforza and Edwards, 1967] L Cavalli-Sforza and A Edwards. Phylogenetic analysis models and estimation procedures. *Evolution*, 32:550–570, 1967. Cited on p. 325

[Cavender and Felsenstein, 1987] J Cavender and J Felsenstein. Invariants of phylogenies in a simple case with discrete states. *Journal of Classification*, 4:57–71, 1987. Cited on p. 348

[Chargaff, 1950] E Chargaff. Chemical specificity of nucleic acids and mechanism for the enzymatic degradation. *Experientia*, 6:201–209, 1950. Cited on p. 126

[Chazelle, 1993] B Chazelle. An optimal convex hull algorithm in any fixed dimension. *Discrete Computational Geometry*, 10:377–409, 1993. Cited on p. 177

[Cohen and Rothblum, 1993] JE Cohen and UG Rothblum. Nonnegative ranks, decompositions, and factorizations of nonnegative matrices. *Linear Algebra Appl.*, 190:149–168, 1993. Cited on p. 116

[Cohen, 2004] JE Cohen. Mathematics is biology's next microscope, only better; biology is mathematics' next physics, only better. *PLoS Biol.* 2(12):e439, 2004. Cited on p. 2

[Consortium, 2004] ENCODE Project Consortium. The ENCODE (ENCyclopedia Of DNA Elements) Project. *Science*, 306(5696):636–40, 2004. Cited on p. 141, 378

[Cowell et al., 1999] RG Cowell, AP Dawid, SL Lauritzen, and DJ Spiegelhalter. *Probabilistic Networks and Expert Systems*. Statistics for Engineering and Information Sciences. Springer-Verlag, New York, 1999. Cited on p. 41

[Cox et al., 1997] D Cox, J Little, and D O'Shea. *Ideals, Varieties, and Algorithms*. Undergraduate Texts in Mathematics. Springer-Verlag, New York, second edition, 1997. An introduction to computational algebraic geometry and commutative algebra. Cited on p. 2, 88, 89, 91, 95, 97, 282, 283, 285

[Craciun and Feinberg, 2005] G Craciun and M Feinberg. Multiple equilibria in complex chemical reaction networks: I. the injectivity property. *SIAM Journal of Applied Mathematics*, 2005. To appear. Cited on p. 94

[Cuyt et al., 2001] A Cuyt, B Verdonk, S Becuwe, and P Kuterna. A remarkable example of catastrophic cancellation unraveled. *Computing*, 66:309–320, 2001. Cited on p. 360

[Darling et al., 2004] ACE Darling, B Mau, FR Blattner, and NT Perna. Mauve: multiple alignment of conserved genomic sequence with rearrangements. *Genome Research*, 14:1394–1403, 2004. Cited on p. 137

[Darwin, 1859] C Darwin. *On the Origin of Species by Means of Natural Selection, or the Preservation of Favoured Races in the Struggle for Life*. John Murray, London, 1859. Cited on p. 67

[DeConcini et al., 1982] C DeConcini, D Eisenbud, and C Procesi. *Hodge algebras*, volume 91 of *Astérisque*. Société Mathématique de France, Paris, 1982. With a French summary. Cited on p. 101

[Demmel, 1997] JW Demmel. *Applied Numerical Linear Algebra*. Society for Industrial and Applied Mathematics (SIAM), Philadelphia, PA, 1997. Cited on p. 351, 352, 353

[Dermitzakis et al., 2003] ET Dermitzakis, A Reymond, N Scamuffa, C Ucla, E Kirkness, C Rossier, and SE Antonarakis. Evolutionary discrimination of mammalian conserved non-genic sequences (CNGs). *Science*, 302:1033–1035, 2003. Cited on p. 130

[Desper et al., 1999] R Desper, F Jiang, O-P Kallioniemi, H Moch, CH Papadimitriou, and AA Schäffer. Inferring tree models for oncogenesis from comparative genome hybridization data. *Journal of Computational Biology*, 6(1):37–51, 1999. Cited on p. 278

[Develin and Sturmfels, 2004] M Develin and B Sturmfels. Tropical convexity. *Documenta Mathematica*, 9:1–27 (electronic), 2004. Cited on p. 116

[Develin et al., 2003] M Develin, F Santos, and B Sturmfels. On the tropical rank of a matrix. *MSRI Proceedings*, 2003. To appear. Cited on p. 114, 116

[Dewey et al., 2004] C Dewey, JQ Wu, S Cawley, M Alexandersson, R Gibbs, and L Pachter. Accurate identification of novel human genes through simultaneous gene prediction in human, mouse, and rat. *Genome Research*, 14(4):661–4, 2004. Cited on p. 129

[Dewey, 2005] C Dewey. MERCATOR: multiple whole-genome orthology map construction. Available at http://hanuman.math.berkeley.edu/~cdewey/mercator/, 2005. Cited on p. 378, 387

[Deza and Laurent, 1997] MM Deza and M Laurent. *Geometry of Cuts and Metrics*, volume 15 of *Algorithms and Combinatorics*. Springer-Verlag, Berlin, 1997. Cited on p. 68

[Douzery et al., 2003] EJP Douzery, F Delsuc, MJ Stanhope, and D Huchon. Local molecular clocks in three nuclear genes: divergence ages of rodents and other mammals, and incompatibility between fossil calibrations. *Molecular Biology and Evolution*, 57:201–213, 2003. Cited on p. 386

[Dress and Huson, 2004] A Dress and D Huson. Constructing splits graphs. *IEEE/ACM Transactions in Computational Biology and Bioinformatics*, 2004. Cited on p. 323

[Dress and Terhalle, 1998] A Dress and W Terhalle. The tree of life and other affine buildings. In *Proceedings of the International Congress of Mathematicians*, number Extra Vol. III in Vol. III (Berlin, 1998), pages 565–574 (electronic), 1998. Cited on p. 117

[Dress et al., 2002] A Dress, JH Koolen, and V Moulton. On line arrangements in the hyperbolic plane. *European Journal of Combinatorics*, 23(5):549–557, 2002. Cited on p. 101

[Durbin et al., 1998] R Durbin, S Eddy, A Korgh, and G Mitchison. *Biological Sequence Analysis: Probabilistic Models of Proteins and Nucleic Acids*. Cambridge University Press, 1998. Cited on p. 2, 3, 4, 24, 27, 30, 142, 253

[Eichler and Sankoff, 2003] EE Eichler and D Sankoff. Structural dynamics of eukaryotic chromosome evolution. *Science*, 301:793–797, 2003. Cited on p. 130

[Evans and Speed, 1993] S Evans and T Speed. Invariants of some probability models used in phylogenetic inference. *The Annals of Statistics*, 21:355–377, 1993. Cited on p. 152, 296, 306, 309

[Ewens and Grant, 2005] WJ Ewens and GR Grant. *Statistical Methods in Bioinformatics: An Introduction*. Statistics for Biology and Health. Springer-Verlag, New York, second edition, 2005. Cited on p. 2

[Faith, 1992] DP Faith. Conservation evaluation and phylogenetic diversity. *Biological Conservation*, 61:1–10, 1992. Cited on p. 336

[Farris, 1972] JS Farris. Estimating phylogenetic trees from distance matrices. *American Naturalist*, 106:645–668, 1972. Cited on p. 325

[Felsenstein, 1978] J. Felsenstein. Cases in which parsimony or compatibility methods will be positively misleading. *Syst. Zool.*, 22:240–249, 1978. Cited on p. 385

[Felsenstein, 1981] J Felsenstein. Evolutionary trees from DNA sequences: a maximum likelihood approach. *Journal of Molecular Evolution*, 17:368–376, 1981. Cited on p. 57, 155, 325, 366

[Felsenstein, 1989] J Felsenstein. PHYLIP – Phylogeny Inference Package (Version 3.2). *Cladistics*, 5:164–166, 1989. Cited on p. 155

[Felsenstein, 2003] J Felsenstein. *Inferring Phylogenies*. Sinauer Associates, Inc., 2003. Cited on p. ix, 67, 147, 152, 326, 334

[Felsenstein, 2004] J Felsenstein. PHYLIP (Phylogeny Inference Package) version 3.6. Distributed by the author, Department of Genome Sciences, University of Washington, Seattle, 2004. Cited on p. 83, 155, 291, 374, 375

[Fernández-Baca et al., 2002] D Fernández-Baca, T Seppäläinen, and G Slutzki. Bounds for parametric sequence comparison. *Discrete Applied Mathematics*, 118:181–198, 2002. Cited on p. 211, 212, 213

[Fernández-Baca et al., 2004] D Fernández-Baca, T Seppäläinen, and G Slutzki. Parametric multiple sequence alignment and phylogeny construction. *Journal of Discrete Algorithms*, 2(2):271–287, 2004. Cited on p. 211

[Ferrari et al., 1995] P Ferrari, A Frigessi, and P de Sa. Fast approximate maximum a posteriori restoration of multi-color images. *Journal of the Royal Statistical Society*, B,57, 1995. Cited on p. 275

[Fitch and Smith, 1983] WM Fitch and TF Smith. Optimal sequence alignments. *PNAS*, 80(5):1382–1386, 1983. Cited on p. 202, 203, 204

[Fleischmann et al., 1995] RD Fleischmann, MD Adams, O White, RA Clayton, EF Kirkness, AR Kerlavage, CJ Bult, JF Tomb, BA Dougherty, and JM Merrick et al. Whole-genome random sequencing and assembly of *Haemophilus influenza* Rd. *Science*, 269(5223):496–512, 1995. Cited on p. 125

[Floyd, 1962] RW Floyd. Algorithm 97: shortest path. *Communications of ACM*, 5(6):345, 1962.

[Forney, 1973] GD Forney. The Viterbi algorithm. *Procedings of the IEEE*, 61(3):268–278, 1973. Cited on p. 46 Cited on p. 227

[Frieze *et al.*, 1998] A Frieze, R Kannan, and S Vempala. Fast Monte Carlo algorithms for low rank approximation. In *39th Symposium on Foundations of Computing*, pages 370–378, 1998. Cited on p. 353

[Fukuda, 2003] K Fukuda. cddlib 0.39d. Available at http://www.ifor.math.ethz.ch/~fukuda/cdd_home/cdd.html, 2003. Cited on p. 190

[Fukuda, 2004] K Fukuda. From the zonotope construction to the Minkowski addition of convex polytopes. *Journal of Symbolic Computation*, 38(4):1261–1272, 2004. Cited on p. 186

[Galtier and Gouy, 1998] N Galtier and M Gouy. Inferring pattern and process: maximum likelihood implementation of a non-homogeneous model of DNA sequence evolution for phylogenetic analysis. *Molecular Biology and Evolution*, 154(4):871–879, 1998. Cited on p. 385, 386

[Garcia *et al.*, 2004] L. D. Garcia, M. Stillman, and B. Sturmfels. Algebraic geometry of Bayesian networks. *Journal of Symbolic Computation*, 39/3-4:331–355, 2004. Special issue on the occasion of MEGA 2003. Cited on p. 40, 102, 284, 351

[Garcia, 2004] LD Garcia. Algebraic statistics in model selection. In M Chickering and J Halpern, editors, *Proceedings of the 20th Conference on Uncertainty in Artificial Intelligence*, pages 177–184. AUAI Press, Arlington, VA, 2004. Cited on p. 288

[Gatermann and Wolfrum, 2005] K Gatermann and M Wolfrum. Bernstein's second theorem and Viro's method for sparse polynomial systems in chemistry. *Advances in Applied Mathematics*, 34(2):252–294, 2005. Cited on p. 94

[Gawrilow and Joswig, 2000] E Gawrilow and M Joswig. Polymake: a framework for analyzing convex polytopes. In G Kalai and GM Ziegler, editors, *Polytopes — Combinatorics and Computation*, pages 43–74. Birkhäuser, 2000. Cited on p. 79, 186

[Gawrilow and Joswig, 2001] E Gawrilow and M Joswig. Polymake: an approach to modular software design in computational geometry. In *Proceedings of the 17th Annual Symposium on Computational Geometry*, pages 222–231. ACM, 2001. June 3-5, 2001, Medford, MA. Cited on p. 79, 186

[Geiger *et al.*, 2001] D Geiger, D Heckerman, H King, and C Meek. Stratified exponential families: graphical models and model selection. *The Annals of Statist.*, 29(2):505–529, 2001. Cited on p. 290

[Geiger *et al.*, 2005] D Geiger, C Meek, and B Sturmfels. On the toric algebra of graphical models. *The Annals of Statistics*, 2005. To appear. Cited on p. 38, 267, 283

[Gentleman *et al.*, 2004] RC Gentleman, VJ Carey, DM Bates, B Bolstad, M Dettling, S Dudoit, B Ellis, L Gautier, Y Ge, J Gentry, et al. Bioconductor: Open software development for computational biology and bioinformatics. *Genome Biology*, 5:R80, 2004. Cited on p. 80

[Gentles and Karlin, 2001] AJ Gentles and S Karlin. Genome-scale compositional comparisons in eukaryotes. *Genome Research*, 4:540–546, 2001. Cited on p. 128

[Gibbs *et al.*, 2004] RA Gibbs, GM Weinstock, ML Metzker, DM Muzny, EJ Sodergren, S Scherer, G Scott, D Steffen, KC Worley, PE Burch, et al. Genome sequence of the brown norway rat yields insights into mammalian evolution. *Nature*, 428(6982):493–521, 2004. Cited on p. 131, 384

[Gorban and Zinovyev, 2004] AN Gorban and AY Zinovyev. The mystery of two straight lines in bacterial genome statistics. *arXiv.org:q-bio.GN/0412015*, 2004. Cited on p. 129

[Grayson and Stillman, 2002] DR Grayson and ME Stillman. Macaulay 2, a software system for research in algebraic geometry. Available at http://www.math.uiuc.edu/Macaulay2/, 2002. Cited on p. 76, 239, 310, 311, 313

[Greig *et al.*, 1989] DM Greig, BT Porteous, and AH Seheult. Exact maximum a posteriori estimation for binary images. *Journal of the Royal Statistical Society*, Series B, 51:271–279, 1989. Cited on p. 274

[Greuel and Pfister, 2002] GM Greuel and G Pfister. *A Singular Introduction to Com-*

mutative Algebra. Springer-Verlag, Berlin and Heidelberg, 2002. Cited on p. 79

[Greuel *et al.*, 2003] GM Greuel, G. Pfister, and H. Schoenemann. Singular: A computer algebra system for polynomial computations. Available at `http://www.singular.uni-kl.de/`, 2003. Cited on p. 239

[Gritzmann and Sturmfels, 1993] P Gritzmann and B Sturmfels. Minkowski addition of polytopes: Computational complexity and applications to Gröbner bases. *SIAM Journal of Discrete Mathematics*, 6:246–269, 1993. Cited on p. 217

[Grötschel *et al.*, 1993] M Grötschel, L Lovász, and A Schrijver. *Geometric Algorithms and Combinatorial Optimization*, volume 2 of Algorithms and Combinatorics. Springer-Verlag, 1993. Cited on p. 176, 184

[Grünbaum, 2003] B Grünbaum. *Convex polytopes*, volume 221 of Graduate Texts in Mathematics. Springer-Verlag, New York, second edition, 2003. Prepared and with a preface by Volker Kaibel, Victor Klee and Günter M. Ziegler. Cited on p. 60

[Guigó *et al.*, 2004] R Guigó, E Birbey, M Brent, E Dermitzakis, L Pachter, H Roest Crollius, V Solovyev, and MQ Zhang. Needed for completion of the human genome: hypothesis driven experiments and biologically realistic mathematical models. *arXiv.org:q-bio.GN/0410008*, 2004. Cited on p. 130

[Gusfield *et al.*, 1994] D Gusfield, K Balasubramanian, and D Naor. Parametric optimization of sequence alignment. *Algorithmica*, 12:312–326, 1994. Cited on p. 55, 194, 209, 211

[Gusfield, 1997] D Gusfield. *Algorithms on Strings, Trees, and Sequences*. Cambridge University Press, 1997. Cited on p. 49, 55, 194, 206

[Hallgrímsdóttir and Sturmfels, 2005] I Hallgrímsdóttir and B Sturmfels. Resultants in genetic linkage analysis. *Journal of Symbolic Computation*, 2005. To appear. Cited on p. 126

[Hammer *et al.*, 1995] R Hammer, M Hocks, U Kulisch, and D Ratz. *C++ Toolbox for Verified Computing: Basic Numerical Problems*. Springer-Verlag, Berlin, 1995. Cited on p. 364, 366

[Hannenhalli and Pevzner, 1999] S Hannenhalli and PA Pevzner. Transforming cabbage into turnip: polynomial algorithm for sorting signed permutations by reversals. *Journal of the ACM*, 46(1):1–27, 1999. Cited on p. 130

[Hansen and Sengupta, 1981] E Hansen and S Sengupta. Bounding solutions of systems of equations using interval analysis. *BIT*, 21:203–211, 1981. Cited on p. 364

[Hansen, 1980] E Hansen. Global optimization using interval analysis – the multidimensional case. *Numerische Mathematik*, 34:247–270, 1980. Cited on p. 360

[Hansen, 1992] E Hansen. *Global Optimization using Interval Analysis*. Marcel Dekker, New York, 1992. Cited on p. 365, 369

[Hartwell *et al.*, 2003] L Hartwell, L Hood, ML Goldberg, LM Silver, RC Veres, and A Reynolds. *Genetics: From Genes to Genomes*. McGraw-Hill Science/Engineering/Math, second edition, 2003. Cited on p. 126

[Hasegawa *et al.*, 1985] M Hasegawa, H Kishino, and T Yano. Dating of the human-ape splitting by a molecular clock of mitochondrial DNA. *Journal of Molecular Evolution*, 22:160–174, 1985. Cited on p. 155

[Hendy and Penny, 1989] M Hendy and D Penny. A framework for the quantitative study of evolutionary trees. *Systematic Zoology*, 38(4), 1989. Cited on p. 332, 334, 385

[Hendy and Penny, 1993] MD Hendy and D Penny. Spectral analysis of phylogenetic data. *Journal of Classification*, 10:5–24, 1993. Cited on p. 152

[Hibi, 1987] T Hibi. Distributive lattices, affine semigroup rings and algebras with straightening laws. *Advanced Studies in Pure Mathematics*, 11:93–109, 1987. Cited on p. 283

[Hillier *et al.*, 2004] LW Hillier, W Miller, E Birney, W Warren, RC Hardison, CP Ponting, P Bork, DW Burt, MAM Groenen, ME Delany, et al. Sequence and comparative analysis of the chicken genome provide unique perspectives on vertebrate evolution. *Nature*, 432(7018):695–716, 2004. Cited on p. 130, 384

[Holland *et al.*, 2003] BR Holland, D Penny, and MD Hendy. Outgroup misplacement and phylogenetic inaccuracy under a molecular clock– a simulation study. *Systematic Biology*, 52(2):229–238, 2003. Cited on p. 385

[Holland *et al.*, 2004] B Holland, KT Huber, V Moulton, and P Lockhart. Using consensus networks to visualize contradictory evidence for species phylogeny. *Molecular Biology and Evolution*, 21(7):1459–1461, 2004. Cited on p. 323

[Hoşten *et al.*, 2005] S Hoşten, A Khetan, and B Sturmfels. Solving the likelihood equations. *Foundations of Computational Mathematics*, 2005. To appear. Cited on p. 108, 335, 343, 344

[Huber *et al.*, 2002] KT Huber, M Langton, D Penny, V Moulton, and M Hendy. Spectronet: A package for computing spectra and median networks. *Applied Bioinformatics*, 1(3):2041–2059, 2002. Cited on p. 323

[Huelsenbeck *et al.*, 2000] JP Huelsenbeck, B Larget, and DL Swofford. A compound poisson process for relaxing the molecular clock. *Genetics*, 154(4):1879–1892, 2000. Cited on p. 385

[Human Genome Sequencing Consortium, 2001] Human Genome Sequencing Consortium. Initial sequencing and analysis of the human genome. *Nature*, 409(6822):860–921, February 2001. Cited on p. 384

[Human Genome Sequencing Consortium, 2004] International Human Genome Sequencing Consortium. Finishing the euchromatic sequence of the human genome. *Nature*, 431(7011):931–945, 2004. Cited on p. 125, 126, 129, 135

[Huson and Bryant, 2005] D Huson and D Bryant. Estimating phylogenetic trees and networks using SplitsTree4. in preparation, 2005. Cited on p. 323

[Huson *et al.*, 2004] D Huson, T Dezulian, T Kloepper, and M Steel. Phylogenetic super-networks from partial trees. *IEEE Transactions on Computational Biology and Bioinformatics*, 1(4):151–158, 2004. Cited on p. 323

[IEEE Task P754, 1985] IEEE, New York. *ANSI/IEEE 754-1985, Standard for Binary Floating-Point Arithmetic*, 1985. A preliminary draft was published in the January 1980 issue of IEEE Computer, together with several companion articles. Available from the IEEE Service Center, Piscataway, NJ, USA. Cited on p. 359

[Imrich and Klavžar, 2000] W Imrich and S Klavžar. *Product Graphs*. Wiley–Interscience Series in Discrete Mathematics and Optimization. Wiley–Interscience, New York, 2000. Structure and recognition, With a foreword by Peter Winkler. Cited on p. 323

[John *et al.*, 2003] K St. John, T Warnow, B Moret, and L Vawter. Performance study of phylogenetic methods: (unweighted) quartet methods and neighbor joining. *Journal of Algorithms*, 48:174–193, 2003. Cited on p. 303

[Jordan, 2005] MI Jordan. *An Introduction to Probabilistic Graphical Models*. In preparation, 2005. Cited on p. 41

[Jukes and Cantor, 1969] TH Jukes and C Cantor. Evolution of protein molecules. In HN Munro, editor, *Mammalian Protein Metabolism*, pages 21–32. New York Academic Press, 1969. Cited on p. 153, 366

[Karlin and Altschul, 1990] S Karlin and SF Altschul. Methods for assessing the statistical significance of molecular sequence features by using general scoring schemes. *Proceedings of the National Academy of Sciences, USA*, 87:2264–2268, 1990. Cited on p. 80

[Kellis *et al.*, 2004] M Kellis, B Birren, and E Lander. Proof and evolutionary analysis of ancient genome duplication in the yeast saccharomyces cerevisae. *Nature*, 8:617–624, 2004. Cited on p. 130

[Kent *et al.*, 2002] WJ Kent, CW Sugnet, TS Furey, KM Roskin, TH Pringle, AM Zhaler, and D Haussler. The human genome browser at UCSC. *Genome Research*, 12(6):996–1006, 2002. Cited on p. 378

[Kent, 2002] WJ Kent. Blat- the blast like alignment tool. *Genome Biology*, 12(4):656–664, 2002. Cited on p. 81, 276

[Khachiyan, 1980] LG Khachiyan. Polynomial algorithms in linear programming. *USSR Computational Mathematics and Mathematical Physics*, 20:53–72, 1980. Cited on p. 176, 184

[Kimura, 1980] M Kimura. A simple method for estimating evolutionary rates of base substitution through comparative studies of nucleotide sequences. *Journal of Molecular Evolution*, 16:111–120, 1980. Cited on p. 154

[Kimura, 1981] M Kimura. Estimation of evolutionary sequences between homologous nucleotide sequences. *Proceedings of the National Academy of Sciences, USA*, 78:454–458, 1981. Cited on p. 154, 329

[Kolmogorov and Zabih, 2003] V Kolmogorov and R Zabih. Multi-camera scene reconstruction via graph cuts. *European Conference on Computer Vision (ECCV)*, 2003. Cited on p. 274, 276

[Korf et al., 2001] I Korf, P Flicek, D Duan, and MR Brent. Integrating genomic homology into gene structure prediction. *Bioinformatics*, 17 Suppl 1:S140–8, 2001. Cited on p. 129

[Korf et al., 2003] I Korf, M Yandell, and J Bedell. *BLAST*. O'Reilly & Associates, Sebastopol, CA, 2003. Cited on p. 81

[Körner, 1989] TW Körner. *Fourier Analysis*. Cambridge University Press, Cambridge, second edition, 1989. Cited on p. 328, 333

[Kschischang et al., 2001] F Kschischang, H Loeliger, and B Frey. Factor graphs and the sum-product algorithm. *IEEE Transactions on Information Theory*, 47(2), February 2001. Cited on p. 41

[Kuhn, 1955] HW Kuhn. The Hungarian method for the assignment problem. *Naval Research Logistics Quarterly*, 2:83–97, 1955. Cited on p. 49

[Kulisch et al., 2001] U Kulisch, R Lohner, and A Facius, editors. *Perspectives on Enclosure Methods*. Springer-Verlag, New York, 2001. Cited on p. 366

[Kulp et al., 1996] D Kulp, D Haussler, MG Reese, and FH Eeckman. A generalized hidden Markov model for the recognition of human genes in DNA. In *Proceedings of the Fourth International Conference on Intelligent Systems for Molecular Biology*, pages 134–142. AAAI Press, 1996. Cited on p. 145

[Kuo, 2005] E Kuo. Viterbi sequences and polytopes. *Journal of Symbolic Computation*, 2005. To appear. Cited on p. 229, 230, 235

[Lake, 1987] JA Lake. A rate-independent technique for analysis of nucleaic acid sequences: evolutionary parsimony. *Molecular Biology and Evolution*, 4:167–191, 1987. Cited on p. 348

[Lanave et al., 1984] CG Lanave, G Preparata, C Saccone, and G Serio. A new method for calculating evolutionary substitution rates. *Journal of Molecular Evolution*, 20:86–93, 1984. Cited on p. 155

[Lander and Waterman, 1988] ES Lander and MS Waterman. Genomic mapping by fingerprinting random clones: a mathematical analysis. *Genomics*, 2:231–239, 1988. Cited on p. 134

[Landsberg and Manivel, 2004] JM Landsberg and L Manivel. On the ideals of secant varieties of Segre varieties. *Foundations of Computational Mathematics*, 4(4):397–422, 2004. Cited on p. 351

[Laubenbacher and Stigler, 2004] R Laubenbacher and B Stigler. A computational algebra approach to the reverse engineering of gene regulatory networks. *Journal of Theoretical Biology*, 229:523–537, 2004. Cited on p. 94

[Lauritzen, 1996] SL Lauritzen. *Graphical models*, volume 17 of Oxford Statistical Science Series. The Clarendon Press Oxford University Press, New York, 1996. Oxford Science Publications. Cited on p. 36, 38, 40, 281, 283

[Lenstra, 1983] HW Lenstra. Integer programming with a fixed number of variables. *Mathematics of Operations Research*, 8(4):538–548, 1983. Cited on p. 48

[Levinson and Gutman, 1987] G Levinson and GA Gutman. Slipped-strand mispairing: a major mechanism for DNA sequence evolution. *Molecular Biology and Evolution*, 4:203–221, 1987. Cited on p. 157

[Levy et al., 2005] D Levy, R Yoshida, and L Pachter. Neighbor joining with subtree weights. Preprint, 2005. Cited on p. 335, 344, 373

[Lin et al., 2002] YH Lin, PA McLenachan, AR Gore, MJ Phillips, R Ota, MD Hendy, and D Penny. Four new mitochondrial genomes and the increased stability of evolutionary trees of mammals from improved taxon sampling. *Molecular Phylo-*

genetics and Evolution, 19:2060–2070, 2002. Cited on p. 382

[Litvinov, 2005] G Litvinov. The Maslov dequantization. idempotent and tropical mathematics: a very brief introduction. *arXiv.org:math/0501038,* 2005. Cited on p. 49

[Loh and Walster, 2002] E Loh and GW Walster. Rump's example revisited. *Reliable Computing,* 8:245–248, 2002. Cited on p. 360

[Madsen *et al.,* 2001] O Madsen, M Scally, CJ Douady, DJ Kao, RW Debry, R Adkins, HM Ambrine, MJ Stanhope, WW DeJong, and MS Springer. Parallel adaptive radiations in two major clades of placental mammals. *Nature,* 409:610–614, 2001. Cited on p. 382

[Marcotte *et al.,* 1999] EM Marcotte, M Pellegrini, MJ Thompson, T Yeates, and D Eisenberg. A combined algorithm for genome-wide prediction of protein function. *Nature,* 402:83–86, 1999. Cited on p. 265

[McAuliffe *et al.,* 2004] JD McAuliffe, L Pachter, and MI Jordan. Multiple-sequence functional annotation and the generalized hidden Markov phylogeny. *Bioinformatics,* 20(12):1850–60, 2004. Cited on p. 157, 402

[Megiddo, 1984] N Megiddo. Linear programming in linear time when the dimension is fixed. *Journal of the Association for Computing Machinery,* 31(1):114–127, 1984. Cited on p. 176, 177, 184

[Mihaescu, 2005] R Mihaescu. The toric ideal of the unhidden Markov model. In preparation, 2005. Cited on p. 238, 246, 247

[Mindell and Honeycutt, 1990] DP Mindell and RL Honeycutt. Ribosomal RNA in vertebrates: evolution and phylogenetic applications. *Annual Review of Ecology and Systematics,* 21:541–566, 1990. Cited on p. 386

[Mond *et al.,* 2003] DMQ Mond, JQ Smith. and D Van Straten. Stochastic factorisations, sandwiched simplices and the topology of the space of explanations. *Proceedings of the Royal Society of London, Series A,* 459:2821–2845, 2003. Cited on p. 24

[Moore, 1967] RE Moore. *Interval Analysis.* Prentice-Hall, Englewood Cliffs, New Jersey, 1967. Cited on p. 360, 363

[Moore, 1979] RE Moore. *Methods and Applications of Interval analysis.* SIAM, Philadelphia, Pennsylvania, 1979. Cited on p. 362

[Mount, 1982] SM Mount. A catalogue of splice junction sequence. *Nucleic Acids Research,* 10(2):459–472, 1982. Cited on p. 144

[Murphy *et al.,* 2001] WJ Murphy, E Eizirik, WE Johnson, YP Zhang, OA Ryder, and SJ O'Brien. Molecular phylogenetics and the origins of placental mammals. *Nature,* 409:614–618, 2001. Cited on p. 382

[Myers, 1999] E Myers. Whole-genome DNA sequencing. *IEEE Computational Engineering and Science,* 3(1):33–43, 1999. Cited on p. 127

[Nasrallah, 2002] JB Nasrallah. Recognition and rejection of self in plant reproduction. *Science,* 296:305–308, 2002. Cited on p. 374

[Needleman and Wunsch, 1970] SB Needleman and CD Wunsch. A general method applicable to the search for similarities in the amino acid sequence of two proteins. *Journal of Molecular Biology,* 48:443–445, 1970. Cited on p. 50

[Neumaier, 1990] A Neumaier. *Interval Methods for Systems of Equations.* Cambridge university press, 1990. Cited on p. 365

[Neyman, 1971] J Neyman. Molecular studies of evolution: A source of novel statistical problems. In S Gupta and Y Jackel, editors, *Statistical Decision Theory and Related Topics,* pages 1–27. Academic Press, New York, 1971. Cited on p. 298

[Ng *et al.,* 1985] SY Ng, P Gunning, R Eddy, P Ponte, J Leavitt, T Shows, and L Kedes. Evolution of the functional human beta-actin gene and its multi-pseudogene family: conservation of noncoding regions and chromosomal dispersion of pseudogenes. *Molecular and Cellular Biology,* 5:2720–2732, 1985. Cited on p. 396

[Nobrega *et al.,* 2003] MA Nobrega, I Ovcharenko, V Afzal, and EM Rubin. Scanning human gene deserts for long-range enhancers. *Science,* 302(5644):413–, 2003. Cited on p. 393

[Ota and Li, 2000] S Ota and WH Li. Njml: A hybrid algorithm for the neighbor-joining and maximum likelihood methods. *Molecular Biology and Evolution*, 17(9):1401–1409, 2000. Cited on p. 356

[Pachter and Speyer, 2004] L Pachter and D Speyer. Reconstructing trees from subtree weights. *Applied Mathematics Letters*, 17(6):615–621, 2004. Cited on p. 335

[Pachter and Sturmfels, 2004a] L Pachter and B Sturmfels. Parametric inference for biological sequence analysis. *Proc Natl Acad Sci U S A*, 101(46):16138–43, 2004. Cited on p. 165, 181, 194

[Pachter and Sturmfels, 2004b] L Pachter and B Sturmfels. Tropical geometry of statistical models. *Proceedings of the National Academy of Sciences, USA*, 101(46):16132–7, 2004. Cited on p. 115, 239

[Pachter and Sturmfels, 2005] L Pachter and B Sturmfels. The mathematics of phylogenomics. *Submitted*, 2005. Cited on p. 215, 345, 390, 400

[Pachter et al., 2002] L Pachter, M Alexandersson, and S Cawley. Applications of generalized pair hidden Markov models to alignment and gene finding problems. *J Comput Biol*, 9(2):389–99, 2002. Cited on p. 145

[Parra et al., 2000] G Parra, E Blanco, and R Guigó. GENEID in drosophila. *Genome Research*, 10(4):511–515, 2000. Cited on p. 236

[Pearl, 1988] J Pearl. *Probabilistic Reasoning in Intelligent Systems*. Morgan Kaufmann, San Francisco, 1988. Cited on p. 41

[Peters et al., 2000] T Peters, R Dildrop, K Ausmeier, and U Ruther. Organization of mouse iroquois homeobox genes in two clusters suggests a conserved regulation and function in vertebrate development. *Genome Research*, 10:1453–1462, 2000. Cited on p. 395

[Pevzner and Tesler, 2003] P Pevzner and G Tesler. Genome rearrangements in mammalian evolution: lessons from human and mouse genomes. *Genome Research*, 13(1):37–45, 2003. Cited on p. 265

[Phillips and Penny, 2003] MJ Phillips and D Penny. The root of the mammalian tree inferred from whole mitochondrial genomes. *Molecular Phylogenetics and Evolution*, 28:171–185, 2003. Cited on p. 382

[Pin, 1998] Jean-Eric Pin. Tropical semirings. In *Idempotency (Bristol, 1994)*, volume 11 of *Publ. Newton Inst.*, pages 50–69. Cambridge Univ. Press, Cambridge, 1998. Cited on p. 49

[Pistone et al., 2000] G Pistone, E Riccomagno, and HP Wynn. *Algebraic Statistics: Computational Commutative Algebra in Statistics*. Chapman & Hall/CRC, December 2000. Cited on p. 3

[Pollard, 2001] TD Pollard. Genomics, the cytoskeleton and motility. *Nature*, 409:842–843, 2001. Cited on p. 396

[Preparata and Shamos, 1985] F Preparata and MI Shamos. *Computational Geometry: An Introduction*. Texts and Monographs in Computer Science. Springer Verlag, New York, 1985. Cited on p. 176

[Radmacher et al., 2001] MD Radmacher, R Simon, R Desper, R Taetle, AA Schäffer, and MA Nelson. Graph models of oncogenesis with an application to melanoma. *Journal of Theoretical Biology*, 212:535–548, 2001. Cited on p. 278

[Rahnenführer et al., 2005] J Rahnenführer, N Beerenwinkel, WA Schulz, C Hartmann, A von Deimling, B Wullich, and T Lengauer. Estimating cancer survival and clinical outcome based on genetic tumor progression scores. *Bioinformatics*, 2005. To appear. Cited on p. 278

[Rall, 1981] LB Rall. *Automatic Differentiation, Techniques and Applications*, volume 120 of Springer Lecture Notes in Computer Science. Springer-Verlag, New York, 1981. Cited on p. 362

[Rambaut and Grassly, 1997] A Rambaut and NC Grassly. Seq-Gen: An application for the Monte Carlo simulation of DNA sequence evolution along phylogenetic trees. *Comput. Appl. Biosci.*, 13:235–238, 1997. Cited on p. 343, 355

[Raphael and Pevzner, 2004] B Raphael and P Pevzner. Reconstructing tumor amplisomes. *Bioinformatics*. 20 Suppl 1, Special ISMB/ECCB 2004 issue:I265–I273, 2004. Cited on p. 131

[Ratz, 1992] D Ratz. *Automatische Ergebnisverifikation bei globalen Optimierungsproblemen.* Ph.D. dissertation, Universität Karlsruhe, Karlsruhe, Germany, 1992. Cited on p. 364, 369

[Richter-Gebert *et al.*, 2003] J Richter-Gebert, B Sturmfels, and T Theobald. First steps in tropical geometry. In GL Litvinov and VP Maslov, editors, *Proceedings of the Conference on Idempotent Mathematics and Mathematical Physics*, 2003. Cited on p. 112

[Ross and Gentleman, 1996] I Ross and R Gentleman. R: A language for data analysis and graphics. *Journal of Computational and Graphical Statistics*, 5(3):299-314, 1996. Cited on p. 79, 303

[Sainudiin *et al.*, 2005] R Sainudiin, SW Wong, K Yogeeswaran, J Nasrallah, Z Yang, and R Nielsen. Detecting site-specific physicochemical selective pressures: applications to the class-I HLA of the human major histocompatibility complex and the SRK of the plant sporophytic self-incompatibility system. *Journal of Molecular Evolution*, in press, 2005. Cited on p. 374

[Sainudiin, 2004] R Sainudiin. Enclosing the maximum likelihood of the simplest DNA model evolving on fixed topologies: towards a rigorous framework for phylogenetic inference. Technical Report BU1653-M, Department of Biol. Stats. and Comp. Bio., Cornell University, 2004. Cited on p. 371

[Saitou and Nei, 1987] N Saitou and M Nei. The neighbor joining method: a new method for reconstructing phylogenetic trees. *Molecular Biology and Evolution*, 4(4):406–425, 1987. Cited on p. 72, 335, 374

[Salakhutdinov *et al.*, 2003] R Salakhutdinov, S Roweis, and Z Ghahramani. Optimization with em and expectation-conjugate-gradient. In *Proceedings of the Twentieth International Conference on Machine Learning (ICML-2003)*, 2003. Cited on p. 261

[Salakhutdinov *et al.*, 2004] R Salakhutdinov, S Roweis, and Z Ghahramani. Relationship between gradient and EM steps in latent variable models. *in preparation*, 2004. Cited on p. 261

[Sandelin *et al.*, 2004] A Sandelin, P Bailey, S Bruce, PG Engström, JM Klos, WW Wasserman, J Ericson, and B Lenhard. Arrays of ultraconserved non-coding regions span the loci of key developmental genes in vertebrate genomes. *BMC Genomics*, 5:99, 2004. Cited on p. 394, 395, 400

[Sankoff and Blanchette, 2000] D Sankoff and M Blanchette. Comparative genomics via phylogenetic invariants for Jukes-Cantor semigroups. In *Stochastic models (Ottawa, ON, 1998)*, volume 26 of *Proceedings of the International Conference on Stochstic Models*, pages 399–418. American Mathematical Society, Providence, RI, 2000. Cited on p. 348

[Sankoff and Nadeau, 2003] D Sankoff and JH Nadeau. Chromosome rearrangements in evolution: from gene order to genome sequence and back. *Proceedings of the National Academy of Sciences, USA*, 100:11188-11189, 2003. Cited on p. 130

[Schenck, 2003] H Schenck. *Computational Algebraic Geometry.* London Mathematical Society Student Texts. Cambridge University Press, 2003. Cited on p. 89

[Schmitz and Zischler, 2003] J Schmitz and H Zischler. A novel family of trna-derived sines in the colugo and two new retrotransposable markers separating dermopterans from primates. *Molecular Phylogenetics and Evolution*, 28:341-349, 2003. Cited on p. 382

[Schrijver, 1986] A Schrijver. *Theory of Linear and Integer Programming.* Wiley-Interscience Series in Discrete Mathematics. John Wiley & Sons Ltd., Chichester, 1986. A Wiley–Interscience Publication. Cited on p. 48

[Schwartz *et al.*, 2003] S Schwartz, WJ Kent, A Smit, Z Zhang, R Baertsch, RC Hardison, D Haussler, and W Miller. Human–mouse alignments with BLASTZ. *Genome Research*, 13:103–107, 2003. Cited on p. 55, 81, 137

[Seidel, 2004] R Seidel. Convex hull computations. In Jacob E. Goodman and Joseph O'Rourke, editors, *Handbook of Discrete and Computational Geometry*. Discrete Mathematics and its Applications (Boca Raton). chapter 22. Chapman & Hall/CRC, Boca Raton, FL, second edition, 2004. Cited on p. 185

[Semple and Steel, 2003] C Semple and M Steel. *Phylogenetics*, volume 24 of Oxford Lecture Series in Mathematics and its Applications. Oxford University Press, Oxford, 2003. Cited on p. ix, 67, 152, 306

[Shoup, 2004] V Shoup. NTL, A Library for doing Number Theory. Available at http://shoup.net/ntl/, 2004. Cited on p. 240

[Siepel and Haussler, 2004] A Siepel and D Haussler. Combining phylogenetic and hidden markov models in biosequence analysis. *Journal of Computational Biology*, 11:413 428, 2004. Cited on p. 157, 402

[Simon et al., 2000] R Simon, R Desper, CH Papadimitriou, A Peng, DS Alberts, R Taetle, JM Trent, and AA Schäffer. Chromosome abnormalities in ovarian adenocarcinoma: III. Using breakpoint data to infer and test mathematical models for oncogenesis. *Genes, Chromosomes & Cancer*, 28:106–120, 2000. Cited on p. 278

[Smith, 1998] JM Smith. *Evolutionary Genetics*. Oxford University Press, second edition, March 1998. Cited on p. ii

[Sneath and Sokal, 1973] PHA Sneath and RR Sokal. *Numerical Taxonomy: the Principles and Practice of Numerical Classification*. W.H. Freeman, San Francisco, 1973. Cited on p. 335

[Speyer and Sturmfels, 2004] D Speyer and B Sturmfels. The tropical Grassmannian. *Adv. Geom.*, 4(3):389 411, 2004. Cited on p. 118, 122

[Speyer and Williams, 2004] D Speyer and L Williams. The tropical totally positive Grassmanian. *Journal of Algebraic Combinatorics, in press*, 2004. Cited on p. 115

[Speyer, 2004] D Speyer. Tropical linear spaces. *Submitted*, 2004. Cited on p. 121

[Stanley, 1999] RP Stanley. *Enumerative combinatorics. Vol. 2*, volume 62 of *Cambridge Studies in Advanced Mathematics*. Cambridge University Press, Cambridge, 1999. With a foreword by Gian-Carlo Rota and appendix 1 by Sergey Fomin. Cited on p. 51, 100, 279

[Steel et al., 1992] MA Steel, MD Hendy, LA Székely, and PL Erdős. Spectral analysis and a closest tree method for genetic sequences. *Applied Mathematics Letters*, 5(6):63–67, 1992. Cited on p. 332, 334

[Strachan and Read, 2004] T Strachan and AP Read. *Human Molecular Genetics*. Garland Press, third edition, June 2004. Cited on p. 131

[Strassen, 1983] V Strassen. Rank and optimal computation of generic tensors. *Linear Algebra Appl.*, 52/53:645 685, 1983. Cited on p. 351

[Strimmer and Moulton, 2000] K Strimmer and V Moulton. Likelihood analysis of phylogenetic networks using directed graphical models. *Molecular Biology and Evolution*, 17:875–881, 2000. Cited on p. 326

[Strimmer and von Haeseler, 1996] K Strimmer and A von Haeseler. Quartet puzzling: A quartet maximum likelihood method for reconstructing tree topologies. *Molecular Biology and Evolution*, 13:964 969, 1996. Cited on p. 356

[Strimmer et al., 2001] K Strimmer, C Wiuf, and V Moulton. Recombination analysis using directed graphical models. *Molecular Biology and Evolution*, 18:97–99, 2001. Cited on p. 326, 327

[Studier and Keppler, 1988] JA Studier and KJ Keppler. A note on the neighbor-joining method of saitou and nei. *Molecular Biology and Evolution*, 5:729 731, 1988. Cited on p. 72

[Sturmfels and Sullivant, 2005] B Sturmfels and S Sullivant. Toric ideals of phylogenetic invariants. *Journal of Computational Biology*, 12:204 228, 2005. Cited on p. 152, 153, 296, 302, 303, 306, 348

[Sturmfels, 1990] B Sturmfels. Gröbner bases and Stanley decompositions of determinantal rings. *Mathematische Zeitschrift*, 205(1):137–144, 1990. Cited on p. 92

[Sturmfels, 1993] B Sturmfels. *Algorithms in Invariant Theory*. Texts and Monographs in Symbolic Computation. Springer-Verlag, Vienna, 1993. Cited on p. 99

[Sturmfels, 2002] B Sturmfels. *Solving Systems of Polynomial Equations*, volume 97 of *CBMS Regional Conference Series in Mathematics*. Published for the Conference

Board of the Mathematical Sciences, Washington, DC, 2002. Cited on p. 88, 112, 285

[Susko, 2003] E Susko. Confidence regions and hypothesis tests for topologies using generalized least squares. *Molecular Biology and Evolution*, 2003. Cited on p. 325, 326

[Swofford, 1998] DL Swofford. *PAUP*. Phylogenetic Analysis using Parsimony (* and other Methods)*. Sunderland Mass., 1998. Cited on p. 375

[Swox, 2004] AB Swox. GMP, the GNU Multiple Precision Arithmetic Library. Available at http://swox.com/gmp/, 2004. Cited on p. 188, 240

[Szabo and Boucher, 2002] A Szabo and K Boucher. Estimating an oncogenetic tree when false negatives and positives are present. *Mathematical Biosciences*, 176:219–240, 2002. Cited on p. 289

[Székely et al., 1993] LA Székely, MA Steel, and PL Erdős. Fourier calculus on evolutionary trees. *Advances in Applied Mathematics*, 14(2):200–210, 1993. Cited on p. 296, 306, 309, 332, 334

[Tamura and Nei, 1993] K Tamura and M Nei. Estimation of the number of nucleotide substitutions in the control region of mitochondrial DNA in humans and chimpanzees. *Molecular Biology and Evolution*, 10:512–526, 1993. Cited on p. 155

[Tavaré, 1986] S Tavaré. Some probabilistic and statistical problems in the analysis of DNA sequences. *Lectures on the Mathematics in the Life Sciences*, 17:57–86, 1986. Cited on p. 155

[Tesler, 2002] G Tesler. Efficient algorithms for multichromosomal genome rearrangements. *Journal of Computer and System Sciences*, 65(3):587–609, 2002. Cited on p. 130, 131

[Thomas et al., 2003] JW Thomas, JW Touchman, RW Blakesley, GG Bouffard, SM Beckstrom-Sternberg, EH Margulies, M Blanchette, AC Siepel, PJ Thomas, JC McDowell, et al. Comparative analyses of multi-species sequences from targeted genomic regions. *Nature*, 424(6950):788–93, 2003. Cited on p. 276, 384, 385

[Tompa et al., 2005] M Tompa, N Li, TL Bailey, GM Church, B De Moor, E Eskin, AV Favorov, MC Frith, Y Fu, WJ Kent, et al. Assessing computational tools for the discovery of transcription factor binding sites. *Nature Biotechnology*, 23(1):137–144, 2005. Cited on p. 147

[Valiant, 1979] L Valiant. The complexity of computing the permanent. *Theoretical Computer Science*, 8:189–201, 1979. Cited on p. 49

[Varchenko, 1995] A Varchenko. Critical points of the product of powers of linear functions and families of bases of singular vectors. *Compositio Mathematica*, 97(3):385–401, 1995. Cited on p. 10

[Venter et al., 2001] JC Venter, MD Adams, EW Myers, PW Li, RJ Mural, GG Sutton, HO Smith, M Yandell, CA Evans, RA Holt, et al. The sequence of the human genome. *Science*, 291(5507):1304–51, 2001. Cited on p. 126

[Viterbi, 1967] AJ Viterbi. Error bounds for convolutional codes and an asymptotically optimum decoding algorithm. *IEEE Transactions on Information Theory*, 13:260–269, 1967. Cited on p. 57, 227

[Vogelstein et al., 1988] B Vogelstein, E Fearon, and S Hamilton. Genetic alterations during colorectal-tumor development. *New England Journal of Medicine*, 319:525–532, 1988. Cited on p. 278

[von Heydebreck et al., 2004] A von Heydebreck, B Gunawan, and L Füzesi. Maximum likelihood estimation of oncogenetic tree models. *Biostatistics*, 5(4):545–556, 2004. Cited on p. 278

[Warshall, 1962] S Warshall. A theorem on boolean matrices. *Journal of the ACM*, 9(1):18, 1962. Cited on p. 46

[Waterman et al., 1992] MS Waterman, M Eggert, and ES Lander. Parametric sequence comparisons. *Proceedings of the National Academy of Sciences, USA*, 89:6090–6093, 1992. Cited on p. 55

[Waterman, 1995] MS Waterman. *Introduction to Computational Biology: Maps, Sequences and Genomes*. Chapman & Hall/CRC, June 1995. Cited on p. 49

[Waterston *et al.*, 2002] RH Waterston, K Lindblad-Toh, E Birney, J Rogers, JF Abril, P Agarwal, R Agarwala, R Ainscough, M Alexandersson, P An, et al. Initial sequencing and comparative analysis of the mouse genome. *Nature*, 420(6915):520–62, 2002. Cited on p. 130, 384

[Watson and Crick, 1953] J Watson and F Crick. A structure for deoxyribose nucleic acid. *Nature*, 171:964–967, 1953. Cited on p. 126

[Wetzel, 1995] R Wetzel. Zur Visualisierung abstrakter Ähnlichkeitsbeziehungen. Master's thesis, Fakultät Mathematik, Universität Bielefeld, 1995. Cited on p. 323, 324

[Winkler, 1984] P Winkler. Isometric embeddings in products of complete graphs. *Discrete Applied Mathematics*, 7:221–225, 1984. Cited on p. 323

[Wolf *et al.*, 2000] MJ Wolf, S Easteal, M Kahn, BD McKay, and LS Jermiin. Trexml: A maximum likelihood program for extensive tree-space exploration. *Bioinformatics*, 16:383–394, 2000. Cited on p. 374

[Woolfe *et al.*, 2005] A Woolfe, M Goodson, DK Goode, P Snell, GK McEwen, T Vavouri, SF Smith, P North, H Callaway, K Kelly, et al. Highly conserved non-coding sequences are associated with vertebrate development. *PLoS Biology*, 3:7, 2005. Cited on p. 394, 400

[Wu and Li, 1985] CI Wu and WH Li. Evidence for higher rates of nucleotide substitution in rodents than in man. *PNAS*, 82(6):1741–1745, 1985. Cited on p. 384

[Yang and Roberts, 1995] Z Yang and D Roberts. On the use of nucleic acid sequences to infer early branchings in the tree of life. *Molecular Biology and Evolution*, 12:451–458, 1995. Cited on p. 385

[Yang, 1997] Z Yang. PAML: A program package for phylogenetic analysis by maximum likelihood. *CABIOS*, 15:555–556, 1997. Cited on p. 303, 375, 400

[Yap and Pachter, 2004] VB Yap and L Pachter. Identification of evolutionary hotspots in the rodent genomes. *Genome Research*, 14(4):574–9, 2004. Cited on p. 154, 305

[Yoder and Yang, 2000] AD Yoder and Z Yang. Estimation of primate speciation dates using local molecular clocks. *Molecular Biology and Evolution*, 17:1081–1090, 2000. Cited on p. 386

[Yu *et al.*, 2002] J Yu, S Hu, J Wang, GKS Wong, S Li, B Liu, Y Deng, L Dai, Y Zhou, X Zhang, et al. A draft sequence of the rice genome (Oryza sativa L. ssp. indica). *Science*, 296(5565):79–92, 2002. Cited on p. 134

[Yu *et al.*, 2005] J Yu, J Wang, W Lin, S Li, H Li, J Zhou, P Ni, W Dong, S Hu, C Zeng, et al. The genomes of Oryza sativa: A history of duplications, 2005. Cited on p. 134

[Zang, 2001] KD Zang. Meningioma: a cytogenetic model of a complex benign human tumor, including data on 394 karyotyped cases. *Cytogenet. Cell Genet.*, 93:207–220, 2001. Cited on p. 278

[Zharkikh, 1994] A Zharkikh. Estimation of evolutionary distances between nucleotide sequences. *Journal of Molecular Evolution*, 39:315–329, 1994. Cited on p. 155

[Ziegler, 1995] GM Ziegler. *Lectures on polytopes*, volume 152 of Graduate Texts in Mathematics. Springer-Verlag, New York, 1995. Cited on p. 60, 183, 217

Index

Printed in the United States
By Bookmasters